Problems and Solutions in Stochastic Calculus with Applications

PROBLEMS AND SOLUTIONS IN STOCHASTIC CALCULUS WITH APPLICATIONS

Patrik Albin
Chalmers University of Technology, Sweden

Kais Hamza
Monash University, Australia

Fima C. Klebaner
Monash University, Australia

EW JERSEY • LONDON • SINGAPORE • BEIJING • SHANGHAI • HONG KONG • TAIPEI • CHENNAI • TOKYO

Published by

World Scientific Publishing Europe Ltd.
57 Shelton Street, Covent Garden, London WC2H 9HE
Head office: 5 Toh Tuck Link, Singapore 596224
USA office: 27 Warren Street, Suite 401-402, Hackensack, NJ 07601

Library of Congress Cataloging-in-Publication Data
Names: Albin, Patrik, author. | Hamza, Kais, author. | Klebaner, Fima C., author.
Title: Problems and solutions in stochastic calculus with applications /
Patrik Albin, Chalmers University of Technology, Sweden, Kais Hamza,
Monash University, Australia, Fima C. Klebaner, Monash University, Australia.
Description: New Jersey : World Scientific, [2025] | Includes index.
Identifiers: LCCN 2023057770 | ISBN 9781800615571 (hardcover) |
ISBN 9781800615601 (paperback) | ISBN 9781800615588 (ebook for institutions) |
ISBN 9781800615595 (ebook for individuals)
Subjects: LCSH: Stochastic analysis. | Stochastic processes. | Calculus.
Classification: LCC QA274.2 .A44 2025 | DDC 519.2/2--dc23/eng/20240326
LC record available at https://lccn.loc.gov/2023057770

British Library Cataloguing-in-Publication Data
A catalogue record for this book is available from the British Library.

For any available supplementary material, please visit
https://www.worldscientific.com/worldscibooks/10.1142/Q0460#t=suppl

Desk Editors: Nandha Kumar Krishnan/Rosie Williamson/Shi Ying Koe

Typeset by Stallion Press
Email: enquiries@stallionpress.com

To my 100 year old mother
— PA

To my wife and children
— KH

In memory of my teacher, Professor Robert Lipster
— FK

About the Authors

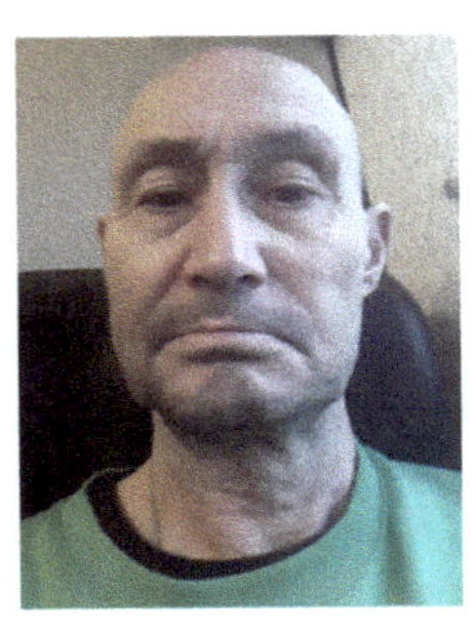

Patrik Albin (MSc 1984, PhD 1987 in Mathematical Statistics from Lund University). Patrik has been an Associate Professor in Analysis and Probability Theory, Mathematical Sciences at Chalmers University of Technology, Gothenburg, Sweden since 1993. He is an active researcher in stochastic processes especially in extreme value theory, Gaussian and stable random processes, Lévy processes and diffusion processes/SDE. He has more than 30 publications and 30 years of experience in teaching Probability and Stochastic Processes. He supervised four PhD students. Patrik has been visiting researcher at UNC Chapel Hill four times, Technion Haifa twice, Cornell Ithaca and Wroclaw Poland.

Kais Hamza (PhD 1988 in Probability from Paris 6, France). Kais is a Professor in the School of Mathematics at Monash University. He is an active researcher with more than 60 publications and 35 years of experience in teaching Probability and Stochastic Processes. Kais has supervised (or co-supervised) 24 PhD students. He is Associate Editor for *Asia-Pacific Financial Markets*.

Fima Klebaner (MSc 1980, PhD 1983 in Stochastic Processes from University of Melbourne). Fima holds a Chair of Statistics in the School of Mathematics at Monash University. He is an active researcher with more than 95 publications and 40 years of experience in teaching Probability, Stochastic Processes, and Financial Mathematics. He supervised (co-supervised) 14 PhD students. Fima is Associate Editor for the *Applied Probability Journals*; the *Australian and New Zealand Journal of Statistics*; *Stochastics: An International Journal Probability and Stochastic Processes.* He is a Fellow of the Institute of Mathematical Statistics and a Lady Davis Professorial Fellow.

Contents

About the Authors vii

Introduction xi

1. Preliminaries from Calculus 1

2. Concepts of Probability Theory 15

3. Basic Stochastic Processes 61

4. Brownian Motion Calculus 99

5. Stochastic Differential Equations 143

6. Diffusion Processes 175

7. Martingales 223

8. Semimartingales 259

9. Pure Jump Processes 283

10. Change of Probability Measure 349

11. Applications in Finance 381

12. Applications in Biology 445

Index 467

Introduction

Just as *the proof of the pudding is in the eating*, the essence of mathematics is in proving. Proofs consist of simple ideas put together. This book tries to do just that: expose the reader to distinct simple ideas and proofs. While many questions here are straightforward, some are more challenging. Some questions are focused on filling gaps and some are given to complement and expand the companion text *Introduction to Stochastic Calculus with Applications* (Third Edition) by Fima Klebaner. All in all, they will help the reader progress in the mathematics of stochastic calculus and its applications.

Typical texts on stochastic calculus are very technical and not accessible to students without a solid mathematical background. Yet, the usual calculus is done by many. This is because foundational questions are notoriously hard, and while they are important for mathematics, they are less so for applications. In this book, we avoid hard foundational questions, but do not sacrifice everything; we try to find the middle road.

For example, Itô's formula is widely used and presented. While this formula might be one of the main results (corresponding to the change of variables in calculus) in stochastic calculus, there are many other important techniques, such as change of time, change of measure and especially martingales, well worthy of a thorough presentation.

The book contains a few challenging questions, but most of the material is quite straightforward.

The book is authored by three active researchers in the areas of probability, stochastic processes and their applications to financial mathematics, mathematical biology, and others, with many problems rooted in their own research. We hope that it will inspire the next generation of researchers in the field.

Regarding Notations

There are interchangeable notations for some things. Indicator function of a set A is denoted by $I(A), I_A, I_A(x), 1_A, 1_A(x)$. A stochastic process is denoted by X, (X_t), $(X_t)_{t\in\mathrm{T}}$, $\{X_t\}$ as well as "process X_t". The random variable X_t may also be denoted by $X(t)$. Asterisks (*) appearing in the equation numbers correspond to the equations stated in the companion book.

Chapter 1

Preliminaries from Calculus

Variation of a Function

The variation $V_g([a,b])$ of a function $g:[a,b]\to\mathbb{R}$ over the interval $[a,b]$ is defined as

$$V_g([a,b]) = \sup\left\{\sum_{i=1}^{n} |g(t_i^n) - g(t_{i-1}^n)| : a = t_0^n < t_1^n < \cdots < t_n^n = b,\ n\in\mathbb{N}\right\}.$$

When $V_g([a,b]) < \infty$, we say that g has a finite variation (FV). Functions (and random processes) encountered in stochastic calculus are always either continuous from the right with existing left limits or continuous from the left with existing right limits,[1] and for such functions (processes), it holds that

$$V_g([a,b]) = \lim\left\{\sum_{i=1}^{n} |g(t_i^n) - g(t_{i-1}^n)| : a = t_0^n < \cdots < t_n^n = b,\ n\in\mathbb{N},\ \max_{1\le i\le n} t_i^n - t_{i-1}^n \downarrow 0\right\}.$$

Hence, this can be considered as the definition of variation when doing stochastic calculus. For a function $g:[0,\infty)\to\mathbb{R}$, we use the shorthand notation $V_g(t)$ for $V_g([0,t])$.

It is easy to see that

$$V_g([a,b]) = |g(a) - g(b)|$$

[1] Commonly called *càdlàg* which is French for "*continue à droite, limite à gauche*".

if and only if g is (not necessarily strictly) increasing or decreasing. Further, we have

$$V_g([a,b]) = \int_a^b |g'(t)|\, dt < \infty$$

when g is continuously differentiable. The Brownian motion, the stochastic process that will be introduced in Chapter 3, is an example of a continuous process with infinite variation over every interval.

One important but still basic result is that g is FV if and only if g can be expressed as the difference between two increasing functions, with the latter expression then being called a Jordan decomposition. One such expression is

$$g(t) = V_g([a,t]) - (V_g([a,t]) - g(t)) \quad \text{for } t \in [a,b].$$

The class of functions with FV over the interval $[a,b]$ is the dual space in functional analysis to the Banach space of continuous functions on $[a,b]$ with the supremum norm. The representation of each functional in the dual space is through the Riemann–Stieltjes (RS) integral we encounter later in this chapter with respect to an appropriate FV function.

Quadratic Variation of a Function

The quadratic covariation (or simply covariation) $[f,g]([a,b])$ between two functions $f,g : [a,b] \to \mathbb{R}$ over the interval $[a,b]$ is defined as

$$\begin{aligned} &[f,g]([a,b]) \\ &\quad = \lim \left\{ \sum_{i=1}^{n} (f(t_i^n) - f(t_{i-1}^n))\,(g(t_i^n) - g(t_{i-1}^n)) : a = t_0^n < \cdots < t_n^n = b, \right. \\ &\qquad\qquad \left. \max_{1 \le i \le n} t_i^n - t_{i-1}^n \downarrow 0 \right\} \end{aligned}$$

whenever this limit exists. Functions and random processes encountered in stochastic calculus always have a well-defined covariation between them. This, however, is a deep result which requires sophisticated tools from martingale theory to prove. We will see a lot of martingale theory later in this book. For functions $f,g : [0,\infty) \to \mathbb{R}$, we use the shorthand notation $[f,g](t)$ for $[f,g]([0,t])$.

The quadratic variation $[g]([a,b])$ of a function $g : [a,b] \to \mathbb{R}$ over the interval $[a,b]$ is defined as

$$[g]([a,b]) = [g,g]([a,b]),$$

and for a function $g : [0,\infty) \to \mathbb{R}$, we use the shorthand notation $[g](t)$ for $[g]([0,t])$. Quadratic variation is very important in the so-called Brownian motion stochastic calculus (see Chapters 4–6). Further, it can be considered the single most important tool to build the whole theory for the so-called semimartingale stochastic calculus (see Chapter 8).

Covariation is an inner product and a positive semi-definite symmetric bilinear form. Therefore, it has the same properties as covariance. In particular, we have

$$0 \le [g] \qquad \text{(positivity)},$$
$$[f,g] = [g,f] \qquad \text{(symmetry)},$$
$$[\alpha f_1 + \beta f_2, g] = \alpha\,[f_1,g] + \beta\,[f_2,g] \qquad \text{(linearity)},$$
$$[f,g] = \frac{1}{2}([f+g] - [f] - [g]) \qquad \text{(polarisation)},$$
$$[f,g] = \frac{1}{4}([f-g] - [f-g]) \qquad \text{(polarisation)}$$

and

$$|[f,g]| \le \sqrt{[f]\,[g]} \qquad \text{(Cauchy–Schwarz inequality)}.$$

The first five of these relations are true whenever the right-hand side is well-defined, while the sixth requires all involved covariations to exist. The proof is more or less by inspection.

Another very important yet basic result is that the covariation between a continuous function and an FV function is zero. Therefore, for example, all continuously differentiable functions have zero quadratic variation. This is arguably the reason that quadratic variation is of little, if any, importance in ordinary (non-stochastic) calculus.

Riemann Integral and Riemann–Stieltjes Integral

The Riemann–Stieltjes (RS) integral

$$\int_a^b f\,dg = \int_a^b f(t)\,dg(t)$$

over the interval $[a,b]$ of a continuous function $f : [a,b] \to \mathbb{R}$ with respect to an increasing function $g : [a,b] \to \mathbb{R}$ is defined as

$$\int_a^b f\,dg = \lim \left\{ \sum_{i=1}^n f(\xi_i)\,(g(t_i^n) - g(t_{i-1}^n)) : \; a = t_0^n < \cdots < t_n^n = b, \right.$$

$$\left. \xi_i \in [t_{i-1}^n, t_i^n],\ \max_{1\le i\le n} t_i^n - t_{i-1}^n \downarrow 0 \right\}.$$

Under the mentioned conditions on f and g, this limit can be shown to exist. The continuity requirement on f can be omitted and the integral can be defined for all functions f and increasing functions g such that the limit exists. However, this leads to a theory with substantially more complicated formulae and proofs; therefore, f is usually assumed to be continuous.

Of course, the ordinary Riemann integral is the special case of the RS integral with $g(x) = x$. This can be viewed as the Euclidian case, while more general (than linear or affine) choices of g correspond to a non-Euclidian measure of length $g(t) - g(s)$ of intervals $[s,t] \subseteq [a,b]$.

By means of the Jordan decomposition, the definition of the RS integral is immediately extended to continuous functions f and functions g that are FV.

For f and g both being continuous and of FV, we have

$$\begin{aligned}\int_a^b f\,dg + \int_a^b g\,df &\leftarrow \sum_{i=1}^n f(t_{i-1}^n)\,(g(t_i^n) - g(t_{i-1}^n)) \\ &\quad + \sum_{i=1}^n g(t_i^n)(f(t_i^n) - f(t_{i-1}^n)) \\ &= \sum_{i=1}^n (f(t_i^n)g(t_i^n) - f(t_{i-1}^n)g(t_{i-1}^n)) \\ &= f(b)g(b) - f(a)g(a).\end{aligned}$$

This is the integration by parts formula for the RS integral. Likewise, if f is continuously differentiable and g is both continuous and of FV, it is not hard to show that

$$\int_a^b f'(g(t))\,dg(t) = \int_{g(a)}^{g(b)} f'(s)\,ds = f(g(b)) - f(g(a)).$$

This is the change of variables formula for the RS integral. Finally, when f is continuous and g continuously differentiable, we have

$$\int_a^b f\,dg = \int_a^b f(t)\,\frac{dg(t)}{dt}\,dt = \int_a^b f(t)g'(t)\,dt.$$

One important feature of the RS integral is that it generalises the concepts of the Riemann integral and summation to a more general concept that includes both the previously mentioned as special cases. To see this, we note that

$$\sum_{i=1}^{n} a_i = \int_{1/2}^{n+1/2} f\,dg,$$

for $f(t) = a_i$ for $t \in [i-1/2, i+1/2)$, $i = 1, \ldots, n+1$, and $g(t) = \lfloor t \rfloor = i-1$ for $t \in [i-1, i)$, $i = 1, \ldots, n+1$. Alternatively, one can use a continuous f, as any continuous f with $f(i) = a_i$ gives the same value for the integral.

The RS integral can be extended to an integral over infinite intervals in the same way as is done with the Riemann integral. Thus extended, we mention another important property of the RS integral: it gives the expected value of a random variable X as

$$E(X) = \int_{-\infty}^{\infty} x\,dF_X(x)$$

for both discrete and continuous random variables, where F_X is the cumulative distribution function (CDF) of X.

Problems

Problem 1.1: Show that if g has an FV on $[a, b]$, then it is bounded on $[a, b]$.

Solution. Let x be a point in $[a, b]$. The variation of g on $[a, x]$ does not exceed the variation of g on $[a, b]$, which is $V < \infty$. By taking the partition $[a, x]$, we have

$$|g(x) - g(a)| \leq V.$$

It now follows that $|g(x)| \leq |g(a)| + V$.

Problem 1.2: Let $g(t)$, $t \geq 0$, be an increasing right-continuous function. Let $a = t_0 < t_1 < \cdots < t_n = b$ be the points of a partition of $[a, b]$. Show that

$$\sum_{i=1}^{n} \Delta g(t_i) = \sum_{i=1}^{n} (g(t_i) - g(t_i-)) \leq g(b) - g(a).$$

Solution. Choose x_i in the interval (t_i, t_{i+1}). Then, $g(t_i) \le g(x_i)$ and $g(t_i-) \ge g(x_{i-1})$. Hence,

$$\Delta g(t_i) = g(t_i) - g(t_i-) \le g(x_i) - g(x_{i-1}).$$

Adding these inequalities, we obtain the required inequality.

Problem 1.3: Let $g(t)$, $t \ge 0$, be an increasing right-continuous function. Show that the number of jumps of size greater than δ is bounded.

Solution. Assume that there exist at least n jumps t_i, $i = 1, \ldots, n$, in $[a, b]$ with size greater than δ. Then, $\sum_{i=1}^n \Delta g(t_i) \ge n\delta$. On the other hand, it follows from the above that

$$n\delta \le \sum_{i=1}^n \Delta g(t_i) \le g(b) - g(a).$$

Hence, $n \le (g(b) - g(a))/\delta$.

Problem 1.4: Let $g(t)$, $t \ge 0$, be an increasing right-continuous function. Show that the number of discontinuities is at most countable. Moreover, show that the sum of jumps is bounded by

$$\sum_{t \in (a,b]} \Delta g(t_i) = \sum_{t \in (a,b]} (g(t_i) - g(t_i-)) \le g(b) - g(a).$$

Solution. Let J_k denote the set of points $t \in (a, b]$ with $\Delta g(t) \ge 1/k$. Then, by the results of Problem 1.3, the number of points in J_k is bounded by $k(g(b) - g(a))$. Since the set of all discontinuities $J = \cup_k J_k$, it follows that it is at most countable as a union of finite sets. The desired inequality follows from the inequality

$$\sum_{t \in J_k} \Delta g(t_i) \le g(b) - g(a)$$

by taking the limit as $k \to \infty$. Note that J_k is an increasing sequence $(J_k \subset J_{k+1})$.

Problem 1.5: Let $g(t)$, $t \ge 0$, be an increasing right-continuous function. Define the discontinuous part g^d of g by Equation (1.11*),

$$g^d(t) = \sum_{s \le t} (g(s) - g(s-)) = \sum_{0 < s \le t} \Delta g(s).$$

Then, $g^c(t) = g(t) - g^d(t)$ is called the continuous part of g. Show that g^c is increasing and continuous.

Solution. Let $a \le x < y \le b$. Applying the inequality from the statement of Problem 1.4 to the interval $[x, y]$, we obtain

$$g^d(y) - g^d(x) \le g(y) - g(x).$$

This implies $g^c(x) \le g^c(y)$ so that g^c is increasing.

Note that for any t, we have

$$g^d(t-) = \lim_{s \uparrow t} g^d(s) = \lim_{s \uparrow t} \sum_{r \le s} \Delta g(r) = \sum_{r < s} \Delta g(r).$$

Hence,

$$g^c(t) = g(t) - g^d(t) = g(t) - \sum_{r \le s} \Delta g(r),$$

and

$$g^c(t-) = g(t-) - g^d(t-) = -\sum_{r \le s} \Delta g(r) + g(t).$$

So, $g^c(t-) = g^c(t)$ and g^c is continuous at t.

Problem 1.6: Let $g(t)$, $t \ge 0$, be a function of FV continuous from the right. Show that g can have at most countably many discontinuities, all of which are jumps $\{t_k\}$, and that, moreover, the sum of the jumps is finite over finite time intervals.

Solution. By Theorem 1.6, g is a difference of two increasing functions,

$$g(t) = a(t) - b(t) = V_g(t) - (V_g - g)(t),$$

both of which are right continuous. Since discontinuities of an increasing function are jumps, all discontinuities of g are jumps. By this decomposition, the set jumps of g are included in the union of the sets of jumps of V_g and $V_g - g$. Since these functions are increasing, these sets are at most countable, hence the set of jumps of g is at most countable.

Let a^d and b^d be the discrete parts of functions a and b, respectively, that is V_g^d and $(V_g - g)^d$,

$$a^d(t) = \sum_{t_k \le t} (a(t_k) - a(t_k-)), \quad b^d(t) = \sum_{t_k \le t} (b(t_k) - b(t_k-)),$$

where t_k's are discontinuities of either function a or b. Now, the function $a^d(t) - b^d(t)$ can be written as

$$a^d(t) - b^d(t) = \sum_{t_k \leq t} (g(t_k) - g(t_k-)) =: g^d(t)$$

and is the discontinuous part of g. (It is clear that the above sum does not change if we remove points t_k's at which g is continuous. Therefore, we may take only the point of discontinuity of g.) Further, the sum of jumps of g is bounded by the sum of a plus the sum of jumps of b that are finite by inspection of Problem 1.5.

Problem 1.7: Let $g(t)$, $t \geq 0$, be a right-continuous function of FV. The discontinuous part g^d of g is defined by $g^d(t) = \sum_{t_k<t}(g(t_k) - g(t_k-)) + g(t) - g(t-)$. Show that $g^c(t) = g(t) - g^d(t)$ is continuous (it is the continuous part of g). Deduce that every function of an FV can be written as a sum of its discontinuous part g^d and a continuous function of the FV.

Solution. This follows from the fact that the continuous part of an increasing function is continuous and from Problem 1.5 and Jordan decomposition.

Problem 1.8: Explain why the definition of variation of a function as the supremum over all partitions (Equation (1.7*)) is more generally valid than that according to Equation (1.9*), which is the limit over shrinking partitions.

Solution. Equation (1.7*) defines the variation as the supremum of a set of real numbers, namely the set of all values the sum in (1.7*) can take for different partitions, and a supremum of a set of real numbers is always a well-defined quantity. Now, we may consider the function $g(\frac{1}{2}) = 1$ and $g(x) = 0$ for $x \in [0,1]$, $x \neq \frac{1}{2}$, which has a well-defined variation $V_g([0,1]) = 2$ according to (1.7*), but which does not have a well-defined variation $V_g([0,1])$ according to (1.9*).

Problem 1.9: Show that for continuous functions, the definition of variation as the supremum over all partitions, Equation (1.7*), is the same as the limit over shrinking partitions, according to Equation (1.9*).

Solution. Let f have an FV V according to the classical definition, Equation (1.7*), which expresses the variation of f as a supremum over all partitions of $[a, b]$. We show that for a given ε, there is a δ such that for *any partition* with maximal distance between points δ, the variation S_1 on this partition satisfies $V - 3\varepsilon \leq S_1 \leq V$.

By definition of supremum, for any $\varepsilon > 0$, there is a partition $t_0, t_1, \ldots, t_n$ such that $S = \sum_{i=0}^{n-1} |f(t_{i+1}) - f(t_i)|$ satisfies $S \geq V - \varepsilon$.

Since f is continuous on $[a, b]$, it is uniformly continuous on $[a, b]$, and for ε as above,

$$\exists \eta > 0 \text{ such that } \forall (s, t) \in [a, b], \ |t - s| < \eta \Rightarrow |f(t) - f(s)| < \frac{\varepsilon}{n+1}.$$

Let

$$\delta = \min\left\{\eta, \min_{i=0,\ldots n-1} (t_{i+1} - t_i)\right\}.$$

Take now any partition with maximum distance δ between points. Denote the points in this partition by x_j, $j = 0, \ldots, m$, and variation on this partition

$$S_1 = \sum_{j=0}^{m-1} |f(x_{j+1}) - f(x_j)|.$$

Among the intervals $[x_j, x_{j+1})$, there are those that don't contain any of the points t_i and those that do. Among those that do, they will contain only one of t_i's by the definition of δ. Denote by J the set of indices such that $[x_j, x_{j+1})$ don't contain any of the points t_i, and by I the set of indices such that $[x_j, x_{j+1})$ contain exactly one of the points t_i; call it t'_j.

Consider the partition formed by t_i's and x_j's together. Using the definitions of I and J, we have that the variation of f corresponding to the joint partition is

$$\tilde{S} = \sum_{j \in J} |f(x_{j+1}) - f(x_j)| + \sum_{j \in I} (|f(x_{j+1}) - f(t'_j)| + |f(t'_j) - f(x_j)|).$$

Note next that the number of terms in the second sum (the number of elements in I) is $n + 1$. This is because each of the intervals contains one of the t_i's and there are $n + 1$ of those. Since $x_{j+1} - t'_j < \eta$ and $t'_j - x_j < \eta$,

$$|f(x_{j+1}) - f(t'_j)| < \frac{\varepsilon}{n+1}, \quad |f(t'_j) - f(x_j)| < \frac{\varepsilon}{n+1},$$

we have that the second sum $\sum_{j\in I} \leq 2\varepsilon$. Next, we have the obvious inequalities

$$S_1 = \sum_{j=0}^{m-1} |f(x_{j+1}) - f(x_j)| \geq \sum_{j\in J} |f(x_{j+1}) - f(x_j)|.$$

But

$$\sum_{j\in J} |f(x_{j+1}) - f(x_j)| = \tilde{S} - \sum_{j\in I} (|f(x_{j+1}) - f(t'_j)| + |f(t'_j) - f(x_j)|)$$
$$\geq \tilde{S} - 2\varepsilon.$$

Since the joint partition is a sub-partition of the original t_i's, its variation can only increase, $\tilde{S} \geq S$, and we finally obtain

$$S_1 \geq S - 2\varepsilon \geq V - \varepsilon - 2\varepsilon = V - 3\varepsilon.$$

This implies that V is a limit on any partition with maximal distance $\delta \to 0$.

Problem 1.10: Repeat the previous problem for càdlàg functions.

Solution. Let f be a càdlàg function on $[a, b]$ with FV $V = V_f([a, b])$ according to the classical definition, Equation (1.7*), which is a supremum over all partitions of $[a, b]$. We show that for a given ε, there is a δ, such that for *any partition* with maximal distance between points δ, the variation S_1 on this partition satisfies $V - 4\varepsilon \leq S_1 \leq V$.

Fix $\varepsilon > 0$. By definition of supremum, there is a partition $t_0, t_1, \ldots, t_n$ such that $S = \sum_{i=1}^{n-1} |f(t_{i+1} - f(t_i)| \geq V - \varepsilon$. Let f^d be a pure jump function with jumps at $\{x \in [a, b] \colon |\Delta f(x)| > \frac{\varepsilon}{n+1}\}$. Note that there are only a finite number of such jumps. Let these jump points be $(u_1, u_2, \ldots, u_p)$. Let $f^c = f - f^d$.

Now, take a partition $\{x_0, x_1, \ldots, x_m\}$ that contains $\{u_1, \ldots, u_p\}$ with $x_i - x_{i-1} < \delta$ for all $i = 1, \ldots, m$, where δ is small enough so that there is at most one t in each subinterval and $|f^c(t) - f^c(s)| < \frac{\varepsilon}{n+1}$ if $|t - s| < \delta$. Let $S_1 = \sum_{j=1}^{m} |f(x_j) - f(x_{j-1})|$. Denote by J, the set of indices such that $[x_{j-1}, x_j)$ does not contain any of the points t_i. Let

$$\tilde{S} = \sum_{j\in J} |f(x_j) - f(x_{j-1})| + \sum_{j\in J^c} (|f(x_j) - f(t_{\ell(j)})| + |f(t_{\ell(j)}) - f(x_{j-1})|),$$

where $\ell(j) = \{i : t_i \in [x_{j-1}, x_j)\}$. Then,

$$\begin{aligned}\tilde{S} \le \sum_{j\in J} |f(x_j) - f(x_{j-1})| + \sum_{j\in J^c} (|f^c(x_j) - f^c(t_{\ell(j)})| \\ + |f^d(x_j) - f^d(t_{\ell(j)})| + |f^c(t_{\ell(j)}) - f^c(x_{j-1})| \\ + |f^d(t_{\ell(j)}) - f^d(x_{j-1})|).\end{aligned}$$

Observe that $|f^d(t_{\ell(j)}) - f^d(x_{j-1})| = 0$ by construction and that f is right continuous. Note also that for $j \in J^c$, $f^d(t_{\ell(j)}) = f^d(x_{j-1})$ and thus

$$\begin{aligned}\sum_{j\in J} |f(x_j) - f(x_{j-1})| + \sum_{j\in J^c} |f^d(x_j) - f^d(t_{\ell(j)})| \\ \le S_1 + \sum_{j\in J^c} |f^c(x_j) - f^c(x_{j-1})|.\end{aligned}$$

Therefore, we have

$$\begin{aligned}\tilde{S} \le S_1 + \sum_{j\in J^c} (|f^c(x_j) - f^c(x_{j-1})| + |f^c(x_j) - f^c(t_{\ell(j)})| \\ - |f^c(t_{\ell(j)}) - f^c(x_{j-1})|),\end{aligned}$$

which is not larger than $S_1 + 3\varepsilon$, giving

$$S_1 \ge \tilde{S} - 3\varepsilon \ge S - 3\varepsilon \ge V - 4\varepsilon.$$

But we also have $S_1 \le V$. This implies that V is a limit on any partition with maximal distance $\delta \to 0$.

Problem 1.11: Prove the polarisation identity and linearity (Equations (1.16*) and (1.17*)) for quadratic variation.

Solution. Clearly, (1.17*) follows readily from using the definition (1.15*) of quadratic variation together with a few simple algebraic manipulations. Further, we may derive (1.16*) from (1.17*) together with symmetry as

$$\begin{aligned}
&\frac{1}{2}([f+g,f+g]-[f,f]-[g,g]) \\
&\quad=\frac{1}{2}([f,f+g]+[g,f+g]-[f,f]-[g,g]) \\
&\quad=\frac{1}{2}([f,f]+[f,g]+[g,f]+[g,g]-[f,f]-[g,g]) \\
&\quad=\frac{1}{2}([f,g]+[g,f]) \\
&\quad=[f,g]=[g,f].
\end{aligned}$$

Problem 1.12: Calculate the RS integral of a continuous function $f : [0,\infty) \to \mathbb{R}$ with respect to the function $g(x) = \lfloor x \rfloor$, the integer part of x, over an interval $(a,b] \subseteq [0,\infty)$.

Solution. Selecting integers $k,\ell \in \mathbb{N}$ such that $\lfloor a \rfloor = k$ and $\lfloor b \rfloor = \ell$, we either have $k=\ell$, in which case $k \le a < b < k+1$ and $\int_{(a,b]} f\,dg = 0$, or else $k<\ell$, in which case $k \le a < k+1 \le \ell \le b < \ell+1$ and $\int_{(a,b]} f\,dg = \sum_{n=\lfloor a \rfloor+1}^{\lfloor b \rfloor} f(n)$.

Problem 1.13: Explain why the RS integral $\int f\,dg$ is not well-defined when $[f,g] \neq 0$.

Solution. $[f,g]$ is the limit over the shrinking partitions of the sums of the products of increments:

$$\begin{aligned}
[f,g] &= \lim_{\delta\to 0}\sum_{i=0}^{n-1}(f(t_{i+1})-f(t_i))(g(t_{i+1})-g(t_i)) \\
&= \lim_{\delta\to 0}\sum_{i=0}^{n-1} f(t_{i+1})(g(t_{i+1})-g(t_i)) \\
&\quad - \lim_{\delta\to 0}\sum_{i=0}^{n-1} f(t_i)(g(t_{i+1})-g(t_i)).
\end{aligned}$$

If the RS integral $\int f\,dg$ exists, both limits are the same, giving $[f,g]=0$. On the other hand, if $[f,g]\neq 0$, the limits are not the same, contradicting the definition which states that the limit

$$\lim_{\delta\to 0}\sum_{i=0}^{n-1} f(\xi_i)(g(t_{i+1})-g(t_i)) = \int f dg$$

should be the same for any choice of $\xi_i \in [t_i, t_{i+1}]$.

Problem 1.14: For two functions $f, g : [0, \infty) \to \mathbb{R}$, we know the values of the quadratic variations $[f+g, f+g](t)$ and $[f-g, f-g](t)$ at a time $t > 0$. Based on this information alone, which of the quantities $[f, f](t)$, $[g, g](t)$, and $[f, g](t)$ can be determined?

Solution. To know the values of $[f+g, f+g](t)$ and $[f-g, f-g](t)$ is equivalent to knowing the values of $\frac{1}{4}([f+g, f+g](t) - [f-g, f-g](t)) = [f, g](t)$ and $\frac{1}{2}([f+g, f+g](t) + [f-g, f-g](t)) = [f, f](t) + [g, g](t)$. Hence, we can say what the value of $[f, g](t)$ is, but not what the value of $[f, f](t)$ or $[g, g](t)$ is (as we know the value of the sum of these two only).

Problem 1.15: Prove the Gronwall inequality: if $0 \le f(t) \le a + b\int_0^t f(s)ds$, $t \ge 0$, then $f(t) \le ae^{bt}$. *Hint*: Let $g(t) = a + b\int_0^t f(s)ds$. (Note that differentiation does not preserve an inequality but integration does.)

Solution. Note that $0 \le f(0) \le a$ and that the given inequality can be written as $0 \le f(t) \le g(t)$, $t \ge 0$. Let $t_0 = \inf\{t \ge 0 : g(t) > 0\}$. If $t_0 = +\infty$, then $f \equiv 0$ and the result immediately holds.

Suppose $t_0 < +\infty$. We have $g'(t) = bf(t)$, and on $(t_0, +\infty)$, the given inequality becomes $g'(t)/g(t) \le b$. Integrating this inequality, we get

$$\ln g(t) - \ln g(t_0) \le b(t - t_0), \text{ i.e. } g(t) \le g(t_0)e^{b(t-t_0)}.$$

If $t_0 > 0$, then $g(t_0) = 0$, and g and consequently f are identically nil. The result again follows immediately.

Suppose $t_0 = 0$. Then, $f(t) \le g(t) \le g(0)e^{bt} = ae^{bt}$.

Chapter 2

Concepts of Probability Theory

Continuous Probability Model

In this section, we teach the scientific probability model that is used by all professional probabilists and also by math students from the graduate level and upwards (and sometimes even earlier). The model is often called the Kolmogorov probability model after the greatest probabilist of all time, A.N. Kolmogorov (1903–1987).

Let Ω be a non-empty set. This set will be the mathematical model for all the possible outcomes of a random experiment and is called the sample space. Elements ω of Ω are thus called outcomes.

A family $\mathcal{F}$ of subsets of Ω is called a σ-field if the following axioms hold:

- $\Omega \in \mathcal{F}$.
- $A \in \mathcal{F} \Rightarrow \overline{A} = A^c \in \mathcal{F}$.
- $A_1, A_2, \ldots \in \mathcal{F} \Rightarrow \bigcup_{n=1}^{\infty} A_n \in \mathcal{F}$.

Members of $\mathcal{F}$ are called measurable sets or events, and the pair $(\Omega, \mathcal{F})$ is called a measurable space. A measurable space is the mathematical model for a random experiment with outcomes in Ω and events in $\mathcal{F}$ (but without assigned probabilities as of yet). The important difference as compared with elementary probability theory is that, in general, not all subsets of Ω are events. (In fact, had they been assumed to be so, one can show that it leads to contradictions for non-trivial continuous probability setups.)

As in elementary probability theory, an event happens when we do the random experiment if the outcome ω of the random experiment is a member of the event.

From the above three axioms for a σ-field, several additional closure properties can easily be derived, such as $\emptyset \in \mathcal{F}$, closedness under finite unions, closedness under countable and finite intersections, and closedness under set differences.

Arbitrary intersections of σ-fields are σ-fields. A union $\bigcup_n \mathcal{F}_n$ of σ-fields is usually not a σ-field, but there is a smallest σ-field denoted $\bigvee_n \mathcal{F}_n$ containing $\bigcup_n \mathcal{F}_n$ (by the usage of the closedness under intersections) which is called the σ-field generated by $\{\mathcal{F}_n\}$. Likewise, for any family $\mathcal{G}$ of subsets of Ω, there is a smallest σ-field $\sigma(\mathcal{G})$ that contains the family which is called the σ-field generated by $\mathcal{G}$.

Three basic examples of σ-fields are $\{\emptyset, \Omega\}$, the family of all subsets of Ω and $\{\emptyset, A, \overline{A}, \Omega\}$ for any $A \subseteq \Omega$. One substantially less basic and particularly important example of a σ-field is the σ-field $\mathcal{B}$ of subsets of $\mathbb{R}$ generated by all intervals, which is called the Borel σ-field. Similarly, the Borel σ-field of subsets of any interval is the σ-field generated by all subintervals of the interval, which in turn simply boils down to $\mathcal{B}$ intersected with the interval.

A probability measure P on $(\Omega, \mathcal{F})$ is a function $\mathrm{P} : \mathcal{F} \to [0, 1]$ satisfying the following axioms:

- $\mathrm{P}(\Omega) = 1$.
- $\mathrm{P}(\overline{A}) = 1 - \mathrm{P}(A) \quad \text{for } A \in \mathcal{F}$.
- $\mathrm{P}(\bigcup_{n=1}^{\infty} A_n) = \sum_{n=1}^{\infty} \mathrm{P}(A_n) \quad \text{for disjoint } A_1, A_2, \ldots \in \mathcal{F}$.

The important difference as compared with elementary probability theory (besides that not necessarily all subsets of Ω are supposed to have assigned probabilities) is the third axiom, which is only assumed to hold for finite unions (that is, $A_{n+1} = A_{n+2} = \cdots = \emptyset$ for some $n \in \mathbb{N}$) in elementary probability. This, one can say, gives the new probability model the same advantage over the elementary one as the Lebesgue integral has over the Riemann integral (see the following). The triple $(\Omega, \mathcal{F}, \mathrm{P})$ is called a probability space and is the mathematical model for a random experiment with assigned probabilities.

From the three axioms for a probability measure, several additional properties can be easily derived. Besides the properties we are familiar with from elementary probability theory, two important new ones (not generally valid in elementary probability theory) are that

$\mathrm{P}(\bigcup_{n=1}^{\infty} A_n) = \lim_{n\to\infty} \mathrm{P}(A_n)$ for events $A_1 \subseteq A_2 \subseteq \cdots$ and $\mathrm{P}(\bigcap_{n=1}^{\infty} A_n) = \lim_{n\to\infty} \mathrm{P}(A_n)$ for events $A_1 \supseteq A_2 \supseteq \cdots$.

Given a sequence of events A_n, $n = 1, 2, \ldots$, there are two sets defined from them: the first is the set that contains all ω's which belong to infinitely many of A_n's, expressed as (spelled infinitely often) $\{A_n \ i.o.\} = \cap_{m=1}^{\infty} \cup_{n=m}^{\infty} A_n$, also known as $\limsup A_n$. The second is the set that contains all ω's which belong to all of the sets from some index onwards, expressed as (spelled eventually) $\{A_n \ ev.\} = \cup_{m=1}^{\infty} \cap_{n=m}^{\infty} A_n$, also known as $\liminf A_n$. An important result regarding the probability of such sets is given by the Borel–Cantelli lemma. It states that if the series converges $\sum_{n=1}^{\infty} \mathrm{P}(A_n) < \infty$, then $\mathrm{P}(A_n \ i.o.) = 0$. If A_n's are independent, then the divergence of the series $\sum_{n=1}^{\infty} \mathrm{P}(A_n) = \infty$ implies that $\mathrm{P}(A_n \ i.o.) = 1$.

A (real-valued) random variable X is a function $X : \Omega \to \mathbb{R}$ such that

$$\{X \in B\} = \{\omega \in \Omega : X(\omega) \in B\} = X^{-1}(B) \in \mathcal{F} \quad \text{for } B \in \mathcal{B}.$$

A random variable X is also called a measurable function from Ω to $\mathbb{R}$. Symbolically, this can be written $X^{-1}(\mathcal{B}) \subseteq \mathcal{F}$. The important difference as compared with elementary probability theory is that not all functions from Ω to $\mathbb{R}$ are random variables but only the measurable ones.

As the family of sets $B \in \mathcal{B}$ such that $\{X \in B\} \in \mathcal{F}$ is a σ-field, it is not hard to see that X is a random variable if and only if

$$\{X \leq x\} = \{\omega \in \Omega : X(\omega) \leq x\} = X^{-1}((-\infty, x]) \in \mathcal{F} \quad \text{for } x \in \mathbb{R}.$$

This of course in turn means that the CDF $F_X(x) = \mathrm{P}(X \leq x)$ makes sense (i.e. it is well-defined). Of course, facts that are trivially true in elementary probability, such as that the sum of random variables is a random variable, must remain valid in the new theory (as it would otherwise not been in use). Additional important properties are that point-wise limits of random variables (measurable functions) are random variables (measurable) and that the decomposition $g(X)$ of a measurable function $g : \mathbb{R} \to \mathbb{R}$ (where $\mathbb{R}$ is equipped with the Borel σ-field $\mathcal{B}$) with a random variable X is a random variable. In particular, the decomposition $g(X)$ of a continuous function $g : \mathbb{R} \to \mathbb{R}$ with a random variable X is a random variable. The proofs of such facts require a few technical manipulations of varying difficulty levels.

The σ-field $\sigma(X)$ generated by a random variable X is the smallest sub-σ-field to $\mathcal{F}$ (that is, σ-field contained in $\mathcal{F}$) that satisfies $X^{-1}(\mathcal{B}) \in \sigma(X)$. Clearly, $\sigma(X)$ is the smallest σ-field of subsets of Ω that makes X a random variable (measurable).

σ-fields have an interpretation as information. This will be important when we discuss conditional expectations later in this chapter. To know exactly which of the events in $\mathcal{F}$ that happens when the random experiment is carried out means that we know the value of all random variables. Less knowledge than that means that we do not know the value of all random variables. Likewise, to know exactly which of the events in $\sigma(X)$ that happens for a random variable X when the random experiment is carried out means that we know the value of X.

One basic non-trivial example of a random variable is the indicator I_A of any $A \in \mathcal{F}$ given by $I_A(\omega) = 1$ for $\omega \in A$ and $I_A(\omega) = 0$ for $\omega \in \overline{A}$. This is simply a Bernoulli-distributed random variable. For I_A, we have $\sigma(I_A) = \{\emptyset, A, \overline{A}, \Omega\}$.

Convergence

Convergence of functions is a more versatile and complicated concept than convergence of real numbers as there are many ways in which a sequence of functions can converge to a limit function. As random variables are functions, the same applies to them. There are four main ways in which a sequence $\{X_n\}_{n=1}^{\infty}$ of random variables can converge to a limit random variable X:

- X_n converges to X in distribution ($X_n \to_D X$) if the CDFs satisfy $\lim_{n\to\infty} F_n(x) = F(x)$ for all continuity points x of F.
- X_n converges to X in probability ($X_n \to_{\mathrm{P}} X$) if $\lim_{n\to\infty} \mathrm{P}(|X_n - X| > \varepsilon) = 0$ for $\varepsilon > 0$.
- X_n converges to X almost surely ($X_n \to_{\text{a.s.}} X$) if $\lim_{n\to\infty} X_n(\omega) \to X(\omega)$ for ω outside an event with probability zero.
- X_n converges to X in r'th mean for an $r \geq 1$ ($X_n \to_{L^r} X$) if $\mathrm{E}(|X|^r) < \infty$ and $\lim_{n\to\infty} \mathrm{E}(|X_n - X|^r) = 0$.

Convergence in distribution is also called weak convergence. It is qualitatively different from the other modes of convergence in that it does not require the random variables $\{X_n\}$ and X to be defined on the same probability space, but just to have similar CDFs. Each of the other three modes of convergence implies convergence in distribution. We have $X_n \to_D X$ if and only if $\lim_{n\to\infty} \mathrm{E}(e^{itX_n}) = \mathrm{E}(e^{itX})$ for $t \in \mathbb{R}$, that is, pointwise convergence of the corresponding characteristic functions.

Convergence in probability takes place if and only if $\{X_n\}$ is a Cauchy sequence in probability, which is to say $\lim_{m,n\to\infty} \mathrm{P}(|X_m - X_n| > \varepsilon) = 0$

for $\varepsilon > 0$. Either of almost sure convergence or convergence in r'th mean implies convergence in probability.

No implications exist between convergence almost surely and convergence in r'th mean as neither of them implies the other in general.

Convergence in r'th mean is also called convergence in (the Banach space) L^r. Convergence in L^r takes place if and only if $\{X_n\}$ is a Cauchy sequence in r'th mean, which is to say that $\mathrm{E}(|X_n|^r) < \infty$ for n sufficiently large and $\lim_{m,n\to\infty} \mathrm{E}(|X_m - X_n|^r) = 0$. When $X_n \to_{L^r} X$, it holds that $\lim_{n\to\infty} \mathrm{E}(X_n^s) = \mathrm{E}(X^s)$ for $s \in [0, r]$ (with s integer valued if the X_n's are not positive). Convergence in L^1 and L^2 are the most important cases of convergence in the mean. By the so-called Loève criterion, convergence in (the Hilbert space) L^2 takes place if and only if $\lim_{m,n\to\infty} \mathrm{E}(X_m X_n)$ exists.

The above-mentioned so-called Cauchy criteria as well as the Loève criterion are very useful tools to establish convergence of a sequence as they do not require that the limit random variable is known but guarantee its existence without that knowledge. We make crucial use of this feature when we construct the so-called Itô stochastic integral in Chapter 4.

Expectation and Lebesgue Integral

The Lebesgue integral (as an improvement of the Riemann integral) emerged around the turn of the twentieth century. It is arguably the most important tool of modern mathematical analysis/calculus. It is also virtually indispensable for modern rigorous probability theory.

We present the integral from a probabilistic perspective as the expected value $\mathrm{E}(X)$ of a random variable X defined on a probability space $(\Omega, \mathcal{F}, \mathrm{P})$ since that is what we will use it for. But our presentation is in fact also valid for any measurable function $f : \Omega \to \mathbb{R}$ from a measurable space $(\Omega, \mathcal{F})$ with a measure P [that need not necessarily be a probability measure with total mass $\mathrm{P}(\Omega) = 1$] to $\mathbb{R}$. The integral that we label

$$\mathrm{E}(X) = \int_\Omega X\,dP = \int_{\omega\in\Omega} X(\omega)\,dP(\omega)$$

has exactly the same building blocks as the math integral $\int_\Omega f\,dP$, where f, Ω, and P may have a less probabilistic interpretation than that indicated above.

The Riemann integral is done by dividing the interval (x-domain) $[a, b]$ that $f : [a, b] \to \mathbb{R}$ should be integrated over into small subintervals (see Chapter 1) and then sampling the (y-) value of f in each subinterval. The approximate value of the integral is the sum of the length of each subinterval times the sampled values of f. For the Lebesgue integral, one instead divides the range of (y-) values of f into small y-subintervals. The approximate value of the integral is then the sum of the length of each x-interval in which f takes values in one of the y-subintervals times the lower endpoint of the corresponding y-subinterval. Written on paper, this becomes

$$\int_{[a,b]} f(x)\,dx \leftarrow \sum_{i=1}^{n} \text{length}(\{x \in [a,b] : f(x) \in [y_{i-1}^n, y_i^n)\})\, y_{i-1}^n,$$

where $\min(f) = y_0^n < y_1^n < \cdots < y_n^n = \max(f)$ and $\max_{1\le i\le n} y_i^n - y_{i-1}^n \downarrow 0$. Now, f could have an infinite range of values. And we want to make sure that the approximating sum (unlike an approximating Riemann or RS sum) always converges. Also, we want to be able to integrate over abstract measurable spaces $(\Omega, \mathcal{F})$ and not just $(\mathbb{R}, \mathcal{B})$. And we are primarily interested in random variables X and their expectation $\mathrm{E}(X)$ rather than math functions f and their integral $\int_\Omega f\,dP$.

So, let X be a random variable defined on a probability space $(\Omega, \mathcal{F}, \mathrm{P})$. We can express X as the difference between two positive random variables $X = X^+ - X^-$, where $X^+ = \max(X, 0)$ and $X^- = \max(-X, 0)$. By defining $\mathrm{E}(X) = \mathrm{E}(X^+) - \mathrm{E}(X^-)$ whenever at least one of $\mathrm{E}(X^+)$ and $\mathrm{E}(X^-)$ is finite, it is sufficient to define the expectation (integral) $\mathrm{E}(X)$ for positive random variables. That definition can be given as

$$\begin{aligned}\mathrm{E}(X) &= \lim_{n\to\infty} \sum_{k=0}^{2^n n - 1} \frac{k}{2^n}\, \mathrm{P}\left(X \in \left[\frac{k}{2^n}, \frac{k+1}{2^n}\right)\right) \\ &= \lim_{n\to\infty} \sum_{k=0}^{2^n n - 1} \frac{k}{2^n}\, \mathrm{P}\left(\left\{\omega \in \Omega : X(\omega) \in \left[\frac{k}{2^n}, \frac{k+1}{2^n}\right)\right\}\right).\end{aligned}$$

Note that the limit and hence the expectation always exist (although they can be infinite) as the sum increases with n.

A random variable is called simple if it is a linear combination of Bernoulli random variables

$$X(\omega) = \sum_{i=1}^{m} c_i I_{A_i}(\omega),$$

where $A_1, \ldots, A_m \in \mathcal{F}$ are disjoint and $c_1, \ldots, c_m \in \mathbb{R}$. It is not hard to see that the expectation of this random variable according to the definition above is

$$\mathrm{E}(X) = \sum_{i=1}^{m} c_i \mathrm{P}(X \in A_i).$$

This in turn is also exactly what we expect from intuition that the expectation of such a random variable should be. Any positive random variable X can be approximated from below with positive simple random variables $\{X_n\}_{n=1}^{\infty}$ such that $X_n(\omega) \uparrow X(\omega)$ as $n \to \infty$. One example of such an approximation is the one we have seen already:

$$X_n(\omega) = \sum_{k=0}^{2^n n-1} \frac{k}{2^n} I_{[\frac{k}{2^n}, \frac{k+1}{2^n})}(X(\omega)).$$

Of course, the limit $\lim_{n\to\infty} \mathrm{E}(X_n)$ exists as it is an increasing limit. It can be shown (and in fact easily proven) that this limit is not dependent on the particular choice of the approximating sequence of simple random variables. Thus, we can alternatively first define the expectation for simple random variables in the mentioned intuitive appealing manner. The expectation of any positive random variable is then obtained by approximating it from below with positive simple random variables. Finally, we define $\mathrm{E}(X)$ for a not necessarily positive random variable as $\mathrm{E}(X^+) - \mathrm{E}(X^-)$ whenever at least one of $\mathrm{E}(X^+)$ and $\mathrm{E}(X^-)$ is finite.

Not only does the Lebesgue integral exist (albeit possibly infinite) for all positive random variables, but it also has a unified definition and appearance regardless of whether X is discrete, continuous or neither of those alternatives. It also has superior theoretical properties. Two obvious properties that it shares with the Riemann integral is linearity and positivity. The following three more sophisticated properties though are not valid for the Riemann integral in general:

- (Monotone Convergence) If $X_n \geq 0$ (pointwise) and X_n increases to a limit X (pointwise), then $\lim_{n\to\infty} \mathrm{E}(X_n) = \mathrm{E}(X)$.
- (Fatou's Lemma) If $X_n \geq 0$, then $\mathrm{E}(\liminf_{n\to\infty} X_n) \leq \liminf_{n\to\infty} \mathrm{E}(X_n)$.
- (Dominated Convergence) If $X_n \to_P X$ and $\mathrm{E}(\sup_{n\geq 1} |X_n|) < \infty$, then $\lim_{n\to\infty} \mathrm{E}(X_n) = \mathrm{E}(X)$.

The proofs of these results are not very difficult.

A random variable X on a probability space $(\Omega, \mathcal{F}, \mathrm{P})$ generates a so-called Stieltjes probability measure dF_X on $(\mathbb{R}, \mathcal{B})$ by $dF_X(B) = \mathrm{P}(X \in B)$ for $B \in \mathcal{B}$. Obviously, the CDF F_X of X satisfies $F_X(x) = dF_X((-\infty, x])$, which is of course the motivation for the notation dF_X. The following important formula holds for a measurable function $g : \mathbb{R} \to \mathbb{R}$:

$$\mathrm{E}(g(X)) = \int_\Omega g(X)\, dP = \int_\mathbb{R} g(x)\, dF_X(x).$$

This in turn follows from an abstract version of the change of variable formula, which we leave out from this treatment.

Conditional Expectation

Stochastic calculus uses a lot of conditional expectations. The elementary notion of conditional expectation turns out to be insufficient here, and the more abstract notion of conditional expectation used in measure-theoretic probability theory (that is, probability theory making use of the abstract Lebesgue integral instead of the Riemann integral or RS integral) is required.

Let X be a random variable on a probability space $(\Omega, \mathcal{F}, \mathrm{P})$ with finite expectation $\mathrm{E}(|X|) < \infty$. Such an X is called integrable by the way. The abstract conditional expectation $\mathrm{E}(X|\mathcal{G})$ of X with respect to a σ-field $\mathcal{G} \subseteq \mathcal{F}$ is a random variable. This is of course unlike the elementary conditional expectation $\mathrm{E}(X|Y = y)$ with respect to another random variable Y which is non-random. The relation between the abstract conditional expectation and the elementary one is that $\mathrm{E}(X|\sigma(Y)) = g(Y)$, where $g : \mathbb{R} \to \mathbb{R}$ is a measurable function satisfying $\mathrm{E}(X|Y = y) = g(y)$. This is not so hard to prove once the theory for the abstract conditional expectation has been developed. Hence, it makes sense to use $\mathrm{E}(X|Y)$ as shorthand notation for $\mathrm{E}(X|\sigma(Y))$.

With the above assumptions on X and $\mathcal{G}$, it follows from the so-called Radon–Nikodym theorem that there exists a $\mathcal{G}$-measurable random variable that we label $\mathrm{E}(X|\mathcal{G})$ which is such that

$$\mathrm{E}(I_B\, \mathrm{E}(X|\mathcal{G})) = \mathrm{E}(I_B X) \quad \text{for all } \mathcal{G}\text{-measurable events } B \in \mathcal{G}.$$

The random variable $\mathrm{E}(X|\mathcal{G})$ is uniquely determined except for its values on an event with probability zero. In other words, two versions of the conditional expectations only differ on such a so-called null event. One can replace $\mathcal{G}$-measurable indicator random variables I_B with bounded (by a constant) $\mathcal{G}$-measurable random variables ξ in the above theorem and definition, as that gives the same conditional expectation $\mathrm{E}(X|\mathcal{G})$.

Besides linearity and positivity, the important properties of the abstract conditional expectation are as follows:

- $\mathrm{E}(X|\{\emptyset,\Omega\}) = \mathrm{E}(X)$ and $\mathrm{E}(X|\mathcal{F}) = X$.
- $\mathrm{E}(XY|\mathcal{G}) = X\,\mathrm{E}(Y|\mathcal{G})$ when X is $\mathcal{G}$-measurable.
- (TOWERING/SMOOTHING) $\mathrm{E}(\mathrm{E}(X|\mathcal{G}_2)|\mathcal{G}_1) = \mathrm{E}(X|\mathcal{G}_1)$ for $\mathcal{G}_1 \subseteq \mathcal{G}_2$.
- $\mathrm{E}(\mathrm{E}(X|\mathcal{G})) = \mathrm{E}(X)$.
- $\mathrm{E}(X|\mathcal{G}) = \mathrm{E}(X)$ when $\mathcal{G}$ and $\sigma(X)$ are independent.
- $\mathrm{E}(X|\mathcal{G}_1 \vee \mathcal{G}_2) = \mathrm{E}(X|\mathcal{G}_1)$ when $\mathcal{G}_2$ is independent of $\sigma(X)$ and $\mathcal{G}_1$.
- (JENSEN'S INEQUALITY) $g(\mathrm{E}(X|\mathcal{G})) \leq \mathrm{E}(g(X)|\mathcal{G})$ when g is convex.
- $\mathrm{E}(g(X,Y)|\sigma(Y)) = \mathrm{E}(g(X,y))\big|_{y=Y}$ when X and Y are independent.
- (MONOTONE CONVERGENCE) $\lim_{n\to\infty} \mathrm{E}(X_n|\mathcal{G}) = \mathrm{E}(X|\mathcal{G})$ when $0 \leq X_n \uparrow X$.
- (FATOU'S LEMMA) $\mathrm{E}(\liminf_{n\to\infty} X_n|\mathcal{G}) \leq \liminf_{n\to\infty} \mathrm{E}(X_n|\mathcal{G})$ when $X_n \geq 0$.
- (DOMINATED CONVERGENCE) $\lim_{n\to\infty} \mathrm{E}(X_n|\mathcal{G}) = \mathrm{E}(X|\mathcal{G})$ when $X_n \to_{\text{a.s.}} X$ and $\mathrm{E}(\sup_{n\geq 1} |X_n|) < \infty$.

Most of these properties are more or less immediate from the definition, while the remaining few are not hard to prove either but their proofs require some technical manipulations.

Abstract conditional probability is defined using conditional expectation (instead of the other way around) as $\mathrm{P}(A|\mathcal{G}) = \mathrm{E}(I_A|\mathcal{G})$ for $A \in \mathcal{F}$.

Stochastic Processes

A stochastic process with (time) parameter set $T \subseteq \mathbb{R}$ is a family $X = \{X(t)\}_{t\in T}$ of random variables defined on a common probability space $(\Omega, \mathcal{F}, \mathrm{P})$. In stochastic calculus, processes with time parameter set $[0,\infty)$ or $[0,T]$, $T > 0$, are the ones that are of interest. We present some basic definitions for the time parameter set $[0,\infty)$ in the following, but they could be rephrased for the time parameter set $[0,T]$ through quite obvious modifications.

A filtration $\{\mathcal{F}_t\}_{\geq 0}$ is an increasing family of σ-fields all contained in $\mathcal{F}$. When $(\Omega, \mathcal{F}, \mathrm{P})$ is equipped with a filtration, we call $(\Omega, \mathcal{F}, \{\mathcal{F}_t\}_{\geq 0}, \mathrm{P})$ a filtered probability space. A stochastic process $\{X(t)\}_{t\geq 0}$ is called adapted to a filtration $\{\mathcal{F}_t\}_{\geq 0}$ if $X(t)$ is $\mathcal{F}_t$-measurable for $t \geq 0$. The smallest filtration that X is adapted to is the filtration generated by X itself $\{\mathcal{F}_t^X\}_{t\geq 0}$ given by $\mathcal{F}_t^X = \bigvee_{s\in[0,t]} \sigma(X(s))$ for $t \geq 0$.

An adapted stochastic process $\{X(t)\}_{t\geq 0}$ is a martingale if $\mathrm{E}(|X(t)|) < \infty$ for $t \geq 0$ and $\mathrm{E}(X(t)|\mathcal{F}_s) = X(s)$ for $0 \leq s \leq t$. If instead $\mathrm{E}(X(t)|\mathcal{F}_s) \geq (\leq)\, X(s)$ for $0 \leq s \leq t$, we call X a submartingale (supermartingale). These definitions at first glance might seem to not lead to very much of interest. On the contrary, it turns out that martingale theory is one of the richest subdisciplines of probability theory. The deeper aspects of that theory though are very difficult and involve some of the arguably most difficult proofs in probability.

From the properties of conditional expectation, it is easy to see that martingales have constant mean (not depending on t) while submartingales (supermartingales) have increasing (decreasing) mean.

The canonical non-trivial example of a martingale is the family of Doob–Lévy martingales $X(t) = \mathrm{E}(Y|\mathcal{F}_t)$ for $t \geq 0$ for any choice of an integrable random variable Y and a filtration $\{\mathcal{F}_t\}_{\geq 0}$. This is established by means of towering. One of the easiest ways to find a submartingale (supermartingale) is to apply a convex (concave) function to a martingale. This is established by means of Jensen's inequality.

A stochastic process $\{X(t)\}_{t\geq 0}$ is a Markov process if $\mathrm{P}(X(t) \in A|\mathcal{F}_s^X) = \mathrm{P}(X(t) \in A|X(s))$ for $0 \leq s \leq t$ and $A \in \mathcal{B}$. The canonical non-trivial example of a Markov process is an independent increment process. The transition CDF of a Markov process is given by

$$P(y, t, x, s) = \mathrm{P}(X(t) \leq y|X(s) = x) \quad \text{for } 0 \leq s \leq t \text{ and } x, y \in \mathbb{R}.$$

Usually (for the Markov processes we employ at least), the corresponding probability density function (PDF) exists and is denoted

$$p(y, t, x, s) = \frac{d}{dy} P(y, t, s, x).$$

Most Markov processes encountered in applied math are time-homogeneous, meaning that $P(y, t+s, x, s)$ does not depend on s. In that case, we write $P(t, x, y) = P(y, t+s, x, s)$ and $p(t, x, y) = p(y, t+s, x, s)$. Also, the theory of Markov processes is very rich and technically very complicated.

Both martingale theory and Markov process theory are indispensable tools in stochastic calculus. Therefore, a really deep understanding of stochastic calculus is not easily acquired. This applies in particular to non-Brownian motion stochastic calculus (semimartingales), which we discuss in Chapter 8.

A random variable $\tau \geq 0$ is a stopping time with respect to a filtration $\{\mathcal{F}_t\}_{\geq 0}$ if

$$\{\tau \leq t\} = \{\omega \in \Omega : \tau(\omega) \leq t\} \in \mathcal{F}_t \quad \text{for } t \geq 0.$$

Here, it is often suitable to allow τ to also take the value ∞ for reasons that become apparent in the following paragraph. If $\mathrm{P}(\tau < \infty) = 1$, we say that τ is finite.

Any non-random time $T \geq 0$ is a stopping time as is $\tau_1 \wedge \tau_2$ when τ_1 and τ_2 are stopping times.

If X is a cádlág adapted process to a right-continuous filtration $\{\mathcal{F}_t\}_{t\geq 0}$ (that is, $\bigcap_{s>t} \mathcal{F}_s = \mathcal{F}_{t+} = \mathcal{F}_t$ for $t \geq 0$), then the hitting time $T_D = \inf\{t \geq 0 : X(t) \in D\}$ is a stopping time for an open set D. If X is continuous, this is also true for D closed.

If $\tau \geq 0$ is a stopping time with respect to a right-continuous filtration $\{\mathcal{F}_t\}_{\geq 0}$, the σ-field

$$\mathcal{F}_\tau = \{A \in \mathcal{F} : A \cap \{\tau \leq t\} \in \mathcal{F}_t \text{ for } t \geq 0\}$$

turns out to be of particular interest in both martingale theory and Markov theory. A Markov process is a strong Markov process if $\mathrm{P}(X(t+\tau) \in A|\mathcal{F}_\tau^X) = \mathrm{P}(X(t+\tau) \in A|X(\tau))$ for $t \geq 0$ and $A \in \mathcal{B}$ for any stopping time τ with respect to the filtration $\{\mathcal{F}_t^X\}_{\geq 0}$. Virtually all Markov processes encountered are strong Markov processes, although counter examples can be constructed.

Problems

Problem 2.1:

(a) Show that mutually exclusive events both of positive probability cannot be independent.

(b) If A and B are independent events and I_A, I_B are their indicators, show that I_A and I_B are independent by showing that for any two functions $h(x)$ and $g(y)$,

$$\mathrm{E}(h(I_A)g(I_B)) = \mathrm{E}(h(I_A))\mathrm{E}(g(I_B)).$$

Solution.

(a) In this case, $A \cap B = \emptyset$ and $\mathrm{P}(A \cap B) = 0$. However, $\mathrm{P}(A)\mathrm{P}(B) > 0$ so that $\mathrm{P}(A \cap B) \neq \mathrm{P}(A)\mathrm{P}(B)$.

(b) Since the quantity $\mathrm{E}(h(I_A)g(I_B)) - \mathrm{E}(h(I_A))\mathrm{E}(g(I_B))$ is unchanged when constants are added to h and g, one can assume without loss of generality that $h(0) = g(0) = 0$ (check it!). Since I_A and I_B take values 0 and 1 only, we let $h(0) = 0$, $h(1) = a$, $g(0) = 0$, $g(1) = b$. Then, the variable $h(I_A)g(I_B)$ takes values 0 and ab, and recalling that $\{I_A = 1\} = A$, $\{I_A = 0\} = A^c$ and similarly for B,

$$\begin{aligned}\mathrm{E}(h(I_A)g(I_B)) &= h(0)g(0)\mathrm{P}(A^c \cap B^c) + h(1)g(0)\mathrm{P}(A \cap B^c)\\ &\quad + h(0)g(1)\mathrm{P}(A^c \cap B) + h(1)g(1)\mathrm{P}(A \cap B)\\ &= ab\mathrm{P}(A \cap B) = ab\mathrm{P}(A)\mathrm{P}(B),\end{aligned}$$

by the independence of A and B. Therefore,

$$\mathrm{E}(h(I_A)g(I_B)) = ab\mathrm{P}(A)\mathrm{P}(B) = \mathrm{E}h(I_A)\mathrm{E}g(I_B).$$

Problem 2.2: Let $\Omega = [0, 1]$ and P denote the uniform probability on $[0, 1]$. Show that $\sin(2\pi x)$ and $\cos(2\pi x)$ are uncorrelated but not independent.

Solution. $\mathrm{E}\sin(2\pi x) = \int_0^1 \sin(2\pi x)dx = [-\cos(2\pi) + \cos(0)]/(2\pi) = 0.$

$\mathrm{E}\cos(2\pi x) = \int_0^1 \cos(2\pi x)dx = [\sin(2\pi) - \sin(0)]/(2\pi) = 0.$

$\mathrm{E}\sin(2\pi x)\cos(2\pi x) = \int_0^1 \sin(2\pi x)\cos(2\pi x)dx.$

Formula for double angle gives $2\sin(2\pi x)\cos(2\pi x) = \sin(4\pi x)$. Therefore,

$$\int_0^1 \sin(2\pi x)\cos(2\pi x)dx = \frac{1}{2}\int_0^1 \sin(4\pi x)dx = -\frac{1}{8\pi}[\cos(4\pi) - \cos(0) = 0.$$

Hence, $\mathrm{E}\sin(2\pi x)\cos(2\pi x) = \mathrm{E}\sin(2\pi x)\mathrm{E}\cos(2\pi x)$, and they are uncorrelated. Since $\sin^2(2\pi x) + \cos^2(2\pi x) = 1$, they are not independent.

Problem 2.3: Find the characteristic function of a discrete uniform distribution on a set of two numbers $\{-b, b\}$ and the characteristic function of a continuous uniform distribution on an interval $[-a, a]$.

Solution. $\mathrm{E}e^{itX} = \frac{1}{2}e^{it(-b)} + \frac{1}{2}e^{itb} = \cos(tb)$.
$\mathrm{E}e^{itX} = \int_{-a}^{a} \frac{1}{2a} e^{itx} dx = \int_{-a}^{a} \frac{1}{2a} \cos(tx) dx + i \int_{-a}^{a} \frac{1}{2a} \sin(tx) dx = \frac{\sin(at)}{at}$.

Problem 2.4: Let X and Y be two independent random variables. X is continuous and has a uniform $[-1, 1]$ distribution. Y is discrete and has a uniform $\{-1, 1\}$ distribution. Show that the distributions of $X + Y$ are continuous uniform $[-2, 2]$.

Solution. Let $\phi_X(t)$ and $\phi_Y(t)$ be characteristic functions of X and Y. Then, the characteristic function of $X + Y$ is $\phi_{X+Y}(t) = \phi_X(t)\phi_Y(t)$.

From the question above, $\phi_X(t) = \frac{\sin(t)}{t}$, $\phi_Y(t) = \cos(t)$. Hence, $\phi_{X+Y}(t) = \frac{\sin(t)\cos(t)}{t} = \frac{\sin(2t)}{2t}$. This corresponds to $U[-2, 2]$ distribution.

Problem 2.5: Let I_n be Bernoulli random variables with parameters p_n, $n = 1, 2, \ldots$, such that the series $\sum_{n=1}^{\infty} p_n < \infty$. Show that, with probability one, $\sum_{n=1}^{\infty} I_n$ is a convergent (random) series. You may interchange expectation and summation due to the positiveness of all terms by the monotone convergence theorem.

Solution. Note that $I_n \geq 0$ and therefore that $\sum_{n=1}^{N} I_n$ is non-decreasing in N; we deduce by the monotone convergence theorem, that

$$\mathrm{E}\sum_{n=1}^{\infty} I_n = \mathrm{E}\lim_N \sum_{n=1}^{N} I_n = \lim_N \mathrm{E}\sum_{n=1}^{N} I_n = \lim_N \sum_{n=1}^{N} \mathrm{E}I_n = \sum_{n=1}^{\infty} \mathrm{E}I_n.$$

Now, $\mathrm{E}I_n = p_n$, hence

$$\mathrm{E}\sum_{n=1}^{\infty} I_n = \sum_{n=1}^{\infty} p_n < \infty.$$

Letting $X = \sum_{n=1}^{\infty} I_n$, we have $X \geq 0$ and by above, $\mathrm{E}X < \infty$.
Hence, $\mathrm{P}(X < \infty) = 1$.

Problem 2.6: Let x be a number in $[0, 1]$ and ε_k be its k-th digit in its binary expansion, $x = 0.\varepsilon_1\varepsilon_2 \ldots \varepsilon_k \ldots$. Let P denote the uniform probability on $[0, 1]$. A normal number is a number in which the proportion of 0's in its binary expansion is half (and the proportion of 1's is half). Give an example of a number that is not normal.

Solution. Any number with a finite expansion has a proportion of all 0's one and is not normal. For example, 0.1. Here, all $\varepsilon_k = 0$ for $k \geq 2$.

Problem 2.7: Let x be a number in $[0,1]$ and ε_k be its k-th digit in its binary expansion, $k = 1, 2$. Looking at ε_k as a random variable by letting x be the outcome of a random experiment in which x is uniform $[0,1]$ gives the distributions of ε_1 and ε_2. Show that they are independent. In a similar way, all ε_k's are independent, $k = 1, 2, \ldots$, Bernoulli random variables with parameters $p = \frac{1}{2}$.

Solution. $\varepsilon_1(x) = 0$ for $x \in [0, 1/2)$ and $\varepsilon_1(x) = 1$ for $x \in [1/2, 1]$. $\varepsilon_2(x) = 0$ for $x \in [0, 1/4)$ and $x \in [1/2, 3/4)$. For other x, $\varepsilon_2(x) = 1$. Hence, $\mathrm{P}(\varepsilon_1 = 0) = \mathrm{P}([0, 1/2)) = 1/2$ and $\mathrm{P}(\varepsilon_1 = 1) = 1/2$. $\mathrm{P}(\varepsilon_2 = 0) = \mathrm{P}([0, 1/4) \cup [1/2, 3/4)) = 1/2$; and $\mathrm{P}(\varepsilon_2 = 1) = 1/2$.

The sets $A = [1/2, 1]$ and $B = [1/4, 1/2) \cup [3/4, 1]$ are independent because $\mathrm{P}(A) = 1/2 = \mathrm{P}(B)$, $A \cap B = [1/2, 3/4]$. $\mathrm{P}(A \cap B) = 1/4 = \mathrm{P}(A)\mathrm{P}(B)$. But $\varepsilon_1 = I_A$ and $\varepsilon_2 = I_B$. Hence, they are independent.

Problem 2.8: Let $\varepsilon_k(x)$ be as above and $r_k(x) = 1 - 2\varepsilon_k(x)$, $k = 1, 2, \ldots$. The latter are called Rademacher functions. State the strong law of large numbers (SLLN) for the r_k's. Deduce the SLLN for ε_k's. Explain what this means for normal numbers in $[0,1]$. Recall that a number in $[0,1]$ is normal if the proportion of 1's in its binary expansion ($x = \sum_{n=1}^{\infty} \varepsilon_k(x) 2^{-k}$) is half.

Solution. Since all ε_k's are independent Bernoulli random variables with parameters $p = \frac{1}{2}$, $r_k(x)$ are independent and take values 1, -1 and have mean 0. The SLLN states that for almost all $x \in [0,1]$,

$$\lim_{n\to\infty} \frac{1}{n} \sum_{k=1}^{n} r_k(x) = 0.$$

"For almost all x" means that the set of all x for which it holds has probability one. In other words, the SLLN

$$\mathrm{P}\left(\left\{x : \lim_{n\to\infty} \frac{1}{n} \sum_{k=1}^{n} r_k(x) = 0\right\}\right) = 1.$$

On the other hand,

$$\frac{1}{n} \sum_{k=1}^{n} r_k(x) = \frac{1}{n} \sum_{k=1}^{n} (1 - 2\varepsilon_k(x)) = 1 - \frac{2}{n} \sum_{k=1}^{n} \varepsilon_k(x),$$

so that, almost surely,

$$0 = \lim_{n\to\infty} \frac{1}{n}\sum_{k=1}^{n} r_k(x) = 1 - \lim_{n\to\infty} \frac{2}{n}\sum_{k=1}^{n} \varepsilon_k(x), \text{ i.e. } \lim_{n\to\infty} \frac{1}{n}\sum_{k=1}^{n} \varepsilon_k(x) = \frac{1}{2}.$$

In other words, the SLLN states that for almost all x, the proportion of 1's is half or that almost every number is normal (P(x is normal) = 1).

Problem 2.9 (Borel–Cantelli lemma): Prove that (a) if $\sum_{n=1}^{\infty} \mathrm{P}(A_n) < \infty$ then $\mathrm{P}(A_n\ i.o.) = 0$, and (b) assuming that A_n's are independent, then $\sum_{n=1}^{\infty} \mathrm{P}(A_n) = \infty$ implies that $\mathrm{P}(A_n\ i.o.) = 1$.

Solution.

(a) By continuity of probability and the fact that sets $\cup_{n=m}^{\infty} A_n$ are decreasing, $\cup_{n=m+1}^{\infty} A_n \subset \cup_{n=m}^{\infty} A_n$,

$$\mathrm{P}(A_n\ i.o.) = \mathrm{P}(\cap_{m=1}^{\infty} \cup_{n=m}^{\infty} A_n) = \lim_{m\to\infty} \mathrm{P}(\cup_{n=m}^{\infty} A_n) = 0.$$

This is because the probability of the union is less than the sum of probabilities, $\mathrm{P}(\cup_{n=m}^{\infty} A_n) \le \sum_{n=m}^{\infty} \mathrm{P}(A_n)$, and since the series converges, the remainder converges to zero.

(b) Using the De Morgan laws (complimentary of a union of sets is the intersection of complimentary sets and similar for intersections), we have

$$\mathrm{P}(A_n\ i.o.) = 1 - \mathrm{P}(A_n^c\ ev.),$$

where A_n^c denotes the complimentary set,

$$\mathrm{P}(A_n^c\ ev.) = \mathrm{P}(\cup_{m=1}^{\infty} \cap_{n=m}^{\infty} A_n^c) = \lim_{m\to\infty} \mathrm{P}(\cap_{n=m}^{\infty} A_n^c)$$

because the sets $\cap_{n=m}^{\infty} A_n^c$ are increasing, $\cap_{n=m}^{\infty} A_n^c \subset \cap_{n=m+1}^{\infty} A_n^c$. Next, by independence,

$$\mathrm{P}(\cap_{n=m}^{\infty} A_n^c) = \prod_{n=m}^{\infty} \mathrm{P}(A_n^c) = \prod_{n=m}^{\infty} (1 - \mathrm{P}(A_n)).$$

The relation between infinite products and series states that if the series of $0 \le c_n \le 1$ diverges, $\sum_{n=1}^{\infty} c_n = \infty$, then the infinite product is zero, $\prod_{n=1}^{\infty}(1 - c_n) = 0$. This implies that $\mathrm{P}(\cap_{n=m}^{\infty} A_n^c) = 0$, $\mathrm{P}(A_n^c\ ev.) = 0$, and $\mathrm{P}(A_n\ i.o.) = 1$.

Problem 2.10:

(a) Continuing with the same notations as above, prove that

$$E\left(\frac{1}{n}\sum_{k=1}^{n} r_k\right)^4 \le \frac{C}{n^2},$$

for some constant C.

(b) Let $X_n = \left(\frac{1}{n}\sum_{k=1}^{n} r_k\right)^4$. Show that, almost surely, $\sum_{n=1}^{\infty} X_n < \infty$.

(c) Deduce the SLLN for Rademacher functions.

Solution.

(a) $E\left(\left(\sum_{k=1}^{n} r_k\right)^4\right)$ is obtained by direct calculations. The main idea is that by the independence of the Rademacher functions, if i, j, k is not equal to l (one index is different from all others), then

$$E(r_i r_j r_k r_l) = E(r_l)E(r_i r_j r_k) = 0.$$

More precisely, the only cases where $E(r_i r_j r_k r_l) \neq 0$ are when

- $i = j \neq k = l$, $i = k \neq j = l$, $i = l \neq j = k$ and
- $i = j = k = l$.

In the latter case, $E(r_i r_j r_k r_l) = E(r_i^4) = E(r_i) = 1/2$, and in the first case, $E(r_i r_j r_k r_l) = E(r_i r_j)E(r_k r_l) = E(r_i^2)E(r_k^2) = E(r_i)E(r_k) = 1/4$, and similarly for the other two cases.

Writing

$$\left(\sum_{k=1}^{n} r_k\right)^4 = \sum_i \sum_j \sum_k \sum_l r_i r_j r_k r_l,$$

we get

$$E\left(\left(\sum_{k=1}^{n} r_k\right)^4\right) = \sum_i \sum_j \sum_k \sum_l E r_i r_j r_k r_l$$

$$= 3n(n-1)\frac{1}{4} + n\frac{1}{2} = \frac{3}{4}n^2 - \frac{1}{4}n.$$

Therefore,

$$E\left(\left(\sum_{k=1}^{n} r_k\right)^4\right) \le Cn^2 \text{ and } E\left(\left(\frac{1}{n}\sum_{k=1}^{n} r_k(x)\right)^4\right) \le \frac{Cn^2}{n^4} = \frac{C}{n^2},$$

for $C = 3/4$.

(b) It now follows, again by monotone convergence, that

$$\mathrm{E}\left(\sum_{n=1}^{\infty} X_n\right) = \sum_{n=1}^{\infty} \mathrm{E}(X_n) \le \sum_{n=1}^{\infty} \frac{C}{n^2} < \infty.$$

Hence, $\sum_{n=1}^{\infty} X_n < \infty$ with probability one.

(c) It follows that, almost surely (i.e. for almost all x),

$$\lim_{n\to\infty} X_n = \lim_{n\to\infty} \left(\frac{1}{n}\sum_{k=1}^{n} r_k\right)^4 = 0.$$

But this is equivalent to: for almost all x,

$$\lim_{n\to\infty} \frac{1}{n}\sum_{k=1}^{n} r_k(x) = 0.$$

Problem 2.11: Show the Cauchy–Schwarz inequality for integrals

$$\left|\int_0^t f(s)g(s)ds\right| \le \sqrt{\int_0^t f^2(s)ds \int_0^t g^2(s)ds}$$

by using such inequality in probability $|\mathrm{E}(XY)| \le \sqrt{\mathrm{E}X^2\mathrm{E}Y^2}$.

Solution. The idea is to make f and g into random variables X and Y using the uniform distribution on $[0, t]$. Indeed, let U be uniform on $[0, t]$. Then, after dividing both sides by t, the desired inequality becomes

$$\left|\int_0^t f(s)g(s)ds/t\right| = |\mathrm{E}(f(U)g(U))|$$

$$\le \sqrt{\mathrm{E}f(U)^2\mathrm{E}g(U)^2} = \sqrt{\int_0^t f^2(s)ds/t \int_0^t g^2(s)ds/t}.$$

Problem 2.12: Construct a probability space $(\Omega, \mathcal{F}, \mathrm{P})$ on which the canonical map $(X(\omega) = \omega)$ has a standard normal distribution.

Solution. Let $\Omega = \mathbb{R}$, $\mathcal{F}$ be the Borel σ-field generated by the intervals on $\mathbb{R}$ and

$$\mathrm{P}(A) = \int_A \frac{1}{\sqrt{2\pi}} e^{-x^2/2} dx,$$

where $A \in \mathcal{F}$.

The random variable $X(\omega) = \omega$ for $\omega \in \Omega$ indeed has a standard normal distribution,

$$\begin{aligned}\mathrm{P}(X \le x) &= \mathrm{P}(X \in (-\infty, x]) = \mathrm{P}(\{\omega : \omega \in (-\infty, x]\}) \\ &= \int_{-\infty}^{x} \frac{1}{\sqrt{2\pi}} e^{-x^2/2} dx.\end{aligned}$$

Problem 2.13: Calculate the expectation $\mathrm{E}(X^+)$ using the definition of the Lebesgue integral for the random variable in the previous problem.

Solution. Using Example 2.9 together with the result in p. 33 in Klebaner's book, we see that

$$\begin{aligned}\mathrm{E}(X^+) &= \lim_{n\to\infty} \sum_{k=0}^{n2^n-1} \frac{k}{2^n} \mathrm{P}\left(\frac{k}{2^n}, \frac{k+1}{2^n}\right) \\ &= \lim_{n\to\infty} \sum_{k=1}^{n2^n-1} \frac{k}{2^n} \int_{k/2^n}^{(k+1)/2^n} \frac{1}{\sqrt{2\pi}} e^{-x^2/2} dx \\ &= \lim_{n\to\infty} \sum_{k=1}^{n2^n-1} \frac{k}{2^n} \frac{1}{\sqrt{2\pi}} e^{-(k/2^n)^2/2} \frac{1}{2^n} \\ &= \int_0^{\infty} x \frac{1}{\sqrt{2\pi}} e^{-x^2/2} dx = \frac{1}{\sqrt{2\pi}}.\end{aligned}$$

Problem 2.14: Let $\mathcal{G}$ be the trivial σ-field, $\mathcal{G} = \{\emptyset, \Omega\}$. Show that if X is $\mathcal{G}$-measurable, then X is almost surely constant; that is, there exists $c \in \mathbb{R}$ such that $\mathrm{P}(X = c) = 1$.

Solution. By definition, for any measurable A ($A \in \mathcal{B}(\mathbb{R})$, the set $\{X \in A\} \in \mathcal{G}$. It follows that for any A, $\{X \in A\} = \Omega$ or $\{X \in A\} = \emptyset$. Let c be one of the values of X, $c \in X(\Omega)$). Then, $\{X = c\}$ is not an empty set, and letting $A = \{c\}$, we find that $\{X \in A\} = \{X = c\} = \Omega$. In other words, $\mathrm{P}(X = c) = \mathrm{P}(\Omega) = 1$.

Problem 2.15: Let X be a random variable defined on a probability space $(\Omega, \mathcal{F}, \mathrm{P})$, where Ω is the sample space of all possible outcomes of a random experiment, $\mathcal{F}$ is the σ-field of those subsets of Ω that are events and P is a probability measure defined on $\mathcal{F}$. Prove that $\mathrm{E}\left(\mathrm{E}(X|\mathcal{G})\right) = \mathrm{E}(X)$ for any σ-field $\mathcal{G}$ that is contained in $\mathcal{F}$ and that $\mathrm{E}(X|\mathcal{G}) = \mathrm{E}(X)$ in the particular case when $\mathcal{G}$ is the trivial σ-field $\mathcal{G} = \{\emptyset, \Omega\}$.

Solution. By the definition of conditional expectation, we have

$$\mathrm{E}(I_A E(X|\mathcal{G})) = \mathrm{E}(I_A X)$$

for any event $A \in \mathcal{G}$. Taking $A = \Omega$, so that $I_A = 1$, we get

$$\mathrm{E}\left(\mathrm{E}(X|\mathcal{G})\right) = \mathrm{E}(X).$$

By definition, $E(X|\mathcal{G})$, where $\mathcal{G}$ is trivial, is the unique $\{\emptyset, \Omega\}$-measurable random variable that satisfies $\mathrm{E}(I_A E(X|\mathcal{G})) = \mathrm{E}(I_A X)$ for $A \in \{\emptyset, \Omega\}$. But the only measurable random variables with resect to the trivial σ-field are constants (see Problem 2.14). Therefore, $\mathrm{E}(X|\mathcal{G}) = c$ (a.s.), for some constant c. Using $\mathrm{E}\left(\mathrm{E}(X|\mathcal{G})\right) = \mathrm{E}(X)$, we find that the left-hand side is c and the right-hand side is $\mathrm{E}(X)$, so that $c = \mathrm{E}(X)$.

Problem 2.16: Consider a finite sample space $\Omega = \{1, \ldots, 2n\}$ with the σ-field $\mathcal{F}$ consisting of all subsets of Ω, equipped with the uniform probability measure P, assigning probability $1/(2n)$ to each outcome $\omega \in \Omega$. Consider the random variable $X(\omega) = \omega$.

(a) Find $\mathrm{E}(X|\mathcal{G})$ for the σ-field $\mathcal{G} = \{\emptyset, A, A^c, \Omega\}$, where $A = \{1, \ldots, n\}$.
(b) Verify that $\mathrm{E}\big(\mathrm{E}(X|\mathcal{G})\big) = \mathrm{E}(X)$.

Solution.

(a) By definition of $\mathcal{G}$-measurability, $\mathrm{E}(X|\mathcal{G}) = c_1$ for all $\omega \in A$ and $\mathrm{E}(X|\mathcal{G})\omega) = c_2$ for all $\omega \in A^c$. Further,

$$\int_B \mathrm{E}(X\ \mathcal{G})d\mathrm{P} = \int_B X d\mathrm{P},\ B \in \{\emptyset, A, A^c, \Omega\}.$$

By taking $B = A$, we obtain

$$\int_A \mathrm{E}(X|\mathcal{G})d\mathrm{P} = \int_A X d\mathrm{P}.$$

But

$$\int_A X d\mathrm{P} = \sum_{\omega \in A} X(\omega)\frac{1}{2n} = \frac{1}{2n}\sum_{\omega \in A} \omega = \frac{1}{2n}\frac{n(n+1)}{2} = \frac{n+1}{4}.$$

On the other hand,

$$\int_A \mathrm{E}(X|\mathcal{G})d\mathrm{P} = c_1 \mathrm{P}(A) = \frac{1}{2}c_1.$$

Hence, $c_1 = \frac{n+1}{2}$. Similarly, we obtain c_2 by replacing A with A^c:

$$\mathrm{E}(X|\mathcal{G}) = \begin{cases} (n+1)/2 & \text{for } \omega \in A, \\ (3n+1)/2 & \text{for } \omega \in A^c. \end{cases}$$

(b)

$$\mathrm{E}(X) = \sum_{\omega=1}^{2n} \omega \frac{1}{2n} = \frac{1}{2n} \sum_{\omega=1}^{2n} \omega = \frac{1}{2n} \frac{2n(2n+1)}{2} = n + \frac{1}{2}.$$

Since $\mathrm{P}(A) = \mathrm{P}(A^c) = \frac{1}{2}$,

$$\mathrm{E}(\mathrm{E}(X|\mathcal{G})) = \frac{1}{2}\frac{n+1}{2} + \frac{1}{2}\frac{3n+1}{2} = n + \frac{1}{2}.$$

Problem 2.17: Consider a finite sample space $\Omega = \{\omega_1, \ldots, \omega_n\}$ with the σ-field $\mathcal{F}$ consisting of all subsets of Ω. The probability measure P, assigning probability $p_i > 0$ to each outcome ω_i, $i = 1, 2, \ldots, n$, with $\sum_{i=1}^{n} p_i = 1$. Let A be a proper subset of Ω and $\mathcal{G} = \{\emptyset, A, A^c, \Omega\}$. Let X be an arbitrary random variable, and denote $x_i = X(\omega_i)$. Show that

$$\mathrm{E}(X|\mathcal{G}) = \left(\frac{1}{P(A)} \sum_{\omega_i \in A} p_i x_i\right) 1_A + \left(\frac{1}{P(A^c)} \sum_{\omega_i \in A^c} p_i x_i\right) 1_{A^c}.$$

Solution. Any $\mathcal{G}$-measurable random variable is of the form $\xi = c_1 1_A + c_2 1_{A^c}$. Thus, $\mathrm{E}(X|\mathcal{G})$ being $\mathcal{G}$-measurable is also of this form. Taking $\xi = 1_A$, we have

$$\mathrm{E}(X1_A) = \mathrm{E}(1_A \mathrm{E}(X|\mathcal{G})) = \mathrm{E}(1_A c_1 1_A + 1_A c_2 1_{A^c}) = c_1 \mathrm{E} 1_A = c_1 P(A).$$

Since $\mathrm{E}(X1_A) = \sum_{\omega_i \in A} p_i x_i$, we obtain $c_1 = \frac{1}{P(A)} \sum_{\omega_i \in A} p_i x_i$. Taking $\xi = 1_{A^c}$, we have $c_2 = \frac{1}{P(A^c)} \sum_{\omega_i \in A^c} p_i x_i$.

Problem 2.18:

(a) Show that, for arbitrary (square-integrable) random variables, X and Y, $X - \mathrm{E}(X|Y)$ and Y are uncorrelated.
(b) Show that, for arbitrary (square-integrable) random variables, X and Y, with means μ_X, μ_Y, variances σ_X^2, σ_Y^2 and correlation ρ, $X - \rho(\sigma_X/\sigma_Y)Y$ and Y are uncorrelated.
(c) Let (X, Y) have a bivariate normal distribution with means μ_X, μ_Y, variances σ_X^2, σ_Y^2 and correlation ρ.

i. Give the conditional distribution of X given $Y = y$, and use it to find $\mathrm{E}(X|Y)$ and $\mathrm{E}(Y|X)$. Specify these for the case of zero means.
ii. Find $\mathrm{E}(X|Y)$ without using the conditional distribution of X given $Y = y$.

Solution.

(a) First, note that $X - \mathrm{E}(X|Y)$ has mean zero: $\mathrm{E}\big(X - \mathrm{E}(X|Y)\big) = \mathrm{E}(X) - \mathrm{E}\big(\mathrm{E}(X|Y)\big) = 0$. Now, $\mathrm{E}\big[\big(X - \mathrm{E}(X|Y)\big)Y\big] = \mathrm{E}(XY) - \mathrm{E}\big(\mathrm{E}(X|Y)Y\big) = \mathrm{E}(XY) - \mathrm{E}\big(\mathrm{E}(XY|Y)\big) = \mathrm{E}(XY) - \mathrm{E}(XY) = 0$. The result follows.

(b) Similarly, $\mathrm{E}\big[\big(X - \rho(\sigma_X/\sigma_Y)Y\big)Y\big] - \big(\mu_X - \rho(\sigma_X/\sigma_Y)\mu_Y\big)\mu_Y = \mathrm{E}(XY) - \rho(\sigma_X/\sigma_Y)\mathrm{E}(Y^2) - \mu_X\mu_Y + \rho(\sigma_X/\sigma_Y)\mu_Y^2 = \mathrm{Cov}(X,Y) - \rho(\sigma_X/\sigma_Y)\,\mathrm{Var}(Y) = \rho\sigma_X\sigma_Y - \rho\sigma_X\sigma_Y = 0$.

(c) i. Direct calculation using the densities $f_{X,Y}(x,y)$ and $f_Y(y)$ yields the conditional distribution of X given $Y = y$ is normal with mean $\mu_X + \rho\frac{\sigma_X}{\sigma_Y}(y - \mu_Y)$ and variance $(1-\rho^2)\sigma_X^2$.
Consequently, $\mathrm{E}(X|Y = y) = \mu_X + \rho\frac{\sigma_X}{\sigma_Y}(y - \mu_Y)$.
Thus, $\mathrm{E}(X|Y) = \mu_X + \rho\frac{\sigma_X}{\sigma_Y}(Y - \mu_Y)$.
Exchanging the roles of X and Y, the conditional distribution of Y given $X = x$ is normal with mean $\mu_Y + \rho\frac{\sigma_Y}{\sigma_X}(x - \mu_X)$ and variance $(1-\rho^2)\sigma_Y^2$. $\mathrm{E}(Y|X) = \mu_Y + \rho\frac{\sigma_Y}{\sigma_X}(X - \mu_X)$.
When the means are zero, $\mathrm{E}(X|Y) = \rho\frac{\sigma_X}{\sigma_Y}Y$.

ii. As an uncorrelated bivariate normal pair, $X - \rho\frac{\sigma_X}{\sigma_Y}Y$ and Y are independent. It follows that $\mathrm{E}(X|Y) = \mathrm{E}(X - \rho\frac{\sigma_X}{\sigma_Y}Y|Y) + \mathrm{E}(\rho\frac{\sigma_X}{\sigma_Y}Y|Y) = \mathrm{E}(X - \rho\frac{\sigma_X}{\sigma_Y}Y) + \rho\frac{\sigma_X}{\sigma_Y}Y = \mu_X + \rho\frac{\sigma_X}{\sigma_Y}(Y - \mu_Y)$. Similarly, $E(Y|X) = \mu_Y + \rho\frac{\sigma_Y}{\sigma_X}(X - \mu_X)$. Further, $E(X|Y) = \rho\frac{\sigma_X}{\sigma_Y}$ and $E(Y|X) = \rho\frac{\sigma_Y}{\sigma_X}X$ for $\mu_X = \mu_Y = 0$.

Problem 2.19: The next two questions show examples of conditioning for the calculation of probabilities and expectations. In later chapters, we shall see a connection to the theory of Markov processes and their generators. The treatment here is elementary. Let $X_n = X_0 + \sum_{i=1}^n Y_i$ be a simple biased random walk with probability p of going up in one step and probability $q = 1 - p$ of going down in one step. Denote by $u(x)$ the probability that a random walk hits b before it hits a, when $X_0 = x \in [a,b]$.

(a) Show that u solves the following second-order difference equation in y:

$$y(x) = py(x+1) + qy(x-1). \qquad (*)$$

(b) Show that, if y solves $(*)$, then $z = \Delta y$, $z(x) = y(x) - y(x-1)$, solves a first-order difference equation and find its general solution.
(c) Show that $y(x)$ has a general solution $y(x) = \lambda(q/p)^x + \mu$ for some constants λ and μ.
(d) Find $u(x)$, using the boundary conditions $u(b) = 1$ and $u(a) = 0$.

Solution.

(a) Let A be the event that random walk (RW) hits b before a. By conditioning on Y_1,

$$u(x) = P(A|X_0 = x) = E(P(A|Y_1, X_0 = x)).$$

Hence,

$$\begin{aligned} u(x) &= \mathrm{P}(A|Y_1 = 1, X_0 = x)\mathrm{P}(Y_1 = 1) \\ &\quad + \mathrm{P}(A|Y_1 = -1, X_0 = x)\mathrm{P}(Y_1 = -1) \\ &= p\mathrm{P}(A|X_0 = x+1) + q\mathrm{P}(A|X_0 = x-1) \\ &= pu(x+1) + qu(x-1). \end{aligned}$$

(b) $(*)$ can be written as $py(x) + qy(x) = py(x+1) + qy(x-1)$ or equivalently $pz(x+1) = qz(x)$. It immediately follows that $z(x) = z(a+1)(q/p)^{x-a-1}$.
(c) From (b),

$$\begin{aligned} y(x) - y(x-1) &= z(a+1)(q/p)^{x-a-1} \\ y(x-1) - y(x-2) &= z(a+1)(q/p)^{x-a-2} \\ &\vdots \\ y(a+1) - y(a) &= z(a+1). \end{aligned}$$

Summing the left-hand expressions, we get $y(x) - y(a)$; summing the right-hand expressions, we get $z(a+1)\big(1 + \cdots + (q/p)^{x-a-1}\big)$. We deduce that

$$y(x) = y(a) + z(a+1)\frac{(q/p)^{x-a} - 1}{(q/p) - 1}.$$

Recall that the random walk is biased so that $p \neq q$ or equivalently that $q/p \neq 1$.

(d) If $X_n = b$, then the probability to hit b before a is 1, and if the RW is in a, then this probability is 0. Therefore,

$$u(x) = u(a)+(u(a+1)-u(a))\frac{(q/p)^{x-a}-1}{(q/p)-1} = u(a+1)\frac{(q/p)^{x-a}-1}{(q/p)-1},$$

and in particular,

$$1 = u(b) = u(a+1)\frac{(q/p)^{b-a}-1}{(q/p)-1}.$$

We deduce that

$$u(a+1) = \frac{(q/p)-1}{(q/p)^{b-a}-1},$$

and finally that

$$u(x) = \frac{(q/p)-1}{(q/p)^{b-a}-1}\,\frac{(q/p)^{x-a}-1}{(q/p)-1} = \frac{(q/p)^{x-a}-1}{(q/p)^{b-a}-1}$$
$$= \frac{(q/p)^{x}-(q/p)^{a}}{(q/p)^{b}-(q/p)^{a}}.$$

Problem 2.20: In gambler's ruin, you start with \$$a$ and your opponent starts with \$$b$, $a, b \in \mathbb{N}$. In each flip of the coin, you win \$1 with probability p when heads comes up, and lose 1 with probability $q = 1 - p$ when tails comes up, $p \in [0, 1]$. The game stops when one party has no money left. Find the expected length of the game in gambler's ruin when

(a) betting is done on a fair coin;
(b) betting is done on a biased coin.

Solution. Since the total amount of money between the players is always constant and equal to $a + b$ (every time you play, the money just gets redistributed), the game stops when you have \$0 or \$$(a + b)$. We now vary the initial fortune and denote by $\theta(x)$ the expected time to the end of the game when your capital is x. By conditioning, on the outcome of the first toss, we obtain that $\theta(x)$ is 1 (the first play) plus the time to get absorbed from $x+1$, which happens with probability p plus the time to get absorbed from $x-1$, which happens with probability q. In other words, it satisfies the recurrence relation

$$\theta(x) = 1 + p\theta(x+1) + q\theta(x-1).$$

When you have no or all the money, the game cannot continue and the time to absorption is zero, $\theta(0) = 0$ and $\theta(a+b) = 0$. Note that the time to absorption (reaching 0 or $a+b$) only depends on the current distribution of the money and not on any previous events — this is the Markov property.

(a) If $p = q = \frac{1}{2}$, then $\theta(x) = 1 + \frac{1}{2}\theta(x+1) + \frac{1}{2}\theta(x-1)$. Rearranging, we have

$$\theta(x+1) - \theta(x) = \theta(x) - \theta(x-1) - 2.$$

Iterating the previous recursive relation and noting that $\theta(0) = 0$ give

$$\theta(x+1) - \theta(x) = \theta(1) - 2x, \quad x = 0, 2, \ldots, a+b-1.$$

Summing over the first n terms gives

$$\begin{aligned}\theta(n) - \theta(0) &= \sum_{x=0}^{n-1} (\theta(x+1) - \theta(x)) = \sum_{x=0}^{n-1} (\theta(1) - 2x) \\ &= n\theta(1) - n(n-1).\end{aligned}$$

Using the boundary conditions $\theta(0) = 0$ and $\theta(a+b) = 0$, we obtain $\theta(1) = a + b - 1$. Substituting this back, we get

$$\theta(n) = n(a+b-1) - n(n-1) = n(a+b-n).$$

Thus, the expected length of the game starting from a is $\theta(a) = ab$.

(b) Let now $p \neq q$. The recurrence equation above is equivalent to

$$\theta(x+1) - \theta(x) = \frac{q}{p}(\theta(x) - \theta(x-1)) - \frac{1}{p}.$$

Letting $r = q/p$ (assuming $p \neq 0$ as in this case, the game takes b deterministic steps to terminate and the answer is trivially $\theta(a) = b$), we can write

$$\theta(x+1) - \theta(x) = r(\theta(x) - \theta(x-1)) - (1+r).$$

Dividing by r^{x+1}, we have

$$\frac{1}{r^{x+1}}(\theta(x+1) - \theta(x)) = \frac{1}{r^x}(\theta(x) - \theta(x-1)) - \frac{1+r}{r^{x+1}}.$$

Iterating gives

$$\frac{1}{r^{x+1}}(\theta(x+1) - \theta(x)) = \frac{1}{r}(\theta(1) - \theta(0)) - \sum_{j=2}^{x+1} \frac{1+r}{r^j}.$$

Hence, recalling that $\theta(0) = 0$ and that $r \neq 1$,

$$\theta(x+1) - \theta(x) = r^x\theta(1) - r^{x+1}\sum_{j=2}^{x+1}\frac{1+r}{r^j} = \theta(1)r^x - (1+r)\sum_{i=0}^{x-1} r^i$$

$$= \theta(1)r^x - (1+r)\frac{1-r^x}{1-r}$$

$$= \left(\theta(1) + \frac{1+r}{1-r}\right)r^x - \frac{1+r}{1-r}.$$

Summing over the first n terms and using the boundary conditions, we obtain that

$$\theta(n) = \left(\theta(1) + \frac{1+r}{1-r}\right)\frac{1-r^n}{1-r} - \frac{1+r}{1-r}n$$

and that

$$\theta(1) = (a+b)\frac{1+r}{1-r^{a+b}} - \frac{1+r}{1-r},$$

and consequently that

$$\theta(a) = \frac{1}{2p-1}\left((a+b)\frac{1-r^a}{1-r^{a+b}} - a\right).$$

Problem 2.21: Let X be a random variable such that it has a finite moment-generating function $m(u) = \mathrm{E}(e^{uX})$, for all $u : |u| \leq a$, $a > 0$.

(a) Show that $|X|$ has a finite moment-generating function on the interval $[-a, a]$.
(b) Deduce that m has a power series expansion, $m(u) = \sum_{n=0}^{\infty} c_n u^n$, and show that the n-th moment of X is given by $\mathrm{E}(X^n) = c_n n!$.

Solution.

(a) Since $e^{u|x|} \leq e^{ux} + e^{-ux}$, $\mathrm{E}(e^{u|X|}) \leq \mathrm{E}(e^{uX}) + \mathrm{E}(e^{-uX}) = m(u) + m(-u) < +\infty$ for $u \in [-a, a]$.
(b) Using the power series expansion for e^x, we obtain by monotone convergence,

$$\mathrm{E}(e^{|uX|}) = \mathrm{E}\left(\sum_{n=0}^{\infty}\frac{|uX|^n}{n!}\right) = \sum_{n=0}^{\infty}\frac{\mathrm{E}(|X|^n)}{n!}|u|^n.$$

We deduce that X has absolute moments of all orders, $\mathrm{E}(|X|^n) < +\infty$ (and the series $\sum_{n=0}^{\infty}\mathrm{E}(X^n)u^n/n!$ is absolutely convergent).

Repeating with X (instead of $|X|$) and using dominated convergence, we obtain

$$m(u) = \mathrm{E}\big(e^{uX}\big) = \mathrm{E}\left(\sum_{n=0}^{\infty} \frac{(uX)^n}{n!}\right) = \sum_{n=0}^{\infty} \frac{\mathrm{E}(X^n)}{n!} u^n.$$

Since the power series expansion is unique, it follows that $\mathrm{E}(X^n) = c_n n!$.

Problem 2.22: Let Z be a standard normal random variable $N(0,1)$. Find the n-th moment of Z, $\mathrm{E}(Z^n)$.

Solution. Using Problem 2.21, $\mathrm{E}(Z^n) = c_n n!$, where c_n is the coefficient of u^n in the power series expansion of $m(u) = \mathrm{E}\big(e^{uZ}\big)$. Since the moment-generating function of Z is given by $m(u) = \mathrm{E}\big(e^{uZ}\big) = e^{u^2/2}$, we obtain c_n by identifying the coefficients of the power series of $e^{u^2/2}$:

$$m(u) = e^{u^2/2} = \sum_{n=0}^{\infty} \frac{\left(\frac{u^2}{2}\right)^n}{n!} = \sum_{n=0}^{\infty} \frac{u^{2n}}{2^n n!}.$$

Since this series consists of only even powers of u, $k = 2n$, the coefficients of odd powers are zero, $c_k = 0$ for odd k. If k is even, $k = 2n$, then $c_k = c_{2n} = \frac{1}{2^n n!}$. Hence, $\mathrm{E}(Z)^k = 0$ for odd k, and for even $k = 2n$, $n = 1, 2, 3, \ldots,$

$$\mathrm{E}\big(Z^{2n}\big) = \frac{(2n)!}{2^n n!}.$$

Problem 2.23: Let X be a random variable such that it has a finite moment-generating function $m(u) = \mathrm{E}\big(e^{uX}\big)$, for all $u : |u| < a$.

(a) Show that m is infinitely differentiable in $(-a, a)$, so that

$$m^{(k)}(u) = \sum_{n=0}^{\infty} \frac{\mathrm{E}\big(X^{n+k}\big)}{n!} u^n$$

and that $\mathrm{E}\big(X^k\big) = m^{(k)}(0)$.

(b) Show that $m^{(k)}(u) = \mathrm{E}\big(X^k e^{uX}\big)$, for all $u : |u| < a$.

Solution.

(a) This is a direct consequence of the general theory: a power series, $f(u) = \sum_{n=0}^{\infty} c_n u^n$, with radius of convergence R is infinitely differentiable in $(-R, R)$, $m^{(n)}(0) = n!c_n$ and, for any $u \in (-R, R)$,

$$f^{(k)}(u) = \sum_{n=k}^{\infty} \frac{c_n n!}{(n-k)!} u^{n-k} = \sum_{n=0}^{\infty} \frac{c_{n+k}(n+k)!}{n!} u^n.$$

As a power series with a radius of convergence greater than or equal to a and with coefficients $c_n = \mathrm{E}(X^n)/n!$ (see Problem 2.21), m is infinitely differentiable in $(-a, a)$, $m^{(k)}(0) = k!c_k = \mathrm{E}(X^k)$ and, for any $u \in (-a, a)$,

$$m^{(k)}(u) = \sum_{n=0}^{\infty} \frac{c_{n+k}(n+k)!}{n!} u^n = \sum_{n=0}^{\infty} \frac{\mathrm{E}(X^{k+n})}{n!} u^n.$$

(b) For any $\varepsilon > 0$, u, $k \in \mathbb{N}$ and x,

$$\begin{aligned} |x|^k e^{|ux|} &= \frac{k!}{\varepsilon^k} \frac{(\varepsilon|x|)^k}{k!} e^{|ux|} < \frac{k!}{\varepsilon^k} e^{(|u|+\varepsilon)|x|} \\ &< \frac{k!}{\varepsilon^k} \left(e^{(|u|+\varepsilon)x} + e^{-(|u|+\varepsilon)x}\right). \end{aligned}$$

For $u \in (-a, a)$, choose $\varepsilon < a - |u|$, then for any $k \in \mathbb{N}$,

$$\mathrm{E}(|X|^k e^{|uX|}) \leq \frac{k!}{\varepsilon^k} (\mathrm{E}(e^{(|u|+\varepsilon)X}) + \mathrm{E}(e^{-(|u|+\varepsilon)X})) < +\infty.$$

Repeating the same reasoning as in Problem 2.21, including a dominated convergence argument, we find that, for any $u \in (-a, a)$,

$$\mathrm{E}(X^k e^{uX}) = \mathrm{E}\left(\sum_{n=0}^{\infty} \frac{X^{k+n}}{n!} u^n\right) = \sum_{n=0}^{\infty} \frac{\mathrm{E}(X^{k+n})}{n!} u^n = m^{(k)}(u).$$

Problem 2.24: Interchanging expectation and derivative. Let $f(u, x)$ be a continuous function with the first partial derivative in u, $\frac{\partial f}{\partial u}$. Let u be fixed, and suppose that for some $\delta > 0$ and any h, $-\delta < h < \delta$,

$$|f(u+h, x) - f(u, x)| \leq H(x)|h|,$$

for some function H, which may depend on δ and u. Let X be a random variable, and suppose that the random variables $f(u+h,X)$, $\frac{\partial}{\partial u}f(u,X)$ and $H(X)$ are integrable.

(a) Show that

$$\frac{\partial}{\partial u}(\mathrm{E}f(u,X)) = \mathrm{E}\left(\frac{\partial}{\partial u}f(u,X)\right).$$

(b) Show that the assumptions in (a) are satisfied if

   i. the functions f and $\frac{\partial f}{\partial u}$ are bounded.
   ii. $f(u,x) = e^{ux}$ and X is a random variable with a finite moment-generating function: $m(u) = \mathrm{E}(e^{uX})$, for all $u \in (-a,a)$, for some $a > 0$ (or infinity).

Solution.

(a) Since for any x, $\lim_{h\to 0}\frac{f(u+h,x)-f(u,x)}{h} = \frac{\partial}{\partial u}f(u,x)$, and the random variables $\frac{f(u+h,X)-f(u,X)}{h}$ are dominated by the integrable random variable $H(X)$, by the dominated convergence theorem,

$$\mathrm{E}\left(\frac{\partial}{\partial u}f(u,X)\right) = \mathrm{E}\lim_{h\to 0}\frac{f(u+h,X)-f(u,X)}{h}$$

$$= \lim_{h\to 0}\mathrm{E}\frac{f(u+h,X)-f(u,X)}{h} = \frac{\partial}{\partial u}(\mathrm{E}f(u,X)).$$

(b) i. If $\frac{\partial f}{\partial u}$ is bounded, then by the mean value theorem for some $\xi \in (u-|h|, u+|h|)$,

$$\left|\frac{f(u+h,x)-f(u,x)}{h}\right| = \left|\frac{\partial}{\partial u}f(\xi,x)\right| \le C,$$

and the assumptions in (a) are clearly satisfied with $H \equiv C$.

ii. Fix u, $|u| < a$, and take $\delta > 0$ such that $|u| + \delta < a$. Then, by the mean value theorem, for any $h \in (-\delta,\delta)$, there exists $\xi \in (-|h|,|h|)$ such that,

$$|f(u+h,x) - f(u,x)| = e^{ux}|e^{hx} - 1| = e^{ux}|x|e^{\xi x}h.$$

Hence,

$$f(u+h,x) - f(u,x) = \frac{\partial}{\partial u}f(u+\xi,x)h = xe^{(u+\xi)x}h.$$

It follows that, for such u, δ, h and ξ,

$$|f(u+h,x)-f(u,x)| = |x|e^{(u+\xi)x}|h| \le |x|e^{|u+\xi||x|}|h|$$
$$\le |x|e^{(|u|+\delta)|x|}|h|.$$

Next, we show that $H(x) = |x|e^{(|u|+\delta)|x|}$ satisfies the assumption in (a): $H(X)$ is integrable. We do so in two steps.

- If X has finite a moment-generating function on $(-a,a)$, then so does $|X|$. Indeed, we observe that $e^{u|X|} \le e^{uX} + e^{-uX}$ and write

$$\mathrm{E}(e^{u|X|}) \le \mathrm{E}(e^{uX}) + \mathrm{E}(e^{(-u)X}),$$

  which is finite for any $u \in (-a,a)$.
- By the theorem of dominated convergence, differentiating in u yields that if $\mathrm{E}(e^{uX}) < +\infty$, for all $u \in (-a,a)$, then $\mathrm{E}(|X|e^{uX}) < +\infty$, for all $u \in (-a,a)$ (see (a)).

Remark: It might be tempting to use the mean value theorem directly on the random quantity

$$f(u+h,X) - f(u,X) = \frac{\partial}{\partial u} f(u+\xi, X)h.$$

This is correct; however, ξ in this case is also random, without a guarantee that it is itself a random variable. Consequently, one may not write the expectation of $\frac{\partial}{\partial u} f(u+\xi, X)$. Using the inequalities as above bypasses this problem.

Problem 2.25: Recall that if a (multivariate) random variable $X = (X_1,\ldots,X_n)$ has moment-generating function $m(v) = m_1(v_1)\ldots m_n(v_n)$, for all $v = (v_1,\ldots,v_n)$ in a ball of $\mathbb{R}^n$ that contains the origin, then the random variables $X_1,\ldots,X_n$ are independent.

Let X be as in Problem 2.23 and $\mathcal{G}$ be a σ-field such that

$$\forall u \in (-a,a),\ \mathrm{E}(e^{uX}|\mathcal{G}) = \mathrm{E}(e^{uX}).$$

Show that X is independent of $\mathcal{G}$.

Solution. Let Y be a bounded $\mathcal{G}$-measurable random variable. Then, for any $u \in (-a,a)$ and any $v \in \mathbb{R}$,

$$\mathrm{E}(e^{uX+vY}) = \mathrm{E}(\mathrm{E}(e^{uX}|\mathcal{G})e^{vY}) = \mathrm{E}(\mathrm{E}(e^{uX})e^{vY}) = \mathrm{E}(e^{uX})\mathrm{E}(e^{vY}).$$

It follows that X and Y are independent. As this is true for any bounded $\mathcal{G}$-measurable random variable, it is true for any $A \in \mathcal{G}$, and the result immediately follows.

Problem 2.26: Let X be a square-integrable random variable. Show that the value of the constant c for which $\mathrm{E}\big((X-c)^2\big)$ is the smallest is given by $\mathrm{E}(X)$.

Solution. Simply look at $\mathrm{E}\big((X-c)^2\big)$ as a quadratic function of c: $\mathrm{E}\big((X-c)^2\big) = \mathrm{E}\big(X^2-2cX+c^2\big) = \mathrm{E}(X^2)-2c\mathrm{E}(X)+c^2$. Differentiating and letting the derivative equal to zero, we immediately get that $\mathrm{E}\big((X-c)^2\big)$ is critical at $c = \mathrm{E}(X)$. By inspection of either the sign of the coefficient of c^2 or the second derivative ($\frac{d^2}{dc^2}\{\mathrm{E}(X-c)^2\} = 2 > 0$), we see that the critical point is a minimum point, i.e. $c = \mathrm{E}(X)$ gives the minimum value of $\mathrm{E}\big((X-c)^2\big)$.

Another approach is to add and subtract $\mathrm{E}(X)$ to $X-c$:

$$\begin{aligned}
\mathrm{E}\big((X-c)^2\big) &= \mathrm{E}\Big(\big((X-\mathrm{E}(X)) + (\mathrm{E}(X)-c)\big)^2\Big) \\
&= \mathrm{E}\Big(\big(X-\mathrm{E}(X)\big)^2 + 2\big(X-\mathrm{E}(X)\big)\big(\mathrm{E}(X)-c\big) + \big(\mathrm{E}(X)-c\big)^2\Big) \\
&= \mathrm{E}\Big(\big(X-\mathrm{E}(X)\big)^2\Big) + 2\big(\mathrm{E}(X)-c\big)\mathrm{E}\big(X-\mathrm{E}(X)\big) + \big(\mathrm{E}(X)-c\big)^2 \\
&= \mathrm{E}\Big(\big(X-\mathrm{E}(X)\big)^2\Big) + \big(\mathrm{E}(X)-c\big)^2 \geq \mathrm{E}\Big(\big(X-\mathrm{E}(X)\big)^2\Big).
\end{aligned}$$

Problem 2.27: Let X, Y be square-integrable random variables. Show that

$$\mathrm{E}(X-\mathrm{E}(X|Y))^2 \leq \mathrm{E}(X-Z)^2$$

for any $\sigma(Y)$-measurable random variable Z.
Hint: Show that $X-\mathrm{E}(X|Y)$ and Z are uncorrelated, and write $X-Z = \big(X-\mathrm{E}(X|Y)\big) + \big(\mathrm{E}(X|Y)-Z\big)$.

Solution. Let $\widehat{X} = \mathrm{E}(X|Y)$. First, we observe that $X-\widehat{X}$ and Z are uncorrelated for any $\sigma(Y)$-measurable random variable Z. Indeed, since $\mathrm{E}(X-\widehat{X}) = \mathrm{E}(X) - \mathrm{E}(\mathrm{E}(X|Y)) = \mathrm{E}(X) - \mathrm{E}(X) = 0$, we have

$$\begin{aligned}
\mathrm{Cov}(X-\widehat{X}, Z) &= \mathrm{E}(Z(X-\widehat{X})) - \mathrm{E}(Z)\mathrm{E}(X-\widehat{X}) \\
&= \mathrm{E}(ZX) - \mathrm{E}(Z\widehat{X}) = \mathrm{E}(ZX) - \mathrm{E}(Z\mathrm{E}(X|Y)) \\
&= \mathrm{E}(ZX) - \mathrm{E}(ZX) = 0.
\end{aligned}$$

Thus (recall that $\widehat{X}$ is $\sigma(Y)$-measurable), $\mathrm{E}\big((X-\widehat{X})(\widehat{X}-Z)\big) = \mathrm{Cov}(X-\widehat{X}, \widehat{X}-Z) = 0$ and

$$\begin{aligned}\mathrm{E}\big((X-Z)^2\big) &= \mathrm{E}\big((X-\widehat{X}+\widehat{X}-Z)^2\big)\\ &= \mathrm{E}\big((X-\widehat{X})^2 + 2(X-\widehat{X})(\widehat{X}-Z) + (\widehat{X}-Z)^2\big)\\ &= \mathrm{E}\big((X-\widehat{X})^2\big) + \mathrm{E}\big((\widehat{X}-Z)^2\big)\\ &\geq \mathrm{E}\big((X-\widehat{X})^2\big).\end{aligned}$$

Problem 2.28: Consider the following definitions for a Gaussian vector $X \in \mathbb{R}^n$ with zero mean, $n \geq 2$. We assume that X is non-degenerate or equivalently that its covariance matrix Σ is non-singular.

(a) $X = AZ$, where A is a (non-random) non-singular matrix and Z is a random vector in $\mathbb{R}^n$ with independent and standard normal $N(0,1)$ components.
(b) $yX = \sum_{i=1}^n y_i X_i$ is a normal random variable with variance $y\Sigma y^T$ (and zero mean) for any non-random vector $y \in \mathbb{R}^n$.
(c) The moment-generating function of X is $m(v) = e^{v\Sigma v^T/2}$, where $\Sigma = \mathrm{Cov}(X, X)$ is the covariance matrix of X and $v \in \mathbb{R}^n$.

Use your knowledge of (1D) normal random variables to show that these definitions are equivalent.

Solution.

(a)⇒(b) Suppose X can be written as AZ. Then, for any y, $yX = yAZ = (yA)Z = bZ$, where $b = yA$. As the sum of independent normal random variables, bZ has a normal distribution with mean zero and variance $\sum_{i=1}^n b_i^2 = bb^T = yAA^Ty^T$. Indeed, with $u \in \mathbb{R}$, $\mathrm{E}\big(e^{u(bZ)}\big) = \mathrm{E}\big(e^{(ub)Z}\big) = e^{(bb^T)u^2/2}$.

$$\begin{aligned}\mathrm{E}\big(e^{u(bZ)}\big) &= \mathrm{E}\big(e^{\sum_{i=1}^n ub_iZ_i}\big) = \mathrm{E}\left(\prod_{i=1}^n e^{ub_iZ_i}\right) = \prod_{i=1}^n \mathrm{E}\big(e^{ub_iZ_i}\big)\\ &= \prod_{i=1}^n e^{(ub_i)^2/2} = e^{\sum_{i=1}^n (ub_i)^2/2}\\ &= e^{(\sum_{i=1}^n b_i^2)u^2/2} = e^{bb^Tu^2/2}.\end{aligned}$$

(b)⇒(a) Here, one needs a result from linear algebra that states that any symmetric positive definite matrix V can be written as a product of a non-singular matrix A and its transpose A^T, and A can be taken to be symmetric, so that $V = A^2$. Since the covariance matrix of a non-degenerate random vector X is symmetric and positive definite, there is such matrix A. Since yX is normal with variance $y\Sigma y^T$, then $\mathrm{E}(e^{yX}) = e^{y\Sigma y^T/2}$. Letting $Z = A^{-1}X$, it is easy to see that Z is a random vector in $\mathbb{R}^n$ with independent and standard normal $N(0,1)$ components:

$$\mathrm{E}(e^{vZ}) = \mathrm{E}(e^{vA^{-1}X}) = e^{(vA^{-1})\Sigma(vA^{-1})^T/2}$$
$$= e^{vA^{-1}AA^T(A^T)^{-1}v^T/2} = e^{vv^T/2}.$$

We have used here the well-known algebraic fact that $(A^{-1})^T = (A^T)^{-1}$.

(b)⇒(c) Since yX is a normal random variable, $\mathrm{E}(e^{yX}) = e^{Var(yX)/2}$ (mgf at 1). But $\mathrm{Var}(yX) = \mathrm{Cov}(yX, yX) = y\,\mathrm{Cov}(X,X)y^T = y\Sigma y^T$, and (c) follows.

(c)⇒(b) This is immediate since for $u \in \mathbb{R}$,

$$\mathrm{E}(e^{u(yX)}) = \mathrm{E}(e^{(uy)X}) = e^{(uy)\Sigma(uy)^T/2} = e^{u^2(y\Sigma y^T)/2},$$

which shows that yX is $N(0, y\Sigma y^T)$.

Problem 2.29:

(a) Show that for a non-negative random variable X,

$$\mathrm{E}(X) = \int_0^\infty \mathrm{P}(X \geq x)dx = \int_0^\infty \mathrm{P}(X > x)dx.$$

Hint: Use $\mathrm{E}(X) = \int_0^\infty x dF(x) = \int_0^\infty \int_0^x dt dF(x)$ and change the order of integration.

(b) Show that if $X \geq 0$, $\mathrm{E}(X) \leq \sum_{n=0}^\infty \mathrm{P}(X > n)$.

Solution.

(a) This is a direct application of Fubini's theorem:

$$\mathrm{E}(X) = \int_0^\infty x dF(x) = \int_0^\infty \int_0^\infty 1_{t\in[0,x]} dt dF(x)$$
$$= \int_0^\infty \int_0^\infty 1_{x\in[t,\infty)} dF(x) dt = \int_0^\infty \mathrm{P}(X \geq t)dt.$$

Replacing $1_{t\in[0,x]}$ with $1_{t\in[0,x)}$ leads to the second equality.

(b) This is a direct consequence of the monotonicity of the distribution function F:

$$\begin{aligned}\mathrm{E}(X) &= \sum_{n=0}^{\infty} \int_{n}^{n+1} \mathrm{P}(X > x)dx = \sum_{n=0}^{\infty} \int_{n}^{n+1} (1 - F(x))dx \\ &\leq \sum_{n=0}^{\infty} \int_{n}^{n+1} (1 - F(n))dx = \sum_{n=0}^{\infty} \mathrm{P}(X > n).\end{aligned}$$

Using the identity $\mathrm{E}(X) = \int_0^\infty \mathrm{P}(X \geq x)dx$ leads to a weaker statement in cases where F has jumps at integer values:

$$\mathrm{E}(X) = \int_0^\infty \mathrm{P}(X \geq x)dx \leq \sum_{n=0}^{\infty} \mathrm{P}(X > n) \leq \sum_{n=0}^{\infty} \mathrm{P}(X \geq n).$$

Problem 2.30: Let U, V, and W be three independent standard normal random variables.

(a) Use these random variables to construct an example of two dependent random variables that, conditionally on a third, are independent.
(b) Use the same random variables to construct an example for which the opposite is true; that is two independent random variables that, conditionally on a third, are dependent.

Solution.

(a) Let $X = U + V$ and $Y = U + W$. Then, (U, X, Y) is a multivariate normal triple with zero mean and covariance matrix

$$\Sigma = \begin{pmatrix} 1 & 1 & 1 \\ 1 & 2 & 1 \\ 1 & 1 & 2 \end{pmatrix}.$$

It follows that conditional on $U = u$, X and Y are independent normal random variables with mean u and variance 1. On the other hand, (X, Y) is a bivariate normal pair with covariance $\mathrm{Cov}(X, Y) = 1 \neq 0$; in other words, X and Y are dependent.
(b) Let $X = U + V$ and $Y = U - V$. Then, (X, Y) is a bivariate normal pair with zero covariance. It follows that X and Y are independent normal random variables. However, conditional on $U = u$, $X = u + V$ and $Y = u - V$ are clearly dependent.

Problem 2.31: Let X be a random variable with moment-generating function m. Suppose that m is finite in an open interval that contains 0. In the following, we explore some basic properties of the behaviour of m as a function. In particular, we try to understand how small m can get.

(a) Show that, if m is finite for $u > 0$, then $m(u) \geq \mathrm{P}(X \geq 0)$ and that, if m is finite for $u < 0$, then $m(u) \geq \mathrm{P}(X \leq 0)$.
(b) Show that if $\mathrm{E}(X) = 0$, then $m(u) \geq 1$.
(c) Show that m has a unique global minimum (for X not identically nil).
(d) Suppose that X has $N(\mu, \sigma^2)$ distribution. Find the global minimum u^* of m and obtain $m(u^*)$. Discuss the solutions of the equation $m(u) = 1$.
(e) Repeat (d) for $X = c - Y$, where Y has an exponential distribution with parameter λ and $c > 0$.

Solution.

(a) Take $u > 0$ for which $m(u)$ is finite. For such u, $e^{uX} > 1$ on $\{X \geq 0\}$ and $e^{uX} > 0$ on $\{X < 0\}$ (in fact everywhere). Therefore,

$$\begin{aligned} m(u) &= \mathrm{E}\big(e^{uX}\big) = \mathrm{E}\big(e^{uX} 1_{X \geq 0} + e^{uX} 1_{X<0}\big) \\ &\geq \mathrm{E}\big(e^{uX} 1_{X \geq 0}\big) \geq \mathrm{P}(X \geq 0). \end{aligned}$$

Similarly for $u < 0$.

(b) This is a direct application of Jensen's inequality:

$$m(u) = \mathrm{E}\big(e^{uX}\big) \geq e^{\mathrm{E}(uX)} = 1.$$

(c) We know from Problem 2.23 that $m''(u) = \mathrm{E}\big(X^2 e^{uX}\big) > 0$ so that m is strictly convex and admits a single global minimum.

(d) In this case, $\mathrm{E}\big(e^{uX}\big) = e^{u\mu + u^2\sigma^2/2}$ and m is minimum when the quadratic function $\mu u + (\sigma^2/2)u^2$ is minimum. This clearly occurs for $u^* = -\mu/\sigma^2$ and then, $m(u^*) = e^{-\mu^2/(2\sigma^2)}$ (which is smaller than 1).

Here, the equation $m(u) = 1$ has two distinct solutions and is actually solvable (algebraically): $m(u) = 1 \Leftrightarrow \mu u + (\sigma^2/2)u^2 = 0 \Leftrightarrow u = 0$ or $u = -2\mu/\sigma^2$.

(e) Here, for $u > -\lambda$,

$$m(u) = \frac{\lambda e^{cu}}{\lambda + u} \text{ and } m'(u) = \frac{\lambda e^{cu}(c(\lambda + u) - 1)}{(\lambda + u)^2}.$$

Furthermore,

$$m''(u) = \frac{\lambda e^{cu}\left((c(\lambda + u) - 1)^2 + 1\right)}{(\lambda + u)^3} \geq 0.$$

Hence, m is minimum at $u^* = (1/c) - \lambda$ and $m(u^*) = c\lambda e^{1-c\lambda}$.

Note that the function $xe^{1-x} \leq 1$ for all x. It follows that $m(u^*) \leq 1$; at its minimum, m is at most 1.

As a strictly convex function, the equation $m(u) = 1$ has at most two solutions. Obviously, 0 is one of them. A second solution exists if and only if $c\lambda \neq 1$. If $c\lambda > 1$, then $u^* = (1/c) - \lambda < 0$, and the second solution of $m(u) = 1$ belongs to the interval $(-\lambda, (1/c) - \lambda)$. If $c\lambda < 1$, then $u^* = (1/c) - \lambda > 0$, and the second solution of $m(u) = 1$ is positive.

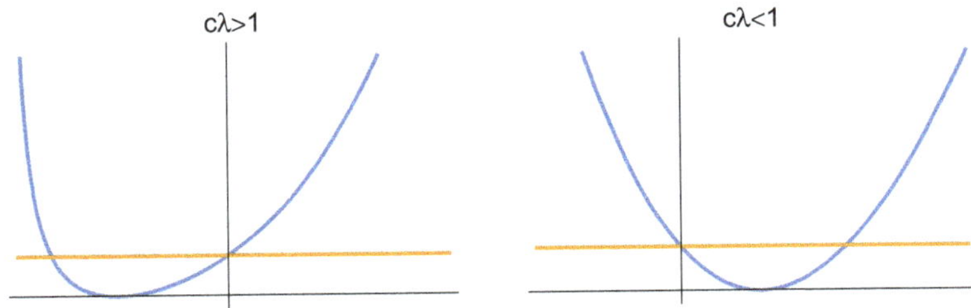

Problem 2.32: Let $X_n = x + \sum_{k=1}^{n} \xi_k$, $n = 0, 1, 2, \ldots$ denote a random walk, with $X_0 = x$ and $(\xi_k)_{k\geq 1}$ being a sequence of independent and identically distributed random variables with distribution $\mathrm{P}(\xi_1 \leq x) = e^{\lambda(x-c)}$ for $x \leq c$ and $\mathrm{P}(\xi_1 \leq x) = 1$ for $x \geq c$, for some positive constants λ, c. (It is easily checked that $\xi_1 = c - Y$, where Y is an exponential λ random variable.) This problem gives results on the random time for this random walk to become negative. Let $T_x = \inf\{n \geq 0 : X_n < 0\}$. We look at the cases of positive and negative drifts, $\mathrm{E}(\xi_1)$. Note that the index x in T_x refers to the starting point and not the target value.

The case of a negative drift, $c\lambda < 1$:

(a) Show that $\mathrm{P}(T_x < \infty) = 1$.
(b) Use Problem 2.31 to show that $\mathrm{E}(T_x) < \infty$, for any x.
Hint: $T_x > n \Rightarrow X_n > 0$.

The case of a positive drift, $c\lambda > 1$:

(c) Show that, if $x \geq 0$, $\mathrm{P}(T_x < \infty | X_1) = \Psi(X_1)$.
(d) Write an integral equation for $\Psi(x) = \mathrm{P}(T_x < \infty)$ using a first-step analysis and specify the domain where it holds.
(e) Let $u^* = -R$ be the (strictly) negative solution of $m(u) = 1$, where m is the moment-generating function of ξ_1 — see Problem 2.31.

Obtain an exponentially decaying in x upper bound of $\Psi(x)$ by stopping the martingale $M_n = e^{-RX_n}$ at an appropriate time.

(f) The integral equation found in (d) has at least one solution, Ψ. Show that (in the case of positive drift) the equation has in fact infinitely many solutions H such that $0 \le H(x) \le 1$, $\forall x$. What conclusion do you draw?
Hint: Show that the space of solutions is convex; in other words, if H_1 and H_2 are solutions, then, for any $0 < \alpha < 1$, $\alpha H_1 + (1-\alpha)H_2$ is also a solution.

Solution.

The case of a negative drift, $c\lambda < 1$:

(a) In this case, $\mathrm{E}(\xi_1) = c - 1/\lambda < 0$. By the law of large numbers,

$$\frac{1}{n}\sum_{k=1}^{n} \xi_k \longrightarrow \mathrm{E}(\xi_1) < 0 \ (a.s.)$$

and therefore, with probability 1, $X_n \longrightarrow -\infty$, i.e. for any ω outside a null event,

$$\forall a > 0, \exists n^*(\omega) \text{ such that } \forall n \ge n^*(\omega),\ X_n(\omega) < -a \text{ and } T_x(\omega) \le n.$$

It immediately follows that, $\Psi(x) = \mathrm{P}(T_x < \infty) = 1$.

(b) We start by observing that $\mathrm{P}(T_x > n+1) = \mathrm{P}(X_1 > 0, X_2 > 0, \ldots, X_n > 0) \le \mathrm{P}(X_n > 0)$. Next, we use the generating function of X_n to bound $\mathrm{P}(X_n > 0)$ — see Problem 2.31. More specifically, we write $\mathrm{P}(X_n > 0) \le m_{X_n}(u)$, where we choose u to be the minimiser of m_{ξ_1}, $u^* = (1/c) - \lambda$. In this case, $m_{\xi_1}(u^*) = \alpha < 1$.

Since $X_n = x + \sum_{k=1}^n \xi_k$, and by the independence of the increments ξ_k,

$$\begin{aligned} m_{X_n}(u) &= \mathrm{E}\big(e^{uX_n}\big) = \mathrm{E}\big(e^{u(x+\sum_{k=1}^n \xi_k)}\big) = e^{ux}\mathrm{E}\big(e^{\sum_{k=1}^n u\xi_k}\big) \\ &= e^{ux}\mathrm{E}\left(\prod_{k=1}^n e^{u\xi_k}\right) = e^{ux}\prod_{k=1}^n \mathrm{E}\big(e^{u\xi_k}\big) = e^{ux}\prod_{k=1}^n m_{\xi_k}(u) \\ &= e^{ux}\prod_{k=1}^n m_{\xi_1}(u) = e^{ux}(m_{\xi_1}(u))^n = e^{ux}(m_{\xi_1}(u))^n. \end{aligned}$$

Hence,

$$m_{X_n}(u^*) = e^{u^*x}(m_{\xi_1}(u^*))^n = e^{u^*x}\alpha^n$$

and with $C = e^{u^*x}/\alpha$,

$$\mathrm{P}(T_x > n+1) \leq e^{u^*x}\alpha^n = C\alpha^{n+1}.$$

It immediately follows that

$$\mathrm{E}(T_x) = \sum_{n=0}^{\infty} \mathrm{P}(T_x > n) \leq C \sum_{k=0}^{\infty} \alpha^n = \frac{C}{1-\alpha} < \infty.$$

The case of a positive drift, $c\lambda > 1$:

(c) Suppose $x \geq 0$. Then, $T_x \geq 1$, $\{T_x = 1\} = \{X_1 < 0\}$ and for $n \geq 2$, $\{T_x = n\} = \{X_1 \geq 0, \ldots, X_{n-1} \geq 0, X_n < 0\}$. Therefore, on $\{X_1 < 0\}$, $T_x = 1$, and on $\{X_1 \geq 0\}$, T_x is the time the random walk started at X_1 and with increments $(\xi_{k+1})_{k\geq 1}$ plus 1: $T_x = 1 + \inf\{n \geq 0 : X_1 + \sum_{k=1}^{n} \xi_{k+1} < 0\}$. Since the conditional distribution of $(\xi_{k+1})_{k\geq 1}$ given X_1 (note that the two objects are in fact independent) is identical to the (unconditional) distribution of $(\xi_k)_{k\geq 1}$, it follows that on $\{X_1 \geq 0\}$, $\mathrm{P}(T_x < \infty | X_1) = \Psi(X_1)$. Combining the two cases, we get

$$\mathrm{P}(T_x < \infty | X_1) = 1_{X_1 < 0} + \Psi(X_1) 1_{X_1 \geq 0} = \Psi(X_1).$$

(d) If $x < 0$, then clearly $T_x = 0$ and $\Psi(x) = 1$. Suppose $x \geq 0$. By conditioning on X_1,

$$\Psi(x) = \mathrm{E}\big(\mathrm{P}(T_x < \infty | X_1)\big) = \mathrm{E}\big(\Psi(X_1)\big).$$

For $x \geq 0$,

$$\begin{aligned}\Psi(x) &= \int_{-\infty}^{\infty} \Psi(y) f_{X_1}(y)dy = \int_{-\infty}^{0} \Psi(y) f_{X_1}(y)dy + \int_{0}^{\infty} \Psi(y) f_{X_1}(y)dy \\ &= \int_{-\infty}^{0} f_{X_1}(y)dy + \int_{0}^{\infty} \Psi(y) f_{X_1}(y)dy.\end{aligned}$$

To find the density of X_1, we first find its distribution function:

$$F_{X_1}(y) = \mathrm{P}(X_1 \leq y) = P(\xi_1 \leq y - x) = \begin{cases} e^{\lambda(y-x-c)} & y \leq x + c, \\ 1 & y > x + c. \end{cases}$$

Differentiating in y, we get

$$f_{X_1}(y) = \begin{cases} \lambda e^{\lambda(y-x-c)} & y \leq x + c, \\ 0 & y > x + c. \end{cases}$$

Proceeding, we obtain for $x \geq 0$,

$$\Psi(x) = e^{-\lambda(x+c)} + \lambda \int_0^{x+c} \Psi(y) e^{\lambda(y-x-c)} dy. \qquad (*)$$

(e) Recall that $-R$ is the non-zero solution of $m(u) = 1$. Then, $M_n = e^{-RX_n}$ is a martingale:

$$\begin{aligned} \mathrm{E}(M_{n+1}|\mathcal{F}_n) &= \mathrm{E}\big(e^{-RX_{n+1}}|\mathcal{F}_n\big) = M_n \mathrm{E}\big(e^{-R\xi_{n+1}}\big) \\ &= M_n m(-R) = M_n. \end{aligned}$$

Hence, by stopping the martingale at a finite stopping time, $T \wedge N$ yields

$$\mathrm{E}\big(M_{T_x \wedge N}\big) = M_0 = e^{-Rx}.$$

On the other hand, since $X_{T_x} < 0$ and $R > 0$, $M_{T_x} = e^{-RX_{T_x}} > 1$ and

$$\begin{aligned} \mathrm{E}\big(M_{T_x \wedge N}\big) &= \mathrm{E}\big(M_{T_x} 1_{T_x < N}\big) + \mathrm{E}\big(M_N 1_{T_x \geq N}\big) \\ &\geq \mathrm{E}\big(M_{T_x} 1_{T_x < N}\big) \geq \mathrm{P}(T_x < N). \end{aligned}$$

Thus, for any $N > 0$, $\mathrm{P}(T_x < N) \leq e^{-Rx}$. Taking $N \to \infty$, we obtain the bound

$$\mathrm{P}(T_x < \infty) = \lim_{N \to \infty} \mathrm{P}(T_x < N) \leq e^{-Rx}.$$

(f) The integral equation found in (d) always has a solution $\Psi(x)$ because we showed that the hitting probabilities satisfy the equation. It is also clear that $H(x) = 1$ is a solution. In the case of positive drift, we know that $\Psi(x) \leq e^{-Rx}$ for $x > 0$ is also a solution in addition to $H(x) = 1$.

It is easily checked that $\alpha\Psi(x) + (1 - \alpha)H(x)$ is also a solution for any α. Hence, there are infinitely many solutions.

The fact that $(*)$ has more than one solution makes the equation on its own insufficient to identify Ψ. Often, the best one can do is obtain a bound on Ψ.

Problem 2.33:

(a) Let X_n be $\mathrm{N}(\mu_n, \sigma_n^2)$. Use characteristic functions to show that X_n converges in distribution, to say X, if and only if μ_n converges, to say μ, and σ_n^2 converges, to say σ^2. Furthermore, in this case, X is $\mathrm{N}(\mu, \sigma^2)$. Here, we extend the definition of the normal distribution to include the degenerate case, $\sigma = 0$. In this case, the random variable is simply a constant equal to μ.

The reader is reminded that, for a periodic function f, the convergence of $f(x_n)$ does not imply that of x_n. In particular, it may be that $e^{i\mu_n t}$ converges for a given t without μ_n converging. However, the convergence of $e^{i\mu_n t}$, for sufficiently many t's, is sufficient to guarantee that of μ_n. (b) illustrates this point.

(b) Let $(x_n)_n$ be a sequence of real numbers. If for all $t \geq 0$, $x_n t - \lfloor x_n t \rfloor$ converges, then so does x_n.

Solution.

(a) If $\mu_n \to \mu$ and $\sigma_n^2 \to \sigma^2$, then the characteristic functions φ_n of X_n converge pointwise:

$$\varphi_n(t) = e^{i\mu_n t - \sigma_n^2 t^2/2} \to e^{i\mu t - \sigma^2 t^2/2}.$$

Since the limit is a characteristic function of a $\mathrm{N}(\mu, \sigma^2)$, then $X_n \xrightarrow{d} X$, where $X \stackrel{d}{=} \mathrm{N}(\mu, \sigma^2)$.

Conversely, suppose $X_n \xrightarrow{d} X$. Then, $\varphi_n(t) = e^{i\mu_n t - \sigma_n^2 t^2/2}$ converges pointwise, say to the characteristic function $\varphi(t)$. It follows that $\sigma_n^2 = (-2/t^2) \ln e^{-\sigma_n^2 t^2/2} = (-2/t^2) \ln |\varphi_n(t)|$ converges and $\lim_n \sigma_n^2 = (-2/t^2) \ln |\varphi(t)|$. Call this limit σ^2. It follows that $e^{-\sigma_n^2 t^2/2} \varphi_n(t) = e^{i\mu_n t}$ converges to $e^{-\sigma^2 t^2/2} \varphi(t)$, which is a continuous function of t, and therefore a characteristic function (by Lévy's continuity theorem). In other words, $(\mu_n)_n$, as a sequence of (constant) random variables, converges weakly. Therefore, $\lim_n \mu_n = \lim_n \mathbb{E}[f(\mu_n)] = \mathbb{E}[f(\mu)] = \mu$, where f is any given bounded function that coincides with the identity in a neighbourhood of μ.

The convergence of both μ_n and σ_n in turn implies that of $\varphi_n(t) = e^{i\mu_n t - \sigma_n^2 t^2/2}$ to $\varphi(t) = e^{i\mu t - \sigma^2 t^2/2}$, the characteristic function of $\mathrm{N}(\mu, \sigma^2)$.

As stated above, the convergence of $e^{i\mu_n t}$, for a fixed t, is not sufficient to imply that of μ_n. However, for a sufficiently rich set (e.g. an open interval containing 0), the convergence of $e^{i\mu_n t}$ is sufficient to guarantee that of μ_n. See Problem 2.34.

Alternatively, and in the case $\inf_n \sigma_n > 0$, let Φ be the distribution function of the standard normal distribution and F that of X. Then, for any pair of points of continuity of F, x and y, $\Phi((x-\mu_n)/\sigma_n)$ and $\Phi((y-\mu_n)/\sigma_n)$ converge, and, since Φ is one-to-one, so do $(x-\mu_n)/\sigma_n$ and $(y-\mu_n)/\sigma_n$. It follows that $(x-y)/\sigma_n$ and therefore σ_n converge. That of μ_n follows immediately.

The convergence of both μ_n and σ_n to μ and $\sigma > 0$, respectively, implies that of $\Phi((x-\mu_n)/\sigma_n)$ to $\Phi((x-\mu)/\sigma)$ (Φ is continuous), the distribution function of $\mathrm{N}(\mu, \sigma^2)$.

(b) Since $x_n - \lfloor x_n \rfloor$ and $x_n - \lfloor x_n t \rfloor / t$ converge, then so does $\lfloor x_n \rfloor - \lfloor x_n t \rfloor / t$. Since $\mathbb{Z} - (1/t)\mathbb{Z}$ is a discrete space, for any $t \in \mathbb{R}_+ \setminus \mathbb{Q}$, there exists $N_t \in \mathbb{N}$ such that for any $n \geq N_t$, $\lfloor x_n \rfloor - \lfloor x_n t \rfloor / t$ is constant, equal to say $p - q/t \in \mathbb{Z} - (1/t)\mathbb{Z}$. Therefore, for any $t \in \mathbb{R}_+ \setminus \mathbb{Q}$ and any $n \geq N_t$, $\lfloor x_n \rfloor - \lfloor x_n t \rfloor / t = p - q/t$ or equivalently that $\lfloor x_n \rfloor - p = \lfloor x_n t \rfloor / t - q/t = (\lfloor x_n t \rfloor - q)/t$. It follows that for any $t \in \mathbb{R}_+ \setminus \mathbb{Q}$ and any $n \geq N_t$, $\lfloor x_n \rfloor - p \in \mathbb{Z} \cap (1/t)\mathbb{Z} = \{0\}$ and $\lfloor x_n \rfloor = p$. We deduce that $x_n - p$, and therefore x_n itself, converges.

Problem 2.34:

(a) Let U be a random variable taking values in $\{-1, 1\}$ and $\mathcal{H}$ be a σ-field. Show that U is independent of $\mathcal{H}$ if and only if $\mathrm{E}(U|\mathcal{H}) = \mathrm{E}(U)$.

(b) Let $(\xi_n)_{n\geq 1}$ be a sequence of random variables taking values in $\{-1, 1\}$ and for $n \geq 0$, $X_n = \sum_{k=1}^{n} \xi_k$ ($X_0 = 0$). Show that $(X_n)_{n\geq 0}$ is a martingale if and only if the ξ_n's are independent and identically distributed with zero mean.

(c) Let $(\eta_n)_{n\geq 1}$ be a sequence of random variables taking values in $\{a, b\}$ ($a < 1 < b$), and for $n \geq 0$, $Y_n = \prod_{k=1}^{n} \eta_k$ ($Y_0 = 1$). Show that $(Y_n)_{n\geq 0}$ is a martingale if and only if the η_n's are independent and identically distributed with mean one.

(d) Explain why in the last two statements, the assumption that the random variables ξ_n and η_n take values in a two-point set is crucial.

Solution.

(a) Without loss of generality, we assume that $\mathrm{E}(U) = 0$ (otherwise consider the difference $U - \mathrm{E}(U)$). Suppose $\mathrm{E}(U|\mathcal{H}) = 0$, then $\mathrm{P}(U = 1|\mathcal{H}) = \mathrm{P}(U = 1) = 1/2$. It follows that for any $A \in \mathcal{H}$,

$$\mathrm{P}(\{U = 1\} \cap A) = \mathrm{E}\big(\mathrm{P}(U = 1|\mathcal{H})1_A\big) = \frac{1}{2}\mathrm{P}(A) = \mathrm{P}(U = 1)\mathrm{P}(A)$$

and that U and $\mathcal{H}$ are independent.

(b) Note that, for random variables taking values in $\{-1, 1\}$, the mean defines the distribution: $\mathrm{P}(\xi = 1) = (1 + \mathrm{E}(\xi))/2$. Clearly, if the ξ_n's are independent and have zero mean, then $(X_n)_n$ is a symmetric simple random walk and a martingale. Next, we consider the converse and suppose that $(X_n)_n$ is a martingale with respect to its natural filtration, $\mathcal{F}_n = \sigma(X_k, k \leq n) = \sigma(\xi_k, k \leq n)$.

The martingale property for $n = 1$ gives $\mathrm{E}(\xi_1) = \mathrm{E}(\xi_1|\mathcal{F}_0) = \mathrm{E}(X_1|\mathcal{F}_0) = X_0 = 0$, from which we deduce that $\mathrm{P}(\xi_1 = -1) = \mathrm{P}(\xi_1 = 1) = 1/2$.

For $n = 2$, we have $X_1 = \mathrm{E}(X_2|\mathcal{F}_1) = X_1 + \mathrm{E}(\xi_2|\xi_1)$ or equivalently that $\mathrm{E}(\xi_2|\xi_1) = 0$. We deduce from (a) that ξ_1 and ξ_2 are independent. Furthermore, since $\mathrm{E}(\xi_2) = 0$, ξ_1 and ξ_2 have the same symmetric distribution.

We proceed by induction and assume that $\xi_1, \ldots, \xi_n$ are independent and have the same symmetric distribution. Writing the martingale property for $n+1$, we get $X_n = \mathrm{E}(X_{n+1}|\mathcal{F}_n) = X_n + \mathrm{E}(\xi_{n+1}|\mathcal{F}_n)$ or equivalently that $\mathrm{E}(\xi_{n+1}|\mathcal{F}_n) = 0$. Applying (a) again, we deduce that ξ_{n+1} is independent of $\mathcal{F}_n$ and consequently that $\xi_1, \ldots, \xi_{n+1}$ are independent. Recall that $\xi_1, \ldots, \xi_n$ are independent by the induction assumption.

(c) We repeat the same proof as in (b). Suppose that $(Y_n)_n$ is a martingale with respect to its natural filtration, $\mathcal{G}_n = \sigma(Y_k, k \leq n) = \sigma(\eta_k, k \leq n)$.

Clearly $\mathrm{E}(\eta_1) = \mathrm{E}(Y_1) = 1$ or equivalently $\mathrm{P}(\eta_1 = b) = (1 - a)/(b - a)$, and $Y_1 = \mathrm{E}(Y_2|\mathcal{G}_1)$ leads to $\mathrm{E}(\eta_2|\eta_1) = 1$, from which we deduce that η_1 and η_2 are independent and identically distributed. Assuming that $\eta_1, \ldots, \eta_n$ are independent and identically distributed, we deduce from the martingale property that $\mathrm{E}(\eta_{n+1}|\mathcal{G}_n) = 0$ and therefore that $\eta_1, \ldots, \eta_{n+1}$ are independent.

(d) The proofs of (b) and (c) rely on the fact that, for random variables taking values in a two-point set, the mean defines the distribution.

When the random variables take three or more values, the situation is completely changed. Suppose ξ_1 and ξ_2 take values in $\{-1, 0, 1\}$. Imposing that $\mathrm{E}(\xi_2|\xi_1) = 0$, leads to $\mathrm{P}(\xi_2 = -1|\xi_1) = \mathrm{P}(\xi_2 = 1|\xi_1) = (1 - \mathrm{P}(\xi_2 = 0|\xi_1))/2$, and in general, this quantity depends on ξ_1, from which we conclude that ξ_1 and ξ_2 do not have to be independent.

Problem 2.35: Recall that, by definition, $A \Delta B$ is the symmetric difference $(A \setminus B) \cup (B \setminus A) = (A \cup B) \setminus (A \cap B) = (A \cup B) \cap (A^c \cup B^c)$ and that for two σ-algebras $\mathcal{G}_1$ and $\mathcal{G}_2$, $\mathcal{G}_1 \vee \mathcal{G}_2 = \sigma(\mathcal{G}_1 \cup \mathcal{G}_2)$.

Let $(\Omega, \mathcal{F}, \mathbb{P})$ be a probability space.

(a) Suppose $\mathcal{C}_1, \mathcal{C}_2 \subset \mathcal{F}$. Show that $\sigma(\mathcal{C}_1 \cup \mathcal{C}_2) = \sigma(\mathcal{C}_1) \vee \sigma(\mathcal{C}_2)$.
(b) Suppose $\mathcal{G}$ is a σ-algebra such that $\mathcal{G} \subset \mathcal{F}$. Show that $\mathcal{H} = \{A \in \mathcal{F} : \exists B \in \mathcal{G}$ such that $\mathbb{P}(A \Delta B) = 0\}$ is a σ-algebra.
(c) Suppose $\mathcal{G}$ is a σ-algebra and let $\mathcal{N}$ be the set of $\mathbb{P}$-null sets: $\mathcal{N} = \{A \in \mathcal{F} : \exists B \in \mathcal{G}$ such that $A \subset B$ and $\mathbb{P}(B) = 0\}$. Show that $\sigma(\mathcal{G} \cup \mathcal{N}) = \mathcal{G} \vee \sigma(\mathcal{N}) = \{A \in \mathcal{F} : \exists B \in \mathcal{G},\ N \in \mathcal{N}$ such that $A = B \cup N\}$.

Solutions

(a) From $\mathcal{C}_1 \cup \mathcal{C}_2 \subset \sigma(\mathcal{C}_1) \cup \sigma(\mathcal{C}_2) \subset \sigma(\mathcal{C}_1) \vee \sigma(\mathcal{C}_2)$, we deduce that $\sigma(\mathcal{C}_1 \cup \mathcal{C}_2) \subset \sigma(\mathcal{C}_1) \vee \sigma(\mathcal{C}_2)$. Conversely, $\mathcal{C}_i \subset \mathcal{C}_1 \cup \mathcal{C}_2$ leads to $\sigma(\mathcal{C}_i) \subset \sigma(\mathcal{C}_1 \cup \mathcal{C}_2)$, to $\sigma(\mathcal{C}_1) \cup \sigma(\mathcal{C}_2) \subset \sigma(\mathcal{C}_1 \cup \mathcal{C}_2)$ and finally to $\sigma(\mathcal{C}_1) \vee \sigma(\mathcal{C}_2) \subset \sigma(\mathcal{C}_1 \cup \mathcal{C}_2)$.
(b) $\mathcal{H}$ is clearly not empty ($\emptyset \in \mathcal{H}$) and
   i. $\Omega \in \mathcal{H}$;
   ii. if $A \in \mathcal{H}$, then $\exists B \in \mathcal{G}$ such that $A^c \Delta B^c = (A^c \cup B^c) \cap (A \cup B) = A \Delta B \in \mathcal{N}$ and $A^c \in \mathcal{H}$;
   iii. if $A_1, A_2, \ldots \in \mathcal{H}$, then $\exists B_1, B_2, \ldots \in \mathcal{G}$ such that $(\bigcup_n A_n) \Delta (\bigcup_n B_n) = \big((\bigcup_n A_n) \cap (\bigcap_n B_n^c)\big) \cup \big((\bigcup_n B_n) \cap (\bigcap_n A_n^c)\big) \subset \big(\bigcup_n (A_n \cap B_n^c)\big) \cup \big(\bigcup_n (B_n \cap A_n^c)\big) = \bigcup_n (A_n \Delta B_n) \in \mathcal{N}$ and $\bigcup_n A_n \in \mathcal{H}$.

   It follows that $\mathcal{H}$ is a σ-algebra.
(c) Let $\mathcal{K} = \{A \in \mathcal{F} : \exists B \in \mathcal{G},\ N \in \mathcal{N}$ such that $A = B \cup N\}$. Clearly, if $B \in \mathcal{G}$ and $N \in \mathcal{N}$, then $B \cup N \in \sigma(\mathcal{G} \cup \mathcal{N})$. It follows that $\mathcal{K} \subset \sigma(\mathcal{G} \cup \mathcal{N})$.

   To show the converse we first observe that $\mathcal{G} \subset \mathcal{K}$ ($B = B \cup \emptyset$ and $\emptyset \in \mathcal{N}$) and $\mathcal{N} \subset \mathcal{K}$ ($N = \emptyset \cup N$), and then show that $\mathcal{K}$ is a

σ-algebra. Indeed,

- $\Omega \in \mathcal{G} \subset \mathcal{K}$.
- Let $A = B \cup N \in \mathcal{K}$. Then $A^c = B^c \cap N^c = B^c \setminus (B^c \cap N)$. Also, since $N \in \mathcal{N}$, there exists $M \in \mathcal{G}$ such that $N \subset M$ and $\mathbb{P}(M) = 0$. Therefore, $B^c \cap N \subset B^c \cap M \subset B^c$ and

  $$A^c = B^c \setminus (B^c \cap N) = \big(B^c \setminus (B^c \cap M)\big) \cup \big((B^c \cap M) \setminus (B^c \cap N)\big).$$

  Since $B, M \in \mathcal{G}$, $B^c \setminus (B^c \cap M) \in \mathcal{G}$. On the other hand, $(B^c \cap M) \setminus (B^c \cap N) \subset M$, from which we deduce that $A^c \in \mathcal{K}$.
- Closure under countable union for $\mathcal{K}$ immediately follows from the same property for $\mathcal{G}$ and the fact that the probability of a countable union of $\mathbb{P}$-null sets is a $\mathbb{P}$-null set.

Problem 2.36: Let $\mathcal{G}$ be a σ-field and g be a bounded measurable function.

(a) Show that if $A \in \mathcal{G}$ is independent of $\mathcal{G}$, then $\mathbb{P}(A) = 0$ or $\mathbb{P}(A) = 1$.
(b) Show that if a $\mathcal{G}$-measurable X is independent of $\mathcal{G}$, then X is (almost surely) constant.

Solution.

(a) Since A is independent of $\mathcal{G}$, $\mathbb{P}(A|\mathcal{G}) = \mathbb{P}(A)$, and since $A \in \mathcal{G}$, $\mathbb{P}(A|\mathcal{G}) = 1_A$ (a.s.). It follows that $\mathbb{P}(A) = 1_A$ (a.s.) and $\mathbb{P}(A)$ can only be either 0 or 1.
(b) We apply (a) to $A = \{X \leq x\}$ to conclude that $\mathbb{P}(X \leq x) \in \{0, 1\}$. Since $F(x) = \mathbb{P}(X \leq x)$ is a non-decreasing and right-continuous function such that $\lim_{x \downarrow -\infty} F(x) = 0$ and $\lim_{x \uparrow +\infty} F(x) = 1$. Therefore, there exists c such that $F(x) = 0$ for $x < c$ and $F(x) = 1$ for $x \geq c$. This is equivalent to saying that $\mathbb{P}(X = c) = F(c) - F(c-) = F(c) - \lim_{x \uparrow c} F(x) = 1$.

Problem 2.37: Let X, Y be independent random variables.

(a) Show that if c is a number such that $0 < \mathbb{P}(X > c) < 1$, then $\{X > c\} \notin \sigma(Y)$.
(b) Show that if $\{X > Y\} \in \sigma(Y)$, then there exists a and b such that $\mathbb{P}(a \leq X \leq b) = 1$ and either $a = b$ (X is constant) or $\mathbb{P}(a < Y < b) = 0$.
(c) Deduce that if Y has a continuous distribution such that $\mathbb{P}(Y \in \operatorname{supp}(X)) > 0$, then $\{X > Y\} \notin \sigma(Y)$.

Solution.

(a) Suppose that $\{X > c\} \in \sigma(Y)$. Hence, there is a measurable set B such that $\{X > c\} = \{Y \in B\}$. On the one hand, it equals $\mathbb{P}(X > c, Y \in B) = \mathbb{P}(X > c)$, and on the other hand, by the independence of X and Y, $\mathbb{P}(X > c, Y \in B) = \mathbb{P}(X > c)\mathbb{P}(Y \in B) = \mathbb{P}(X > c)^2$. This yields that either $\mathbb{P}(X > c) = 0$ or $\mathbb{P}(X > c) = 1$, which contradicts the assumption that $0 < \mathbb{P}(X > c) < 1$. This proves that $\{X > c\} \notin \sigma(Y)$.

(b) Suppose that $\{X > Y\} \in \sigma(Y)$ and therefore that $\{X \leq Y\} \in \sigma(Y)$. It follows that there exists a measurable set B such that $\{X \leq Y\} = \{Y \in B\}$ and also $\{X > Y\} = \{Y \in B^c\}$. Using the independence of X and Y, we get

$$\begin{aligned}\int_B \mathbb{P}(X > y)dF_Y(y) &= \int_B (1 - \mathbb{P}(X \leq y))dF_Y(y)\\ &= \mathbb{P}(Y \in B) - \int_B \mathbb{P}(X \leq y)dF_Y(y)\\ &= \mathbb{P}(Y \in B) - \mathbb{P}(X \leq Y, Y \in B) = 0.\end{aligned}$$

Similarly,

$$\begin{aligned}\int_{B^c} \mathbb{P}(X \leq y)dF_Y(y) &= \int_{B^c} (1 - \mathbb{P}(X > y))dF_Y(y)\\ &= \mathbb{P}(Y \in B^c) - \int_{B^c} \mathbb{P}(X > y)dF_Y(y)\\ &= \mathbb{P}(Y \in B^c) - \mathbb{P}(X > Y, Y \in B^c) = 0.\end{aligned}$$

Therefore, there exists a ($\mathbb{P}_Y$-negligible) set N such that $\mathbb{P}(Y \in N) = 0$ and

$$\forall y \in B \setminus N,\ \mathbb{P}(X > y) = 0 \text{ and } \forall y \in B^c \setminus N,\ \mathbb{P}(X \leq y) = 0.$$

Let $a = \sup\{y : y \in B^c \setminus N\}$ and $b = \inf\{y : y \in B \setminus N\}$. Then, on the one hand, $\mathbb{P}(X < a) = 0$ and $\mathbb{P}(X \leq b) = 1$, and on the other hand, $B \backslash N \subset [b, +\infty)$ and $B^c \backslash N \subset (-\infty, a]$. It immediately follows that $\mathbb{P}(a \leq X \leq b) = 1$ and $(a, b) \subset N$ or that $\mathbb{P}(a < Y < b) = 0$.

(c) We reason by contradiction and suppose that $\{X > Y\} \in \sigma(Y)$. From (b), we know that there exists $a \leq b$ such that $\mathbb{P}(a < Y < b) = 0$ (or $a = b$) and $\text{supp}(X) \subset [a, b]$. Since Y has a continuous distribution,

$$\mathbb{P}(Y \in \text{supp}(X)) \leq \mathbb{P}(a \leq Y \leq b) = \mathbb{P}(a < Y < b) = 0.$$

This contradicts the assumption that $\mathbb{P}(Y \in \text{supp}(X)) > 0$. As a consequence, $\{X > Y\} \notin \sigma(Y)$.

Problem 2.38: Let A be a set and $\mathcal{C}$ be a family of subsets of Ω. Show that $\sigma(\mathcal{C}) \cap A = \sigma_A(\mathcal{C} \cap A)$, the smallest σ-algebra of subsets of A that contains the family $\mathcal{C} \cap A = \{B \cap A,\ B \in \mathcal{C}\}$.

Solution. This is a useful result on the trace on A of the σ-algebra generated by the family $\mathcal{C}$.

(a) $\sigma(\mathcal{C}) \cap A$ is a σ-algebra of subsets of A. Furthermore, $\mathcal{C} \cap A \subset \sigma(\mathcal{C}) \cap A$. It immediately follows that $\sigma_A(\mathcal{C} \cap A) \subset \sigma(\mathcal{C}) \cap A$.

(b) Let $\mathcal{A} = \{B \subset \Omega,\ B \cap A \in \sigma_A(\mathcal{C} \cap A)\}$. Then $\mathcal{A}$ is σ-algebra (of subsets of Ω):

- $\Omega \cap A = A \in \sigma_A(\mathcal{C} \cap A)$, that is $\Omega \in \mathcal{A}$;
- for $B \in \mathcal{A}$, $B \cap A \in \sigma_A(\mathcal{C} \cap A)$ and $B^c \cap A = (B \cap A)^c \cap A \in \sigma_A(\mathcal{C} \cap A)$, that is $B^c \in \mathcal{A}$;
- for $B_n \in \mathcal{A}$, $B_n \cap A \in \sigma_A(\mathcal{C} \cap A)$ and $\left(\bigcup_n B_n\right) \cap A = \bigcup_n (B_n \cap A) \in \sigma_A(\mathcal{C} \cap A)$, that is $\bigcup_n B_n \in \mathcal{A}$.

Since $\mathcal{C} \subset \mathcal{A}$, $\sigma(\mathcal{C}) \subset \mathcal{A}$ and $\sigma(\mathcal{C}) \cap A \subset \mathcal{A} \cap A \subset \sigma_A(\mathcal{C} \cap A)$.

Chapter 3

Basic Stochastic Processes

Brownian Motion

Brownian motion (BM), named after the botanist R. Brown (and also called the Wiener process after the mathematician N. Wiener), is a stochastic process $\{B(t)\}_{t \geq 0}$ with the following properties:

- INDEPENDENT INCREMENTS for all $s, t \geq 0$, $B(t+s) - B(s)$ is independent of $\mathcal{F}_s^B$, i.e. all $B(u)$, $u \leq s$.
- STATIONARY INCREMENTS $B(t+s) - B(s)$ is normal $N(0,t)$ distributed for $s, t \geq 0$.
- CONTINUITY $\{B(t)\}_{t \geq 0}$ have continuous sample paths with probability one (see also the following section).

The existence of a process with the above properties can be established in a few different ways, but none of them are really elementary and each relies on some advanced results in stochastic process theory.

As for all independent increment processes (including also, for example, the Poisson process), many calculations with BM are best done by expressing the BM values involved as sums of independent increments, for example,

$$\begin{aligned} &\mathrm{P}(B(1) \leq x,\, B(2) \leq y,\, B(3) \leq z) \\ &\quad = \mathrm{P}(\xi_1 \leq 1,\, \xi_1 + \xi_2 \leq y,\, \xi_1 + \xi_2 + \xi_3 \leq z), \end{aligned}$$

for $x, y, z \in \mathbb{R}$, where ξ_1, ξ_2 and ξ_3 are independent standard normal distributed. The reason is that this resolves the problem with the dependence of the different process values and reduces that problem to a calculation involving only independent random variables.

BM can alternatively be started at $B(0) = x$ with the other three axioms above kept unchanged. It is customary to write $\{B^x(t)\}_{t\geq 0}$ for a BM started at x. Alternatively, one can write $P_x(B(t) \leq y)$ for $P(B^x(t) \leq y)$, etc. to clarify what starting point is used. It follows from the axioms above that the probability law (all probabilities calculated) for $\{B^x(t)\}_{t\geq 0}$ coincide with those of $\{B^0(t) + x\}_{t\geq 0}$ (so-called space homogeneity). If nothing else is said, the starting point is $x = 0$.

There is a tradition to be deliberately unspecific about the starting point not only for BM but also for time-homogeneous Markov processes in general (as we shall see, BM is one). The reason for that is that the transition probabilities for time-homogeneous Markov processes have a sort of life of their own (being stochastic semi-groups), making them meaningful to talk about, for example,

$$P(y, t, x, s) = P(y, t - s, x, 0) = P(t - s, x, y),$$

even if the Markov process was not at all started at x. This in turn can be a little less confusing if one is deliberately unspecific about the starting point. However, in many calculations for BM, the starting point is needed, in which case it is 0 unless something else is specifically indicated.

A stochastic process $\{X(t)\}_{t\in T}$ is Gaussian if $\sum_{i=1}^{n} a_i X(t_i)$ is univariate normal distributed for each choice of $a_1, \ldots, a_n \in \mathbb{R}$, $t_1, \ldots, t_n \in T$ and $n \in \mathbb{N}$. A Gaussian process obviously has a normal distributed process value $X(t)$ at each time $t \in T$, but the converse is not true: the requirement that a process is Gaussian is in fact a much deeper and more demanding requirement than just individual process values being Gaussian.

The probability distribution for any vector of $(X(t_1), \ldots, X(t_n))$ of Gaussian process values is completely determined by the mean function $T \ni t \curvearrowright E(X(t)) \in \mathbb{R}$ and the covariance function $T^2 \ni (s,t) \curvearrowright \mathrm{Cov}(X(s), X(t)) \in \mathbb{R}$ of the process. This follows from that the same applies to the (multivariate) characteristic function of the vector by usage of the definition of a Gaussian process. Further, a collection of Gaussian process values with mutual covariances zero are independent. This follows by a quite similar argument.

Gaussian processes are also called normal processes, but they are in fact very far from normal in the sense that their properties are atypical for processes in general. This is an important fact that sometimes seems to be less known than it should be among some applied stochastic process users. For example, the property that everything is determined by the mean and covariance functions is more or less unique for Gaussian processes and so on.

It is an exercise to establish that BM $\{B(t)\}_{t\geq 0}$ is a Gaussian process with mean function $\mathrm{E}(B(t))=0$ and covariance function $\mathrm{Cov}(X(s),X(t))= E(X(s)X(t)) = \min(s,t)$. As usual, this involves using the independence of the increments. As all probabilities for Gaussian processes are determined using the mean and covariance functions, we can alternatively define BM as (a continuous version of) a Gaussian process with the above-mentioned mean function and covariance function.

Properties of BM Paths

A sample path of BM $\{B(t)\}_{t\geq 0} = \{B(\omega,t)\}_{t\geq 0}$ (where now also the dependence on the outcome $\omega \in \Omega$ is present in the notation) is the function

$$B(\omega,\cdot) : [0,\infty) \to \mathbb{R}$$

for an (one) outcome $\omega \in \Omega$ of the random experiment. With probability one (that is, for all $\omega \in \Omega$ outside a null event), the properties of BM sample paths are as follows:

- continuous functions,
- non-monotone in every interval,
- non-differentiable in every point,
- infinite variation over every interval,
- linear quadratic variation $[B](t) = t$ for $t \geq 0$.

While continuity stems from the very definition (construction) of BM, the claims about non-monotonicity and non-differentiability are somewhat technically complicated to establish, it can be done with more or less elementary methods. Claims about the variation and quadratic variation (valid at each single $t \geq 0$ but not jointly for the whole sample path) can be quite easily established using moment methods involving just the first two moments.

Three Martingales of BM

Let $\{\mathcal{F}_t\}_{t\geq 0}$ be the filtration generated by BM $\{B(t)\}_{t\geq 0}$ itself $\{\mathcal{F}_t^B\}_{t\geq 0}$. That filtration can be shown to be continuous, that is, $\mathcal{F}_{t+} = \mathcal{F}_t$ for $t \geq 0$ and $\mathcal{F}_{t-} = \mathcal{F}_t$ for $t > 0$. The following three stochastic processes are all martingales with respect to $\{\mathcal{F}_t\}_{t\geq 0}$:

- $\{B(t)\}_{t\geq 0}$.
- $\{B(t)^2 - t\}_{t\geq 0}$.
- $\{e^{uB(t)-u^2t/2}\}_{t\geq 0}$ for any $u \in \mathbb{R}$.

The proofs of these statements follows the usage of independent normal distributed increments together with some of the basic rules of conditional expectations.

In Chapter 4, we will learn a cleverer way to judge, more or less just by inspection, whether a function $\{f(B(t),t)\}_{t\geq 0}$ of BM is a martingale based on usage of the fundamentally important Itô formula.

Markov Property of BM

It is not hard to see that BM $\{B(t)\}_{t\geq 0}$ is a time-homogeneous Markov process with transition PDF

$$p(t,x,y) = \frac{1}{\sqrt{2\pi t}}\, e^{(y-x)^2/t} \quad \text{for } t>0 \text{ and } x,y\in\mathbb{R}.$$

BM is also a strong Markov process, and one transparent way to express the strong Markov property is the fact that $\{B(\tau+t)-B(\tau)\}_{t\geq 0}$ is a BM for any stopping time τ with respect to $\{\mathcal{F}_t^B\}_{t\geq 0}$.

For any time-homogeneous Markov process $\{X(t)\}_{t\geq 0}$ with transition PDF $p(t,x,y)$, it is easy to establish that the joint PDF of $(X(t_1),\ldots,X(t_n))$ for $0\leq t_1<\cdots<t_n$ is given by

$$\begin{aligned} &f_{X(t_1),\ldots,X(t_n)}(x_1,\ldots,x_n)\\ &\quad = f_{X(t_1)}(x_1)\prod_{i=2}^{n} p(t_i-t_{i-1},x_{i-1},x_i) \quad \text{for } x_1,\ldots,x_n\in\mathbb{R}, \end{aligned}$$

where $f_{X(t_1)}$ is the PDF of $X(t_1)$ of course (and provided that this density exists). In particular, this formula applies to BM with its transition PDF mentioned above.

Hitting Times, Maximum and Minimum, Joint Distributions and Zeros of BM

Consider the first time BM $\{B(t)\}_{t\geq 0}$ takes the value $x\neq 0$:

$$T_x = \inf\{t>0 : B(t)=x\}.$$

Then, T_x is a stopping time (with respect to the filtration generated by BM itself), and for $a < 0 < b$ the stopping time $\tau = \min(T_a, T_b)$ satisfies $E(\tau) < \infty$, while on the other hand, $E(T_a) = E(T_b) = \infty$. However, we do have so-called recurrence in that $P(T_a < \infty) = P(T_b < \infty) = 1$ (see also the end of this chapter).

The so-called reflection principle for BM states that the process $\{\hat{B}(t)\}_{t\geq 0}$ given by $\hat{B}(t) = B(t)$ for $t \in [0, \tau]$ and $\hat{B}(t) = 2\,B(\tau) - B(t)$ for $t \geq \tau$ is a BM for every stopping time τ. This result has a famous consequence that is also sometimes called the reflection principle, namely that

$$\mathrm{P}\Big(\max_{s\in[0,t]} B(s) > x\Big) = \mathrm{P}\Big(\min_{s\in[0,t]} B(s) < -x\Big) = 2\mathrm{P}(B(t) > x) \quad \text{for } t, x > 0.$$

To see this, take $\tau = T_x$, and note that by the first version of the reflection principle,

$$\mathrm{P}\Big(B(t) < x \;\Big|\; \max_{s\in[0,t]} B(s) > x\Big) = \mathrm{P}\Big(B(t) > x \;\Big|\; \max_{s\in[0,t]} B(s) > x\Big) = \frac{1}{2},$$

so that

$$\begin{aligned}
&\mathrm{P}\Big(\max_{s\in[0,t]} B(s) > x\Big) \\
&\quad = \mathrm{P}(B(t) > x) + \mathrm{P}\Big(\max_{s\in[0,t]} B(s) > x,\ B(t) < x\Big) \\
&\quad = \mathrm{P}(B(t) \geq x) + \mathrm{P}\Big(B(t) < x \;\Big|\; \max_{s\in[0,t]} B(s) > x\Big)\mathrm{P}\Big(\max_{s\in[0,t]} B(s) > x\Big) \\
&\quad = \mathrm{P}(B(t) \geq x) + \frac{1}{2}\mathrm{P}\Big(\max_{s\in[0,t]} B(s) > x\Big).
\end{aligned}$$

Note that despite perhaps looking plain, it is in fact quite rare that it is possible to give a closed-form expression as above for the distribution of maxima and minima for a non-degenerate continuous time random process.

From the above-mentioned results on maxima and minima, it follows readily that the stopping time T_x has a so-called inverse Gamma distribution with PDF

$$f_{T_x}(t) = \frac{|x|}{\sqrt{2\pi t^3}}\, e^{-x^2/(2t)} \quad \text{for } x \neq 0 \text{ and } t > 0.$$

Further, it follows from the strong Markov property that $T_b - T_a$ is independent of $\{B(t)\}_{t\in[0,T_A]}$ (or $\mathcal{F}^B_{T_a}$ really) with the same distribution as T_{b-a} for $0 < a < b$.

The joint distribution of BM and its maximum has PDF given by

$$f_{B(t),\max_{s\in[0,t]} B(s)}(x,y)$$
$$= \sqrt{\frac{2}{\pi t^3}}\,(2y-x)\,e^{-(2y-x)^2/(2t)} \quad \text{for } x\in\mathbb{R},\ y\geq x^+ \text{ and } t>0.$$

There is a famous result about the occurrence of zeros of BM called the arcsine law. It is a more or less immediate consequence of the result about the distribution of maxima and reads as follows:

$$\mathrm{P}\big(B(t)\neq 0 \text{ for } t\in(a,b)\big) = \frac{2}{\pi}\arcsin(\sqrt{a/b}) \quad \text{for } 0<a<b.$$

Size of Increments of BM

BM can be viewed as a continuous version of a simple symmetric random walk. In fact, properly normalised and time scaled, such a random walk converges weakly to a BM (when the normalisation and scaling are suitable and subsequently modified). The relation between BM and a simple random walk is illustrated by the law of the iterated logarithm for BM which has exactly the same appearance as it has for the random walk:

$$\mathrm{P}\left(\limsup_{t\to\infty}\frac{B(t)}{\sqrt{2t\,\ln\ln(t)}}=1\right) = \mathrm{P}\left(\liminf_{t\to\infty}\frac{B(t)}{\sqrt{2t\,\ln\ln(t)}}=-1\right)=1.$$

BM in Higher Dimensions

A higher (than one) dimensional BM in $\mathbb{R}^n$ is simply defined as an $\mathbb{R}^n$-valued process $(B_1(t),\ldots,B_n(t))$ for $t\geq 0$, where $\{B_1(t)\}_{t\geq0},\ldots,$ $\{B_n(t)\}_{t\geq0}$ are independent ordinary BM's. By polarisation together with the fact that $(B_i\pm B_j)/\sqrt{2}$ both are BM for $i\neq j$, it follows that $[B_i,B_j]=0$ for $i\neq j$.

A topic that turns out to be of surprisingly large interest is what kind of sets BM in different dimensions visits and does not visit with different likelihoods. The basic results here are as follows: For $n=1$, BM visits any point eventually with probability 1 (recurrence). For $n=2$, BM visits any open set eventually with probability 1 (also called recurrence), but the hitting probability of single individual points is 0. For $n\geq 3$, BM eventually leaves any bounded set for good with probability 1 (transience).

Problems

Problem 3.1: Let Y_1, Y_2, Y_3, Y_4 be i.i.d. $N(0,1)$ random variables. Let $X_1 = (Y_1 + Y_2)/\sqrt{2}$, $X_2 = (Y_1 + Y_2 + Y_3 + Y_4)/2$, and $X_3 = (Y_1 - Y_2)/\sqrt{2}$. State with reason if the following is true.

(a) X_1, X_2 and X_3 are identically distributed $N(0,1)$.
(b) (X_1, X_2, X_3) is a Gaussian vector.
(c) X_1 and X_2 are independent.
(d) X_1 and X_3 are independent.

Solution.

(a) The X_i's are linear combinations of independent normal random variables and therefore normal themselves. Calculations of means and variances show that the X_i's are standard normal (zero mean and unit variance).
(b) If X and Y are the (column) vectors (X_1, X_2, X_3) and (Y_1, Y_2, Y_3, Y_4), respectively, then $X = AY$ with

$$A = \begin{pmatrix} 1/\sqrt{2} & 1/\sqrt{2} & 0 & 0 \\ 1/2 & 1/2 & 1/2 & 1/2 \\ 1/\sqrt{2} & -1/\sqrt{2} & 0 & 0 \end{pmatrix},$$

and as such, (X_1, X_2, X_3) is a Gaussian vector.
(c) X_1 and X_2 are NOT independent because $\text{Cov}(X_1, X_2) = 1/\sqrt{2} \neq 0$.
(d) X_1 and X_3 are independent because they are jointly normal with covariance zero, $\text{Cov}(X_1, X_3) = 0$.

Problem 3.2:

(a) Obtain, for an arbitrary random vector Z with covariance matrix Ξ, the covariance matrix of AZ.
(b) Suppose $X = (X_1, X_2, \ldots, X_d)$ has a multivariate normal distribution, such that all its components are identically distributed with zero mean and unit variance, and for $i \neq j$, $\text{Cov}(X_i, X_j) = \rho$, where ρ is appropriately chosen in $(-1, 1)$.
   i. Give an expression for Σ_{ij}, the row i and column j entry of Σ.
   ii. Show that for any $\rho \in (-1, 1)$, a two-dimensional normal X with covariance Σ exists. Give a representation of the form $X = AZ$,

where Z is made up of two independent standard normal random variables.

iii. Investigate the case $d = 3$. In particular, show that for some values of ρ, there are no random vectors X that meet the above requirements.

iv. Next, we tackle the general case. In view of the highly structured nature of Σ, it is intuitively clear that an equally structured solution should exist. Construct such a solution when $\rho \in [-1/(d-1), 1)$.

Solution.

(a) Since the covariance is linear in each argument (bilinear),

$$\mathrm{Cov}((AZ)_i, (AZ)_j)$$

$$= \mathrm{Cov}\left(\sum_k A_{ik} Z_k, \sum_\ell A_{j\ell} Z_\ell\right) = \sum_{k,\ell} A_{ik} A_{j\ell} \Lambda_{k\ell} = (A\Xi A^T)_{ij}.$$

In other words, the covariance matrix of AX is $A\Xi A^T$.

(b) i. $\Sigma_{ii} = 1$ and for $i \neq j$, $\Sigma_{ij} = \rho$.

ii. For Σ to be a covariance matrix, it must be symmetric, which it is, and positive semi-definite or equivalently that the eigenvalues of Σ are non-negative. Here, the characteristic polynomial is

$$\det(\Sigma - \lambda I) = \lambda^2 - 2\lambda + 1 - \rho^2$$

and the eigenvalues of Σ are $1 \pm \rho$. Therefore, for any $\rho \in (-1, 1)$, Σ is positive semi-definite and a covariance matrix.

A lower triangular solution to the equation $AA^T = \Sigma$ is

$$A = \begin{pmatrix} 1 & 0 \\ \rho & \sqrt{1-\rho^2} \end{pmatrix}.$$

A is of course not unique, as can be seen in iv. below. In conclusion,

$$X = AZ = \begin{pmatrix} 1 & 0 \\ \rho & \sqrt{1-\rho^2} \end{pmatrix} \begin{pmatrix} Z_1 \\ Z_2 \end{pmatrix} = \begin{pmatrix} Z_1 \\ \rho Z_1 + \sqrt{1-\rho^2} Z_2 \end{pmatrix}$$

has a multivariate normal distribution with a covariance matrix Σ.

iii. In this case, the characteristic polynomial is

$$\begin{aligned}\det(\Sigma - \lambda I) &= -\lambda^3 + 3\lambda^2 - 3(1-\rho^2)\lambda + 2\rho^3 - 3\rho^2 + 1 \\ &= -(\lambda - (1-\rho))^2(\lambda - (1+2\rho)),\end{aligned}$$

and the eigenvalues of Σ are $1-\rho$ and $1+2\rho$. Therefore, Σ is positive semi-definite if and only if $\rho \in [-1/2, 1)$.

A lower triangular solution to the equation $AA^T = \Sigma$ is

$$\begin{pmatrix} 1 & 0 & 0 \\ \rho & \sqrt{1-\rho^2} & 0 \\ \rho & \frac{\rho\sqrt{1-\rho}}{\sqrt{1+\rho}} & \sqrt{\frac{(1-\rho)(1+2\rho)}{1+\rho}} \end{pmatrix}.$$

It follows that

$$X = AZ = \begin{pmatrix} Z_1 \\ \rho Z_1 + \sqrt{1-\rho^2}\, Z_2 \\ \rho Z_1 + \frac{\rho\sqrt{1-\rho}}{\sqrt{1+\rho}} Z_2 + \sqrt{\frac{(1-\rho)(1+2\rho)}{1+\rho}}\, Z_3 \end{pmatrix}$$

has a multivariate normal distribution with covariance matrix Σ.

Again, other solutions exist. Indeed, allowing the dimension of Z to increase leads to a "simpler" construction:

$$X_i = \sqrt{\rho} Z_0 + \sqrt{1-\rho} Z_i, \quad i = 1, 2, 3,$$

or equivalently that

$$A = \begin{pmatrix} \sqrt{\rho} & \sqrt{1-\rho} & 0 & 0 \\ \sqrt{\rho} & 0 & \sqrt{1-\rho} & 0 \\ \sqrt{\rho} & 0 & 0 & \sqrt{1-\rho} \end{pmatrix}.$$

Note however that this construction is more restrictive in that ρ must be non-negative rather than greater than $-1/2$.

iv. We look for a $d \times d$ matrix A of the form: $A_{ii} = \beta$ and $j \neq i$, $A_{ij} = \alpha$. In this case,

$$(AA^T)_{ii} = (A^2)_{ii} = \sum_{k=1}^{d} A_{ik} A_{ki} = \beta^2 + (d-1)\alpha^2,$$

and for $j \neq i$,

$$(AA^T)_{ij} = (A^2)_{ij} = \sum_{k=1}^{d} A_{ik}A_{kj} = 2\alpha\beta + (d-2)\alpha^2.$$

Equating the first quantity to 1 and the second to ρ, we end up with the following system of equations in α and β:

$$\begin{cases} (d-1)\alpha^2 + \beta^2 = 1, \\ (d-2)\alpha^2 + 2\alpha\beta = \rho. \end{cases}$$

This system is easily solved and the (four) solutions are found to be

$$\alpha = \frac{1}{d}\left(\varepsilon_1\sqrt{1-\rho} + \varepsilon_2\sqrt{1+(d-1)\rho}\right) \text{ and } \beta = \alpha + \varepsilon_1\sqrt{1-\rho},$$

where $\varepsilon_i \in \{-1, 1\}$. Therefore, for any $\rho \geq [-1/(d-1), 1)$ and any of the four solutions above, $X_i = \beta Z_i + \alpha \sum_{j\neq i} Z_j$ defines a multivariate normal vector with covariance matrix Σ.

Problem 3.3: Derive the moment-generating function of the multivariate normal distribution $N(\boldsymbol{\mu}, \Sigma)$.

Solution. The probability density of a random vector $\mathbf{X} = (X_1, X_2, \ldots, X_n) \sim N(\boldsymbol{\mu}, \Sigma)$ is

$$f_{\mathbf{X}}(\boldsymbol{x}) = \frac{1}{(2\pi)^{n/2}|\Sigma|^{1/2}} e^{-\frac{1}{2}(\boldsymbol{x}-\boldsymbol{\mu})\Sigma^{-1}(\boldsymbol{x}-\boldsymbol{\mu})^T},$$

and its moment-generating function can be obtained directly from the definition:

$$\begin{aligned} M_{\mathbf{X}}(\mathbf{t}) &= \mathrm{E}(e^{\mathbf{t}\mathbf{X}^T}) = \int_{-\infty}^{\infty} e^{\boldsymbol{t}\boldsymbol{x}^T} f_{\mathbf{X}}(\boldsymbol{x})d\boldsymbol{x} \\ &= \int_{-\infty}^{\infty} \frac{1}{(2\pi)^{n/2}|\Sigma|^{1/2}} e^{\boldsymbol{t}\boldsymbol{x}^T - \frac{1}{2}(\boldsymbol{x}-\boldsymbol{\mu})\Sigma^{-1}(\boldsymbol{x}-\boldsymbol{\mu})^T} d\boldsymbol{x} \\ &= e^{\boldsymbol{\mu}\boldsymbol{t}^T + \frac{1}{2}\boldsymbol{t}\Sigma\boldsymbol{t}^T} \int_{-\infty}^{\infty} \frac{1}{(2\pi)^{n/2}|\Sigma|^{1/2}} e^{-\frac{1}{2}(\boldsymbol{x}-\boldsymbol{\mu}-\boldsymbol{t}\Sigma)\Sigma^{-1}(\boldsymbol{x}-\boldsymbol{\mu}-\boldsymbol{t}\Sigma)^T} d\boldsymbol{x} \\ &= e^{\boldsymbol{\mu}\boldsymbol{t}^T + \frac{1}{2}\boldsymbol{t}\Sigma\boldsymbol{t}^T}, \end{aligned}$$

where in order to obtain the third line, recall that Σ is symmetric, $\Sigma = \Sigma^T$ as $\Sigma \equiv AA^T$.

Problem 3.4:

(a) Let $X_1, \ldots, X_n$ be independent discrete random variables, and let $Y_k = X_1 + \cdots + X_k$, $k = 1, \ldots, n$. Express $\mathrm{P}(Y_1 = y_1, \ldots, Y_n = y_n)$ in terms of the probability mass functions p_k of X_k, $k = 1, \ldots, n$.

(b) Suppose now that $X_1, \ldots, X_n$ are independent and jointly absolutely continuous random variables and as before, $Y_k = X_1 + \cdots + X_k$, $k = 1, \ldots, n$. Express the joint density of $(Y_1, \ldots, Y_n)$ in terms of the densities f_k of X_k, $k = 1, \ldots, n$.

(c) Let B be a BM. Write the joint density of $(B_1, \ldots, B_n)$ in as explicit a form as possible.

(d) Obtain the determinant of the n-by-n matrix

$$\Sigma = \begin{pmatrix} 1 & 1 & 1 & 1 & \cdots & 1 \\ 1 & 2 & 2 & 2 & \cdots & 2 \\ 1 & 2 & 3 & 3 & \cdots & 3 \\ \vdots & \vdots & \vdots & \vdots & \ddots & \vdots \\ 1 & 2 & 3 & 4 & \cdots & n \end{pmatrix}$$

from the expression in (c).

Solution.

(a) We easily see that $Y_1 = y_1, Y_2 = y_2, \ldots, Y_n = y_n$ if and only if, letting $y_0 = 0$, $X_k = y_k - y_{k-1}$, $k = 1, \ldots, n$. It immediately follows, from the independence of X_k's, that

$$\begin{aligned} &\mathrm{P}(Y_1 = y_1, \ldots, Y_n = y_n) \\ &= \mathrm{P}(X_1 = y_1 - y_0, \ldots, X_n = y_n - y_0) = \prod_{k=1}^{n} p_k(y_k - y_{k-1}). \end{aligned}$$

(b) Here, we proceed by induction. Let g_k be the joint density of $(Y_1, \ldots, Y_k)$ and G_k be its distribution function. Informed by (a), we show that $g_n(y_1, \ldots, y_n) = \prod_{k=1}^{n} f_k(y_k - y_{k-1})$. Clearly, $g_1(y_1) = f_1(y_1 - y_0)$; recall that $y_0 = 0$. Suppose the statement is true up to $n-1$. Then, using the induction assumption and letting $v_1 = u_0 = 0$,

$$G_n(y_1,\ldots,y_n)$$
$$= \mathrm{P}(Y_1 \le y_1,\ldots,Y_n \le y_n)$$
$$= \int_{-\infty}^{y_1} \mathrm{P}(X_2 \le y_2 - u_1,\ldots,X_2+\cdots+X_n \le y_n - u_1) f_1(u_1)du_1$$
$$= \int_{-\infty}^{y_1} \left(\int_{-\infty}^{y_2-u_1} \cdots \int_{-\infty}^{y_n-u_1} \prod_{k=2}^{n} f_k(v_k - v_{k-1}) dv_n \ldots dv_2 \right) f_1(u_1)du_1$$
$$= \int_{-\infty}^{y_1} \left(\int_{-\infty}^{y_2} \cdots \int_{-\infty}^{y_n} \prod_{k=2}^{n} f_k(u_k - u_{k-1}) du_n \ldots du_2 \right) f_1(u_1)du_1$$
$$= \int_{-\infty}^{y_1} \cdots \int_{-\infty}^{y_n} \prod_{k=1}^{n} f_k(u_k - u_{k-1}) du_n \cdots du_1,$$

where we have operated the change of variables $u_k = v_k + u_1$, $k = 2,\ldots,n$.

(c) Let g be the density of $(B_1, B_2 \ldots, B_n)$. Since $B_1, B_2 - B_1, \ldots, B_n - B_{n-1}$ are independent (as increments of a BM) and $N(0,1)$, we can apply (b) to get, with $y_0 = 0$,

$$g(y_1,\ldots,y_n) = \prod_{k=1}^{n} \frac{1}{\sqrt{2\pi}} e^{-(y_k - y_{k-1})^2/2}$$
$$= \frac{1}{(2\pi)^{n/2}} e^{-\frac{1}{2}\sum_{k=1}^{n}(y_k - y_{k-1})^2}.$$

(d) Σ is in fact the covariance matrix of a BM at times $1, 2, \ldots, n$. As a random vector which has an n-variate Gaussian distribution with mean zero and covariance matrix Σ, $(B_1, B_2, \ldots, B_n)$ has the following joint probability density function:

$$\frac{1}{(2\pi)^{n/2}\sqrt{|\Sigma|}} e^{-\frac{1}{2}\mathbf{y}\Sigma^{-1}\mathbf{y}^T}. \qquad (*)$$

Identifying the terms in the above expression and those in that of g, we immediately deduce that $|\Sigma| = 1$ (and that $\mathbf{y}\Sigma^{-1}\mathbf{y}^T = \sum_{k=1}^{n}(y_k - y_{k-1})^2$).

Problem 3.5: Let $B(t)$ be a BM. Show that the following processes are BMs on $[0, T]$:

(a) $X(t) = -B(t)$.
(b) $X(t) = B(T-t) - B(T)$, where $T < \infty$.
(c) $X(t) = cB(t/c^2)$, where $c \neq 0$.
(d) $X(t) = tB(1/t)$, for $t > 0$, and $X(0) = 0$.

Solution.

(a) $X = -B$ is clearly a Gaussian process. Indeed, for $t_1 < \cdots < t_n \leq T$, $(X(t_1), \ldots, X(t_n)) = (-B(t_1), \ldots, -B(t_n)) = -(B(t_1), \ldots, B(t_n))$ is Gaussian. Furthermore,

$$\begin{aligned}\operatorname{Cov}(X(s), X(t)) &= \operatorname{Cov}(-B(s), -B(t)) = \operatorname{Cov}(B(s), B(t)) \\ &= \min(s, t).\end{aligned}$$

(b) $X(t) = B(T-t) - B(T)$ is again clearly Gaussian since for $t_1 < \cdots < t_n \leq T$, $(X(t_1), \ldots, X(t_n)) = (B(T-t_1) - B(T), \ldots, B(T-t_n) - B(T))$ is Gaussian as a linear transformation of the Gaussian vector $(B(T-t_1), \ldots, B(T-t_n), B(T))$. Computing the covariance function, we find

$$\begin{aligned}&\operatorname{Cov}(X(s), X(t)) \\ &\quad = \operatorname{Cov}(B(T-s) - B(T), B(T-t) - B(T)) \\ &\quad = \min(T-s, T-t) - \min(T-s, T) - \min(T, T-t) + T \\ &\quad = \min(T-s, T-t) - (T-s) - (T-t) + T \\ &\quad = \min(T-s, T-t) + s + t - T = \min(t, s).\end{aligned}$$

(c) For $t_1 < \cdots < t_n \leq T$, $(X(t_1), \ldots, X(t_n)) = (cB(t_1/c^2), \ldots, cB(t_n/c^2)) = c(B(t_1/c^2), \ldots, B(t_n/c^2))$ is obviously Gaussian and

$$\begin{aligned}&\operatorname{Cov}(X(s), X(t)) \\ &\quad = \operatorname{Cov}(cB(s/c^2), cB(t/c^2)) = c^2 \operatorname{Cov}(B(s/c^2), B(t/c^2)) \\ &\quad = c^2 \min(s/c^2, t/c^2) = \min(s, t).\end{aligned}$$

(d) Again, $X(t)$ is a Gaussian process, has zero mean, and is such that, for $s, t \neq 0$,

$$\begin{aligned}&\operatorname{Cov}(X(s), X(t)) \\ &\quad = \operatorname{Cov}(sB(1/s), tB(1/t)) = st \operatorname{Cov}(B(1/s), B(1/t)) \\ &\quad = st \min(1/s, 1/t) = \min(t, s).\end{aligned}$$

Problem 3.6:

(a) Let $B(t)$ and $W(t)$ be two independent BMs. Show that $X(t) = (B(t)+W(t))/\sqrt{2}$ is also a BM.
(b) Construct a counterexample to show the necessity of the assumption of independence.

Solution.

(a) As the sum of two independent Gaussian processes, X is also Gaussian: for any $t_1 < \cdots < t_n$, any $\lambda_1, \ldots, \lambda_n$, $\lambda_1 X(t_1) + \cdots + \lambda_n X(t_n) = (\lambda_1 B(t_1) + \cdots + \lambda_n B(t_n))/\sqrt{2} + (\lambda_1 W(t_1) + \cdots + \lambda_n W(t_n))/\sqrt{2}$ is normal as the sum of two independent normal random variables. Furthermore, for any s, t, $\mathrm{Cov}(B(s), W(t)) = 0$ and

$$\begin{aligned}\mathrm{Cov}(X(s), X(t)) &= \mathrm{Cov}((B(s) + W(s))/\sqrt{2}, (B(t) + W(t))/\sqrt{2}) \\ &= \mathrm{Cov}(B(s) + W(s), B(t) + W(t))/2 \\ &= \big(\mathrm{Cov}(B(s), B(t)) + \mathrm{Cov}(W(s), W(t))\big)/2 \\ &= \min(s, t).\end{aligned}$$

(b) A simple counterexample consists of letting $W(t) = B(t+1) - B(1)$. B and W are both BMs, but they are not independent and

$$\begin{aligned}&\mathrm{Cov}(X(s), X(t)) - \min(s, t) \\ &= \mathrm{Cov}((B(s) + W(s))/\sqrt{2}, (B(t) + W(t))/\sqrt{2}) - \min(s, t) \\ &= \mathrm{Cov}(B(s) + W(s), B(t) + W(t))/2 - \min(s, t) \\ &= \big(\min(s, t + 1) - \min(s, 1) + \min(s + 1, t) - \min(1, t)\big)/2.\end{aligned}$$

As the latter is not zero (take for example $s = 2$ and $t = 4$), X cannot be a BM.

Problem 3.7:

(a) State with reason whether $B_2 - 2B_1$ is independent of B_u for $u \leq 1$.
(b) Give a condition on the numbers a, b so that $aB_t + bB_s$ is independent of B_u for $u \leq s < t$.
(c) Let $X_t = B_{t+1} - B_1$ for $t \geq 0$. Show that the process X_t is a BM.
(d) Let $X_t = B_{t+1} - B_1$ for $0 \leq t \leq 1$ and $X_t = B_{t-1} + B_2 - B_1$ for $1 < t \leq 2$. Is it a BM on $[0, 2]$?

Solution.

(a) $\text{Cov}(B_2 - 2B_1, B_u) = \text{Cov}(B_2, B_u) - 2\,\text{Cov}(B_1, B_u) = u - 2u = -u \neq 0$. Hence, $B_2 - 2B_1$ and B_u for $0 < u \leq 1$ are not independent.

(b) $\text{Cov}(aB_t + bB_s, B_u) = au + bu = (a+b)u$. Since the vector $(aB_t + bB_s, B_u)$ is Gaussian, zero correlation implies independence. Hence, the necessary (and sufficient) condition for independence is $b = -a$.

(c) The increments of the process X_t are normal, $X_t - X_s = B_{t+1} - B_1 - (B_{s+1} - B_1) = B_{t+1} - B_{s+1}$, which is $N(0, t-s)$. It is also independent of B_u for $u \leq s+1$, meaning $X_t - X_s$ is independent of X_v, $v \leq s$. X_t is continuous. Thus, it fulfils the definition of BM.

(d) X_t is obtained by cutting BM on $[0,2]$ at the point $t = 1$ and swapping the pieces between $[0,1]$ and $[1,2]$. The process X_t is Gaussian, continuous, zero mean. Let's check the covariance function.

When $s \leq 1 < t \leq 2$, $t - 1 \leq 1 \leq s + 1 \leq 2$.
Hence,

$$\begin{aligned}
\text{Cov}(X_s, X_t) &= \text{Cov}(B_{s+1} - B_1, B_{t-1} + B_2 - B_1) \\
&= \text{Cov}(B_{s+1}, B_{t-1}) - \text{Cov}(B_1, B_{t-1}) \\
&\quad + \text{Cov}(B_{s+1}, B_2) - \text{Cov}(B_{s+1}, B_1) \\
&\quad - \text{Cov}(B_1, B_2) + \text{Cov}(B_1, B_1) \\
&= t - 1 - (t-1) + (s+1) - 1 = s.
\end{aligned}$$

By symmetry of covariance, it is t when $t < s$, hence it is $\min(s,t)$ for any choice of s and t. Thus, X_t is a BM.

Problem 3.8:

(a) Let $0 \leq a < b \leq c < d$. Give the joint distribution of the increments $B_c - B_a$ and $B_d - B_b$. State when these are independent.

(b) State with reasons whether the following process $V_t = B_{2t} - B_t$, $t \geq 0$, is Gaussian and is a BM.

Solution.

(a) The column vector $Y = (B_a, B_b, B_c, B_d)^T$ has a normal distribution with 0 mean vector and covariance matrix

$$\Sigma = \begin{pmatrix} a & a & a & a \\ a & b & b & b \\ a & b & c & c \\ a & b & c & d \end{pmatrix}.$$

Let $U = \begin{pmatrix} B_c - B_a \\ B_d - B_b \end{pmatrix}$.

Then, $U = AY$ for matrix

$$A = \begin{pmatrix} -1 & 0 & 1 & 0 \\ 0 & -1 & 0 & 1 \end{pmatrix}.$$

Since U is a linear transformation of the Gaussian vector Y, it has normal distribution with mean vector $AE(Y) = 0$ and covariance function ($A\Sigma A^T$ or compute covariances directly)

$$\Sigma_U = \begin{pmatrix} c-a & c-b \\ c-b & d-b \end{pmatrix}.$$

They are independent iff covariance zero, $b = c$.

(b) It is a Gaussian process, V_t has $N(0,t)$ distribution. $(V_{t_1}, V_{t_2}) = (B_{2t_1} - B_{t_1}, B_{2t_2} - B_{t_2})$ has a bivariate normal distribution as a joint distribution of increments of BM (even on overlapping intervals see (a)). Similarly, all finite-dimensional distributions of V_t are multivariate normal. But V_t is not a BM because, for example, its covariance function for $s < t$ is

$$\text{Cov}(V_t, V_s) = \text{Cov}(B_{2t} - B_t, B_{2s} - B_s) = 2s - \min(t, 2s) \neq s.$$

Problem 3.9: Let $X_t = e^{B_t}$. State with reason whether

(a) X_t/X_s is independent of B_u for $u \le s < t$.
(b) X_t^2/X_s is independent of B_u for $u \le s < t$.
(c) $(\log X_t - \log X_s)^2$ is independent of X_u for $u \le s < t$.

Solution.

(a) $X_t/X_s = e^{B_t - B_s}$. It is independent of B_u for $u \le s < t$ because it is a function of $B_t - B_s$, which is independent of B_u for $u \le s < t$.
(b) $X_t^2/X_s = e^{2B_t - B_s} = e^{B_t - B_s}e^{B_t}$ is not independent of B_u, $u \le s < t$ because if that was the case, then $\log(X_t^2/X_s) = 2B_t - B_s$ and B_u would be independent and therefore uncorrelated, which they are not as their covariance is u.
(c) $(\log X_t - \log X_s)^2 = (B_t - B_s)^2$ is independent of B_u for $u \le s < t$ because it is a function of $B_t - B_s$, which is independent of B_u for $u \le s < t$. Hence, it is independent of any function of B_u, such as $X_u = e^{B_u}$.

Problem 3.10 (Random integral, which is not an Itô integral): Let $X_t = \int_0^t B_s ds$. Recall that a main result on Riemann integration states that a continuous function on a compact interval is Riemann integrable. State with reason if the following is true:

(a) X_t is a random process, and it is continuously differentiable as a function of t and $X_t' = B_t$.
(b) X_t is a Gaussian process.
(c) $X_t - X_s$ is independent of B_u, $u \leq s$.

Solution.

(a) Yes, X is obviously random and $X(\omega)$ is a continuously differentiable function of t as an integral of a continuous function. Furthermore, $X_t'(\omega) = B_t(\omega)$.

(b) As a Riemann integral, X_t is the limit of Riemann sums:

$$\int_0^t B_s ds = \lim_n \sum_{k=0}^{n-1} B_{s_k}(s_{k+1} - s_k),$$

where the limit is taken as the mesh $\max_{0 \leq k \leq n-1}(s_{k+1} - s_k)$ of the partition $0 = s_0 < s_1 < \cdots < s_n = t$ approaches zero. It follows that for $t_1 < t_2 < \cdots < t_m$, $(X_{t_1}, \ldots, X_{t_m})$ can be seen as the limit of the vector of Riemann sums:

$$\left(\sum_{k=0}^{n_1-1} B_{s_k}(s_{k+1} - s_k), \sum_{k=n_1}^{n_2-1} B_{s_k}(s_{k+1} - s_k), \ldots, \right.$$
$$\left. \sum_{k=n_{m-1}}^{n_m-1} B_{s_k}(s_{k+1} - s_k) \right),$$

where $0 = s_0 < s_1 < \cdots < s_{n_1} = t_1 < s_{n_1+1} < \cdots < s_{n_m} = t_m$ is a concatenation of partitions of the interval $[0, t_1], \ldots, [t_{m-1}, t_m]$. Since these vectors are Gaussian, so is their limit, from which we deduce that X is a Gaussian process.

(c) $X_t - X_s = \int_s^t B_r dr$. Since B_r for $r > s$ depends on B_s, analysing the dependence of $\int_s^t B_r dr$ on B_u, $u \leq s$, is not immediate. It is possible that the sum (or integral) of random variables U_λ that depend on another random variable V is independent of the latter. For example, B_3 and $-B_2$ are not independent of B_1 but their sum is. Since $X_t - X_s$ and B_u, $u \leq s$, are jointly Gaussian, the question of dependence is settled by looking at the covariance. For $u \leq s$,

$$\begin{aligned}\int_s^t \mathrm{E}(|B_u B_r|)dr &\leq \int_s^t \sqrt{\mathrm{E}((B_u B_r)^2)}dr \\ &= \int_s^t \sqrt{\mathrm{E}\Big(B_u^2(B_u + (B_r - B_u))^2\Big)}dr \\ &= \int_s^t \sqrt{\mathrm{E}\Big(B_u^4 + B_u^2(B_r - B_u)^2 + 2B_u^3(B_r - B_u)\Big)}dr \\ &= \int_s^t \sqrt{3u^2 + u(r-u)}dr \ < \ +\infty,\end{aligned}$$

from which we deduce that Fubini's theorem applies and that

$$\begin{aligned}\mathrm{Cov}\left(\int_s^t B_r dr, B_u\right) &= \mathrm{E}\left(\int_s^t B_u B_r dr\right) = \int_s^t \mathrm{E}(B_u B_r)dr \\ &= \int_s^t \mathrm{Cov}(B_u, B_r)dr = \int_s^t u dr = (t-s)u \neq 0.\end{aligned}$$

In other words, $X_t - X_s$ and B_u, $u \leq s$ are not independent. Alternatively, by writing $B_r = B_s + (B_r - B_s)$ we can see that $X_t - X_s = (t-s)B_s + \int_s^t (B_r - B_s)dr$, a sum of two independent *rv*.

Problem 3.11: Let $B(t)$ be a BM and $0 \leq s < t$. Show that the conditional distribution of $B(s)$ given $B(t) = b$ is normal and gives its mean and variance.

Solution. Since $(B(s), B(t))$ is bivariate normal with zero means and variances s, t, covariance s and correlation $\sqrt{s/t}$, we have the conditional distribution is normal with mean $\frac{\sqrt{s}}{\sqrt{t}}\frac{\sqrt{s}}{\sqrt{t}}b = \frac{s}{t}b$ and variance $(1 - s/t)s = s(t-s)/t$.

Alternative solution: for $s \leq t$, $B(s)$ and $B(s) - (s/t)B(t)$ are uncorrelated and are therefore independent (see Problem 3.33). Therefore, $B(s)$ can be written as the sum of two independent zero-mean random variables with variances $s - s^2/t$ for the first one (and s^2/t for the second one): $B(s) = \big(B(s) - (s/t)B(t)\big) + (s/t)B(t)$. It follows that conditional on $B(t) = b$, $B(s)$ is normal with mean $(s/t)b$ and variance $s - s^2/t$.

Problem 3.12: Show that the random variables $M(t) = \max_{s \le t} B(s)$ and $|B(t)|$ have the same distribution.

Solution. Recall $M(t)$ is the maximum of BM on $[0,t]$: $M(t) = \max_{s \le t} B(s)$. Its distribution is given by the reflection principle of the BM (see Theorem 3.15 of Klebaner's book): for any $x > 0$, $\mathrm{P}(M(t) \ge x) = 2\mathrm{P}(B(t) \ge x)$.

It immediately follows that, for $x \ge 0$,

$$\begin{aligned} \mathrm{P}(|B(t)| > x) &= \mathrm{P}(B(t) > x) + \mathrm{P}(B(t) < -x) = 2\mathrm{P}(B(t) > x) \\ &= \mathrm{P}(M(t) > x). \end{aligned}$$

Problem 3.13: Show that the moments of order r of the hitting time T_x are finite $\mathrm{E}(T_x^r) < \infty$ if and only if $r < 1/2$.

Solution. The density of T_x is

$$f_{T_x}(t) = \frac{|x|}{\sqrt{2\pi}} t^{-3/2} e^{-x^2/2t},$$

see Theorem 3.18 in Klebaner's book. It follows that

$$\begin{aligned} \mathrm{E}(T_x^r) &= \int_0^\infty t^r f_{T_x}(t)dt = \frac{|x|}{\sqrt{2\pi}} \int_0^\infty t^{r-\frac{3}{2}} e^{-\frac{x^2}{2t}} dt \\ &= \frac{|x|}{\sqrt{2\pi}} \int_0^\infty s^{-r-\frac{1}{2}} e^{-\frac{x^2}{2}s} ds. \end{aligned}$$

Up to a change of variable ($u = (x^2/2)s$), the latter is a gamma function which is known to converge for $-r - 1/2 > -1$, that is for $r < 1/2$.

Problem 3.14: Derive the conditional distribution $B(t)$ given the maximum $M(t)$.

Solution. The joint distribution of $(B(t), M(t))$ has the density

$$f_{B,M}(x,y) = \sqrt{\frac{2}{\pi}} \frac{(2y-x)}{t^{3/2}} e^{-(2y-x)^2/(2t)}$$

for $y \ge 0$ and $x \le y$ — see Theorem 3.21 in Klebaner's book.

Either by integrating $f_{B,M}$ in x or by applying the reflection principle (see Problem 3.12), we find that $f_M(y) = 2f_B(y) = \sqrt{2/(\pi t)} e^{-y^2/(2t)}$, from which we deduce that

$$\begin{aligned} f_{B(t)|M(t)}(x|y) &= \sqrt{\frac{2}{\pi}} \frac{(2y-x)}{t^{3/2}} e^{-(2y-x)^2/(2t)} \times \sqrt{\frac{\pi t}{2}} e^{y^2/(2t)} \\ &= \frac{(2y-x)}{t} e^{-\left((2y-x)^2 - y^2\right)/(2t)}. \end{aligned}$$

Problem 3.15: By considering $-B(t)$, derive the joint distribution of $B(t)$ and $m(t) = \min_{s\le t} B(s)$.

Solution. Note that $\min_{s\le t} B(s) = -\max_{s\le t}(-B(s))$ and letting $W(t) = -B(t)$, we conclude that $W(t)$ is also a BM, and

$$\begin{aligned}
\mathrm{P}\big(B(t) \ge x, \min_{s\le t} B(s) \le y\big) &= \mathrm{P}\big(W(t) \le -x, \max_{s\le t} W(s) \ge -y\big) \\
&= 1 - \Phi\left(\frac{-2y + x}{\sqrt{t}}\right),
\end{aligned}$$

where Φ is the standard normal distribution function. Since $\phi(z) = \Phi'(z) = e^{-z^2/2}/\sqrt{2\pi}$ and $\phi'(z) = \Phi''(z) = -ze^{-z^2/2}/\sqrt{2\pi}$, for $y \le 0$ and $x \ge y$, the joint density of $(B(t), m(t))$ is

$$\begin{aligned}
f_{B,m}(x, y) &= -\frac{\partial^2}{\partial x \partial y}\mathrm{P}(B(t) \ge x, m(t) \le y) = \frac{\partial^2}{\partial x \partial y}\Phi\left(\frac{-2y + x}{\sqrt{t}}\right) \\
&= -\frac{2}{t}\phi'\left(\frac{-2y + x}{\sqrt{t}}\right) = \frac{2}{t}\frac{1}{\sqrt{2\pi}}e^{-\frac{(-2y+x)^2}{2t}}\left(\frac{-2y + x}{\sqrt{t}}\right) \\
&= \sqrt{\frac{2}{\pi}}\frac{(x - 2y)}{t^{3/2}}e^{\frac{-(2y-x)^2}{2t}}.
\end{aligned}$$

An alternative approach consist of first observing that $m(t) = \min_{s\le t} B(s) = -\max_{s\le t}(-B(s))$ and then writing that

$$\begin{aligned}
&\mathrm{P}(B(t) \ge -x, M(t) \ge -y) \\
&\quad = 1 - \mathrm{P}(B(t) \le -x) - \mathrm{P}(M(t) \le -y) + \mathrm{P}(B(t) \le -x, M(t) \le -y),
\end{aligned}$$

from which we deduce that $\mathrm{P}(B(t) \ge -x, M(t) \ge -y)$ and $\mathrm{P}(B(t) \le -x, M(t) \le -y)$ have identical second cross derivatives and consequently that

$$\begin{aligned}
f_{B,m}(x, y) &= \frac{\partial^2 \mathrm{P}(B(t) \le x, m(t) \le y)}{\partial x \partial y} = \frac{\partial^2 \mathrm{P}(B(t) \ge -x, M(t) \ge -y)}{\partial x \partial y} \\
&= \frac{\partial^2 \mathrm{P}(B(t) \le -x, M(t) \le -y)}{\partial x \partial y} = f_{B,M}(-x, -y).
\end{aligned}$$

The result is obtained by recalling Theorem 3.21 in Klebaner's book — see also Problem 3.14.

Problem 3.16: Show that the random variables $M(t)$, $|B(t)|$ and $M(t) - B(t)$ have the same distributions — see Problem 3.12.

Solution. Let $a > 0$, and $D = \{(x, y) : y - x > a\}$.
Then,

$$\begin{aligned}
&\mathrm{P}(M(t) - B(t) > a) \\
&= \int\int_D f_{B,M}(x, y)dxdy = \int_0^\infty \int_{-\infty}^{y-a} f_{B,M}(x, y)dxdy \\
&= \int_0^\infty \int_{-\infty}^{y-a} \sqrt{\frac{2}{\pi}} \frac{(2y - x)}{t^{3/2}} e^{\frac{-(2y-x)^2}{2t}} dxdy \\
&= \sqrt{\frac{2}{t\pi}} \int_0^\infty \left(\int_{-\infty}^{y-a} \left(\frac{\partial}{\partial x} e^{\frac{-(2y-x)^2}{2t}} \right) dx \right) dy \\
&= \sqrt{\frac{2}{t\pi}} \int_0^\infty e^{\frac{-(y+a)^2}{2t}} dy = 2 \int_0^\infty \frac{1}{\sqrt{2\pi t}} e^{\frac{-(y+a)^2}{2t}} dy \\
&= 2 \int_a^\infty \frac{1}{\sqrt{2\pi t}} e^{\frac{-u^2}{2t}} du = 2\mathrm{P}(B(t) > a),
\end{aligned}$$

which is the same as was obtained for $M(t)$ and $|B(t)|$ in Problem 3.12.

Problem 3.17: The first zero of BM started at zero is 0. What is the second zero?

Solution. $T_2 = \inf\{t > 0 : B(t) = 0\}$. Since any interval $(0, \epsilon)$ contains a zero of BM, $T_2 = 0$, the second zero is also zero.

Problem 3.18: Let T be the last time before 1 a BM visits 0. Explain why $X(t) = B(t + T) - B(T) = B(t + T)$ is not a BM.

Solution. By the argument in Problem 3.17, the first zero of the BM is a limit from the right of other zeroes. However, by the definition of T, there are no zeroes between T and 1 (T is isolated from the right) in the original BM B. In other words, 0 is a zero for X, but it is not a limit of other zeroes. Thus, X cannot be a BM.

Problem 3.19: The law of large numbers and the law of iterated logarithm for BM are stated in Klebaner's book (see Theorems 3.30 and 3.31) in the neighbourhood of infinity. Mapping infinity to zero with the map $1/t$, formulate them near zero.

Solution. Let $B(t)$ be a BM. Then, the process $W(t) = tB(1/t)$ is also a BM (see Problem 3.5(d)).

Applying the law of large numbers and the law of iterated logarithm for BM at large t to the process W allows us to obtain the behaviour near zero of the original BM B:

The law of large numbers for BM near zero:

$$\lim_{t\to 0} B(t) = \lim_{t\to\infty} B(1/t) = \lim_{t\to\infty} \frac{tB(1/t)}{t} = \lim_{t\to\infty} \frac{W(t)}{t} = 0, \text{ a.s.}$$

The law of iterated logarithm for BM near zero:

$$\begin{aligned}
&\limsup_{t\to 0} \frac{B(t)}{\sqrt{2t\ln\ln(1/t)}} \\
&\quad = \limsup_{t\to\infty} \frac{\sqrt{t}B(1/t)}{\sqrt{2\ln\ln t}} = \limsup_{t\to\infty} \frac{tB(1/t)}{\sqrt{2t\ln\ln t}} = \limsup_{t\to\infty} \frac{W(t)}{\sqrt{2t\ln\ln t}} \\
&\quad = 1 \text{ a.s.}
\end{aligned}$$

Similarly, for lim inf,

$$\liminf_{t\to 0} \frac{B(t)}{\sqrt{2t\ln\ln(1/t)}} = -1 \text{ a.s.}$$

Problem 3.20: Let $B(t)$ be BM and $\alpha > 0$. Show that $X(t) = e^{-\alpha t}B(e^{2\alpha t})$ is a Gaussian process. Find its mean and covariance functions.

Solution. Let $Y(t) = B(e^{2\alpha t})$, then $X(t) = e^{-\alpha t}Y(t)$. The process Y is clearly a Gaussian process: for any n and $t_1 < t_2 < \cdots < t_n$, $(Y(t_1), Y(t_2), \ldots, Y(t_n))$ is a Gaussian vector. The process X is also Gaussian, as $(X(t_1), X(t_2), \ldots, X(t_n))$ is obtained by a linear transformation of the Gaussian vector $(Y(t_1), Y(t_2), \ldots, Y(t_n))$; simply take the diagonal matrix with diagonal entries $(e^{-\alpha t_1}, e^{-\alpha t_2}, \ldots, e^{-\alpha t_n})$.

$X(t)$ has a mean of zero,

$$\mathrm{E}(X(t)) = \mathrm{E}\big(e^{-\alpha t}B(e^{2\alpha t})\big) = e^{-\alpha t}\mathrm{E}\big(B(e^{2\alpha t})\big) = 0,$$

and covariance

$$\begin{aligned}
&\mathrm{Cov}(X(t), X(s)) \\
&\quad = \mathrm{Cov}\left(e^{-\alpha t}Y(t), e^{-\alpha s}Y(s)\right) = e^{-\alpha t}e^{-\alpha s}\,\mathrm{Cov}(B(e^{2\alpha t}), Y(e^{2\alpha s})) \\
&\quad = e^{-\alpha(t+s)} \min\left(e^{2\alpha t}, e^{2\alpha s}\right) = e^{-\alpha(t+s)}e^{2\alpha\min(t,s)} \\
&\quad = e^{-\alpha(t+s-2\min(t,s))} = e^{-\alpha|t-s|}.
\end{aligned}$$

Problem 3.21: Let X be a Gaussian process with zero mean and covariance $\gamma(s,t) = e^{-\alpha|t-s|}$. Show that X has a version with continuous paths.

Solution. Using Problem 3.20 and the fact that the mean and covariance functions uniquely determine a Gaussian process, we can immediately identify X as being of the form $X(t) = e^{-\alpha t}B(e^{2\alpha t})$, from which we conclude that it has continuous paths since B itself has continuous paths.

Alternatively, and following the arguments in Problem 3.20, we see that $W(t) = \sqrt{t}X\big((\ln t)/(2\alpha)\big)$ defines a Gaussian process on $[1, \infty)$ with zero mean and covariance $\min(s, t)$. In other words, W is a BM (on $[1, \infty)$) and $X(t) = e^{-\alpha t}W(e^{2\alpha t})$ is clearly continuous as a product of a continuous function, $e^{-\alpha t}$, and a composite of two continuous functions, W and $e^{2\alpha t}$.

Problem 3.22: Show that in a normal random walk, $S_n = S_0 + \sum_{i=1}^{n} \xi_i$, where the ξ_i's are independent standard normal random variables and S_0 is a constant, $e^{uS_n - nu^2/2}$ is a martingale.

Solution. Let $M_n = e^{uS_n - nu^2/2}$. Since $M_n \geq 0$, integrability follows from the martingale property, which is shown as follows: $S_{n+1} = S_n + \xi_{n+1}$, with ξ_{n+1} independent of $\mathcal{F}_n$, the σ-field generated by ξ_k, $k \leq n$. Therefore,

$$\begin{aligned}\mathrm{E}\big(e^{uS_{n+1}}\big|\mathcal{F}_n\big) &= \mathrm{E}\big(e^{uS_n + u\xi_{n+1}}\big|\mathcal{F}_n\big) = e^{uS_n}\mathrm{E}\big(e^{u\xi_{n+1}}\big|\mathcal{F}_n\big)\\ &= e^{uS_n}\mathrm{E}\big(e^{u\xi_{n+1}}\big) = e^{uS_n + u^2/2}.\end{aligned}$$

Multiplying both sides of the above identity by $e^{-(n+1)u^2/2}$, the martingale property is obtained:

$$\mathrm{E}(M_{n+1}|\mathcal{F}_n) = M_n.$$

Taking expectation on both sides, we get

$$\mathrm{E}(M_{n+1}) = \mathrm{E}(M_n) = \mathrm{E}(M_0) = e^{uS_0} < +\infty,$$

from which the integrability of M_n follows.

Problem 3.23: Let $S_n = S_0 + \sum_{i=1}^{n} \xi_i$ be a random walk, with independent and identically distributed increments with distribution $\mathrm{P}(\xi_1 = 1) = p$, $\mathrm{P}(\xi_1 = -1) = 1 - p$, and S_0 is a constant. Show that for any λ, $e^{\gamma S_n - \lambda n}$ is a martingale for an appropriate value of γ.

Solution. Similarly to the previous question, ξ_{n+1} is independent of S_n and of all previous ξ_i's, $i \leq n$, or independent of $\mathcal{F}_n$. Using the properties of conditional expectation, one gets

$$\begin{aligned}
\mathrm{E}\big(e^{\gamma S_{n+1}}\big|\mathcal{F}_n\big) &= \mathrm{E}\big(e^{\gamma S_n+\gamma\xi_{n+1}}\big|\mathcal{F}_n\big) = e^{\gamma S_n}\mathrm{E}\big(e^{\gamma\xi_{n+1}}\big|\mathcal{F}_n\big) \\
&= e^{\gamma S_n}\mathrm{E}\big(e^{\gamma\xi_{n+1}}\big).
\end{aligned}$$

In the special case where $\mathrm{P}(\xi_1 = 1) = p$, $\mathrm{P}(\xi_1 = -1) = 1 - p$, one gets $\mathrm{E}\big(e^{\gamma\xi_{n+1}}\big) = pe^{\gamma} + (1-p)e^{-\gamma}$.

One can choose γ such that

$$\mathrm{E}\big(e^{\gamma\xi_{n+1}}\big) = pe^{\gamma} + (1-p)e^{-\gamma} = e^{\lambda} \tag{$*$}$$

and then multiply both sides by $e^{-(n+1)\lambda}$ to obtain the martingale property for any λ. To solve $(*)$, let $x = e^{\gamma}$, rewrite $(*)$ as a quadratic equation in x, and solve it to get the two solutions $\big(e^{\lambda} \pm \sqrt{e^{2\lambda} - 4p(1-p)}\big)/(2p)$. The two candidates γ that turn $e^{\gamma S_n - \lambda n}$ into a martingale are obtained by taking logs:

$$\ln\big(e^{\lambda} \pm \sqrt{e^{2\lambda} - 4p(1-p)}\big) - \ln(2p).$$

Problem 3.24: Let $S_n = S_0 + \sum_{i=1}^{n}\xi_i$ be a random walk, with independent and identically distributed increments with common moment-generating function $\phi(u) = \mathrm{E}\big(e^{u\xi_1}\big)$, and S_0 is a constant. Show that for any γ, $e^{\gamma S_n - \lambda n}$ is a martingale for an appropriate value of λ.

Solution. Similarly to the previous question, ξ_{n+1} is independent of S_n and of all previous ξ_i's, $i \leq n$, or independent of $\mathcal{F}_n$. Using the properties of conditional expectation, one gets

$$\begin{aligned}
\mathrm{E}\big(e^{\gamma S_{n+1}}\big|\mathcal{F}_n\big) &= \mathrm{E}\big(e^{\gamma S_n+\gamma\xi_{n+1}}\big|\mathcal{F}_n\big) = e^{\gamma S_n}\mathrm{E}\big(e^{\gamma\xi_{n+1}}\big|\mathcal{F}_n\big) \\
&= e^{\gamma S_n}\mathrm{E}\big(e^{\gamma\xi_{n+1}}\big) = e^{\gamma S_n}\phi(\gamma).
\end{aligned}$$

As in the previous two problems, we multiply both sides by $\phi(\gamma)^{-(n+1)}$ to obtain the martingale property for any γ. It follows that the appropriate λ is $\ln\phi(\gamma)$.

Problem 3.25: Let the wealth of a company at time t be modelled by $X_t = x+\mu t+\sigma B_t$, where x is a positive number. In this question, we consider two different cases: positive drift ($\mu > 0$) and negative drift ($\mu < 0$). Let T_x be the first time when the process X hits 0, $T_x = \inf\{t : X_t = 0\}$. Denote the probability of ruin $\Psi(x) = P(T_x < \infty)$. Note that the index x in T_x refers to the starting point and not the target value.

(a) Let $\mu > 0$. Show that $\Psi(x) \leq e^{-Rx}$ with $R = 2\mu/\sigma^2$.
Hint: Show that $M_t = e^{-RX_t}$ is a martingale. Then, stop it at $T_x \wedge N$, and argue just as you did for ruin probabilities in a random walk (Problem 2.32).

(b) Let $\mu < 0$. Show that $\mathrm{E}(T_x) < \infty$. Deduce that $\Psi(x) = 1$.
Hint: Show that $\mathrm{P}(T_x > t) \leq \mathrm{P}(X_t > 0)$, then use (c) together with the moment generating function (MGF) bound for $P(X_t > 0)$.

Solution.

(a) $\mu > 0$. We start by recalling that for any λ, $\mathcal{E}(\lambda)_t = e^{\lambda B_t - \lambda^2 t/2}$ is a martingale.

M is of the form $\mathcal{E}(\lambda)$ if and only if $R\mu = (R\sigma)^2/2$, that is $R = 2\mu/\sigma^2$. For such an R, $M_t = e^{-RX_t}$ is a martingale.

Next, using optional stopping at the finite stopping time $T_x \wedge N$ and noting that $X_{T_x} = 0$ and $M_{T_x} = e^{-RX_{T_x}} = 1$, we get

$$\begin{aligned} e^{-Rx} &= M_0 = \mathrm{E}\big(M_{T_x \wedge N}\big) = \mathrm{E}\big(M_{T_x} 1_{T_x \leq N}\big) + \mathrm{E}\big(M_N 1_{T_x > N}\big) \\ &\geq \mathrm{E}\big(M_{T_x} 1_{T_x \leq N}\big) = \mathrm{E}\big(e^{-RX_{T_x}} 1_{T_x \leq N}\big) \\ &\geq \mathrm{E}\big(1_{T_x \leq N}\big) = \mathrm{P}\big(T_x \leq N\big). \end{aligned}$$

Hence, $\mathrm{P}(T_x \leq N) \leq e^{-Rx}$ and taking N to infinity, we get that

$$\mathrm{P}(T_x < \infty) = \lim_{N \to \infty} \mathrm{P}(T_x \leq N) \leq e^{-Rx}.$$

(b) $\mu < 0$. Since $\mathrm{P}(T_x > t) = \mathrm{P}(X_s > 0,\ \forall s \leq t) \leq \mathrm{P}(X_t > 0)$,

$$\mathrm{E}(T_x) = \int_0^\infty \mathrm{P}(T_x > t)dt \leq \int_0^\infty \mathrm{P}(X_t > 0)dt.$$

Next, we use the upper bound on $\mathrm{P}(X_t > 0)$ found in Problem 2.31: if m is the moment-generating function of Y, then $\mathrm{P}(Y > 0) \leq m(u)$ for any $u > 0$. In this case, X_t is $N(x + \mu t, \sigma^2 t)$ and the bound becomes

$$\mathrm{P}(X_t > 0) \leq e^{u(x+\mu t) + u^2\sigma^2 t/2} = e^{ux} e^{u(\mu + u\sigma^2/2)t}.$$

Taking $u \in (0, -2\mu/\sigma^2)$ yields $u(\mu + u\sigma^2/2) < 0$ and $e^{u(\mu+u\sigma^2/2)t}$ integrable on $(0, \infty)$. It follows that for such a u,

$$\mathrm{E}(T_x) \leq \int_0^\infty \mathrm{P}(X_t > 0)dt \leq e^{ux} \int_0^\infty e^{u(\mu+u\sigma^2/2)t} < +\infty.$$

An integrable random variable must clearly be finite, and we deduce that $\Psi(x) = \mathrm{P}(T_x < \infty) = 1$.

Problem 3.26: The process X_t is defined for discrete times $t = 1, 2, \ldots$. It can take only three values: 1, 2 and 3. Its behaviour is defined by the following rule: from state 1 it goes to 2, from 2 it goes to 3 and from 3 it goes back to 1. X_1 takes values 1, 2, 3 with equal probabilities. Show that this process is Markov. Show also that

$$\mathrm{P}(X_3 = 3 | X_2 = 1 \text{ or } 2, X_1 = 3) \neq \mathrm{P}(X_3 = 3 | X_2 = 1 \text{ or } 2).$$

This demonstrates that to apply the Markov property, we must know the present state of the process exactly; it is not enough to know that it can take one of the two (or more) possible values.

Solution. Test for the Markov property[1]: from the definition of the process, it is straightforward that the future behaviour of the process depends only on its current state, i.e. it is Markov.

Further, if $X_1 = 3$, then X_2 must be 1, implying that $X_3 = 2$ and cannot be 3, so that $\mathrm{P}(X_3 = 3 | X_2 = 1 \text{ or } 2, X_1 = 3) = 0$.

Using standard calculations of conditional probabilities, one can show that

$$\mathrm{P}(X_3 = 3 | X_2 = 1 \text{ or } 2) = \frac{1}{2},$$

thus $\mathrm{P}(X_3 = 3 | X_2 = 1 \text{ or } 2, X_1 = 3) \neq \mathrm{P}(X_3 = 3 | X_2 = 1 \text{ or } 2)$.

Problem 3.27: A discrete-time process $X(t)$, $t = 0, 1, 2, \ldots$, is said to be autoregressive of order p (AR(p)) if there exists $a_1, \ldots, a_p \in \mathbb{R}$, and a white noise $Z(t)$ ($\mathrm{E}(Z(t)) = 0$, $\mathrm{E}(Z^2(t)) = \sigma^2$ and, for $s > 0$, $\mathrm{E}(Z(t)Z(t+s)) = 0$) such that

$$X(t) = \sum_{s=1}^{p} a_s X(t-s) + Z(t),$$

and $Z(t)$ is independent of $X(t-1), X(t-2), \ldots$.

(a) Show that $X(t)$ is Markovian if and only if $p = 1$.
(b) Show that if $X(t)$ is AR(2), then $\boldsymbol{Y}(t) = (X(2t), X(2t-1))$ is Markovian.
(c) Suppose that $Z(t)$ is a Gaussian process. Write the transition probability function of an AR(1) process $X(t)$.

---

[1](Definition 3.8) $X(t)$ is a Markov process if for any t and $s > 0$, the conditional distribution of $X(t+s)$ given $\mathcal{F}_t$ is the same as the conditional distribution of $X(t+s)$ given $X(t)$, that is, $P(X(t+s)y|\mathcal{F}_t) = P(X(t+s)y|X(t))$, a.s.

Solution.

(a) Let $A \subset \mathbb{R}$.

$$P(X(t) \in A|\mathcal{F}_{t-1}) = P\left(\sum_{s=1}^{p} a_s X(t-s) + Z(t) \in A|\mathcal{F}_{t-1}\right).$$

Clearly, only $X(t-1), \ldots, X(t-p)$ enter the conditional distribution, hence

$$P\left(\sum_{s=1}^{p} a_s X(t-s) + Z(t) \in A|X(t-1), \ldots, X(t-p)\right).$$

If $p = 1$, then $X(t) = a_1 X(t-1) + Z(t)$, and

$$\begin{aligned} P(X(t) \in A|\mathcal{F}_{t-1}) &= P(a_1 X(t-1) + Z(t) \in A|\mathcal{F}_{t-1}) \\ &= P(a_1 X(t-1) + Z(t) \in A|X(t-1)), \end{aligned}$$

and $X(t)$ is Markov. On the other hand, if $p > 1$, then the distribution of $X(t)$ depends not only on $X(t-1)$ but also on $X(t-2)$ or further (up to $X(t-p)$), and it is not Markov.

(b) If $p = 2$, then $X(t) = a_1 X(t-1) + a_2 X(t-2) + Z(t)$. $\mathbf{Y}(t) = (X(2t), X(2t-1))$, and $\mathbf{Y}(t-1) = (X(2t-2), X(2t-3))$. Let $\mathbf{B} \subset \mathbb{R}^2$, and $\mathcal{F}_t = \sigma(X(s), s \leq 2t)$. Then,

$$\begin{aligned} &P(\mathbf{Y}(t) \in \mathbf{B}|\mathcal{F}_{t-1}) \\ &\quad = P(X(2t), X(2t-1) \in \mathbf{B}|\mathcal{F}_{t-1}) \\ &\quad = P((a_1 X(2t-1) + a_2 X(2t-2) + Z(2t), \\ &\qquad a_1 X(2t-2) + a_2 X(2t-3) + Z(2t-1) \in \mathbf{B}|\mathcal{F}_{t-1}) \\ &\quad = P((a_1(a_1 X(2t-2) + a_2 X(2t-3) + Z(2t-1)) \\ &\qquad + a_2 X(2t-2) + Z(2t), \\ &\qquad a_1 X(2t-2) + a_2 X(2t-3) + Z(2t-1) \in \mathbf{B}|\mathcal{F}_{t-1}). \end{aligned}$$

It is now clear that the only values from the past (in $\mathcal{F}_{t-1}$) that enter into this conditional distribution are $X(2t-2)$ and $X(2t-3)$. Thus proceeding,

$$\begin{aligned} P(\mathbf{Y}(t) \in \mathbf{B}|\mathcal{F}_{t-1}) &= P(\mathbf{Y}(t) \in \mathbf{B}|X(2t-2), X(2t-3)) \\ &= P(\mathbf{Y}(t) \in \mathbf{B}|\mathbf{Y}(t-1)). \end{aligned}$$

(c) The one-step transition probability function can be found by

$$\begin{aligned}
&\mathrm{P}(X(t) \le y | X(t-1) = x) \\
&= \mathrm{P}(a_1 X(t-1) + Z(t) \le y | X(t-1) = x) \\
&= \mathrm{P}(a_1 x + Z(t) \le y | X(t-1) = x) \\
&= \mathrm{P}(Z(t) \le y - a_1 x) = \Phi\left(\frac{y - a_1 x}{\sigma}\right).
\end{aligned}$$

The one-step transition probability density function is

$$\frac{\partial}{\partial y}\Phi\left(\frac{y - a_1 x}{\sigma}\right) = \frac{1}{\sigma} f\left(\frac{y - a_1 x}{\sigma}\right).$$

Problem 3.28: The distribution of a random variable $\tau > 0$ has the lack of memory property (memoryless) if for any $a, b > 0$, $\mathrm{P}(\tau > a + b | \tau > a) = \mathrm{P}(\tau > b)$. Verify the lack of memory property for the exponential $\exp(\lambda)$ distribution. Show that if τ has a lack of memory property and has a continuous distribution, then it has an exponential distribution.

Solution. Let τ have an exponential distribution $\exp(\lambda)$, that is, $\mathrm{P}(\tau > a) = e^{-\lambda a}$. Thus, by using that the set $\{\tau > a + b\} \subset \{\tau > a\}$,

$$\begin{aligned}
\mathrm{P}(\tau > a + b | \tau > a) &= \frac{\mathrm{P}(\tau > a + b, \tau > a)}{\mathrm{P}(\tau > a)} \\
&= \frac{\mathrm{P}(\tau > a + b)}{\mathrm{P}(\tau > a)} = e^{-\lambda b} = \mathrm{P}(\tau > b).
\end{aligned}$$

Let now τ have the lack of memory property, and let $G(a) = \mathrm{P}(\tau > a)$. Then, we have from above that the memoryless property is equivalent to

$$\forall a, b \in \mathbb{R}_+, \;\; G(a + b) = G(a)G(b). \tag{$*$}$$

This is a functional equation in G. To find all continuous and strictly positive solutions of this equation, we take logs. Let $g(a) = \ln G(a)$. Then, $(*)$ becomes

$$\forall a, b \in \mathbb{R}_+, \;\; g(a + b) = g(a) + g(b).$$

We now prove that the only continuous functions of this type are the linear functions. The proof is done in a number of steps:

(a) $g(0) = 2g(0)$, from which we deduce that $g(0) = 0$. Let $\alpha = g(1)$.
(b) $\forall m \in \mathbb{N}$, $g(m + 1) = g(m) + g(1)$, and therefore $\forall m \in \mathbb{N}$, $g(m) = mg(1) = \alpha m$.

(c) $\forall m \in \mathbb{N},\ \forall a \in \mathbb{R}_+,\ g(ma) = g((m-1)a) + g(a) = \cdots = mg(a)$. It follows that $\forall m \in \mathbb{N},\ g(1) = mg(1/m)$ and therefore that $g(1/m) = g(1)/m = \alpha m$.
(d) $\forall q = m/n \in \mathbb{Q}_+,\ g(q) = mg(1/n) = \alpha m/n = \alpha q$.
(e) Let $r \in \mathbb{R}_+$. Then, there exists a sequence of rational numbers $(q_n)_n$ such that $r = \lim_n q_n$ ($\mathbb{Q}$ is dense in $\mathbb{R}$) and $g(r) = g(\lim_n q_n) = \lim_n g(q_n) = \lim_n \alpha q_n = \alpha r$.

Reverting back to $G(a) = e^{g(a)}$, we deduce that for any $a \in \mathbb{R}_+$, $G(a) = e^{\alpha a}$. Letting $\lambda = -\alpha$ (recall that $G(a) \leq 1$), we conclude that τ is exponential with parameter $\lambda = -\ln \mathrm{P}(\tau > 1)$.

Problem 3.29: Show that among all zero-mean stochastic processes $\{X(t)\}_{t\geq 0}$ with finite second moments $\mathrm{E}(X(t)^2) < \infty$ for $t \geq 0$, the class of martingales contains all processes with independent increments and are all included among processes with uncorrelated increments.

Solution. For X that is zero mean with independent increments, we have

$$\begin{aligned}\mathrm{E}(X(t)|\mathcal{F}_s^X) &= \mathrm{E}(X(t) - X(s)|\mathcal{F}_s^X) + \mathrm{E}\{X(s)|\mathcal{F}_s^X\} \\ &= \mathrm{E}\{X(t) - X(s)\} + X(s) = X(s)\end{aligned}$$

for $s \leq t$, where we use the independent increments and (2.21) together with the fact that X is adapted to the σ-field $\{\mathcal{F}_t^X\}_{t\geq 0}$. Hence, X is a martingale.

On the other hand, for X a zero-mean martingale, we have

$$\begin{aligned}&\mathrm{E}\{(X(u) - X(t))\,(X(s) - X(r))\} \\ &\quad= \mathrm{E}\{\mathrm{E}\{(X(u) - X(t))\,(X(s) - X(r))\,|\,\mathcal{F}_s^X\}\} \\ &\quad= \mathrm{E}\{(X(s) - X(r))\,\mathrm{E}\{X(u) - X(t)\,|\,\mathcal{F}_s^X\}\} \\ &\quad= \mathrm{E}\{(X(s) - X(r))\,(X(s) - X(s))\} \\ &\quad= 0\end{aligned}$$

for $0 \leq r \leq s \leq t \leq u$, where we made use of Equation (2.20*) in Klebaner's book and the fact that X is adapted together with Equation (2.18*) and the martingale property.

Problem 3.30: Prove Equation (3.4*) in Klebaner's book. (Note that it is assumed that $0 < t_1 < \cdots < t_n$ in this formula.)

Solution. We prove (3.4*) by induction. Note that the property (3.4*) when $n=1$ is just (3.3*).

Now, assume that (3.4) holds for $n=k$. Note that (3.4) for $n=k$ in turn means that $(B^x(t_1),\ldots,B^x(t_k))$ has probability density function

$$f_{(B^x(t_1),\ldots,B^x(t_k))}(y_1,\ldots,y_k)=p_{t_1}(x,y_1)\prod_{i=2}^{k}p_{t_i-t_{i-1}}(y_{i-1},y_i),$$

for $(y_1,\ldots,y_k)\in\mathbb{R}^k$. For the case when $n=k+1$, it therefore follows from conditioning on the value $(y_1,\ldots,y_k)$ of $(B^x(t_1),\ldots,B^x(t_k))$ and using the independence of increments that (by writing $f(y_1,\ldots,y_k)=f_{(B^x(t_1),\ldots,B^x(t_k))}(y_1,\ldots,y_k)$ for the joint density)

$$\begin{aligned}
\mathrm{P}\left\{\bigcap_{i=1}^{k+1}\{B^x(t_i)\le x_i\}\right\} &= \int_{-\infty}^{x_1}\cdots\int_{-\infty}^{x_k}\mathrm{P}\{B^x(t_{k+1})-B^x(t_k)+y_k\le x_{k+1}\}\\
&\quad\times f(y_1,\ldots,y_k)\,dy_1\ldots dy_k\\
&=\int_{-\infty}^{x_1}\cdots\int_{-\infty}^{x_k}\Phi\left(\frac{x_{k+1}-y_k}{\sqrt{t_{k+1}-t_k}}\right)p_{t_1}(x,y_1)\\
&\quad\times\prod_{i=2}^{k}p_{t_i-t_{i-1}}(y_{i-1},y_i)dy_1\ldots dy_k\\
&=\int_{-\infty}^{x_1}\cdots\int_{-\infty}^{x_k}\int_{-\infty}^{x_{k+1}}p_{t_1}(x,y_1)\\
&\quad\times\prod_{i=2}^{k+1}p_{t_i-t_{i-1}}(y_{i-1},y_i)\,dy_1\ldots dy_{k+1},
\end{aligned}$$

as $B^x(t_{k+1})-B^x(t_k)$ is $\mathrm{N}(0,t_{k+1}-t_k)$-distributed. This proves (3.4) by induction.

Problem 3.31: Let ξ and η be independent standard normal random variables. Show that the process $\{X(t)\}_{t\in\{0,1\}}$ given by $X(0)=\operatorname{sign}(\eta)\,\xi$ and $X(1)=\operatorname{sign}(\xi)\,\eta$ is not Gaussian despite each of the process values $X(0)$ and $X(1)$ being standard Gaussian.

Solution. It is an elementary exercise to see that $X(0)$ and $X(1)$ are standard Gaussian (normal) distributed. Also, note that

$$X(0)\,X(1)=\operatorname{sign}(\eta)\,\xi\;\operatorname{sign}(\xi)\,\eta=|\xi|\,|\eta|\ge 0.$$

However, if $(X(0), X(1))$ were bivariate standard Gaussian (as it must be if X is a Gaussian process), then the above non-negativity is possible if and only if $X(0)$ and $X(1)$ have perfect correlation 1. But this is not true, as

$$\begin{aligned}\mathrm{Corr}(X(0), X(1)) &= \mathrm{Cov}(X(0), X(1)) \\ &= \mathrm{E}(X(0)X(1)) = \mathrm{E}(|\xi||\eta|) = (\mathrm{E}|\xi|)^2 = \frac{2}{\pi}.\end{aligned}$$

by elementary calculations [where we used the fact that $X(0)$ and $X(1)$ are standard Gaussian].

Problem 3.32: Prove that the finite-dimensional distributions of a zero-mean Gaussian stochastic process $\{X(t)\}_{t\in T}$ are completely characterised by the covariance function of the process.

Solution. Given $t_1, \ldots, t_n \in T$, the distribution of the random variable $(X(t_1), \ldots, X(t_n))$ is determined by its characteristic function (Fourier transform):

$$\mathrm{E}\{\mathrm{e}^{i\sum_{j=1}^{n} a_j X(t_j)}\} \quad \text{for} \quad (a_1, \ldots, a_n) \in \mathbb{R}^n.$$

As $\sum_{j=1}^{n} a_j X(t_j)$ is a univariate zero-mean Gaussian random variable, whose characteristic function in turn is equal to

$$\exp\left[-\frac{1}{2}\,Var\left(\sum_{j=1}^{n} a_j X(t_j)\right)\right] = \exp\left[-\frac{1}{2}\sum_{i=1}^{n}\sum_{j=1}^{n} a_i\, a_j\ \mathrm{Cov}(X(t_i), X(t_j))\right],$$

which in turn is obviously determined by the covariance function of X.

Problem 3.33: Show that for any t, $0 < t < 1$ the random variables $B(1)$ and $B(t) - tB(1)$ are independent. Show that the processes $B(t) - tB(1)$ and $B(1)$ are independent. The process $B(t) - tB(1)$ is called a Brownian bridge.

Solution. $(B(1), B(t) - tB(1))$ is virtually trivially bivariate normal, which means that every linear combination of the components in the vector is univariate normal. Recall that bivariate normal random variables are independent if and only if they are uncorrelated. Direct calculations give

$$\begin{aligned}\mathrm{Cov}(B(1), B(t) - tB(1)) &= \mathrm{Cov}(B(1), B(t)) - \mathrm{Cov}(B(1), tB(1)) \\ &= t - t = 0.\end{aligned}$$

To show the independence of the process $X(t) = B(t) - tB(1)$ and $B(1)$, it is enough to show that for any linear combination $\sum_{i=1}^n a_i X(t_i)$ and $B(1)$ are independent. Note that in the question above, we established the independence of $X(t)$ and $B(1)$ for any fixed t. Since the vector $B(1), X(t_i)$, $i = 1, \ldots, n$, has a multivariate normal distribution, the independence is equivalent to zero covariance. Thus, by the properties of covariance,

$$\operatorname{Cov}\left(\sum_{i=1}^n a_i X(t_i), B(1)\right) = \sum_{i=1}^n a_i \operatorname{Cov}(X(t_i), B(1)) = 0.$$

Problem 3.34: Show that the process $X(t) = \sqrt{\frac{2}{\pi}} \sum_{j=1}^\infty \frac{\sin(jt)}{j} \xi_j$, where ξ_j are i.i.d. standard normal random variables is a Brownian bridge on $[0, \pi]$.

Solution. One can see as in the previous question that the process $B(t) - \frac{t}{T} B(T)$ is a Brownian bridge on $[0, T]$. Taking $T = \pi$ and using formula (3.7) on p. 62 in the book,

$$B(t) = \frac{t}{\sqrt{\pi}} \xi_0 + \sqrt{\frac{2}{\pi}} \sum_{j=1}^\infty \frac{\sin(jt)}{j} \xi_j,$$

we obtain $B(\pi) = \sqrt{\pi}\xi_0$, $t/\pi B(\pi) = \xi_0 t/\sqrt{\pi}$, hence $X(t) = B(t) - t/\pi B(\pi)$, and it is a Brownian bridge.

Alternatively, one can verify directly that $X(t)$ is a Gaussian process with zero mean and covariance function $s(1 - t/\pi)$ for $0 \le s < t \le \pi$.

Problem 3.35 (Fake BM, by Häggstrom): Let Z be a $N(0, 1)$ random variable and $a > 0$ be arbitrary. $N(t)$ be a Poisson process defined for $t \ge a > 0$ with intensity $\lambda(t) = 1/(4u)$.

(a) Show that the process $X(t) = \sqrt{t} Z(-1)^{N(t)}$ defined for $t \ge a$ is a martingale, has $N(0, t)$ marginal distribution, but is not a BM.

(b) Show that $X(0)$ is not defined by the previous formula. However, show that one can define the process $X(t)$ for all $t \ge 0$ as a Markov jump process with the following transition probabilities $P(y, t, 0, 0) = \mathrm{P}(X(t) \le y | X(0) = 0) = \mathrm{P}(N(0, t) \le y)$ and for $s > 0$,

$$P(y, t, x, s) = I(x\sqrt{t/s} \le y) p_{st} + I(x\sqrt{t/s} \ge -y)(1 - p_{st}),$$

with $p_{st} = P(N(t) - N(s) \text{ even})$. Check the Chapman–Kolmogorov equations.

Solution.

(a) For $a \le s < t$, since $X(t) = X(s)\sqrt{t/s}(-1)^{N(t)-N(s)}$, by independence of Poisson increments,

$$\mathrm{E}(X(t)|X(s)) = X(s)\sqrt{t/s}\mathrm{E}(-1)^{N(t)-N(s)}.$$

Now, by using the expectation of a function of a Poisson random variable,

$$\begin{aligned}\mathrm{E}(-1)^{N(t-s)} &= e^{-\int_s^t \lambda(u)du}\sum_{k=0}^{\infty}(-1)^k\frac{(\int_s^t \lambda(u)du)^k}{k!}\\ &= e^{-2\int_s^t \lambda(u)du} = e^{-\frac{1}{2}\log t/s} = \frac{\sqrt{s}}{\sqrt{t}}.\end{aligned}$$

The martingale property follows. Since for any $t > 0$, $(-1)^{N(t)} = \pm 1$, and $-Z$ has $N(0,1)$ distribution, $X(t)$ has $N(0,t)$ marginals. Clearly, $X(t)$ jumps and takes values on the parabola $\pm\sqrt{t}$, it is not continuous, and cannot be a BM.

(b) Since the rate is not integrable near zero $\int_0^a \lambda(u)du = \infty$, the Poisson process $N(t)$ is not well defined. We check the consistency of the transition probabilities. Only consistency from $t = 0$ needs to be checked, as for $s > 0$, it is obvious because $N(t)$ for $t > s > 0$ is well defined. Denote by $f_s(x)$ the PDF of $N(0,s)$ distribution. Then,

$$\begin{aligned}\mathrm{P}(X(t) \le y) &= \int_{-\infty}^{\infty} \mathrm{P}(X(t) \le y|X(s) = x)f_s(x)dx\\ &= \int P(y,t,x,s)f_s(x)dx = \int_{-\infty}^{\infty} I(x\sqrt{t/s} \le y)p_{st}\\ &\quad + I(x\sqrt{t/s} \ge -y)(1-p_{st})f_s(x)dx.\end{aligned}$$

When we write it as a sum of two integrals, the first one equals to

$$\begin{aligned}\int_{-\infty}^{\infty} I(x\sqrt{t/s} \le y)f_s(x)dx &= \int_{-\infty}^{y\sqrt{s/t}} f_s(x)dx\\ &= \mathrm{P}(N(0,s) \le y\sqrt{s/t}),\end{aligned}$$

which is $\mathrm{P}(N(0,t) \le y)$. Similarly, the second integral equals to $\mathrm{P}(N(0,t) \ge -y)$. Using the symmetry of $N(0,t)$ distribution, it follows that $\mathrm{P}(X(t) \leq y)$ equals to

$p_{st}\mathrm{P}(N(0,t) \leq y) + (1 - p_{st})\mathrm{P}(N(0,t) \geq -y) = \mathrm{P}(N(0,t) \leq y)$,

and the Chapman–Kolmogorov equations hold.

Problem 3.36: Show that for $t \geq 1$, the processes

$$X(t) = \sqrt{t}\cos(B(\log t)), \quad \text{and} \quad Y(t) = \sqrt{t}\sin(B(\log t))$$

are martingales.

Solution. Consider the complex-valued process $Z(t) = \sqrt{t}e^{iB(\log t)}$. It is a martingale, as the following calculations show:

$$\mathrm{E}(e^{iB(\log t)}|B(\log s)) = e^{iB(\log s)}\mathrm{E}e^{iB(\log(t/s))}.$$

Hence,

$$\mathrm{E}(e^{iB(\log t)}|B(\log s)) = e^{iB(\log s)}e^{-\frac{1}{2}\log(t/s)} = e^{iB(\log s)}\frac{\sqrt{s}}{\sqrt{t}}.$$

Therefore,

$$\mathrm{E}(\sqrt{t}e^{iB(\log t)}|B(\log s)) = \sqrt{s}e^{iB(\log s)}.$$

Hence, both the real and imaginary parts are martingales.

Problem 3.37 (Fake BM, by Hamza and Klebaner): Let $B(t)$ be BM and $N(t)$ a Poisson process with parameter λ, independent of each other. Show that with the choice $\lambda = \frac{\sqrt{e}}{2(\sqrt{e}-1)}$, the process $X(t) = \sqrt{t}e^{-\frac{1}{2}N(\log t)}B(e^{N(\log t)})$, $t \geq 1$, is a martingale. Show that for all $t \geq 1$, it has $N(0,t)$ distribution. Show that it is not a BM.

Remark: A non-decreasing process with independent increments is called a subordinator, and when it is used as time in BM, a subordinated BM process results.

Solution. Consider for $0 < a < b$ and $\mathcal{F}_a = \sigma(N(u), u \leq a, B(v), v \leq e^{N(a)})$,

$$\begin{aligned}
&\mathrm{E}\big(e^{-\frac{1}{2}N(b)}B(e^{N(b)})|\mathcal{F}_a\big)\\
&\quad = \mathrm{E}\big(e^{-\frac{1}{2}N(a)}e^{-\frac{1}{2}(N(b)-N(a))}(B(e^{N(a)}) + B(e^{N(b)}) - B(e^{N(a)}))|\mathcal{F}_a\big)\\
&\quad = e^{-\frac{1}{2}N(a)}B(e^{N(a)})\mathrm{E}\big(e^{-\frac{1}{2}(N(b)-N(a))}|\mathcal{F}_a\big)\\
&\qquad + e^{-\frac{1}{2}N(a)}\mathrm{E}\big(e^{-\frac{1}{2}(N(b)-N(a))}(B(e^{N(b)}) - B(e^{N(a)}))|\mathcal{F}_a\big).
\end{aligned}$$

Since $N(b) - N(a)$ is independent of $\mathcal{F}_a$ and $B(e^{N(b)}) - B(e^{N(a)})$ is independent of $\mathcal{F}_a$, the conditional expectation equals to the unconditional one:

$$\begin{aligned}
&\mathrm{E}\big(e^{\frac{1}{2}N(b)}B(e^{N(b)})|\mathcal{F}_a\big)\\
&\quad = e^{-\frac{1}{2}N(a)}B(e^{N(a)})e^{\lambda(b-a)((e^{-\frac{1}{2}}-1)}\\
&\qquad + e^{-\frac{1}{2}N(a)}\mathrm{E}e^{-\frac{1}{2}(N(b)-N(a))}(B(e^{N(b)}) - B(e^{N(a)})),
\end{aligned}$$

where we used that $\mathrm{E}e^{u(N(b)-N(a))} = \mathrm{e}^{\lambda(b-a)(e^u-1)}$. Next, we condition on the Poisson process and use the independence assumption. Since the increments of BM have zero mean, the second term vanishes:

$$\begin{aligned}
&\mathrm{E}e^{-\frac{1}{2}(N(b)-N(a))}\big(B(e^{N(b)}) - B(e^{N(a)})\big)\\
&\quad = \mathrm{E}\big(\mathrm{E}e^{-\frac{1}{2}(N(b)-N(a))}(B(e^{N(b)}) - B(e^{N(a)}))|N(a), N(b)\big)\\
&\quad = \mathrm{E}e^{-\frac{1}{2}(N(b)-N(a))}\big(\mathrm{E}(B(e^{N(b)}) - B(e^{N(a)}))|N(a), N(b)\big) = 0.
\end{aligned}$$

Hence, we have

$$\mathrm{E}\big(e^{-\frac{1}{2}N(b)}B(e^{N(b)})|\mathcal{F}_a\big) = e^{-\frac{1}{2}N(a)}B(e^{N(a)})e^{\lambda(b-a)(e^{-\frac{1}{2}}-1)}.$$

Now, taking $a = \log s$ and $b = \log t$, we obtain

$$e^{\lambda(b-a)((e^{-\frac{1}{2}}-1)} = e^{\lambda \log(t/s)(e^{-\frac{1}{2}}-1)} = \left(\frac{s}{t}\right)^{\lambda(1-e^{-\frac{1}{2}})}.$$

Finally, taking λ such that $\lambda(1 - e^{-\frac{1}{2}}) = \frac{1}{2}$, the required martingale property follows.

To see that $X(t)$ has $N(0,t)$ distribution, note the scaling property of $B(t)$, for any constant c, $B(ct)$ has the same distribution as $\sqrt{c}B(t)$. Hence, by conditioning on $N(t)$, taking $c = e^{N(\log t)}$, $e^{-\frac{1}{2}N(\log t)}B(e^{N(\log t)})$ is distributed as $B(1)$, and $X(t)$ has the distribution of $\sqrt{t}B(1)$. Clearly, $X(t)$ is discontinuous, hence it is not a BM. Furthermore, one can show that $X(t)$ is not a Gaussian process by considering its bivariate distributions.

Problem 3.38 (Fake BM, by Oleszkiewicz): Given $a \geq 0$, for $t \geq e^{-a}$, let $\mathcal{F}_t^a = \sigma(V_1, V_2, (B_s)_{0 \leq s \leq a+\ln t})$, where V_1, V_2 and B are independent, V_1 and V_2 are $N(0,1)$ and B is a BM. Let

$$X_t^a = \sqrt{t}(V_1 \cos B_{a+\ln t} + V_2 \sin B_{a+\ln t}).$$

Show that $(X_t^a, \mathcal{F}_t^a)_{t \geq e^{-a}}$ is a continuous martingale with $N(0,t)$ marginals, but it is non-Gaussian.

Solution. First, we show that X_t^a has $N(0,t)$ marginals:

$$\begin{aligned}
\mathrm{E}(e^{i\lambda X_t^a}) &= \mathrm{E}\big(\mathrm{E}(e^{i\lambda\sqrt{t}(V_1 \cos B_{a+\ln t} + V_2 \sin B_{a+\ln t})} | B_{a+\ln t})\big) \\
&= \mathrm{E}\big(\mathrm{E}(e^{i\lambda\sqrt{t}V_1 \cos B_{a+\ln t}} | B_{a+\ln t})\mathrm{E}(e^{i\lambda\sqrt{t}V_2 \sin B_{a+\ln t}} | B_{a+\ln t})\big) \\
&= \mathrm{E}(e^{-\frac{1}{2}\lambda^2 t \cos^2 B_{a+\ln t}} e^{-\frac{1}{2}\lambda^2 t \sin^2 B_{a+\ln t}}) \\
&= \mathrm{E}(e^{-\frac{1}{2}\lambda^2 t}) = e^{-\frac{1}{2}\lambda^2 t}.
\end{aligned}$$

For the martingale property, consider the process

$$Y_t^a = \sqrt{t}(V_1 - iV_2)e^{iB_{a+\ln t}}.$$

We show that Y_t^a, $t \geq e^{-a}$ is a martingale:

$$\begin{aligned}
\mathrm{E}(Y_{t+s}^a | Y_t^a) &= \mathrm{E}(\sqrt{t+s}(V_1 - iV_2)e^{iB_{a+\ln(t+s)}} | Y_t^a) \\
&= \mathrm{E}\big(\mathrm{E}(\sqrt{t+s}(V_1 - iV_2)e^{iB_{a+\ln(t+s)}} | Y_t^a, V_1, V_2) | Y_t^a\big) \\
&= \sqrt{t+s}\mathrm{E}\big((V_1 - iV_2)e^{iB_{a+\ln t}} \\
&\quad \times \mathrm{E}(e^{i(B_{a+\ln(t+s)} - B_{a+\ln t})} | Y_t^a, V_1, V_2) | Y_t^a\big) \\
&= \sqrt{\frac{t+s}{t}} Y_t^a e^{-\frac{1}{2}\ln(\frac{t+s}{t})} = Y_t^a.
\end{aligned}$$

Hence, both the real and imaginary parts are martingales. Note that

$$\begin{aligned}
Y_t^a = \sqrt{t}\big(&V_1 \cos B_{a+\ln t} + V_2 \sin B_{a+\ln t} \\
&+ i(V_1 \sin B_{a+\ln t} - V_2 \cos B_{a+\ln t})\big).
\end{aligned}$$

Thus, the real part $X_t^a = \sqrt{t}(V_1 \cos B_{a+\ln t} + V_2 \sin B_{a+\ln t})$ is a martingale.

Next, we show that X_t^a is not a Gaussian process. Recall that if X_t is a Gaussian process, then the increment $X_t - X_s$, for any $s < t$, is Gaussian.

However, taking $s = 1$ and $t = e$,

$$\begin{aligned}
&\mathrm{E}[e^{i\lambda(X_e^a - X_1^a)}] \\
&\quad = \mathrm{E}\big(e^{i\lambda[\sqrt{e}(V_1 \cos B_{a+1} + V_2 \sin B_{a+1}) - (V_1 \cos B_a + V_2 \sin B_a)]}\big) \\
&\quad = \mathrm{E}\big(\mathrm{E}\big(e^{i\lambda[V_1(\sqrt{e}\cos B_{a+1} - \cos B_a) + V_2(\sqrt{e}\sin B_{a+1} - \sin B_a)]} \big| B_a, B_{a+1}\big)\big) \\
&\quad = \mathrm{E}\big(e^{-\frac{1}{2}\lambda^2[(\sqrt{e}\cos B_{a+1} - \cos B_a)^2 + (\sqrt{e}\sin B_{a+1} - \sin B_a)^2]}\big) \\
&\quad = \mathrm{E}\big(e^{-\frac{1}{2}\lambda^2[e+1-2\sqrt{e}(\cos B_{a+1}\cos B_a + \sin B_{a+1}\sin B_a)]}\big) \\
&\quad = \mathrm{E}\big(e^{-\frac{1}{2}\lambda^2[e+1-2\sqrt{e}\cos(B_{a+1} - B_a)]}\big) \\
&\quad = \mathrm{E}\big(e^{-\frac{1}{2}\lambda^2[e+1-2\sqrt{e}\cos B_1]}\big).
\end{aligned}$$

This shows that $X_e^a - X_1^a$ is not Gaussian, and hence X_t^a is not a Gaussian process.

Chapter 4

Brownian Motion Calculus

Arguably, the most important purpose of this book is to study stochastic differential equations (SDEs). Recall that an ordinary differential equation (ODE) with initial value is given as

$$x'(t) = \mu(x(t), t) \quad \text{for } t \in [0, T], \quad x(0) = x_0,$$

for a $T \in (0, \infty)$ and an $x_0 \in \mathbb{R}$, where $\mu : \mathbb{R} \times [0, T] \to \mathbb{R}$ is the (measurable) coefficient function of the equation. This can be expressed in differential form as

$$dx(t) = \mu(x(t), t)\, dt \quad \text{for } t \in [0, T], \quad x(0) = x_0.$$

(Unlike what some people seem to believe, there is absolutely nothing non-rigorous with differentials as long as one knows what one means with them.) Alternatively, we can express the equation in integrated form as

$$x(t) = x_0 + \int_0^t \mu(x(s), s)\, ds \quad \text{for } t \in [0, T].$$

Of course, what makes the above expressions equations is that (the solution) $x : [0, T] \to \mathbb{R}$ appears on both sides of the equality.

For SDEs, the second and third ways to express the equation are used, but the first virtually never. This is simply because the solution is virtually never differentiable in the usual sense. An SDE with coefficient functions $\mu, \sigma : \mathbb{R} \times [0, T] \to \mathbb{R}$ is given is differential form by

$$dX(t) = \mu(X(t), t)\, dt + \sigma(X(t), t)\, dB(t) \quad \text{for } t \in [0, T], \quad X(0) = x_0.$$

Here, $\{B(t)\}_{t\geq 0}$ is BM and the solution $X : [0,T] \to \mathbb{R}$ (or really $X : \Omega \times [0,T] \to \mathbb{R}$) is denoted with a capital X as it (unlike in the case of the ODE) is a stochastic process because of the presence of BM. Alternatively, we can express the SDE in integrated form as

$$X(t) = x_0 + \int_0^t \mu(X(s), s)\, ds + \int_0^t \sigma(X(s), s)\, dB(s) \quad \text{for } t \in [0,T].$$

Both these equations mean the same things: they are just two different ways to express the one and same SDE. But the question is, what the equation really means because BM is non-differentiable everywhere so that the differential $dB(t)$ does not exist in the first expression for the SDE. Further, BM is not FV, so the right-most integral does not exist in the usual math sense in the SDE written in integrated form.

This chapter is intended to provide a well-defined meaning to the so-called Itô integral process

$$\left\{\int_0^t X\, dB\right\}_{t\in[0,T]} = \left\{\int_0^t X(s)\, dB(s)\right\}_{t\in[0,T]}$$

and study its properties. The integral is not the same as any integral featured in math and, in particular, is not the same as the RS integral or the (more general) Lebesgue–Stieltjes integral, but is something new and entirely specific for stochastic calculus. It carries the name Itô integral after its creator, K. Itô, who developed this in around 1950. The Itô integral (process) is exactly what is needed to give meaning to the SDE in integrated form and thereby also to the SDE in differential form.

Definition of the Itô Integral and Itô Integral Processes

Throughout this chapter, $\{B(t)\}_{t\geq 0}$ denotes BM and $\{\mathcal{F}_t\}_{\geq 0} = \{\mathcal{F}_t^B\}_{\geq 0}$ is the filtration generated by BM itself.

A stochastic process $\{X(t)\}_{t\in[0,T]}$ is called measurable if it is a measurable function $X : \Omega \times [0,T] \to \mathbb{R}$. This in turn means that $X^{-1}(\mathcal{B}) \subseteq \sigma(\mathcal{F} \times ([0,T] \cap \mathcal{B}))$. The process X for which we define the Itô integral process $\{\int_0^t X\, dB\}_{t\in[0,T]}$ must be measurable — this is required for the construction of the integral.

The concept of a measurable process is slightly too technical to be suitable to be fully investigated and utilised in our treatment so we feel content with informing the reader that all processes with cádlág and/or cáglád sample paths are measurable, as are then, in particular, all continuous processes.

As sums, differences, and products of measurable processes are measurable, it follows that, for example, $\{X(t)\, I_B(t)\}_{t\in[0,T]}$ is measurable when X is a measurable process and B is a subinterval of $[0,T]$ (or any measurable subset of $[0,T]$). Every process X that we will Itô integrate will be included among these examples of measurable processes. Thereby, we consider the topic of measurability of X resolved for our purposes.

For a measurable process, the Fubini theorem ensures that

$$\mathrm{E}\Big(\int_0^T X(t)\,dt\Big) = \int_0^T \mathrm{E}(X(t))\,dt$$

in the sense that both sides are well-defined simultaneously, and when that occurs they agree.

The Itô integral is constructed in three steps for subsequently larger classes of measurable and adapted processes that can conveniently be denoted S_T, E_T and P_T. The class of simple processes S_T consists of processes of the type

$$X(t) = \xi_0 I_{\{0\}}(t) + \sum_{i=0}^{n-1} \xi_i I_{(t_i,t_{i+1}]}(t) \quad \text{for } t\in[0,T],$$

for a grid $0 = t_0 < t_1 < \cdots < t_n = T$ of non-random times and for $\xi_0, \xi_1, \ldots, \xi_{n-1}$ random variables with ξ_i $\mathcal{F}_{t_i}$-measurable and $E(\xi_i^2) < \infty$ for $i = 0, \ldots, n-1$.

The class E_T consists of measurable and adapted processes $\{X(t)\}_{t\in[0,T]}$ such that

$$\mathrm{E}\Big(\int_0^T X(t)^2\,dt\Big) = \int_0^T \mathrm{E}(X(t)^2)\,dt < \infty.$$

The class P_T consists of measurable and adapted processes $\{X(t)\}_{t\in[0,T]}$ such that

$$\mathrm{P}\Big(\int_0^T X(t)^2\,dt < \infty\Big) = 1.$$

Clearly, we have $S_T \subseteq E_T \subseteq P_T$.

For $X \in S_T$, we define the Itô integral process $\{\int_0^t X\,dB\}_{t\in[0,T]}$ by $\int_0^0 X\,dB = 0$ and

$$\int_0^t X\,dB = \sum_{i=0}^{m-1} \xi_i(B(t_{i+1}) - B(t_i)) + \xi_m(B(t) - B(t_m)),$$

for $t \in (t_m, t_{m+1}]$ and $m = 0, \ldots, n-1$. When considering the Itô integral process $\{\int_0^t X\,dB\}_{t\in[0,T]}$ of a process $X \in S_T$ at a finite number of points

$s_1, \ldots, s_j \in [0, T]$, there is no restriction to assume that $s_1, \ldots, s_j$ are members of the grid $0 = t_0 < t_1 < \cdots < t_n = T$ that is used to define X, as otherwise that grid can be enriched to include $s_1, \ldots, s_j$ without changing any values of either X or the Itô integral process. This technique is often useful to make notation less complicated.

For $X \in E_T$, one can prove that there exists an approximating sequence $\{X_n\}_{n=1}^{\infty} \subseteq S_T$ such that

$$\lim_{n\to\infty} \mathrm{E}\left\{\int_0^T (X_n(t) - X(t))^2\, dt\right\} = 0.$$

The proof is quite straightforward in the case of a continuous X but is exceedingly difficult in the general case. For $X \in E_T$, the Itô integral process $\{\int_0^t X\, dB\}_{t\in[0,T]}$ is well defined and defined as a limit in the sense of convergence in L^2 of $\int_0^t X_n\, dB$ as $n \to \infty$ for each $t \in [0, T]$, where $\{X_n\}_{n=1}^{\infty} \subseteq S_T$ is an approximating sequence as above. The proof consists of proving that $\{\int_0^t X_n\, dB\}_{n=1}^{\infty}$ is a Cauchy sequence in L^2 for each $t \in [0, T]$ so that the L^2-limits denoted $\{\int_0^t X\, dB\}_{t\in[0,T]}$ exist. In addition, one must prove that the limits don't depend on which particular approximating sequence that is chosen, so that the integral process is not multiple-valued.

For $X \in P_T$, one can prove that there exists an approximating sequence $\{X_n\}_{n=1}^{\infty} \subseteq E_T$ such that

$$\int_0^T (X_n(t) - X(t))^2\, dt \to_P 0 \quad \text{as } n \to \infty.$$

For $X \in P_T$, the Itô integral process $\{\int_0^t X\, dB\}_{t\in[0,T]}$ is well defined, and defined as a limit in the sense of convergence in probability of $\int_0^t X_n\, dB$ as $n \to \infty$ for each $t \in [0, T]$, where $\{X_n\}_{n=1}^{\infty} \subseteq E_T$ is an approximating sequence as above. The proof consists of proving that $\{\int_0^t X_n\, dB\}_{n=1}^{\infty}$ is a Cauchy sequence in probability for each $t \in [0, T]$ so that the limits in probability denoted $\{\int_0^t X\, dB\}_{t\in[0,T]}$ exist. In addition, one must prove that the limits don't depend on which particular approximating sequence is chosen so that the integral process is not multiple-valued. The proof is not overly difficult but requires some special tools to be developed.

By combining the approximation of processes in P_T with processes in E_T and the approximation of processes in E_T with processes in S_T, it is not hard to see that the approximating sequence $\{X_n\}_{n=1}^{\infty} \subseteq E_T$ of an $X \in P_T$ in the previous paragraph can in fact be selected to belong to S_T.

A very important and useful fact is that for a continuous $X \in P_T$, it holds that

$$\sup_{t\in[0,T]} \left| \int_0^t X\,dB - \int_0^t \sum_{i=1}^n X(t_{i-1}^n) I_{(t_{i-1}^n, t_i^n]}\,dB \right| \to_P 0,$$

for a sequence of partitions $0 = t_0^n < t_1^1 < \cdots < t_n^n = T$ of $[0,T]$ such that $\max_{1\le i\le n} t_i^n - t_{i-1}^n \downarrow 0$. In other words, we have

$$\sum_{i=1}^n X(t_{i-1}^n) I_{[0,t]}(B(t_i^n) - B(t_{i-1}^n)) \to_P \int_0^t X\,dB \quad \text{for } t \in [0,T]$$

and

$$\sum_{i=1}^n X(t_{i-1}^n)(B(t_i^n) - B(t_{i-1}^n)) \to_P \int_0^T X\,dB.$$

As virtually all processes X that we Itô integrate are continuous, this rule virtually always applies for us.

It is important to note that

$$\sum_{i=1}^n X(t_i^n)(B(t_i^n) - B(t_{i-1}^n)) \\ - \sum_{i=1}^n X(t_{i-1}^n)(B(t_i^n) - B(t_{i-1}^n)) \to [X,B](T) \neq 0$$

in general. This emphasises the fact that the Itô integral is not an RS integral. In the particular case when X is FV, we do however have $[X,B](T) = 0$.

For BM itself, for example, we have

$$\sum_{i=1}^n B(t_i^n)(B(t_i^n) - B(t_{i-1}^n)) - \sum_{i=1}^n B(t_{i-1}^n)(B(t_i^n) - B(t_{i-1}^n)) \to [B](T) = T.$$

By "twisting" this example a little, one deduces that

$$\int_0^T B\,dB \leftarrow \sum_{i=1}^n B(t_{i-1}^n)(B(t_i^n) - B(t_{i-1}^n))$$

$$= \frac{1}{2}\sum_{i=1}^n (B(t_i^n) + B(t_{i-1}^n))(B(t_i^n) - B(t_{i-1}^n))$$

$$- \frac{1}{2}\sum_{i=1}^n (B(t_i^n) - B(t_{i-1}^n))(B(t_i^n) - B(t_{i-1}^n))$$

$$\to \frac{1}{2}B(T)^2 - \frac{T}{2}.$$

The Itô integral $\int_a^b X\,dB$ over an interval $(a,b] \subseteq [0,T]$ is defined as $\int_0^T I_{(a,b]} X\,dB$. As the Itô integral process $\int_0^t X\,dB$ is continuous as a function of t (see the following), it does not matter if the interval endpoints a and b are included or not.

Besides linearity, the important properties of the Itô integral process are as follows:

- $\int_0^T I_{(a,b]}\,dB = \int_a^b dB = B(b) - B(a)$ for $(a,b] \subseteq [0,T]$.
- (ZERO-MEAN) $\mathrm{E}(\int_0^t X\,dB) = 0$ for $t \in [0,T]$ and $X \in E_T$.
- (ADAPTEDNESS) $\{\int_0^t X\,dB\}_{t\in[0,T]}$ is adapted.
- (SAMPLE PATH CONTINUITY) $\int_0^t X\,dB$ is continuous as a function of $t \in [0,T]$.
- (MARTINGALE) $\{\int_0^t X\,dB\}_{t\in[0,T]}$ is a martingale for $X \in E_T$.
- (ISOMETRY) $\mathrm{E}((\int_0^t X\,dB)^2) = \int_0^t E(X(s)^2)\,ds$ for $t \in [0,T]$ and $X \in E_T$.
- (QUADRATIC VARIATION) $[\int_0^t X\,dB] = [\int_0^{(\cdot)} X\,dB](t) = \int_0^t X(s)^2\,ds$ for $t \in [0,T]$.
- (COVARIATION) $[\int_0^t X\,dB, \int_0^t Y\,dB] = \int_0^t X(s)Y(s)\,ds$ for $t \in [0,T]$.

All properties are more or less immediate for processes in S_T. Continuity and quadratic variation/covariation are complicated to prove for processes in E_T and P_T. The remaining properties for Itô integrals in these spaces are inherited from the construction by the convergence of approximating sequences. Isometry is obviously invalid for $X \in P_T \backslash E_T$. The zero-mean and martingale properties can both hold and fail for processes $X \in P_T \backslash E_T$ — this depends on the particular choice of X.

Itô Integral and Gaussian Processes

For a measurable non-random process $x : [0, T] \to \mathbb{R}$ (with no dependence on the outcome $\omega \in \Omega$), we have $E_T = P_T$ and $x \in P_T$ if and only if x is square-integrable, $\int_0^T x(t)^2\, dt < \infty$. In that case, $\{\int_0^t x\, dB\}_{t\in[0,T]}$ is a zero-mean Gaussian process and martingale with covariance function

$$\begin{aligned}\operatorname{Cov}\left(\int_0^s x\, dB, \int_0^t x\, dB\right) &= \mathrm{E}\left[\left(\int_0^s x\, dB\right)\left(\int_0^t x\, dB\right)\right]\\ &= \int_0^{\min(s,t)} x(u)^2\, du.\end{aligned}$$

This property is probably easiest to establish by rewriting

$$\int_0^t x\, dB = \int_0^T x I_{[0,t]}(s)\, dB$$

and employing polarisation together with isometry. Similarly, one might consider a family of non-random processes $\{\{x(t,s)\}_{s\in[0,t]}\}_{t\in[0,T]}$ such that $\{x(t,s)\}_{s\in[0,t]} \in P_t$ for $t \in [0,T]$ and establish that $\{\int_0^t x(s,t)\, dB(s)\}_{t\in[0,T]}$ is a zero-mean Gaussian process satisfying an obvious version of the covariance formula above. However, in doing so, the martingale property is lost in general.

Itô's Formula for BM

For a twice continuously differentiable (C^2) function $f : \mathbb{R} \to \mathbb{R}$ together with a grid $0 = t_0^n < t_1^n < \cdots < t_n^n = t$, a second-order Taylor expansion shows that

$$\begin{aligned}f(B(t)) - f(B(0)) &= \sum_{i=0}^{n-1} (f(B(t_{i+1}^n)) - f(B(t_i^n)))\\ &= \sum_{i=0}^{n-1} f'(B(t_i^n))(B(t_{i+1}^n) - B(t_i^n))\\ &\quad + \frac{1}{2}\sum_{i=0}^{n-1} f''(B(t_i^n))(B(t_{i+1}^n) - B(t_i^n))^2 + \text{higher order terms}.\end{aligned}$$

Sending $\max_{1\le i\le n} t_i^n - t_{i-1}^n \downarrow 0$, it is immediate that the first term on the right-hand side converges to $\int_0^t f'(B)\, dB$. By being more specific with the third term on the right-hand side, it is not hard either to show that the term converges to 0. By somewhat more complicated arguments, it follows that the second sum on the right-hand side converges to

$$\int_0^t f''(B(s))\, d[B](s) = \int_0^t f''(B(s))\, ds.$$

The proof goes by proving that (by continuity of $f''(B)$) asymptotically (in the limit) that the sum is the same as

$$\sum_{i=0}^{n-1} f''(B(s_{j_i}^m))(B(t_{i+1}^n) - B(t_i^n))^2,$$

for a suitable choice of $0 \le j_1 \le \cdots \le j_n \le m$, where $0 = s_0^m < s_1^m < \cdots < s_m^m = t$ is a courser grid than $\{t_i^n\}_{i=0}^n$, that is, $m \le n$ and $\{s_i^m\}_{i=0}^m \subseteq \{t_i^n\}_{i=0}^n$. Now, first send $\max_{1\le i\le n} t_i^n - t_{i-1}^n \downarrow 0$ and then $\max_{1\le i\le m} s_i^m - s_{i-1}^m \downarrow 0$ afterwards to obtain the desired using first that $[B](t) = t$ and then the convergence of approximating Riemann sums.

The reasoning in the previous paragraph establishes our first version of the immensely important Itô formula (sometimes also called Itô's lemma):

$$f(B(t)) = f(B(0)) + \int_0^t f'(B)\, dB = \frac{1}{2}\int_0^t f''(B(s))\, ds.$$

Often it is more convenient to use this result written in the differential form:

$$df(B(t)) = f'(B(t))\, dB(t) + \frac{1}{2}\, f''(B(t))\, dt.$$

Note that we here have made rigorous the notation

$$dB(t)^2 = d[B](t) = dt.$$

This kind of calculus we will see much more of in the sequel.

A simple application of Itô's formula is to reestablish the earlier established fact that

$$\int_0^t B\, dB = \frac{1}{2}\, B(t)^2 - \frac{t}{2}.$$

Itô Processes and Stochastic Differentials

An Itô process $\{X(t)\}_{t\in[0,T]}$ (as opposed to the earlier discussed Itô integral process special case thereof) is given by

$$\{X(t)\}_{t\in[0,T]} = \left\{X(0) + \int_0^t \mu(s)\, ds + \int_0^t \sigma\, dB\right\}_{t\in[0,T]}.$$

Here, $X(0)$ is an $\mathcal{F}_0$-measurable random variable (and thus a constant as long as we keep $\mathcal{F}_t = \mathcal{F}_t^B$) and $\sigma \in P_T$ (by necessity), while $\{\mu(t)\}_{t\in[0,T]}$ is a measurable and adapted process with $\int_0^T |\mu(t)|\, dt < \infty$. It follows that the Itô process is adapted and continuous.

An Itô process is often conveniently displayed in differential form as

$$dX(t) = \mu(t)\, dt + \sigma(t)\, dB(t).$$

This expression is called a stochastic differential. To say that a stochastic process X has stochastic differential thus means that it is an Itô process.

By writing $\mu = \mu^+ - \mu^-$, we see that $\{\int_0^t \mu(s)\, ds\}_{t\in[0,T]}$ is FV. And so it follows that

$$[X](t) = \int_0^t \sigma(s)^2\, ds \quad \text{for } t \in [0, T].$$

Using polarisation, we obtain a formula for the covariation between two Itô processes X and Y (with respect to the one and same BM):

$$[X, Y](t) = \int_0^t \sigma_X(s)\sigma_Y(s)\, ds \quad \text{for } t \in [0, T].$$

Written in differential form, this becomes

$$dX(t)dY(t) = d[X, Y](t) = \sigma_X(t)\sigma_Y(t)\, dt.$$

This in turn is arguably one of the most important and useful expressions in stochastic calculus.

The Itô integral of one Itô process X with respect to another Y is defined as

$$\left\{\int_0^t X\, dY\right\}_{t\in[0,T]} = \left\{\int_0^t X(s)\, \mu_Y(s)\, ds + \int_0^t X\sigma_Y\, dB\right\}_{t\in[0,T]}$$

provided that both integrals on the right-hand side are well defined [that is, when $\int_0^t |X(s)\mu_Y(s)|\, ds < \infty$ and $X\sigma_Y \in P_T$].

As X is continuous, it can be shown (but is not entirely immediate) that

$$\sum_{i=1}^{n} X(t_{i-1}^n) I_{[0,t]}(Y(t_i^n) - Y(t_{i-1}^n)) \to_P \int_0^t X\, dY \quad \text{for } t \in [0, T],$$

for a sequence of partitions $0 = t_0^n < \cdots < t_n^n = T$ of $[0, T]$ such that $\max_{1\le i\le n} t_i^n - t_{i-1}^n \downarrow 0$.

Itô's Formula for Itô Processes

By means of replacing BM with an Itô process in the derivation of Itô's formula for BM and using the fact that $dX(t)^2 = d[X](t)$, we obtain Itô's formula for an Itô process:

$$f(X(t)) = f(X(0)) + \int_0^t f'(X)\,dX + \frac{1}{2}\int_0^t f''(X)\,d[X].$$

Written out in full detail, this becomes

$$f(X(t)) = f(X(0)) + \int_0^t \left(f'(X(s))\mu_X(s) + \frac{1}{2} f''(X(s))\sigma_X(s)^2\right) ds + \int_0^t f'(X)\sigma_X\,dB.$$

This famous result together with the following bivariate versions of it are the most important results in stochastic calculus. Often it is convenient to write Itô's formula in differential form as

$$df(X(t)) = f'(X(t))\,dX(t) + \frac{1}{2} f''(X(t))\,d[X](t).$$

In Itô's formula, it is sufficient to require that f is C^2 on the range of values of the Itô process X involved.

For two Itô processes X and Y and a function $f : \mathbb{R}^2 \to \mathbb{R}$ that have continuous partial derivatives up to order two (C^2, that is), we have the following bivariate Itô formula:

$$\begin{aligned} df(X(t),Y(t)) = {} & \frac{\partial f}{\partial x}(X(t),Y(t))\,dX(t) + \frac{\partial f}{\partial y}(X(t),Y(t))\,dY(t) \\ & + \frac{1}{2}\frac{\partial^2 f}{\partial x^2}(X(t),Y(t))\,d[X](t) + \frac{1}{2}\frac{\partial^2 f}{\partial y^2}(X(t),Y(t))\,d[Y](t) \\ & + \frac{\partial^2 f}{\partial x \partial y}(X(t),Y(t))\,d[X,Y](t). \end{aligned}$$

Entirely analogous multivariate formulae hold for three or more Itô processes. The proof is from an obvious extension of the proof for one Itô process making use of the Taylor expansions in several variables. As in one dimension, it is sufficient to require that f is C^2 on the range of values of the Itô processes involved.

One particularly interesting application of the bivariate Itô formula is integration by parts:

$$d(X(t)Y(t)) = X(t)\,dY(t) + Y(t)\,dX(t) + d[X,Y](t).$$

This result can alternatively be established by means of rearranging

$$\begin{aligned}&\sum_{i=0}^{n-1}(X(t_{i+1}^n) - X(t_i^n))(Y(t_{i+1}^n) - Y(t_i^n))\\ &\quad = X(t)Y(t) - X(0)Y(0) - \sum_{i=0}^{n-1} X(t_i^n)(Y(t_{i+1}^n) - Y(t_i^n))\\ &\qquad - \sum_{i=0}^{n-1} Y(t_i^n)(X(t_{i+1}^n) - X(t_i^n)),\end{aligned}$$

for $0 = t_0^n < t_1^n < \cdots < t_n^n = t$ and sending $\max_{1\leq i\leq n} t_i^n - t_{i-1}^n \downarrow 0$ on both sides.

It is very instructive to compare the integration by parts formula above with the integration by parts formula for the RS integral (which the former is a proper extension of), where the last term on the right-hand side is not present (because it vanishes due to the processes involved being both FV and continuous).

Another very important application of the bivariate Itô formula is the special case when $Y(t) = t$:

$$df(X(t),t) = \frac{\partial f}{\partial x}(X(t),t)\,dX(t) + \frac{\partial f}{\partial y}(X(t),t)\,dt + \frac{1}{2}\frac{\partial^2 f}{\partial x^2}(X(t),t)\,d[X](t).$$

In fact, for this formula, the requirements on f can be relaxed to f being C^2 in the first variable and only C^1 in the second ($C^{2,1}$, that is).

By applying the Itô formula of the previous paragraph to BM, we conclude that $\{f(B(t),t)\}_{t\in[0,T]}$ is a martingale whenever $\frac{\partial f}{\partial y} + \frac{1}{2}\frac{\partial^2 f}{\partial x^2} = 0$ and $\{\frac{\partial f}{\partial x}(B(t),t)\}_{t\in[0,T]} \in E_T$. From this, we immediately recover all three martingales of BM listed in Chapter 3 just by inspection.

Itô Processes in Higher Dimensions

Higher-dimensional Itô integral processes with values in $\mathbb{R}^n$ with respect to $\mathbb{R}^d$-valued BM $\{\mathbf{B}(t)\}_{t\geq 0} = \{(B_1(t),\ldots,B_d(t))\}_{t\geq 0}$ (with independent BM components) can be defined: Let $\boldsymbol{\sigma} : [0,T] \to \mathbb{R}^{n\times d}$ satisfy $\{\sigma_{ij}(t)\}_{t\in[0,T]} \in P_T$ for $(i,j) \in \{1,\ldots,n\} \times \{1,\ldots,d\}$, where the filtration $\{\mathcal{F}_t\}_{t\in[0,T]}$ now is $\mathcal{F}_t = \bigvee_{j=1}^{d} \mathcal{F}_t^{B_j}$ for $t \in [0,T]$. Enlarging the filtrations of the component BM in this way will not affect the value of an Itô integral process with respect to any of the component BM except that the space P_T will have more members because more processes are adapted.

The $\mathbb{R}^n$-valued Itô integral process $\{\int_0^t \boldsymbol{\sigma}\, d\mathbf{B}\}_{t\in[0,T]}$ is defined by

$$\left\{\int_0^t \boldsymbol{\sigma}\, d\mathbf{B}\right\}_{t\in[0,T]} = \left\{\left(\sum_{j=1}^{d}\int_0^t \sigma_{1j}\, dB_j, \ldots, \sum_{j=1}^{d}\int_0^t \sigma_{nj}\, dB_j\right)\right\}_{t\in[0,T]}.$$

By adding an FV integral process,

$$\left\{\int_0^t \boldsymbol{\mu}(s)\, ds\right\}_{t\in[0,T]} = \left\{\left(\int_0^t \mu_1(s)\, ds, \ldots, \int_0^t \mu_n(s)\, ds\right)\right\}_{t\in[0,T]},$$

where $\boldsymbol{\mu} : [0,T] \to \mathbb{R}^n$ have measurable and adapted components with $\int_0^T |\mu_i(t)|\, dt < \infty$ for $i = 1, \ldots, n$, we obtain an $\mathbb{R}^n$-valued Itô process

$$\{\mathbf{X}(t)\}_{t\in[0,T]} = \left\{\mathbf{X}(0) + \int_0^t \boldsymbol{\mu}(s)\, ds + \int_0^t \boldsymbol{\sigma}\, d\mathbf{B}\right\}_{t\in[0,T]}.$$

The components of this process are Itô processes in themselves and these components will have covariation

$$\{[X_i, X_j](t)\}_{t\in[0,T]} = \left\{\int_0^t (\boldsymbol{\sigma}\,\boldsymbol{\sigma}^{\mathrm{T}})_{ij}(t)\, dt\right\}_{t\in[0,T]} \quad \text{for } i,j \in \{1,\ldots,n\}.$$

Problems

Problem 4.1: Let $X(t) = 2I_{[0,1]}(t) + 3I_{(1,3]}(t) - 5I_{(3,4]}(t)$. Give the Itô integral $\int_0^4 X(t)dB(t)$ as a sum of random variables, and give its distribution, mean and variance. Show that the process $M(t) = \int_0^t X(s)dB(s)$, $0 \le t \le 4$, is a Gaussian process and give its covariance function.

Solution. Since X is a simple process, the Ito integral, by definition, is a sum:

$$\int_0^4 X(t)dB(t) = 2(B(1) - B(0)) + 3(B(3) - B(1)) - 5(B(4) - B(3)).$$

Since the increments of BM are independent, normal mean zero and variance being the length of the interval, the distribution of this sum is that of

$$2N(0,1) + 3N(0,2) - 5N(0,1) = N(0, 4 + 18 + 25) = N(0, 47).$$

Next, we give $M(t)$.

When $0 \le t \le 1$, $M(t) = 2B(t)$;
when $1 < t \le 3$, $M(t) = 2B(1) + 3(B(t) - B(1)) = 3B(t) - B(1)$;
when $3 < t \le 4$, $M(t) = 3B(3) - B(1) - 5(B(t) - B(3)) = -5B(t) + 8B(3) - B(1)$.

We can also write the same in one line as

$$M(t) = 2B(t)I_{[0,1]}(t) + (3B(t) - B(1))I_{(1,3]}(t) \\ + (-5B(t) + 8B(3) - B(1))I_{(3,4]}(t).$$

A proof of the Gaussian process: by the characterisation property of Gaussian vectors, Problem 2.28, we have to show that $\sum_{i=1}^n a_i \int_0^{t_i} X(u)dB(u)$ is normal for each choice of real $a_1, \ldots, a_n$ and (say) $0 \equiv t_0 < t_1 < \cdots < t_n$. The sum equals $\sum_{i=1}^n (\sum_{j=i}^n a_j) \int_{t_{i-1}}^{t_i} X(u)dB(u)$, where the integrals are independent normal so that sum is normal.

Alternatively, we can see that the process $M(t)$ has independent Gaussian increments and is therefore Gaussian.

Problem 4.2: Give values of α for which the following process is defined $Y(t) = \int_0^t (t-s)^{-\alpha} dB(s)$. (This process is used in the definition of the so-called fractional Brownian motion.)

Solution. Let t be fixed (say $t = 1$). The Itô integral $\int_0^t X(s)dB(s)$ is defined if $\int_0^t X^2(s)ds < \infty$ with probability one. When $X(s)$ is non-random, it is the same as $\int_0^t X^2(s)ds < \infty$. Here,

$$\int_0^t (t-s)^{-2\alpha} ds = \frac{t^{1-2\alpha}}{1-2\alpha} < \infty, \text{ for any } \alpha < \frac{1}{2},$$
$$= \infty, \text{ for any } \alpha > \frac{1}{2}$$

$$\text{and when } \alpha = \frac{1}{2}, \quad \int_0^t (t-s)^{-1} ds = -\log(t-s)|_0^t = \infty.$$

Thus, $Y(t)$ is defined (as a random variable) if and only if $\alpha < \frac{1}{2}$.

Allowing now t to vary from 0 to T, we see that for $\alpha < \frac{1}{2}$ the process $Y(t)$ is defined. Note that this process does not have the familiar properties of the integral because the integrand $X(s)$ depends also on t. This is the same as in the usual integral, e.g. $\int_0^t (t-s)^{-2\alpha} ds$: it does not satisfy the additivity property, $\int_0^a + \int_a^b = \int_0^{a+b}$. Consequently, while the Itô integral process $Y(t)$ is well defined, it is not a martingale.

Problem 4.3: State which of the following Itô integrals exist, and calculate its mean and variance:

(a) $\int_0^T sign(B_t)dB_t$.

(b) $\int_0^T \sin(B_t + 1)dB_t$.

(c) $\int_0^T \sin(B_{t+1})dB_t$.

(d) $\int_0^T e^{B_t^2} dB_t$.

Solution.

(a) $sign(B_t)$ is adapted ($\mathcal{F}_t$-measurable because it is a function of B_t, which is $\mathcal{F}_t$-measurable). $sign(x)$ is a càdlàg function (has left and right limits). Therefore, $sign(B_t)$ is an adapted regular process. $\int_0^T sign^2(B_t)dt = \int_0^T 1dt = T < \infty$. Hence, Itô integral exists, it has zero mean and variance T.

(b) $\sin(B_t + 1)$ is adapted, as a function of B_t, and is a continuous and bounded function $\sin(x + 1)$. Hence, Itô integral exists, it has zero mean and variance $\int_0^T \mathrm{E}\sin^2(B_t+1)dt$. (This is possible to compute using the complex exponential but we need not do it here).

(c) $\int_0^T \sin(B_{t+1})dB_t$. Since B_{t+1} is not adapted (not $\mathcal{F}_t$-measurable), the Itô integral is not defined.

(d) $\int_0^T e^{B_t^2} dB_t$. Since e^{x^2} is a continuous function, the Ito integral is defined. Since $\mathrm{E}e^{2B_t^2} = \infty$ for $t \geq 1/4$, but finite for $t < 1/4$, the Itô integral has mean zero and finite variance for $t < 1/4$, but for $t \geq 1/4$ it may not have finite mean. In fact, it is possible to show by using inequalities that it does not.

Problem 4.4: We have seen that $\int_0^T B_t dB_t = B_T^2/2 - T/2$. Calculate the variance of both sides and show that they are equal. You can use $\mathrm{E}Z^4 = 3$, for $Z = N(0, 1)$.

Solution. By isometry, $\mathrm{E}(\int_0^T B_t dB_t)^2 = \int_0^T \mathrm{E}B_t^2 dt = \int_0^T t dt = T^2/2$. $Var(B_T^2/2 - T/2) = Var(B_T^2/2)$. In distribution, we have, $B_T = \sqrt{T}Z$; therefore, $Var(B_T^2/2) = Var(TZ^2)/4 = (T^2/4)Var(Z^2) = (T^2/4)(\mathrm{E}Z^4 - (\mathrm{E}Z^2)^2) = (T^2/4)(3 - 1) = T^2/2$.

Problem 4.5: Let X_t and Y_t be simple deterministic processes.

(a) Show that $\mathrm{E}(\int_0^T X_t dB_t \int_0^T Y_t dB_t) = \int_0^T X_t Y_t dt$.

(b) Recover the covariance function of Brownian motion from the above result. (Use $B_u = \int_0^T I_{[0,u]}(t)dB_t$.)

Solution.

(a) By taking the partition obtained from both partitions for X and Y, we obtain $X_t = \sum_{i=1}^n x_i I_{(t_i,t_{i+1}]}(t)$, $Y_t = \sum_{i=1}^n y_i I_{(t_i,t_{i+1}]}(t)$. Then, $\int_0^T X_t dB_t = \sum_{i=1}^n x_i(B(t_{i+1}) - B(t_i))$, and $\int_0^T Y_t dB_t = \sum_{i=1}^n y_i(B(t_{i+1}) - B(t_i))$. Hence,

$$\int_0^T X_t dB_t \int_0^T Y_t dB_t = \sum_{i=1}^n \sum_{j=1}^n x_i(B(t_{i+1}) - B(t_i)) \times y_j(B(t_{j+1}) - B(t_j)).$$

Taking expectations, we can see that the terms with $i \neq j$ are zero, $\mathrm{E}(B(t_{i+1}) - B(t_i))(B(t_{j+1}) - B(t_j)) = 0$, and

$$\mathrm{E}\Big(\int_0^T X_t dB_t \int_0^T Y_t dB_t\Big) = \sum_{i=1}^n x_i y_i (t_{i+1} - t_i) = \int_0^T X_t Y_t dt.$$

The proof of a more general result uses isometry.

For any two random variables U, V,

$$\mathrm{E}(UV) = \frac{1}{2}(\mathrm{E}(U+V)^2 - \mathrm{E}U^2 - \mathrm{E}V^2).$$

Letting $U = \int_0^T X_t dB_t$ and $V = \int_0^T Y_t dB_t$, we have

$$\begin{aligned}
&\mathrm{E}\left(\int_0^T X_t dB_t\right)\left(\int_0^T Y_t dB_t\right) \\
&= \frac{1}{2}\left(\mathrm{E}\left(\int_0^T (X_t + Y_t) dB_t\right)^2 \right.\\
&\left. - \mathrm{E}\left(\int_0^T X_t dB_t\right)^2 - \mathrm{E}\left(\int_0^T Y_t dB_t\right)^2\right).
\end{aligned}$$

Using isometry,

$$\mathrm{E}\left(\int_0^T X_t dB_t\right)\left(\int_0^T Y_t dB_t\right) = \int_0^T \mathrm{E}(X_t Y_t) dt.$$

(b) Recover the covariance function of Brownian motion from the above result. (Use $B_u = \int_0^T I_{[0,u]}(t) dB_t$.)

Let $u < v$.

$$\begin{aligned}
\mathrm{E}(B_u B_v) &= \mathrm{E}\int_0^T I_{[0,u]}(t)dB_t \int_0^T I_{[0,v]}(t)dB_t \\
&= \int_0^T I_{[0,u]}(t)I_{[0,v]}(t)dt = \int_0^T I^2_{[0,u]}(t)dt \\
&= \int_0^T I_{[0,u]}(t)dt = \int_0^u dt = u.
\end{aligned}$$

Problem 4.6: Show that if X is a simple adapted and bounded adapted process, then $\int_0^t X(s)dB(s)$ is continuous.

Solution. A simple bounded adapted process can be written as a sum:

$$X(t) = \xi_0 I_{\{0\}}(t) + \sum_{i=0}^{n-1} \xi_i I_{(t_i,t_{i+1}]}(t),$$

where $\xi_0, \xi_1, \ldots, \xi_{n-1}$ are random variables, and each ξ_i is $\mathcal{F}_{t_i}$ measurable and bounded, in particular $\mathrm{E}(\xi_i^2) < \infty$. Therefore, the Itô integral is well defined and can be written as

$$\int_0^t X(s)dB(s) = \int_0^T X(s)I_{(0,t]}(s)dB(s) = \sum_{i=0}^{n-1} \xi_i(B(t_{i+1} \wedge t) - B(t_i \wedge t)),$$

where $t_i \wedge t = \min(t_i, t)$, and is therefore continuous, since all the terms are continuous.

Problem 4.7: Let X_n be a Gaussian sequence convergent in distribution to X. Show that the distribution of X is either normal or degenerate. Deduce that if $\mathrm{E}X_n \to \mu$ and $Var(X_n) \to \sigma^2 > 0$, then the limit is $N(\mu, \sigma^2)$. Since convergence in probability implies convergence in distribution, deduce convergence of the Itô integrals of simple non-random processes to a Gaussian limit.

Solution. The moment-generating function for each X_n in the Gaussian sequence is $e^{\mu_n t + \sigma_n^2 t^2/2}$. Convergence in distribution is equivalent to $\lim_{n\to\infty} e^{\mu_n t+\sigma_n^2 t^2/2} = m(t)$, where $m(t)$ is the moment-generating function of X. This implies that the quadratic polynomials $\mu_n t + \sigma_n^2 t^2/2$ converge for all t. Taking $t = \pm 1$, we have that $\mu_n + \sigma_n^2/2$ and $-\mu_n + \sigma_n^2/2$ converge. Adding them, we obtain σ_n^2 converges, hence both sequences converge $\mu_n \to \mu$ and $\sigma_n^2 \to \sigma^2 \geq 0$.

If $\sigma^2 = 0$, then $m(t) = e^{\mu t}$, and the limit is a constant $X = \mu$.
If $\sigma^2 > 0$, then $m(t) = e^{\mu t + \sigma^2 t^2/2}$ and the limit is $N(\mu, \sigma^2)$.

An Itô integral of a non-random function is a limit of approximating Itô integrals of simple non-random functions $X_n(t)$. Since $X_n(t)$ takes finitely many non-random values, $\int_0^T X_n(t)dB(t)$ has a normal distribution with mean zero and variance $\int_0^T X_n^2(t)dt$ (as a linear combination of a multivariate normal). It now follows by the first statement that the Itô integral, as a limit, $\int_0^T X(t)dB(t)$ has normal distribution with mean zero and variance $\int_0^T X^2(t)dt$.

Problem 4.8: Let X be a non-random (deterministic) continuous process (function).

(a) Using its martingale exponential, show that $\int_0^t X(s)dB(s)$ is a Gaussian random variable. See Problem 4.7.
(b) Deduce that in fact, $\int_0^\cdot X(s)dB(s)$ is a Gaussian process.

Solution.

(a) We know that

$$\mathcal{E}\left(u\int_0^\cdot X(s)dB(s)\right)(t) = \exp\left(u\int_0^t X(s)dB(s) - \frac{u^2}{2}\int_0^t X(s)^2 ds\right)$$

is a martingale for $t \in [0, T]$ as soon as, for example, Novikov's condition (see Chapter 8) is satisfied:

$$\mathrm{E}\left(\exp\left(\frac{u^2}{2}\int_0^T X(t)^2 dt\right)\right) = \exp\left(\frac{u^2}{2}\int_0^T X(t)^2 dt\right) < +\infty.$$

As X is continuous and therefore bounded on $[0, T]$, the above is clearly satisfied and the martingale exponential is a true martingale. It follows that

$$1 = \mathrm{E}\left(\exp\left(u\int_0^t X(s)dB(s) - \frac{u^2}{2}\int_0^t X(s)^2 ds\right)\right)$$

or equivalently that

$$\mathrm{E}\left(\exp\left(u\int_0^t X(s)dB(s)\right)\right) = \exp\left(\frac{u^2}{2}\int_0^t X(s)^2 ds\right),$$

from which we immediately deduce that $\int_0^t X(s)dB(s)$ is normal (with mean zero and variance $\int_0^t X(s)^2 ds$).

(b) To show that $\int_0^t X(s)dB(s)$ is a Gaussian process, we show that for any n, any $t_1 < t_2 < \cdots < t_n$ and any $u_1, u_2, \ldots, u_n$,

$$u_1 \int_0^{t_1} X(s)dB(s) + u_2 \int_0^{t_2} X(s)dB(s) + \cdots$$
$$+ u_n \int_0^{t_n} X(s)dB(s) \qquad (*)$$

is a Gaussian random variable. However, $(*)$ can be rewritten as

$$\int_0^{t_n} (u_1 1_{[0,t_1]}(s) + u_2 1_{[0,t_2]}(s) + \cdots + u_{n-1} 1_{[0,t_{n-1}]}(s) + u_n)$$
$$\times X(s)dB(s),$$

which is nothing but another stochastic integral of the form $\int_0^t X(s)dB(s)$. The result follows from (a).

Problem 4.9: Let $X(t)$ be a deterministic function, with $\int_0^T X^2(t)dt < \infty$. Show that the process $Y(t) = \int_0^t X(s)dB(s)$ is a Gaussian process with independent increments.

Remark: This property is inherited from the Itô integrals of simple deterministic processes (by taking limits).

Solution. It should be (intuitively) clear that the Itô integrals of deterministic functions over disjoint intervals $[a, b]$ and $[c, d]$ are independent due to independence of increments of Brownian motion. This is obvious for simple adapted processes and then by taking limits. This implies that the process $Y(t)$ has independent increments. A more formal proof is as follows.

To prove the process $Y(t)$ is Gaussian, we have to show (see Problem 2.28) that $\sum_{i=1}^n a_i Y(t_i) = \sum_{i=1}^n a_i \int_0^{t_i} X(s)dB(s)$ is normal for each choice of real $a_1, \ldots, a_n$ and $0 \equiv t_0 < t_1 < \cdots < t_n = T$. The sum equals

$$\sum_{i=1}^n \int_{t_{i-1}}^{t_i} \sum_{j=i}^n a_j X(s)dB(s) = \int_0^T \tilde{X}(s)dB(s),$$

where $\tilde{X}(s) = b_i X(s) 1_{(t_{i-1},t_i]}(s)$, with $b_i = \sum_{j=i}^n a_j$. Since $\tilde{X}$ is a deterministic process, the previous Problem 4.7 $\int_0^T \tilde{X}(s)dB(s)$ has a normal distribution. Thus, $Y(t)$ is a Gaussian process. To see that its increments are independent, calculate covariance for disjoint intervals $[a,b]$ and $[c,d]$:

$$\text{Cov}\left(\int_a^b X(s)dB(s), \int_c^d X(s)dB(s)\right)$$
$$= \int_0^T X^2(s)1_{(a,b]}(s)1_{(c,d]}(s)ds = 0.$$

The increments have a joint bivariate normal distribution since Y is a Gaussian process, and zero covariance implies independence.

Problem 4.10: Show that if the function $X(t,s)$, $0 \le t,s \le T$, is non-random and satisfy $\int_0^t X^2(t,s)ds < \infty$, $0 \le t \le T$, then $Y(t) := \int_0^t X(t,s)dB(s)$ is a Gaussian random variable. The collection $Y(t)$, $0 \le t \le T$, is a Gaussian process with zero mean and covariance function for $u \ge 0$ given by $\text{Cov}(Y(t), Y(t+u)) = \int_0^t X(t,s)X(t+u,s)ds$.

Solution. For fixed t, the distribution of $Y(t)$, as an Itô integral of a non-random function, is Gaussian with mean zero and variance $\int_0^t X^2(t,s)ds$ by Theorem 4.11.

To prove the process is Gaussian, we have to show (see Problem 2.28) that $\sum_{i=1}^n a_i \int_0^{t_i} X(t_i,s)dB(s)$ is normal for each choice of real $a_1,\ldots,a_n$ and (say) $0 \equiv t_0 < t_1 < \cdots < t_n$. The sum equals $\sum_{i=1}^n \int_{t_{i-1}}^{t_i} \sum_{j=i}^n a_j X(t_j,s)dB(s)$, where the integrals are normal and independent (according to Problem 4.9) and therefore the sum is normal.

For the covariance function, we first observe that, for $u \ge 0$,

$$Y(t+u) = \int_0^t X(t+u,s)dB(s) + \int_t^{t+u} X(t+u,s)dB(s).$$

Since $X(t+u,s)$ is non-random, the second integral $\int_t^{t+u} X(t+u,s)dB(s)$ is independent of $\mathcal{F}_t$. Therefore,

$$\text{Cov}(Y(t), Y(t+u)) = \text{Cov}\left(\int_0^t X(t,s)dB(s), \int_0^t X(t+u,s)dB(s)\right)$$
$$= \text{E}\left(\int_0^t X(t,s)dB(s) \int_0^t X(t+u,s)dB(s)\right)$$
$$= \int_0^t X(t,s)X(t+u,s)ds,$$

where the last equality follows from Theorem 4.5 as the expectation of the product of two Itô integrals.

Problem 4.11: Show that a Gaussian martingale on a finite time interval $[0, T]$ is a square-integrable martingale with independent increments. Deduce that if X is non-random and $\int_0^t X^2(s)ds < \infty$ then $Y(t) = \int_0^t X(s)dB(s)$ is a Gaussian square-integrable martingale with independent increments.

Solution. Recall that by square integrable, we mean that the second moments are bounded. Clearly, if $M(t)$ is Gaussian on $[0, T]$, $\mathrm{E}M^2(t) < \infty$ for any $t \leq T$.

By Jensen's inequality for conditional expectation with $g(x) = x^2$,

$$\mathrm{E}(M^2(T)|\mathcal{F}_t) \geq (\mathrm{E}(M(T)|\mathcal{F}_t))^2 = M^2(t).$$

Therefore, $M(t)$ is a square-integrable martingale.

The covariance between $M(s)$ and $M(t) - M(s)$ for $s < t$ is zero because by the martingale property, $\mathrm{E}(M(t) - M(s)|\mathcal{F}_s) = 0$ and

$$\begin{aligned}\mathrm{E}(M(s)(M(t) - M(s)) &= \mathrm{E}\mathrm{E}\left(M(s)(M(t) - M(s))|\mathcal{F}_s\right) \\ &= \mathrm{E}\left(M(s)\mathrm{E}(M(t) - M(s)|\mathcal{F}_s)\right) = 0.\end{aligned}$$

Now, if jointly Gaussian variables are uncorrelated, they are independent. Thus, the increment $M(t) - M(s)$ is independent of $M(s)$.

Next, consider $Y(t) = \int_0^t X(s)dB(s)$. Since $X(s)$ is non-random, $\mathrm{E}\int_0^T X^2(s)ds = \int_0^T X^2(s)ds < \infty$, as given in the question. Thus, the Itô integral $Y(t)$ is a martingale. It is also Gaussian, as shown in Problem 4.10 above. Thus, it is a square-integrable martingale with independent increments.

Problem 4.12: Obtain the alternative relation for the quadratic variation of Itô processes, Equation 4.62: $[X, X](t) = X^2(t) - X^2(0) - 2\int_0^t X(s)dX(s)$, by applying Itô's formula to $X^2(t)$.

Solution. Take $f(x) = x^2$, so $f'(x) = 2x$ and $f''(x) = 2$.

Substitute in Itô's formula to get

$$dX^2(t) = 2X(t)dX(t) + d[X, X](t).$$

Writing this in the integral form, we obtain the result.

Problem 4.13: $X(t)$ has a stochastic differential with $\mu(x) = bx + c$ and $\sigma^2(x) = 4x$. Assuming $X(t) > 0$, find the stochastic differential for the process $Y(t) = \sqrt{X(t)}$.

Solution. Let $f(x) = \sqrt{x}$, so $f'(x) = \dfrac{1}{2\sqrt{x}}$ and $f''(x) = -\dfrac{1}{4}x^{-3/2}$. Itô's formula gives

$$df(X(t)) = \frac{1}{2\sqrt{X(t)}}dX(t) - \frac{1}{8X(t)\sqrt{X(t)}}\sigma^2(X(t))dt.$$

Substitute $Y(t) = \sqrt{X(t)}$ and the stochastic differential $dX(t) = (bX(t) + c)dt + \sqrt{4X(t)}dB(t)$ in the equation above to obtain

$$dY(t) = \frac{1}{2}\left[bY(t) + \frac{c-1}{Y(t)}\right]dt + dB(t).$$

Problem 4.14: A process $X(t)$ on $(0, 1)$ has a stochastic differential with coefficient $\sigma(x) = x(1 - x)$. Assuming $0 < X(t) < 1$, show that the process defined by $Y(t) = \ln(X(t)/(1 - X(t)))$ has a constant diffusion coefficient.

Solution. Let $f(x) = \ln\left(\dfrac{x}{1-x}\right)$, so $f'(x) = \dfrac{1}{x(1-x)}$ and $f''(x) = \dfrac{2x-1}{x^2(1-x)^2}$. Itô's formula gives

$$df(X(t)) = \frac{1}{X(t)(1-X(t))}dX(t) + \frac{2X(t)-1}{2X^2(t)(1-X(t))^2}\sigma^2(X(t))dt.$$

Substitute $Y(t)$ and $dX(t) = \mu(X(t))dt + X(t)(1-X(t))dB(t)$ to obtain

$$dY(t) = \left(\frac{\mu(X(t))}{X(t)(1-X(t))} + X(t) - \frac{1}{2}\right)dt + dB(t).$$

Hence, the diffusion coefficient of the process $Y(t)$ is a constant $(= 1)$.

Problem 4.15: $X(t) > 0$ has a stochastic differential with $\mu(x) = cx$ and $\sigma^2(x) = x^a$, $c > 0$ and $a > -1$. Let $Y(t) = X(t)^b$. What choice of b will give a constant diffusion coefficient for Y?

Solution. Let $f(x) = x^b$, so $f'(x) = bx^{b-1}$ and $f''(x) = b(b-1)x^{b-2}$. Itô's formula:

$$df(X(t)) = bX(t)^{b-1}dX(t) + \tfrac{1}{2}b(b-1)X(t)^{b-2}\sigma^2(X(t))dt.$$

Substitute $dX(t) = cX(t)dt + X(t)^{\frac{a}{2}}dB(t)$ and $Y(t) = X(t)^b$:

$$dY(t) = (\dots)dt + bX(t)^{b-1+\frac{a}{2}}dB(t).$$

Hence, the diffusion coefficient is a constant if $b = 0$ or if $b = 1 - \frac{a}{2}$.

Problem 4.16: Let $X(t) = tB(t)$ and $Y(t) = e^{B(t)}$. Find $d\left(\frac{X(t)}{Y(t)}\right)$.

Solution. Let $f(x,t) = txe^{-x}$, so $f_x(x,t) = te^{-x}(1-x)$, $f_{xx}(x,t) = te^{-x}(x-2)$, and $f_t(x,t) = xe^{-x}$.

Then, take $x = B(t)$ and note $\dfrac{X(t)}{Y(t)} = f(B(t),t)$.

Use Itô's formula for a function of two variables to obtain

$$df(B(t),t) = te^{-B(t)}(1 - B(t))dB(t) + \frac{1}{2}te^{-B(t)}(B(t)-2)dt + B(t)e^{-B(t)}dt,$$

$$d(tB(t)e^{-B(t)}) = e^{-B(t)}(B(t) + \frac{1}{2}tB(t) - t)dt + te^{-B(t)}(1 - B(t))dB(t).$$

Problem 4.17: Obtain the differential of a ratio formula $d\left(\frac{X(t)}{Y(t)}\right)$ by taking $f(x,y) = \frac{x}{y}$. Assume that the process Y stays away from 0.

Solution. Take $f(x,y) = \dfrac{x}{y}$, so $f_x(x,y) = \dfrac{1}{y}$, $f_{xx}(x,y) = 0$, $f_y(x,y) = -\dfrac{x}{y^2}$, $f_{yy}(x,y) = \dfrac{2x}{y^3}$ and $f_{xy}(x,y) = -\dfrac{1}{y^2}$.

Itô's formula for a function of two variables gives

$$d\left(\frac{X(t)}{Y(t)}\right) = \frac{1}{Y(t)}dX(t) - \frac{X(t)}{Y^2(t)}dY(t) + \frac{X(t)}{Y^3(t)}d[Y,Y](t) - \frac{1}{Y^2(t)}d[X,Y](t).$$

Problem 4.18: Find $d(M^2(t))$, where $M(t) = e^{B(t)-\frac{t}{2}}$.

Solution. Take $f(x,t) = e^{2x-t}$, so $f_x = 2e^{2x-t}$, $f_{xx} = 4e^{2x-t}$ and $f_t = -e^{2x-t}$. Note that $M^2(t) = f(B(t),t)$ and use Itô's formula to obtain

$$\begin{aligned} d(M^2(t)) &= 2e^{2B(t)-t}dB(t) + \left(2e^{2B(t)-t} - e^{2B(t)-t}\right)dt \\ &= e^{2B(t)-t}dt + e^{2B(t)-t}dB(t). \end{aligned}$$

Problem 4.19: Let $M(t) = B^3(t) - 3tB(t)$. Show that M is a martingale, first directly and then by using the Itô integrals.

Solution. By definition, a stochastic process $\{M(t), t \geq 0\}$ is a martingale if for any t, it is integrable: $\mathrm{E}|M(t)| < \infty$, and for any $s > 0$, it satisfies the martingale property: $\mathrm{E}(M(t+s)|\mathcal{F}_t) = M(t)$ a.s.

$M(t)$ is integrable for any t:

$$\mathrm{E}|M(t)| = \mathrm{E}|B^3(t) - 3tB(t)| \leq \mathrm{E}|B^3(t)| + 3t|B(t)| < \infty,$$

since all moments of a normal distribution are finite.

Use the expansion $(a+b)^3 = a^3 + 3a^2b + 3ab^2 + b^3$ with decomposition $B(t+s) = B(t) + (B(t+s) - B(t))$. Take $a = B(t)$, $b = B(t+s) - B(t)$, and use the facts that $\mathrm{E}(B(t+s) - B(t))^3 = 0$, and $\mathrm{E}(B(t+s) - B(t))^2 = s$, $\mathrm{E}(B(t+s) - B(t)) = 0$, and that $B(t)$ is $\mathcal{F}_t$-measurable and $B(t+s) - B(t)$ is independent of $\mathcal{F}_t$, to get the martingale property.

Alternatively, take $f(x,t) = x^3 - 3xt$, note $M(t) = f(B(t),t)$, and use Itô's formula to obtain

$$\begin{aligned} dM(t) &= 3B^2(t)dB(t) + \frac{1}{2}6B(t)dt - 3B(t)dt - 3tdB(t) \\ &= (3B^2(t) - 3t)dB(t). \end{aligned}$$

Thus, $M(t) = \int_0^t (3B^2(s) - 3s)dB(s)$ is an Itô integral process.

Since $\int_0^T \mathrm{E}(3B^2(t) - 3t)^2 dt < \infty$, the martingale property of the Itô integral holds, and $M(t)$ is a martingale on $[0,T]$ for any T.

Problem 4.20: Show that $M(t) = e^{t/2}\sin(B(t))$ is a martingale by using Itô's formula.

Solution. Take $f(x,t) = e^{t/2}\sin(x)$, so $f_x(x,t) = e^{t/2}\cos(x)$, $f_{xx}(x,t) = -e^{t/2}\sin(x)$ and $f_t(x,t) = \frac{1}{2}e^{t/2}\sin(x)$. Itô's formula gives

$$df(B(t),t) = e^{\frac{t}{2}}\cos(B(t))dB(t) - \frac{1}{2}e^{\frac{t}{2}}\sin(B(t))dt + \frac{1}{2}e^{\frac{t}{2}}\sin(B(t))dt.$$

Thus, $M(t) = f(B(t),t) = \int_0^t e^{s/2}\cos(B(s))dB(s)$ is an Itô integral process. Clearly,

$$\int_0^T \mathrm{E}(e^{s/2}\cos(B(s)))^2 ds < \infty,$$

and the martingale property follows from that of Itô integral.

More advanced knowledge of martingale exponential also yields this martingale from $e^{t/2}\sin(B(t)) = \frac{1}{2i}(\mathcal{E}(iB)(t) - \mathcal{E}(-iB)(t))$.

Problem 4.21: For a function of n variables and n-dimensional Brownian motion, write Itô's formula for $f(B_1(t),\ldots,B_n(t))$ by using the gradient notation $\nabla f = (\frac{\partial}{\partial x_1},\ldots,\frac{\partial}{\partial x_n})$.

Solution. By Itô's formula,

$$df(B_1(t),\ldots,B_n(t)) = \sum_i^n \frac{\partial f}{\partial x_i}(\mathbf{B}(t))dB_i(t) + \frac{1}{2}\sum_i^n \frac{\partial^2 f}{\partial x_i^2}(\mathbf{B}(t))dt$$
$$= \nabla f(\mathbf{B}(t))\cdot d\mathbf{B}(t) + \frac{1}{2}\nabla\cdot\nabla f(\mathbf{B}(t))dt,$$

where "$\cdot$" is the inner product of vectors in $\mathbb{R}^n$ and $\mathbf{B}(t) = (B_1(t),\ldots,B_n(t))$ with $d\mathbf{B}(t) = (dB_1(t),\ldots,dB_n(t))$. Note that $B_i(t)$'s are independent, so all the mixed terms $\frac{\partial^2 f}{\partial x_i \partial x_j}dB_i(t)dB_j(t)$ are zero.

The operator $\nabla\cdot\nabla = \Delta = \sum_i \frac{\partial^2}{\partial x_i^2}$ is the Laplacian, so that Itô's formula becomes

$$df(B(t)) = \frac{1}{2}\Delta f(B(t))dt + \nabla f\cdot dB.$$

Problem 4.22: $\Phi(x)$ is the standard normal distribution function. Show that for a fixed $T > 0$, the process $\Phi(\frac{B(t)}{\sqrt{T-t}})$, $0 \le t \le T$ is a martingale.

Solution. Denote $\phi(x) = \Phi'(x)$. Then,

$$\frac{\partial}{\partial x}\Phi\left(\frac{x}{\sqrt{T-t}}\right) = \phi\left(\frac{x}{\sqrt{T-t}}\right)\frac{1}{\sqrt{T-t}},$$

and

$$\frac{\partial^2}{\partial x^2}\Phi\left(\frac{x}{\sqrt{T-t}}\right) = \phi'\left(\frac{x}{\sqrt{T-t}}\right)\frac{1}{T-t} = -\frac{x}{(T-t)^{3/2}}\phi\left(\frac{x}{\sqrt{T-t}}\right),$$

by $\phi'(x) = -x\phi(x)$. Further,

$$\frac{\partial}{\partial t}\Phi\left(\frac{x}{\sqrt{T-t}}\right) = -\frac{1}{2}\phi\left(\frac{x}{\sqrt{T-t}}\right)\frac{x}{(T-t)^{3/2}}.$$

Thus, by Itô's formula, for all $0 \le t < T$,

$$d\Phi\left(\frac{B(t)}{\sqrt{T-t}}\right) = \phi\left(\frac{B(t)}{\sqrt{T-t}}\right)\frac{1}{\sqrt{T-t}}dB(t),$$

and

$$\Phi\left(\frac{B(t)}{\sqrt{T-t}}\right) = \frac{1}{2} + \int_0^t \phi\left(\frac{B(s)}{\sqrt{T-s}}\right) \frac{1}{\sqrt{T-s}} dB(s).$$

Since $\phi\left(\frac{B(s)}{\sqrt{T-s}}\right)\frac{1}{\sqrt{T-s}} \leq \frac{1}{\sqrt{T-s}}$, the Itô integral above is a martingale (see the martingale property of the Itô integral).

Thus, for all $t < T$ and $s < t$, $X(t) = \Phi\left(\frac{B(t)}{\sqrt{T-t}}\right)$ satisfies the martingale property:

$$E(X(t)|\mathcal{F}_s) = X(s).$$

Next, as $t \to T$, $X(t) \to Y = I(B(T) > 0) + \frac{1}{2}I(B(T) = 0) = I(B(T) > 0)$ a.s.

The martingale property holds also for $t = T$ by dominated convergence. We shall see an alternative derivation by identifying this martingale with a Doob–Lévy martingale $\mathrm{E}(I(B_T > 0)|\mathcal{F}_t)$, Example 7.5.

Problem 4.23: Let $X(t) = (1-t)\int_0^t \frac{dB(s)}{1-s}$, where $0 \leq t < 1$. Find $dX(t)$.

Solution. Let $Y(t) = \int_0^t \frac{dB(s)}{1-s}$, so $dY(t) = \frac{dB(t)}{1-t}$ and $X(t) = (1-t)Y(t)$. Take $f(y,t) = (1-t)y$, so $f_y(y,t) = (1-t)$, $f_{yy}(y,t) = 0$ and $f_t(y,t) = -y$. Use Itô's formula and note $X = f(Y(t),t)$ to obtain

$$df(Y(t),t) = (1-t)dY(t) - Y(t)dt,$$
$$dX(t) = dB(t) - \frac{X(t)}{1-t}dt.$$

An alternative solution which is more or less by inspection involves using integration by parts.

Problem 4.24: Let $X(t) = tB(t)$. Find its quadratic variation $[X,X](t)$.

Solution. Use Itô's formula with $f(x,t) = tx$, so $f_x = t, f_{xx} = 0$, and $f_t = x$.

Thus, $dX(t) = tdB(t) + B(t)dt$, and $d[X,X](t) = (dX(t))^2 = t^2dt$:

$$[X,X](t) = \int_0^t d[X,X](s) = \int_0^t s^2 ds = t^3/3.$$

Problem 4.25: (Gaussian fake Brownian motions.) A family of processes (indexed by r) are all Gaussian processes with Brownian marginals

$N(0,t)$ but not a Brownian motion unless $r=0$. Let $r>-\frac{1}{2}$, $X_0=0$, and for $t>0$,

$$X_t = \sqrt{2r+1}\, t^{-r} \int_0^t s^r dB_s.$$

(a) Show that $(X)_t$ is a Gaussian process with $N(0,t)$ marginals and give its covariance function.

(b) Show that for $t>0$, (X_t) solves the SDE

$$dX_t = -\frac{r}{t}X_t dt + \sqrt{2r+1}dB_t.$$

Solution.

(a) Since $\int_0^t s^{2r}ds = \frac{1}{2r+1}t^{2r+1} < \infty$ for $r > -\frac{1}{2}$, the Itô integral $\int_0^t s^r dB_s = Y_t$ is well defined for all such r. It has mean zero and variance $\frac{1}{2r+1}t^{2r+1}$. It follows that X_t has mean zero and variance

$$\mathrm{E}(X_t^2) = (2r+1)t^{-2r}\mathrm{E}(Y_t^2) = (2r+1)t^{-2r}\frac{1}{2r+1}t^{2r+1} = t.$$

The Itô integral Y_t is a Gaussian process as an Itô integral of a deterministic function. Since X_t is a product of a deterministic function and a Gaussian process, it is also a Gaussian process. This proves (a).

The covariance function $(u<t)$,

$$\begin{aligned}\mathrm{Cov}(X_u, X_t) &= (2r+1)u^{-r}t^{-r}\,\mathrm{Cov}(Y_u, Y_t)\\ &= (2r+1)u^{-r}t^{-r}\,\mathrm{Cov}(Y_u, Y_u) = u^{r+1}t^{-r}.\end{aligned}$$

(b) Use $X_t = \sqrt{2r+1}\; t^{-r}Y_t$ and $dY_t = t^r dB_t$ to derive the SDE for X_t. Using the integration by parts formula and taking into account that covariation between t^r and Y_t is zero,

$$\begin{aligned}dX_t &= \sqrt{2r+1}d(t^{-r}Y_t) = \sqrt{2r+1}t^{-r}dY_t + \sqrt{2r+1}(-r)t^{-r-1}Y_t dt\\ &= \sqrt{2r+1}dB_t - rt^{-1}X_t dt, \text{ since } Y_t = t^r X_t/\sqrt{2r+1}.\end{aligned}$$

This shows the SDE for X_t. This SDE is not defined at $t=0$ due to division of the drift by t. However, due to $\lim_{t\to 0} E(X_t^2) = 0$, the value at $t=0$ of X_t can be defined by continuity, $X_0 = 0$.

Problem 4.26: Let $X(t) = \int_0^t (t-s)dB(s)$. Find $dX(t)$ and its quadratic variation $[X,X](t)$. Compare with the quadratic variation of the Itô integrals.

Solution.

$$\begin{aligned} X(t) &= tB(t) - \int_0^t s dB(s) \\ &= \int_0^t s dB(s) + \int_0^t B(s) ds - \int_0^t s dB(s) \\ &= \int_0^t B(s) ds, \end{aligned}$$

where the second row can be readily obtained by substituting $d(tB(t))$, as obtained in Problem 4.24.

Thus, $X(t)$ is differentiable and of finite variation, so the quadratic variation is zero, $[X, X](t) = 0$.

Problem 4.27: Norbert Wiener defined the stochastic integral $\int_0^T g(t) dB_t$ for continuously differentiable functions g as

$$\int_0^T g(t) dB_t = g(T) B_T - \int_0^T B_t g'(t) dt.$$

Taking $g(t) = t$, show that the Itô integral $\int_0^1 t dB_t$ and the Riemann integral $\int_0^1 B_t dt$ have the same distribution.

Solution. The Itô integral has $N(0, 1/3)$ distribution. If we do Riemann sum approximation of the Riemann integral and show that it has $N(0, 1/3)$ distribution, then they both have $N(0, 1/3)$ distribution.

Alternatively, the following proves that both integrals have the same distribution and gives another way to establish the distribution of $\int_0^1 B_t dt$:

$\int_0^1 t dB_t = B_1 - \int_0^1 B_t dt = \int_0^1 (B_1 - B_t) dt$. But $\{B_1 - B_t\}_{t \geq 0}$ and $\{B_{1-t}\}_{t \geq 0}$ have same finite-dimensional distributions (same covariance function). Finally, by letting $s = 1 - t$, $\int_0^1 B_{1-t} dt = \int_0^1 B_s ds$.

Problem 4.28: Norbert Wiener defined the stochastic integral process $\{\int_0^t g\, dB\}_{t \geq 0}$ with respect to B for continuously differentiable functions $g : [0, \infty) \to \mathbb{R}$ as

$$\int_0^t g\, dB = g(t) B(t) - \int_0^t B\, dg = g(t) B(t) - \int_0^t B(r) g'(r)\, dr \quad \text{for } t \geq 0.$$

Find the covariance function and mean of $\{\int_0^t g\, dB\}_{t \geq 0}$ defined in this way by means of direct calculation (not using the Itô integral theory).

Solution. As mathematical expectation E commutes with integration $\int$ by the Fubini theorem, we have

$$\mathrm{E}\left\{g(t)B(t)-\int_0^t B(r)g'(r)\,dr\right\}$$

$$= g(t)\,\mathrm{E}\{B(t)\} - \int_0^t \mathrm{E}\{B(r)\}\,g'(r)\,dr = 0.$$

Assuming that $0 \le s \le t$, the same reasoning gives

$$\mathrm{Cov}\left\{g(s)B(s)-\int_0^s B(r_1)g'(r_1)\,dr_1, g(t)B(t)-\int_0^t B(r_2)g'(r_2)\,dr_2\right\}$$

$$= \mathrm{E}\left\{\left(g(s)B(s)-\int_0^s B(r_1)g'(r_1)\,dr_1\right)\left(g(t)B(t)-\int_0^t B(r_2)g'(r_2)\,dr_2\right)\right\}$$

$$= g(s)g(t)\,\mathrm{E}\{B(s)B(t)\} - g(s)\int_0^t \mathrm{E}\{B(s)B(r)\}\,g'(r)\,dr$$

$$- g(t)\int_0^s \mathrm{E}\{B(t)B(r)\}\,g'(r)\,dr$$

$$+ \int_0^s\int_0^t \mathrm{E}\{B(r_1)B(r_2)\}\,g'(r_1)g'(r_2)\,dr_1dr_2$$

$$= g(s)g(t)\,s - g(s)\int_0^s r\,g'(r)\,dr - g(s)\int_s^t s\,g'(r)\,dr - g(t)\int_0^s r\,g'(r)\,dr$$

$$+ 2\int_{r_1=0}^{r_1=s}\int_{r_2=r_1}^{r_2=s} r_1\,g'(r_1)g'(r_2)\,dr_1dr_2$$

$$+ \int_{r_1=0}^{r_1=s}\int_{r_2=s}^{r_2=t} r_1\,g'(r_1)g'(r_2)\,dr_1dr_2$$

$$= g(s)^2 s - (g(s)+g(t))\int_0^s r\,g'(r)\,dr + 2\,g(s)\int_0^s r\,g'(r)\,dr$$

$$- 2\int_0^s r\,g'(r)g(r)\,dr + (g(t)-g(s))\int_0^s r\,g'(r)\,dr$$

$$= \int_0^s g(r)^2\,dr,$$

so that the covariance function is

$$\mathrm{Cov}\left\{\int_0^s g\,dB, \int_0^t g\,dB\right\} = \int_0^{\min\{s,t\}} g(r)^2\,dr \quad \text{for } s,t \ge 0.$$

Problem 4.29: Show that convergence in $\mathbb{L}^p$ of random variables for $p \geq 1$ implies convergence in probability as well as that convergence in $\mathbb{L}^p$ of random variables implies convergence of moments of order up to $[p]$.

Solution. If $X_n \to X$ in $\mathbb{L}^p$, then Tjebysjev's inequality shows that

$$\mathbf{P}\{|X_n - X| \geq \varepsilon\} \leq \frac{\mathrm{E}\{|X_n - X|^p\}}{\varepsilon^p} \to 0 \quad \text{as } n \to \infty \text{ for } \varepsilon > 0.$$

Further, we have by Hölder's inequality followed by Jensen's inequality,

$$\begin{aligned}
\big|\mathrm{E}\{X_n^m\} - \mathrm{E}\{X^m\}\big| &= \left|\mathrm{E}\left\{\sum_{i=0}^{m-1}\binom{m}{i}(X_n - X)^{m-i}X^i\right\}\right| \\
&\leq \sum_{i=0}^{m-1}\binom{m}{i}\mathrm{E}\{|X_n - X|^{m-i}|X|^i\} \\
&\leq \sum_{i=0}^{m-1}\binom{m}{i}\big(\mathrm{E}\{|X_n - X|^m\}\big)^{(m-i)/m}\,\mathrm{E}\{|X|^m\}\big)^{i/m} \\
&\leq \sum_{i=0}^{m-1}\binom{m}{i}\big(\mathrm{E}\{|X_n - X|^p\}\big)^{(m-i)/p}\,\mathrm{E}\{|X|^p\}\big)^{i/p} \\
&\to 0 \quad \text{as } n \to \infty \text{ for } m \leq [p].
\end{aligned}$$

Problem 4.30: Show that if $X_n \to X$ in $\mathbb{L}^p$ for $p \geq 1$ as $n \to \infty$, then $\mathrm{E}\{X_n | \mathcal{G}\} \to \mathrm{E}\{X | \mathcal{G}\}$ in $\mathbb{L}^1$ as $n \to \infty$ for any σ-algebra $\mathcal{G} \subseteq \mathcal{F}$.

Solution. This follows using Properties 2.24 and 2.20 of conditional expectation in Klebaner's book together with Hölder's inequality as

$$\begin{aligned}
\mathrm{E}\{|\mathrm{E}\{X_n | \mathcal{G}\} - \mathrm{E}\{X | \mathcal{G}\}|\} &\leq \mathrm{E}\{\mathrm{E}\{|X_n - X| | \mathcal{G}\}\} \\
&= \mathrm{E}\{|X_n - X|\} \\
&\leq \big(\mathrm{E}\{|X_n - X|^p\}\big)^{1/p} \\
&\to 0 \quad \text{as } n \to \infty.
\end{aligned}$$

Problem 4.31: Show that the Itô integral process $\{\int_0^t X\, dB\}_{t \in [0,T]}$ for $X \in S_T$ is a martingale.

Solution. Pick a partition $0 = t_0 < t_1 < \cdots < t_n = T$ of the interval $[0, T]$ and consider an $X \in S_T$ given by

$$X(t) = I_{\{0\}}(t)\,\xi_0 + \sum_{i=0}^{n-1} I_{(t_i, t_{i+1}]}(t)\,\xi_i \quad \text{for } t \in [0, T],$$

where ξ_i is $\mathcal{F}_{t_i}$-measurable with $E(\xi_i^2) < \infty$ for $i = 0, \ldots, n-1$. Recall that

$$\int_0^t X\,dB$$
$$= \begin{cases} \displaystyle\sum_{i=0}^{m-1} \xi_i\,(B(t_{i+1}) - B(t_i)) + \xi_m\,(B(t) - B(t_m)) & \text{for } t \in (t_m, t_{m+1}], \\ 0 & \text{for } t = 0, \end{cases}$$

for $m = 0, \ldots, n-1$. As $\mathcal{F}_0^B = \{\emptyset, \Omega\}$ (as $B(0)$ is a constant), we have

$$\mathrm{E}\Big\{\int_0^t X\,dB \,\Big|\, \mathcal{F}_0^B\Big\} = \mathrm{E}\Big\{\int_0^t X\,dB\Big\} = 0 = \int_0^0 X\,dB,$$

by Equation (2.17*) together with Property 3 of the Itô integral process for S_T in Klebaner's book. Now, pick $0 < s \le t \le T$ together with integers $0 \le k \le m \le n-1$ such that $s \in (t_k, t_{k+1}]$ and $t \in (t_m, t_{m+1}]$. Then, we have

$$\mathrm{E}\Big\{\int_0^t X\,dB \,\Big|\, \mathcal{F}_s^B\Big\}$$
$$= \mathrm{E}\Big\{\sum_{i=0}^{m-1} \xi_i\,(B(t_{i+1}) - B(t_i)) + \xi_m\,(B(t) - B(t_m)) \,\Big|\, \mathcal{F}_s^B\Big\}$$
$$= \sum_{i=0}^{k-1} \xi_i\,(B(t_{i+1}) - B(t_i)) + \xi_k\,\mathrm{E}\big\{B(t_{k+1}) - B(t_k) \big| \mathcal{F}_s^B\big\}$$
$$+ \sum_{i=k+1}^{m-1} \mathrm{E}\big\{\xi_i\,\mathrm{E}\{B(t_{i+1}) - B(t_i) \,|\, \mathcal{F}_{t_i}^B\} \big| \mathcal{F}_s^B\big\}$$
$$+ \mathrm{E}\big\{\xi_m\,\mathrm{E}\{B(t) - B(t_m) \,|\, \mathcal{F}_{t_m}^B\} \big| \mathcal{F}_s^B\big\}$$

$$= \sum_{i=0}^{k-1} \xi_i \left(B(t_{i+1}) - B(t_i)\right) + \xi_k \left(B(s) - B(t_k)\right) + 0 + 0$$

$$= \int_0^s X\,dB \quad \text{for } k < m,$$

while in the same fashion,

$$\mathrm{E}\left\{\int_0^t X\,dB \,\middle|\, \mathcal{F}_s^B\right\}$$

$$= \sum_{i=0}^{m-1} \xi_i \left(B(t_{i+1}) - B(t_i)\right) + \xi_m \,\mathrm{E}\{B(t) - B(t_m) \,|\, \mathcal{F}_s^B\}$$

$$= \sum_{i=0}^{m-1} \xi_i \left(B(t_{i+1}) - B(t_i)\right) + \xi_m \left(B(s) - B(t_m)\right)$$

$$= \int_0^s X\,dB \quad \text{for } k = m,$$

Problem 4.32: Prove Theorem 4.9 in Klebaner's book for $X \in S_T$.

Solution. Pick a partition $0 = s_0 < s_1 < \cdots < s_m = T$ of the interval $[0, T]$ and consider an $X \in S_T$ given by

$$X(t) = I_{\{0\}}(t)\xi_0 + \sum_{i=0}^{m-1} I_{(s_i, s_{i+1}]}(t)\xi_i \quad \text{for } t \in [0, T],$$

where ξ_i is $\mathcal{F}_{t_i}$-measurable with $E(\xi_i^2) < \infty$ for $i = 0, \ldots, m-1$. Now, any given grid $0 = t_0 < t_1 < \cdots < t_n = T$ may be refined to a grid $0 = t'_0 < t'_1 < \cdots < t'_k = T$ with at most $n + m - 1$ members that also include the times $0 < s_1 < \cdots < s_{m-1} < T$. Writing $s_j = t'_{i(j)}$ for $j = 1, \ldots, m-1$, we then have

$$\left|\sum_{i=1}^{n}\left(\int_0^{t_i} X\,dB - \int_0^{t_{i-1}} X\,dB\right)^2 - \sum_{i=1}^{k}\left(\int_0^{t'_i} X\,dB - \int_0^{t'_{i-1}} X\,dB\right)^2\right|$$

$$\leq \sum_{j=1}^{m-1}\left(\int_0^{t'_{i(j)+1}} X\,dB - \int_0^{t'_{i(j)-1}} X\,dB\right)^2$$

$$+ \sum_{j=1}^{m-1}\left(\int_0^{t'_{i(j)+1}} X\,dB - \int_0^{t'_{i(j)}} X\,dB\right)^2$$

$$+\sum_{j=1}^{m-1}\bigg(\int_0^{t'_{i(j)}} X\,dB - \int_0^{t'_{i(j)-1}} X\,dB\bigg)^2$$

$$\leq 3\sum_{j=1}^{m-1}\bigg(\int_0^{t'_{i(j)+1}} X\,dB - \int_0^{t'_{i(j)}} X\,dB\bigg)^2$$

$$+3\sum_{j=1}^{m-1}\bigg(\int_0^{t'_{i(j)}} X\,dB - \int_0^{t'_{i(j)-1}} X\,dB\bigg)^2$$

$$=3\sum_{j=1}^{m-1}\xi_j^2(B(t'_{i(j)+1}) - B(t'_{i(j)}))^2$$

$$+3\sum_{j=1}^{m-1}\xi_{j-1}^2(B(t'_{i(j)}) - B(t'_{i(j)-1}))^2$$

$$\to 0 \quad \text{with probability 1 as} \quad \max_{1\leq i\leq k} t'_i - t'_{i-1} \leq \max_{1\leq i\leq n} t_i - t_{i-1} \downarrow 0,$$

by the continuity of B. To finish the proof, it is therefore sufficient to prove that, in the sense of convergence in probability,

$$\lim_{\max_{1\leq i\leq n} t_i - t_{i-1}\downarrow 0}\sum_{i=1}^{n}\bigg(\int_0^{t_i} X\,dB - \int_0^{t_{i-1}} X\,dB\bigg)^2 = \int_0^T X(r)^2\,dr$$

for grids $0 = t_0 < t_1 < \cdots < t_n = T$ that include the times $0 < s_1 < \cdots < s_{m-1} < T$. That this is so in turn is an immediate consequence of the fact that $[B]([s,t]) = t-s$.

Problem 4.33: Show that the Itô integral process $\{\int_0^t X\,dB\}_{t\in[0,T]}$ for $X \in E_T$ is a martingale.

Solution. Picking $\{X_n\}_{n=1}^{\infty} \subseteq S_T$ such that

$$\mathrm{E}\bigg\{\int_0^T (X_n(t) - X(t))^2\,dt\bigg\} \to 0 \quad \text{as } n \to \infty,$$

the definition of the Itô integral process for E_T together with Problem 4.30 and the martingale property for the Itô integral process for S_T (recall Problem 4.31) show that

$$\mathrm{E}\bigg\{\int_0^t X\,dB\,\bigg|\,\mathcal{F}_s^B\bigg\} \leftarrow \mathrm{E}\bigg\{\int_0^t X_n\,dB\,\bigg|\,\mathcal{F}_s^B\bigg\} = \int_0^s X_n\,dB \to \int_0^s X\,dB$$

for $0 \leq s \leq t \leq T$, in the sense of convergence in $\mathbb{L}^1$ (for the first limit) and in the sense of convergence in $\mathbb{L}^2$ (for the second limit), respectively, as $n \to \infty$. Now, the conditional expectation on the left-hand side in the above equation must be equal to something, which in turn can therefore only be the integral on the right-hand side.

Problem 4.34: For two Itô processes $X = \{X(t)\}_{t\in[0,T]}$ and $Y = \{Y(t)\}_{t\in[0,T]}$, the Stratonovich integral process $\{\int_0^t X\,\partial Y\}_{t\in[0,T]}$ of X wrt. Y is defined as

$$\int_0^t X\,\partial Y \equiv \int_0^t X\,dY + \frac{1}{2}\,[X,Y](t) \quad \text{for } t \in [0,T]$$

(see also Section 5.9 in Klebaner's book). With this notation, show that $df(X(t)) = f'(X(t))\,\partial X(t)$ for f twice continuously differentiable.

Solution. First, we must agree on what is the exact meaning of the statement we are challenged to show, that $df(X(t)) = f'(X(t))\,\partial X(t)$. And that in turn must be

$$f(X(t)) - f(X(0)) = \int_0^t f'(X)\,\partial X.$$

Now, by the definition of the Stratonovich integral, we have

$$\int_0^t f'(X)\,\partial X = \int_0^t f'(X)\,dX + \frac{1}{2}\,[f'(X),X](t).$$

Here, the arguments from Example 4.23 in Klebaner's book carry over with only obvious modifications to show that

$$[f'(X),X](t) = \int_0^t f''(X)\,d[X,X],$$

so that

$$\int_0^t f'(X)\,\partial X = \int_0^t f'(X)\,dX + \frac{1}{2}\int_0^t f''(X)\,d[X,X].$$

But the right-hand side of this in turn equals $f(X(t)) - f(X(0))$ by Itô's formula, Theorem 4.16 in Klebaner's book. (Note that we only require f to be twice continuously differentiable in this exercise, rather than thrice continuously differentiable, as is required in the corresponding Theorem 5.19 in Klebaner's book.)

Problem 4.35: Show that for a process $X \in E_T$, the following process is a martingale:

$$\left\{\left(\int_0^t X\,dB\right)^2 - \int_0^t X(s)^2\,ds\right\}_{t\in[0,T]}.$$

Solution. If we have proved that the above process is a martingale for $X \in S_T$, then given an $X \in E_T$, we may pick a sequence $\{X_n\}_{n=1}^{\infty} \subseteq S_T$ such that

$$\lim_{n\to\infty} \mathrm{E}\left\{\int_0^T (X_n(t) - X(t))^2\,dt\right\} = 0,$$

and

$$\int_0^t X_n\,dB \to \int_0^t X\,dB \quad \text{as } n\to\infty,$$

for $t \in [0,T]$ in the sense of convergence in $\mathbb{L}^2$. From this, in turn, we conclude by means of Hölder's inequality and Jensen's inequality (together with the elementary fact that $(x+y)^2 \leq 2(x^2+y^2)$) that

$$\begin{aligned}
&\mathrm{E}\left\{\left|\int_0^t X_n(s)^2\,ds - \int_0^t X(s)^2\,ds\right|\right\}\\
&\quad = \mathrm{E}\left\{\left|\int_0^t (X_n(s)-X(s))\,(X_n(s)+X(s))\,ds\right|\right\}\\
&\quad \leq \mathrm{E}\left\{\sqrt{\int_0^t (X_n(s)-X(s))^2\,ds}\sqrt{\int_0^t (X_n(s)+X(s))^2\,ds}\right\}\\
&\quad \leq \sqrt{\mathrm{E}\left\{\int_0^t (X_n(s)-X(s))^2\,ds\right\}}\sqrt{\mathrm{E}\left\{\int_0^t (X_n(s)+X(s))^2\,ds\right\}}\\
&\quad \leq \sqrt{\mathrm{E}\left\{\int_0^T (X_n(s)-X(s))^2\,ds\right\}}\\
&\qquad \times \sqrt{2\,\mathrm{E}\left\{\int_0^T (X_n(s)-X(s))^2\,ds\right\} + 2\,\mathrm{E}\left\{\int_0^T (2X(s))^2\,ds\right\}}\\
&\quad \to 0 \quad \text{as } n\to\infty.
\end{aligned}$$

Similarly by using also the isometry property,

$$\begin{aligned}
&\mathrm{E}\bigg\{\bigg|\Big(\int_0^t X_n\,dB\Big)^2-\Big(\int_0^t X\,dB\Big)^2\bigg|\bigg\}\\
&\quad=\mathrm{E}\bigg\{\bigg|\Big(\int_0^t X_n\,dB-\int_0^t X\,dB\Big)\Big(\int_0^t X_n\,dB+\int_0^t X\,dB\Big)\bigg|\bigg\}\\
&\quad\le\sqrt{\mathrm{E}\bigg\{\Big(\int_0^t (X_n-X)\,dB\Big)^2\bigg\}}\sqrt{\mathrm{E}\bigg\{\Big(\int_0^t (X_n+X)\,dB\Big)^2\bigg\}}\\
&\quad=\sqrt{\mathrm{E}\bigg\{\int_0^t (X_n(s)-X(s))^2\,ds\bigg\}}\sqrt{\mathrm{E}\bigg\{\int_0^t (X_n(s)+X(s))^2\,ds\bigg\}}\\
&\quad\to 0\quad\text{as } n\to\infty.
\end{aligned}$$

Thus, we have

$$\int_0^t X_n(s)^2\,ds\to\int_0^t X(s)^2\,ds\quad\text{and}$$

$$\Big(\int_0^t X_n\,dB\Big)^2\to\Big(\int_0^t X\,dB\Big)^2\quad\text{as } n\to\infty$$

for $t\in[0,T]$ in the sense of convergence in $\mathbb{L}^1$. Hence, we may use that their conditional expectations converge in L^1 (proven in Problem 4.30) together with (the assumed proven) martingale property when $X_n\in S_T$ to conclude that

$$\begin{aligned}
\mathrm{E}\bigg\{\Big(\int_0^t X\,dB\Big)^2-\int_0^t X(r)^2\,dr\,\bigg|\,\mathcal{F}_s\bigg\}&\leftarrow\mathrm{E}\bigg\{\Big(\int_0^t X_n\,dB\Big)^2-\int_0^t X_n(r)^2\,dr\,\bigg|\,\mathcal{F}_s\bigg\}\\
&=\Big(\int_0^s X_n\,dB\Big)^2-\int_0^s X_n(r)^2\,dr\\
\text{as } n\to\infty\qquad&\to\Big(\int_0^s X\,dB\Big)^2-\int_0^s X(r)^2\,dr
\end{aligned}$$

for $0\le s<t\le T$ in the sense of convergence in $\mathbb{L}^1$, thereby establishing the requested martingale property for $X\in E_T$.

Pick a grid $0=t_0<t_1<\cdots<t_n=T$ and consider an $X\in S_T$ given by

$$X(t)=I_{\{0\}}(t)\,\xi_0+\sum_{i=0}^{n-1}I_{(t_i,t_{i+1}]}(t)\,\xi_i\quad\text{for } t\in[0,T],$$

where ξ_i is $\mathcal{F}_{t_i}$-measurable with $E(\xi_i^2) < \infty$ for $i = 0, \ldots, n-1$. Recall that

$$\int_0^t X\,dB$$
$$= \begin{cases} \sum_{i=0}^{m-1} \xi_i\,(B(t_{i+1}) - B(t_i)) + \xi_m\,(B(t) - B(t_m)) & \text{for } t \in (t_m, t_{m+1}], \\ 0 & \text{for } t = 0. \end{cases}$$

In order to prove the martingale property

$$\mathrm{E}\bigg\{\bigg(\int_0^t X\,dB\bigg)^2 - \int_0^t X(r)^2\,dr \,\bigg|\, \mathcal{F}_s\bigg\} = \bigg(\int_0^s X\,dB\bigg)^2 - \int_0^s X(r)^2\,dr$$

for $0 \le s < t \le T$, we may without loss of generality assume that $s = t_j$ and $t = t_k$ for some $0 \le j < k \le n$, as the grid $0 = t_0 < t_1 < \cdots < t_n = T$ can otherwise be enriched to accommodate s and t without affecting the values of

$$\bigg(\int_0^t X\,dB\bigg)^2 - \int_0^t X(r)^2\,dr \quad \text{and} \quad \bigg(\int_0^s X\,dB\bigg)^2 - \int_0^s X(r)^2\,dr.$$

Here, the random variable to the right is $\mathcal{F}_s$-measurable; therefore, simple algebraic manipulations show that the martingale property to be established holds if

$$\mathrm{E}\bigg\{\bigg(\int_0^t X\,dB\bigg)^2 - \bigg(\int_0^s X\,dB\bigg)^2 - \int_s^t X(r)^2\,dr \,\bigg|\, \mathcal{F}_s\bigg\}$$
$$= \mathrm{E}\bigg\{\bigg(\int_s^t X\,dB\bigg)^2 + 2\int_0^s X\,dB \int_s^t X\,dB - \int_s^t X(r)^2\,dr \,\bigg|\, \mathcal{F}_s\bigg\}$$
$$= 0.$$

That this identity holds in turn follows from the fact that

$$\mathrm{E}\bigg\{\int_0^s X\,dB \int_s^t X\,dB \,\bigg|\, \mathcal{F}_s\bigg\}$$
$$= \bigg(\int_0^s X\,dB\bigg) \sum_{i=j}^{k-1} \mathrm{E}\{\xi_i\,\mathrm{E}\{(B(t_{i+1}) - B(t_i)) | \mathcal{F}_{t_i}\} | \mathcal{F}_s\} = 0.$$

Similarly,

$$\mathrm{E}\Big\{\Big(\int_s^t X\,dB\Big)^2 \Big| \mathcal{F}_s\Big\} = \sum_{i=j}^{k-1} \mathrm{E}\{\xi_i^2\,\mathrm{E}\{(B(t_{i+1})-B(t_i))^2|\mathcal{F}_{t_i}\}|\mathcal{F}_s\}$$

$$+\,2 \sum_{j\le i_1<i_2\le k-1} \mathrm{E}\{\xi_{i_1}\,\xi_{i_2}\,(B(t_{i_1+1})-B(t_{i_1}))$$

$$\times\,\mathrm{E}\{(B(t_{i_2+1})-B(t_{i_2}))|\mathcal{F}_{t_{i_2}}\}|\mathcal{F}_s\}$$

$$= \sum_{i=j}^{k-1} \mathrm{E}\{\xi_i^2\,(t_{i+1}-t_i)|\mathcal{F}_s\} + 0$$

$$= \mathrm{E}\Big\{\int_s^t X(r)^2\,dr \,\Big|\, \mathcal{F}_s\Big\}.$$

It is tempting to try to solve the exercise by means of applying Itô's formula, which readily gives

$$\Big(\int_0^t X\,dB\Big)^2 - \int_0^t X(s)^2\,ds = 2\int_0^t \Big(\int_0^s X(r)\,dB(r)\Big)\,X(s)\,dB(s).$$

Here, we know that $\int_0^s X(r)\,dB(r)$ and $X(s)$ are both square-integrable. But this only implies that $\big(\int_0^s X(r)\,dB(r)\big)\,X(s)$ is integrable (rather than square-integrable) in general; therefore, we cannot conclude that the process on the right-hand side is a martingale from what we have learned so far.

Problem 4.36: Prove Itô's formula, Theorem 4.13 in Klebaner's book.

Solution. We shall prove that for a twice continuously differentiable function f, it holds that

$$f(B(t)) = f(B(0)) + \int_0^t f'(B(r))\,dB(r)$$

$$+\,\frac{1}{2}\int_0^t f''(B(r))\,dr \quad \text{for } t>0.$$

To that end, we consider partitions $0 = t_0 < t_1 < \cdots < t_n = t$ of the interval $[0,t]$ that becomes finer and finer so that $\max_{1\le i\le n} t_i - t_{i-1} \downarrow 0$.

By the Taylor expansion, we have

$$\begin{aligned}
f(B(t)) - f(B(0)) &= \sum_{i=1}^{n} (f(B(t_i)) - f(B(t_{i-1}))) \\
&= \sum_{i=1}^{n} f'(B(t_{i-1}))\,(B(t_i) - B(t_{i-1})) \\
&\quad + \frac{1}{2}\sum_{i=1}^{n} f''(B(t_{i-1}))\,(B(t_i) - B(t_{i-1}))^2 \\
&\quad + \sum_{i=1}^{n} \int_{B(t_{i-1})}^{B(t_i)} (B(t_i) - r)\,(f''(r) - f''(B(t_{i-1})))dr.
\end{aligned}$$

Here, the first term on the right-hand side converges to $\int_0^t f'(B)\,dB$ in probability as $f(B)$ is a continuous and adapted process. Moreover, recalling that the quadratic variation of B over an interval equals the length of that interval, it follows that the second term on the right-hand side converges to $\frac{1}{2}\int_0^t f''(B(r))\,dr$ by means of introducing a second cruder grid $\{t'_j\}_{j=1}^m$, approximating the value of $f''(B(t_{i-1}))$ by $f''(B(t'_{j-1}))$ for an appropriate j, and sending first $\max_{1\le i\le n} t_i - t_{i-1} \downarrow 0$ and then $\max_{1\le j\le m} t'_j - t'_{j-1} \downarrow 0$ afterwards, as this makes it possible to replace $(B(t_i) - B(t_{i-1}))^2$ with $t_i - t_{i-1}$ in the first limit as $\max_{1\le i\le n} t_i - t_{i-1} \downarrow 0$ and the approximation of $f''(B(t_{i-1}))$-values by $f''(B(t'_{j-1}))$-values is accurate in the second limit as $\max_{1\le j\le m} t'_j - t'_{j-1} \downarrow 0$ by the continuity of $f''(B)$.

Finally, the third term on the right-hand side is bounded by

$$\begin{aligned}
&\sup_{r,s\in[0,t],\,|r-s|\le \max_{1\le i\le n} t_i - t_{i-1}} |f''(B(r)) - f''(B(s))| \left|\sum_{i=1}^{n} \int_{B(t_{i-1})}^{B(t_i)} (B(t_i) - r)\,dr\right| \\
&= \sup_{r,s\in[0,t],\,|r-s|\le \max_{1\le i\le n} t_i - t_{i-1}} |f''(B(r)) - f''(B(s))| \sum_{i=1}^{n} \frac{(B(t_i) - B(t_{i-1}))^2}{2} \\
&\to 0 \times \frac{t}{2}.
\end{aligned}$$

Problem 4.37: Let $(B(t))_{t\ge 0}$ be a Brownian motion.

(a) Show that $\int_{(0,t]} \frac{|B(s)|}{s} ds < \infty$ (a.s.).

(b) Show that $\int_{(0,t]} \frac{B(s)}{s} ds$ is in L^2.

(c) Show that $W(t) = B(t) - \int_{(0,t]} \frac{B(s)}{s} ds$ is a Brownian motion.

(d) Show that the natural filtration of W is strictly smaller than that of B ($\mathcal{F}_t^W \subsetneq \mathcal{F}_t^B$) and that $\mathbb{E}[B(t)|\mathcal{F}_t^W] = 0$.

(e) Compute $\int_\varepsilon^t \frac{1}{s} dW(s)$ for $t \geq \varepsilon > 0$ and deduce that $\mathcal{F}_t^B = \bigcap_{\varepsilon>0} \left(\mathcal{F}_t^W \vee \mathcal{F}_\varepsilon^B\right)$.

Solution.

(a) We prove the stronger statement that $\mathrm{E}\left(\int_{(0,t]} \frac{|B(s)|}{s} ds\right) < \infty$; that is, $\int_{(0,t]} \frac{|B(s)|}{s} ds$ is in L^1 (and so is $\int_{(0,t]} \frac{B(s)}{s} ds$):

$$\begin{aligned}
\mathrm{E}\left(\int_{(0,t]} \frac{|B(s)|}{s} ds\right) &= \int_{(0,t]} \frac{\mathrm{E}\left(|B(s)|\right)}{s} ds \quad \text{(by Fubini)} \\
&= \mathrm{E}\left(|B(1)|\right) \int_{(0,t]} \frac{\sqrt{s}}{s} ds = 2\mathrm{E}\left(|B(1)|\right)\sqrt{t} \\
&= 2\mathrm{E}\left(|B(t)|\right) < \infty.
\end{aligned}$$

Having a finite mean, the random variable $\int_{(0,t]} \frac{|B(s)|}{s} ds$ must be almost surely finite.

(b) We show that $\mathrm{E}\left[\left(\int_{(0,t]} \frac{B(s)}{s} ds\right)^2\right] < +\infty$.

$$\begin{aligned}
\mathrm{E}\left[\left(\int_{(0,t]} \frac{B(s)}{s} ds\right)^2\right] &= \mathrm{E}\left[\left(\left|\int_{(0,t]} \frac{B(s)}{s} ds\right|\right)^2\right] \\
&\leq \mathrm{E}\left[\left(\int_{(0,t]} \frac{|B(s)|}{s} ds\right)^2\right] \\
&= \mathrm{E}\left[\int_{(0,t]} \int_0^t \frac{|B(r)B(s)|}{rs} dr ds\right] \\
&= \int_{(0,t]} \int_{(0,t]} \frac{\mathrm{E}[|B(r)B(s)|]}{rs} dr ds
\end{aligned}$$

$$\leq \int_{(0,t]}\int_{(0,t]} \frac{\sqrt{\mathrm{E}[B(r)^2]\mathrm{E}[B(s)^2]}}{rs}drds$$

(by Hölder's inequality)

$$= \int_{(0,t]}\int_{(0,t]} \frac{\sqrt{rs}}{rs}drds = \left(\int_{(0,t]} \frac{1}{\sqrt{s}}ds\right)^2$$

$$= \frac{t}{4} < +\infty.$$

(c) First, we establish that W is a Gaussian process, that is, for any m and $t_1 < \cdots < t_m$, $(W(t_1), \ldots, W(t_m))$ is a Gaussian vector. Since $W(t)$ can be approximated by the Riemann sums,

$$W^{(n)}(t) = B(t) - \sum_{k=1}^{n} \frac{B(s_k)}{s_k}(s_k - s_{k-1}),$$

for partitions of $(0,t]$, $s_0 = 0 < s_1 < s_2 < \cdots < s_n = t$, we see that $W^{(n)}(t)$, as a linear combination of the Gaussian vector $(B(s_1), \ldots, B(s_n))$, is a Gaussian random variable. We easily deduce that $(W(t_1), \ldots, W(t_m))$ can be approximated by a sequence of m-dimensional Gaussian vectors. As such, it is itself a Gaussian vector.

As a Gaussian process, the law of W is entirely defined by its mean and covariance functions. Clearly, $\mathrm{E}[W(t)] = 0$. For $t < v$,

$$\begin{aligned}
\mathrm{Cov}(W(t), W(v)) &= \mathrm{E}[W(t)W(v)] \\
&= \mathrm{E}\left[\left(B(t) - \int_{(0,t]} \frac{B(s)}{s}ds\right)\left(B(v) - \int_{(0,v]} \frac{B(u)}{u}du\right)\right] \\
&= t - \mathrm{E}\left[B(t)\int_{(0,v]} \frac{B(u)}{u}du\right] - \mathrm{E}\left[B(v)\int_{(0,t]} \frac{B(s)}{s}ds\right] \\
&\quad + \mathrm{E}\left[\int_{(0,t]} \frac{B(s)}{s}ds \int_{(0,v]} \frac{B(u)}{u}du\right].
\end{aligned}$$

Next, we use the Fubini theorem to compute each of the above expectations by swapping the expectation and integral. Note that by Hölder's inequality,

$$\int_{(0,v]} \frac{\mathrm{E}[|B(t)B(u)|]}{u} du \leq \int_{(0,v]} \frac{\sqrt{\mathrm{E}[B(t)^2]}\sqrt{\mathrm{E}[B(u)^2]}}{u} du$$
$$= \int_{(0,v]} \frac{\sqrt{t}\sqrt{u}}{u} du = 2\sqrt{tv} \ < \ +\infty$$

and similarly for the other terms (see (b)),

$$\begin{aligned}
&\mathrm{Cov}(W(t), W(v)) \\
&= t - \int_{(0,v]} \frac{\mathrm{E}[B(t)B(u)]}{u} du - \int_{(0,t]} \frac{\mathrm{E}[B(v)B(s)]}{s} ds \\
&\quad + \int_{(0,v]} \int_{(0,t]} \frac{\mathrm{E}[B(s)B(u)]}{su} ds du \\
&= t - \int_0^v \frac{t \wedge u}{u} du - \int_0^t \frac{v \wedge s}{s} ds + \int_0^v \left(\int_0^t \frac{s \wedge u}{su} ds \right) du \\
&= t - t - t(\ln v - \ln t) - t + \int_0^t (1 + \ln t - \ln u) du + \int_t^v \frac{t}{u} du \\
&= -t \ln v + t \ln t - t + t + t \ln t - (t \ln t - t) + t(\ln v - \ln t) \\
&= t.
\end{aligned}$$

This completes the proof that W is a Gaussian process with continuous paths, mean zero, and covariance $\min(s,t)$; that is, W is a Brownian motion.

Alternatively, we can use Theorem 4.12. Indeed, by the integration by parts formula,

$$(\ln t)B(t) = (\ln \varepsilon)B(\varepsilon) + \int_\varepsilon^t (\ln s) dB(s) + \int_\varepsilon^t \frac{B(s)}{s} ds.$$

Since $\int_0^t (\ln s)^2 ds < +\infty$, $\int_0^t (\ln s) dB(s)$ is well defined (and represents a martingale). Also, by the law of iterated logarithm, almost surely,

$$\limsup_{\varepsilon \downarrow 0} \frac{B(\varepsilon)}{\sqrt{2\varepsilon \ln \ln \varepsilon}} = 1 \text{ and } \liminf_{\varepsilon \downarrow 0} \frac{B(\varepsilon)}{\sqrt{2\varepsilon \ln \ln \varepsilon}} = -1.$$

It follows that, almost surely,

$$\limsup_{\varepsilon\downarrow 0}(\ln\varepsilon)B(\varepsilon) = \limsup_{\varepsilon\downarrow 0}(\ln\varepsilon)\sqrt{2\varepsilon\ln\ln\varepsilon}\frac{B(\varepsilon)}{\sqrt{2\varepsilon\ln\ln\varepsilon}} = 0.$$

Similarly, almost surely,

$$\liminf_{\varepsilon\downarrow 0}(\ln\varepsilon)B(\varepsilon) = \limsup_{\varepsilon\downarrow 0}(\ln\varepsilon)\sqrt{2\varepsilon\ln\ln\varepsilon}\frac{B(\varepsilon)}{\sqrt{2\varepsilon\ln\ln\varepsilon}} = 0;$$

that is, almost surely, $\lim_{\varepsilon\downarrow 0}(\ln\varepsilon)B(\varepsilon) = 0$ and

$$(\ln t)B(t) = \int_0^t (\ln s)dB(s) + \int_0^t \frac{B(s)}{s}ds.$$

From this, we deduce that

$$\begin{aligned} W(t) &= B(t) - (\ln t)B(t) + \int_0^t (\ln s)dB(s) \\ &= \int_0^t (1 - \ln t + \ln s)dB(s). \end{aligned}$$

A direct application of Theorem 4.12 shows that W is a Gaussian process with zero mean and covariance function

$$\begin{aligned} &\mathrm{Cov}(W(t), W(v)) \\ &= \int_0^t (1 - \ln t + \ln s)(1 - \ln(t+u) + \ln s)ds \\ &= (t\ln t - t)(\ln(t+u) - 1) - (\ln t + \ln(t+u) - 2)\int_0^t (\ln s)ds \\ &\quad + \int_0^t (\ln s)^2 ds \\ &= (t\ln t - t)(\ln(t+u) - 1) - (\ln t + \ln(t+u) - 2)(t\ln t - t) \\ &\quad + t(\ln t)^2 - 2t\ln t + 2t \\ &= t. \end{aligned}$$

This completes the proof that W is a Brownian motion.

(d) Note that B and W are both Brownian motions with respect to their own filtrations. W is clearly adapted to the natural filtration of B. If the two filtrations were equal, then $B(t) - W(t) = \int_0^t \frac{B(s)}{s}ds$

would be a continuous martingale of finite variation and must therefore be nil ($W(0) = B(0) = 0$). This leads to a contradiction from which we deduce that the natural filtration of W is strictly smaller than that of B.

Also, for $s < t$,

$$\begin{aligned}
W(s) &= \mathbb{E}\left[W(t)\,|\mathcal{F}_s^W\right] \\
&= \mathbb{E}\left[B(t) - \int_{(0,t]} \frac{B(r)}{r}dr\,|\mathcal{F}_s^W\right] \\
&= \mathbb{E}\left[\mathbb{E}\left[B(t) - \int_{(0,t]} \frac{B(r)}{r}dr\,|\mathcal{F}_s^B\right]|\mathcal{F}_s^W\right] \\
&= \mathbb{E}\left[B(s) - \int_{(0,s]} \frac{B(r)}{r}dr - \mathbb{E}\left[\int_{(s,t]} \frac{B(r)}{r}dr\,|\mathcal{F}_s^B\right]|\mathcal{F}_s^W\right] \\
&= W(s) - \mathbb{E}\left[\int_{(s,t]} \frac{\mathbb{E}\left[B(r)\,|\mathcal{F}_s^B\right]}{r}dr\,|\mathcal{F}_s^W\right] \\
&= W(s) - \mathbb{E}\left[\int_{(s,t]} \frac{B(s)}{r}dr\,|\mathcal{F}_s^W\right] \\
&= W(s) - \mathbb{E}\left[B(s)\,|\mathcal{F}_s^W\right]\int_{(s,t]} \frac{dr}{r}.
\end{aligned}$$

We immediately deduce that for any $s > 0$, $\mathbb{E}[B(s)|\mathcal{F}_s^W] = 0$.

(e) For $t \geq \varepsilon > 0$,

$$\frac{1}{t}B(t) = \frac{1}{\varepsilon}B(\varepsilon) + \int_\varepsilon^t \frac{1}{s}dB(s) - \int_\varepsilon^t \frac{B(s)}{s^2}ds$$

and

$$\int_\varepsilon^t \frac{1}{s}dW(s) = \int_\varepsilon^t \frac{1}{s}\left(dB(s) - \frac{B(s)}{s}ds\right) = \frac{1}{t}B(t) - \frac{1}{\varepsilon}B(\varepsilon),$$

from which we deduce that

$$B(t) = \frac{t}{\varepsilon}B(\varepsilon) + t\int_\varepsilon^t \frac{1}{s}dW(s).$$

It follows that $B(t)$ is $(\mathcal{F}_t^W \vee \mathcal{F}_\varepsilon^B)$-measurable and that $\mathcal{F}_t^B = \mathcal{F}_t^W \vee \mathcal{F}_\varepsilon^B = \bigcap_{\alpha>0}\left(\mathcal{F}_t^W \vee \mathcal{F}_\alpha^B\right)$.

Chapter 5

Stochastic Differential Equations

Consider an ODE with an initial value

$$x'(t) = \mu(x(t), t) \quad \text{for } t \in [0, T], \quad x(0) = x_0,$$

for a $T \in (0, \infty)$ and an $x_0 \in \mathbb{R}$. According to the Peano theorem, it is sufficient to require continuity of the coefficient function $\mu : \mathbb{R} \times [0, T] \to \mathbb{R}$ in order for a solution to exist at least in some sub-interval $[0, S]$ of $[0, T]$. The solution need not be unique.

If the coefficient μ is continuous as well as Lipschitz continuous in the first variable uniformly in time, that is,

$$|\mu(x, t) - \mu(y, t)| \leq K\,|x - y| \quad \text{for } x, y \in \mathbb{R} \text{ and } t \in [0, T],$$

for some constant $K > 0$, then the Pickard–Lindelöf theorem ensures that there exists a unique solution to the ODE.

We shall now discuss stochastic differential equations (SDEs):

$$dX(t) = \mu(X(t), t)\,dt + \sigma(X(t), t)\,dB(t) \quad \text{for } t \in [0, T], \quad X(0) = x_0.$$

We will cite two results about the existence and uniqueness of solutions to SDEs that are not too different from the Peano and Pickard–Lindelöf theorems in appearance. Their proofs however are quite technically complicated and will not be discussed here.

Definition of SDE

Through out this chapter, $\{B(t)\}_{t\geq 0}$ denotes BM and $\{\mathcal{F}_t\}_{\geq 0} = \{\mathcal{F}_t^B\}_{\geq 0}$ is the filtration generated by BM itself. An SDE with coefficient functions $\mu, \sigma : \mathbb{R} \times [0,T] \to \mathbb{R}$ is given in differential form by

$$dX(t) = \mu(X(t),t)\,dt + \sigma(X(t),t)\,dB(t) \quad \text{for } t \in [0,T], \quad X(0) = x_0.$$

The function μ is called the drift coefficient of the SDE, while σ is called the diffusion coefficient. Both these functions must be measurable. The initial value x_0 is any real (non-random) number. Alternatively, the SDE can be expressed in integrated form as

$$X(t) = x_0 + \int_0^t \mu(X(s),s)\,ds + \int_0^t \sigma(X(s),s)\,dB(s) \quad \text{for } t \in [0,T].$$

A solution X to the SDE is called a diffusion process. Obviously, the solution (if it exists) is an Itô process. According to Chapter 4, X must also be adapted and continuous with

$$\int_0^T |\mu(X(t),t)|\,dt < \infty \quad \text{and} \quad \{\sigma(X(t),t)\}_{t\in[0,T]} \in P_T.$$

An SDE can also be started with a random (variable) initial value $X(0) = X_0$ that is independent of $\{\mathcal{F}_t^B\}_{t\geq 0}$. One way to accomplish this is to insert the random X_0 instead of the non-random x_0 in the solution to the non-random initial value SDE when that solution has been constructed. This however leads to a non-adapted solution. So, a more satisfactory way is to work with the enlarged filtration $\mathcal{F}_t = \mathcal{F}_t^B \bigvee \sigma(X_0)$ instead of $\mathcal{F}_t^B$: everything we have done so far and will do in the sequel based on the filtration being $\mathcal{F}_t^B$ will also work without any alterations for $\mathcal{F}_t^B \bigvee \sigma(X_0)$. In fact, there is yet another alteration of the filtration that is always done in more technically oriented and complete treatments than ours, namely to enlarge $\mathcal{F}_t$ with all null events of $\mathcal{F}$ to create a so-called augmented filtration. This has to do with the fact that conditional expectations and stochastic convergence limits are unique only outside a null event. Therefore, to ensure that, for example, the limits of adapted random items are adapted, this modification is required.

A solution to the SDE is called a strong solution if it is a solution with the BM B given (for any such given B). An SDE has strong uniqueness if (given any BM B) any pair of strong solutions agree (except for on a null event). Note that this does not assume existence, so an SDE can display strong uniqueness without having a strong solution. (Such SDEs do exist!) Unless otherwise specifically mentioned, we discuss and deal with strong solutions only.

One might suspect that a solution to an SDE should resemble a solution of the corresponding ODE with σ taken to be zero. This is sometimes the case, but other times the solutions to these SDEs and ODEs are very different.

SDEs with coefficients $\mu(x,t) = \mu(x)$ and $\sigma(x,t) = \sigma(x)$ that do not depend on $t \in [0,T]$ are called (time) homogeneous. More detailed results about many issues are known for homogeneous SDEs than for general SDEs (for example, sharper criteria for the existence and/or uniqueness of solutions). We will see a selection of results for homogeneous SDEs in Chapter 6. Argubly, most SDEs encountered in applied mathematics are homogeneous.

SDEs of the above-discussed type are the ones we shall focus on, and they are more specifically called diffusion-type SDEs. This is opposed to a more general form of SDEs (which we will occasionally encounter as well) given by

$$dX(t) = \mu(t)\,dt + \sigma(t)\,dB(t) \quad \text{for } t \in [0,T], \quad X(0) = x_0,$$

where $\mu(t)$ and $\sigma(t)$ may depend on not only $X(t)$ but on the entire past $\{X(s)\}_{s\in[0,t]}$ of X.

There is no general method to find explicit solutions of (diffusion-type) SDEs expressed in terms of μ, σ, and B. On the other hand, when given a candidate solution, the procedure to validate the solution is often by the use of Itô's formula.

Stochastic Exponential and Logarithm

Let X have a stochastic differential. A solution $\{U(t)\}_{t\in[0,T]}$ to the (usually non-diffusion type) SDE

$$dU(t) = U(t)\,dX(t) \quad \text{for } t \in [0,T], \quad U(0) = 1,$$

is called a stochastic exponential of X and denoted $\mathcal{E}(X)$. [Note that if the SDE were an ODE, the solution would be $U(t) = e^{X(t)-X(0)}$.] It turns out that the stochastic exponential exists and is uniquely given by

$$U(t) = \mathcal{E}(X)(t) = e^{X(t)-X(0)-\frac{1}{2}[X](t)} \quad \text{for } t \in [0,T].$$

Given this expression, it is easy to check by Itô's formula that it solves the requested SDE. (Note that the solution coincides with the formal ODE solution if and only if $[X] = 0$, meaning that the Itô process X has no Itô integral part.)

Let X have a stochastic differential and be strictly positive. A solution $\{U(t)\}_{t\in[0,T]}$ to the (usually non-diffusion type) SDE

$$dX(t) = X(t)\,dU(t) \quad \text{for } t \in [0,T], \quad U(0) = 0,$$

is called a stochastic logarithm of X and denoted $\mathcal{L}(X)$. [Note that if the SDE were an ODE the solution would be $U(t) = \ln(X(t)) - \ln(X(0))$.] It turns out that the stochastic logarithm exists and is uniquely given by

$$U(t) = \mathcal{L}(X)(t) = \ln\left(\frac{X(t)}{X(0)}\right) + \int_0^t \frac{d[X](s)}{2X(s)^2} \quad \text{for } t \in [0,T].$$

Given this expression, it is easy to check by Itô's formula that it solves the requested SDE. The solution can also be derived using the fact that X must be the stochastic exponential of U modulo a multiplicative constant. (Note that again the solution coincides with the formal ODE solution if and only if $[X] = 0$.)

It is a very useful exercise to establish that $\mathcal{L}(\mathcal{E}(X)) = X - X(0)$ and $\mathcal{E}(\mathcal{L}(X)) = X/X(0)$.

Solutions to Linear SDEs

A general linear (in general non-diffusion type) SDE is given by

$$\begin{gathered} dX(t) = (\alpha(t) + \beta(t)X(t))\,dt + (\gamma(t) + \delta(t)X(t))\,dB(t) \\ \text{for } t \in [0,T], \quad X(0) = x_0. \end{gathered}$$

Here, the coefficients α, β, γ, and δ are continuous adapted stochastic processes satisfying appropriate integrability conditions. It is not very hard (albeit a bit notationally complicated) to show that the solution to the general linear SDE is given by

$$X(t) = U(t)\left(x_0 + \int_0^t \frac{\alpha(s) - \gamma(s)\delta(s)}{U(s)}\,ds + \int_0^t \frac{\gamma}{U}\,dB\right)$$

with

$$U(t) = \exp\left(\int_0^t \left(\beta(s) - \frac{1}{2}\delta(s)^2\right) ds + \int_0^t \delta\,dB\right) \quad \text{for } t \in [0,T].$$

One famous example of a linear SDE is the Langevin equation $\alpha(t) = -\mu$, $\beta = \delta = 0$ and $\sigma(t) = \sigma$ for constants $\mu, \sigma > 0$. Another is the stochastic exponential of BM $\alpha = \beta = \sigma = 0$ and $\delta(t) = 1$. Yet another example is offered by the so-called Brownian bridge:

$$dX(t) = \frac{b - X(t)}{T - t}\,dt + dB(t) \quad \text{for } t \in [0, T), \quad X(0) = a.$$

This turns out to be simply the process $X(t) = (B(t) + a | B(T) + a = b)$, that is, a Brownian motion on $[0, T]$ started at a and forced (or conditioned rather) to finish at b. That process is also called pinned BM.

Existence and Uniqueness of Solutions

We will cite two basic results for the existence and/or uniqueness of solutions to SDEs. They can both be somewhat strengthened in different ways at the cost of a more complicated appearance. We do not discuss such improvements.

The first result is the basic existence and uniqueness result for an SDE

$$dX(t) = \mu(X(t), t)\,dt + \sigma(X(t), t)\,dB(t) \quad \text{for } t \in [0, T].$$

Assume that there exist constants $K_1 = K_1(N, T) > 0$ and $K_2 = K_2(T) > 0$ such that

$$|\mu(x,t) - \mu(y,t)| + |\sigma(x,t) - \sigma(y,t)| \le K_1 |x - y| \quad \text{for } t \in [0, T] \text{ and } |x|, |y| \le N$$

for each $N > 0$ and

$$|\mu(x,t)| + |\sigma(x,t)| \le K_2(1 + |x|) \quad \text{for } t \in [0, T] \text{ and } x \in \mathbb{R}.$$

In other words, the coefficients are locally Lipschitz with (global) linear growth in the x-variable uniformly in the t-variable. For any choice of a non-random initial value $X(0) = x_0$ or a random initial value $X(0) = X_0$, there exists a unique strong solution $\{X(t)\}_{t \in [0,T]}$ to the SDE. If in addition, X_0 satisfies $E(X_0^2) < \infty$, then it holds that

$$E\left(\sup_{t \in [0,T]} X(t)^2\right) \le C(1 + E(X_0^2)),$$

where $C = C(K_2, T)$ is a constant that depends on K_2 and T only.

The local Lipschitz condition suffices for uniqueness but not for existence as one can see, for example, with the ODE $dx(t) = x(t)^2\,dt$, $x(0) = 1$, with unique solution $x(t) = 1/(1 - t)$. It is easy to see that a global Lipschitz condition, that is, $K_1 = K_1(T)$ not depending on N, implies linear growth.

Obviously, the canonical application of the above-cited existence and uniqueness result is to a linear SDE with bounded non-random (to make the SDE diffusion-type) coefficients $\alpha, \beta, \sigma, \delta : [0,T] \to \mathbb{R}$.

The second basic result only concerns uniqueness and is attributed to Yamada and Watanabe: assume that the drift coefficient μ is globally Lipschitz and that the diffusion coefficient σ is globally Hölder of order $\alpha \geq 1/2$, that is,

$$|\sigma(x,t) - \sigma(y,t)| \leq K_3|x-y|^\alpha \quad \text{for } t \in [0,T] \text{ and } x, y \in \mathbb{R}$$

for some constant $K_3 = K_3(T)$. Then, the SDE displays strong uniqueness.

The canonical application of the Yamada–Watanabe theorem is to the so-called Girsanov SDE $dX(t) = |X(t)|^r \, dB(t)$, $X(0) = 0$, with $r \in [1/2, 1]$, for which the theorem yields the uniqueness of the solution $X = 0$.

The criteria imposed in the above-cited existence and uniqueness results are far from necessary: SDE can have very "wild" coefficients [such as, $\sigma(x,t) = |x|^{1000}$] and still have a well-defined unique solution.

Markov Property of Solutions

More or less in general, solutions to SDEs are Markov processes as well as strong Markov processes. If the SDE is time-homogeneous, then so is the Markov process solution. In fact, at least historically, some authors called solutions to SDEs (diffusion processes) continuous Markov processes. This indicates that the converse statement that continuous Markov processes are diffusion processes holds. However, we will not discuss this topic.

Construction of Weak Solutions

A solution to the SDE is called a weak solution if the BM that features in (the solution to) the SDE is (not given from the beginning but) constructed together with the solution (can be chosen at liberty). Clearly, any strong solution is a weak solution.

An SDE has weak uniqueness if any pair of weak solutions $\{X_1(t)\}_{t\in[0,T]}$ and $\{X_2(t)\}_{t\in[0,T]}$ have common finite-dimensional distributions, that is,

$$\mathrm{P}(X_1(t_1) \leq x_1, \ldots, X_1(t_n) \leq x_n) = \mathrm{P}(X_2(t_1) \leq x_1, \ldots, X_2(t_n) \leq x_n)$$

for $x_1, \ldots, x_n \in \mathbb{R}$, $t_1, \ldots, t_n \in [0,T]$ and $n \in \mathbb{N}$.

The difference between strong and weak solutions might seem minor, but in fact, it is the other way around: it is, from a theoretical perspective, substantially easier to find weak solutions than to find strong solutions. However, for virtually all specific SDEs encountered in practice, strong solutions exist when weak solutions do.

There is a famous example of an SDE that has a unique weak solution but no strong solution, namely the Tanaka SDE $dX(t) = \text{sign}(X(t))\,dB(t)$, $X(0) = 0$, where $\text{sign}(x) = 1$ for $x \geq 0$ and $\text{sign}(x) = -1$ for $x < 0$. The statement about the weak solution is not too hard to prove while that for (no) strong solutions requires a quite sophisticated extension of Itô's formula to the convex function $\mathbb{R} \ni x \curvearrowright |x| \in \mathbb{R}$.

The basic existence and uniqueness result for weak solutions to an SDE

$$dX(t) = \mu(X(t), t)\,dt + \sigma(X(t), t)\,dB(t) \quad \text{for } t \in [0, T], \quad X(0) = x_0,$$

is as follows: assume that $\sigma(x, t)$ is strictly positive and continuous and that both $\mu(x, t)$ and $\sigma(x, t)$ have linear growth in the x-variable (uniformly in the t-variable). Then, the SDE has a unique solution that is a strong Markov process. Note that the Lipschitz condition for the coefficients featuring in the basic existence and uniqueness result for strong solutions is not needed for weak solutions.

There are many connections between solutions to SDEs and solutions to partial differential equations (PDEs), as we shall see. The generator of an SDE is the partial differential operator

$$L_s f(x, s) = (L_s f)(x, s) = \frac{\sigma(x, s)^2}{2}\,\frac{\partial^2 f(x, s)}{\partial x^2} + \mu(x, s)\,\frac{\partial f(x, s)}{\partial x}.$$

Now, assume that μ and σ are bounded on compact subsets of $[0, T] \times \mathbb{R}$. From Itô's formula, it is easy to see that

$$\left\{ f(X(t)) - f(x_0) - \int_0^t (L_s f)(X(s))\,ds \right\}_{t \in [0,T]}$$

is a zero-mean martingale for any C^2 function $f : \mathbb{R} \to \mathbb{R}$ with compact support.

Given just the coefficients μ and σ of the SDE together with the initial value x_0, a martingale problem for the SDE is the converse issue to find a continuous and adapted stochastic process $\{X(t)\}_{t \in [0,T]}$ such that

$$\left\{ f(X(t)) - f(x_0) - \int_0^t (L_s f)(X(s))\,ds \right\}_{t \in [0,T]}$$

is a zero-mean martingale for any C^2 function $f : \mathbb{R} \to \mathbb{R}$ with compact support.

It turns out that X is a solution to the martingale problem for an SDE only if X is a weak solution to the SDE. This of course requires a construction of a BM B associated with the Markov process X such that X solves the SDE. The proof is not hard to outline but a complete proof requires more sophisticated tools than we have developed so far.

Another approach for obtaining weak solutions is given by a change of measure for a Brownian motion, see Chapter 10.

Backward and Forward Equations

Let L_s be the generator of the SDE

$$dX(t) = \mu(X(t), t)\, dt + \sigma(X(t), t)\, dB(t) \quad \text{for } t \in [0, T].$$

A fundamental solution of the so-called Kolmogorov backward PDE

$$\frac{\partial u(x, s)}{\partial s} + L_s u(x, s) = 0$$

is a non-negative function $p(t, y, x, s)$ for $x, y \in \mathbb{R}$ and $0 \le s < t \le T$ such that

$$u(x, s) = \int_{-\infty}^{\infty} g(y) p(y, t, x, s)\, dy$$

is bounded and [regarded as a function of (x, s)] solves the PDE for any bounded function $g : \mathbb{R} \to \mathbb{R}$ with

$$\lim_{s \uparrow t} u(x, s) = g(x).$$

Under technical conditions on the coefficients μ and σ, a strictly positive fundamental solution $p(y, t, x, s)$ to the backward equation exists. Further, that fundamental solution solves the backward equation [regarded as a function of (x, s)] as well as the so-called forward equation

$$-\frac{\partial p}{\partial t} + L_t^\star p = -\frac{\partial p}{\partial t} + \frac{\partial^2}{\partial y^2}\left(\frac{\sigma(y,t)^2}{2} p\right) - \frac{\partial}{\partial y}\left(\mu(y, t)\, p\right) = 0$$

[regarded as a function of (y, t)]. (Here, the $\star$ in the notation stems from that $L^\star$ is the adjoint differential operator to L in a certain sense.) Moreover, there exists a Markov process $\{X(t)\}_{t \in [0,T]}$ that has $p(y, t, x, s)$ as a transition PDF and is a weak solution to the SDE. (This of course requires the construction of a BM B associated with the Markov process X such that X solves the SDE.)

And so, we have found a method to construct weak solutions to SDEs based on the theory for PDEs and the Markov process.

Stratonovich Stochastic Calculus

There is an alternative to the so-called Itô stochastic calculus we have developed so far, labelled Stratonovich stochastic calculus. The idea here is to alter the definition of the stochastic integral involved so that the rules from ordinary calculus (such as, integration by parts) come into play again. The underlying theory remains the same though, as are the Itô processes involved, being the sum of an Itô integral process and an (absolutely) continuous FV process. It is just a matter of expressing them differently.

The Stratonovich integral process $\{\int_0^t X\,\partial Y\}_{t\in[0,T]}$ of an Itô process $\{X(t)\}_{t\in[0,T]}$ with respect to another Itô process $\{Y(t)\}_{t\in[0,T]}$ (both built with and adapted to the filtration of the one and the same BM B) is defined as

$$\left\{\int_0^t X\,\partial Y\right\}_{t\in[0,T]} = \left\{\int_0^t X\,dY + \frac{1}{2}\,[X,Y](t)\right\}_{t\in[0,T]}.$$

Written in differential form, this becomes

$$X(t)\,\partial Y(t) = X(t)\,dY(t) + \frac{1}{2}\,d[X,Y](t) \quad \text{for } t\in[0,T].$$

As X is continuous, it follows from what we learned about Itô integrals of one Itô process with respect to another, together with the definition of variation, that

$$\frac{1}{2}\sum_{i=1}^{n}(X(t_i^n)+X(t_{i-1}^n))(Y(t_i^n)-Y(t_{i-1}^n)) \to_P \int_0^t X\,\partial Y$$

for a sequence of partitions $0 = t_0^n < \cdots < t_n^n = t$ of $[0,t]$ such that $\max_{1\le i\le n} t_i^n - t_{i-1}^n \downarrow 0$.

Expressed with Stratonovich differentials, the chain rule becomes

$$\begin{aligned} d(X(t)Y(t)) &= X(t)\,dY(t) + Y(t)\,dX(t) + d[X,Y](t) \\ &= X(t)\,\partial Y(t) + Y(t)\,\partial X(t). \end{aligned}$$

Making use of the fact that

$$\begin{aligned} d[f'(X)(t),X(t)] &= \left(f''(X(t))\,dX(t) + \frac{1}{2}\,f'''(X(t))\,d[X](t)\right)dX(t) \\ &= f''(X(t))\,d[X](t) \end{aligned}$$

for $f : \mathbb{R} \to \mathbb{R}$ of class C^3, it further follows that Itô's formula (for $f \in C^3$) simplifies to

$$df(X(t)) = f'(X(t))\,dX(t) + \frac{1}{2} f''(X(t))\,d[X](t) = f'(X(t))\,\partial X(t).$$

By mere insertion in the definition of the differential form of a Stratonovich (integral) diffusion-type SDE,

$$dX(t) = \mu(X(t), t)\,dt + \sigma(X(t), t)\,\partial B(t) \quad \text{for } t \in [0, T]$$

and using that

$$d[\sigma(X(t), t), B(t)] = \sigma(X(t), t)\,\frac{\partial \sigma}{\partial x}(X(t), t)dt,$$

we see that the equation translates to the Itô (integral) SDE

$$dX(t) = \left(\mu(X(t), t) + \frac{1}{2}\,\sigma(X(t), t)\,\frac{\partial \sigma}{\partial x}(X(t), t)\right) dt$$
$$+ \sigma(X(t), t)\,dB(t) \quad \text{for } t \in [0, T].$$

Problems

Problem 5.1 (Gaussian diffusions): Show that if $X(t)$ satisfies the SDE $dX(t) = a(t)dt + b(t)dB(t)$, with deterministic bounded coefficients $a(t)$ and $b(t)$, such that $\int_0^T |a(t)|dt < \infty$, and $\int_0^T b^2(t)dt < \infty$, and $X(0)$ is either a deterministic constant or has a normal distribution independent of B, then $X(t)$ is a Gaussian process with independent Gaussian increments.

Solution. By Problem 4.11, $\int_0^t b(s)dB(s)$ is a Gaussian process with independent increments. But then, that must also be true for $X(t) = X(0) + \int_0^t a(s)ds + \int_0^t b(s)dB(s)$. The only difference is that increments are no longer necessarily zero-mean.

Remark: It is easy to see that any martingale with finite second moments has uncorrelated increments. If it is also a Gaussian process, then the increments are independent. The above process is such an example.

Problem 5.2: For the process (X_t), $t \geq 0$, solve the SDE

$$dX_t = -X_t dt + dB_t, \quad X_0 \in \mathbb{R}.$$

(a) Show that the process $Y_t = e^t X_t$ has independent Gaussian increments, and give the distribution of the increments over the time interval $[s, t]$.
(b) Show that the process (X_t) has Gaussian increments, but they are not independent.
(c) State with reason whether the process (X_t) is a Gaussian process, and give its mean and covariance functions.
(d) Find the conditional distribution of X_t given X_s for $s < t$.
(e) Show that if X_0 has distribution $N(0, 1/2)$ and is independent of the process (B_t), then for any time t, X_t has $N(0, 1/2)$ distribution.

Solution.

(a) Using integration by parts (product rule),

$$dY_t = e^t dX_t + X_t e^t dt = e^t dB_t.$$

Integrating from s to t,

$$Y_t - Y_s = \int_s^t e^u dB_u.$$

Since the increments of Brownian motion are independent, $Y_t - Y_s$ is independent of the past values of $B_r, r \leq s$. This implies that they are independent of all the values of Y_r, $r \leq s$, $Y_r = X_0 + \int_0^r e^u dB_u$:

$$Y_t - Y_s = \int_s^t e^u dB_u \sim N(0, \int_s^t e^{2u} du).$$

(b) From above, by multiplying by e^{-t},

$$X_t = e^{-t+s} X_s + e^{-t} \int_s^t e^u dB_u,$$

$$X_t - X_s = (e^{-t+s} - 1) X_s + e^{-t} \int_s^t e^u dB_u.$$

Also (by taking $t = s$ and $s = 0$),

$$X_s = e^{-s} X_0 + e^{-s} \int_0^s e^u dB_u.$$

Since the stochastic integral of deterministic function e^u is Gaussian, it follows that X_s is Gaussian, and $X_t - X_s$ is a sum of two independent Gaussian, which is Gaussian.

It is clear that $X_t - X_s$ depends on X_s; therefore, the increments are not independent.

(c) The process (Y_t) is a Gaussian process because it has independent Gaussian increments. Since $X_t = e^{-t}Y_t$, it is also Gaussian.

This can be seen as follows. Let $y = (y_1, \ldots, y_n)$ be a non-random vector, and $Y = (Y_{t_1}, \ldots, Y_{t_n})$. We know that for any such y, the dot product

$$yY = \sum_{i=1}^{n} y_i Y_{t_i}$$

is a Gaussian random variable. Consider

$$yX = \sum_{i=1}^{n} y_i X_{t_i} = \sum_{i=1}^{n} y_i e^{-t_i} Y_{t_i} = \sum_{i=1}^{n} \tilde{y}_i Y_{t_i},$$

where $\tilde{y}_i = y_i e^{-t_i}$. Hence, it is Gaussian as a linear combination of Y_{t_i}'s. Since Itô integral has zero mean,

$$\mathrm{E}(X_t) = e^{-t}X_0,$$

$$\begin{aligned}
\mathrm{Cov}(X_s, X_t) &= \mathrm{Cov}\left(X_s, e^{-t+s}X_s + e^{-t}\int_s^t e^u dB_u\right) \\
&= e^{-t+s}\,\mathrm{Cov}(X_s, X_s) = e^{-t+s}Var(X_s) \\
&= e^{-t+s}e^{-2s}\int_0^s e^{2u}du \\
&= \frac{1}{2}e^{-t-s}(e^{2s}-1) = \frac{1}{2}(e^{-t+s} - e^{-t-s}).
\end{aligned}$$

(d) Since

$$X_t = e^{-t+s}X_s + e^{-t}\int_s^t e^u dB_u,$$

the conditional distribution of X_t given X_s is normal with mean $e^{-t+s}X_s$ and variance equal to

$$Var\left(e^{-t}\int_s^t e^u dB_u\right) = e^{-2t}\int_s^t e^{2u}du = \frac{1}{2}(1 - e^{-2(t-s)}).$$

(e) At any time s, X_s is a sum of two independent normals.

$$X_s = e^{-s}X_0 + e^{-s}\int_0^s e^u dB_u.$$

$$e^{-s}X_0 \sim N\left(0, \frac{1}{2}e^{-2s}\right)$$

$$e^{-s}\int_0^s e^u dB_u \sim N\left(0, e^{-2s}\int_0^s e^{2u}du\right).$$

$$e^{-2s}\int_0^s e^{2u}du = \frac{1}{2}(1 - e^{-2s}).$$

Hence, because variances of independent variables add up, we have for any $s > 0$,

$$X_s \sim N\left(0, \frac{1}{2}\right).$$

Problem 5.3: Let $X_t = (1-t)\int_0^t \frac{1}{1-s}dB_s$ and $Y_t = \int_0^t \frac{1}{1-s}dB_s$ for $0 \le t < 1$.

(a) Show that the process (Y_t) is well defined for $t < 1$, and it is a martingale.

(b) Show that the process (X_t), $0 \le t < 1$, solves the stochastic differential equation

$$dX_t = -\frac{X_t}{1-t}dt + dB_t, \quad X_0 = 0.$$

Hint: $X_t = (1-t)Y_t$, then use the product rule.

(c) State with reason whether the process (X_t) is a Gaussian process.

(d) Calculate the mean and covariance functions of the process (X_t).

(e) Let $Z_t = B_t - tB_1$, $0 \le t < 1$. State with reason why processes (X_t) and (Z_t) are the same Gaussian process (called a Brownian bridge). *Hint*: Calculate the mean and covariance functions for (Z_t).

Solution.

(a) We check that for $t < 1$, $\int_0^t (\frac{1}{1-s})^2 ds < \infty$:

$$\int_0^t \left(\frac{1}{1-s}\right)^2 ds = \left(\frac{1}{1-t} - 1\right) < \infty.$$

(b) $dX_t = d((1-t)Y_t) = -Y_t dt + (1-t)dY_t = -\frac{X_t}{1-t}dt + dB_t$
because $dY_t = \frac{1}{1-t}dB_t$, and $Y_t = X_t/(1-t)$.

(c) Y_t is a Gaussian process, as a stochastic integral of a deterministic function with independent Gaussian increments.

X_t is a Gaussian process because it is a deterministic function times Y_t.

To prove it, take $y \in \mathbb{R}^n$ and consider $X = (X_{t_1}, X_{t_2}, \ldots, X_{t_n})$:

$$yX = \sum_{i=1}^{n} y_i X_{t_i} = \sum_{i=1}^{n} y_i(1-t_i)Y_{t_i} = \sum_{i=1}^{n} \tilde{y}_i Y_{t_i},$$

where $\tilde{y}_i = y_i(1-t_i)$.

Since the process Y_t is Gaussian, the random variable $sum_{i=1}^{n}\tilde{y}_i Y_{t_i}$ is Gaussian, as a linear combination of Y_{t_i}'s. Thus, we obtain that for any y, yX is Gaussian. Hence, (X_t) is a Gaussian process.

(d) Since the Itô integral has zero mean (since $\int_0^t (\frac{1}{1-s})^2 ds < \infty$, the condition for integrability is verified)

$$\mathrm{E}(X_t) = 0,$$

$$\mathrm{Cov}(X_t, X_s) = \mathrm{Cov}((1-t)Y_t, (1-s)Y_s) = (1-t)(1-s)\,\mathrm{Cov}(Y_t, Y_s).$$

Take $s < t$. Then, due to the independence of the increments of Y process, and using the variance of Ito integral,

$$\mathrm{Cov}(Y_t, Y_s) = \mathrm{Cov}(Y_s, Y_s) = Var(Y_s) = \int_0^s \left(\frac{1}{1-u}\right)^2 du$$
$$= \frac{1}{1-s} - 1.$$

Finally, from above,

$$\mathrm{Cov}(X_t, X_s) = (1-t)s.$$

(e) $\mathrm{E}(Z_t) = \mathrm{E}(B_t - tB_1) = 0$. Take $s < t$. Then, by using the covariance function of Brownian motion, we obtain

$$\begin{aligned}\mathrm{Cov}(Z_t, Z_s) &= \mathrm{Cov}(B_t - tB_1, B_s - sB_1)\\ &= \mathrm{Cov}(B_t, B_s) - t\,\mathrm{Cov}(B_1, B_s) - s\,\mathrm{Cov}(B_t, B_1)\\ &\quad + st\,\mathrm{Cov}(B_1, B_1)\\ &= s - ts - st + st = s(1-t).\end{aligned}$$

It is easy to see that the process $Z_t = B_t - tB_1$ is Gaussian, for example by taking linear combinations. Thus, Z_t is a Gaussian process with a mean function zero and the same covariance function as the process X_t above. We have established earlier that the mean and covariance functions uniquely determine the Gaussian process, hence the processes Z_t and X_t are the same process. They give different representations of the Brownian bridge process.

Problem 5.4: Give the SDEs for $X(t) = \cos(B(t))$ and $Y(t) = \sin(B(t))$.

Solution. Let $f(x) = \cos(x)$, so $f_x(x) = -\sin(x)$ and $f_{xx}(x) = -\cos(x)$.

By Itô's formula,

$$dX(t) = df(B(t)) = -\sin(B(t))dt - \frac{1}{2}\cos(B(t))dB(t).$$

Similarly, let $g(x) = \sin(x)$, so $g_x(x) = \cos(x)$ and $g_{xx}(x) = -\sin(x)$, and by Ito's formula, we obtain

$$dY(t) = dg(B(t)) = \cos(B(t))dt - \frac{1}{2}\sin(B(t))dB(t).$$

Problem 5.5: Solve the SDE $dX(t) = B(t)X(t)dt + B(t)X(t)dB(t)$, $X(0) = 1$.

Solution. By definition of the stochastic exponential ($U(t) = \mathcal{E}(X)(t)$ if $dU(t) = U(t)dX(t)$ with $U(0) = 1$, $U(t) = e^{X(t)-X(0)-\frac{1}{2}[X,X](t)}$), X is the stochastic exponential of R with $dR(t) = B(t)dt + B(t)dB(t)$.

Consequently, $R(t) = \int_0^t B(s)ds + \int_0^t B(s)dB(s)$, and therefore $X(t) = \mathcal{E}(R)(t) = e^{\int_0^t (B(s)-\frac{1}{2}B^2(s))ds+\int_0^t B(s)dB(s)}$.

Problem 5.6: Solve the SDE $dX(t) = -\alpha X(t)dt + \sigma B(t)dB(t)$. Comment whether it is a diffusion-type SDE.

Solution. In a diffusion-type SDE, one should have $\sigma(X(t), t)$ (σ as a function of $X(t)$ and t only). Here, $\sigma(t) = \sigma B(t)$, thus it is not a diffusion-type SDE.

Let $dM(t) = \sigma B(t)dB(t)$, then $dX(t) = -\alpha X(t)dt + dM(t)$ is a Langevin-type SDE. Follow the solution of a Langevin-type SDE (as in Example 5.6).

Consider the process $Y(t) = X(t)e^{\alpha t}$. Use the differential of the product rule, and note that the covariation of $e^{\alpha t}$ with $X(t)$ is zero, as it is a differentiable function $(d(e^{\alpha t})dX(t) = \alpha e^{\alpha t}dtdX(t) = 0)$. Thus,

$$dY(t) = e^{\alpha t}dX(t) + \alpha e^{\alpha t}X(t)dt.$$

Substitute the SDE for $dX(t)$ to obtain

$$dY(t) = \sigma e^{\alpha t}B(t)dB(t).$$

This gives

$$Y(t) = Y(0) + \int_0^t \sigma e^{\alpha s}B(s)dB(s).$$

And the solution for $X(t)$ is

$$X(t) = e^{-\alpha t}\left(X(0) + \int_0^t \sigma e^{\alpha s}B(s)dB(s)\right).$$

Problem 5.7: Solve the SDE $dX_t = B_t dt + X_t dB_t$, $X_0 = 1$.

Solution. Consider $d(X_t e^{-B_t})$. Then, $d(e^{-B_t}) = -e^{-B_t}dB_t + \frac{1}{2}e^{-B_t}dt$.

$$\begin{aligned} d(X_t e^{-B_t}) &= e^{-B_t}dX_t + X_t d(e^{-B_t}) + dX_t d(e^{-B_t}) \\ &= e^{-B_t}(B_t dt + X_t dB_t) + X_t\left(-e^{-B_t}dB_t + \frac{1}{2}e^{-B_t}dt\right) \\ &\quad - X_t e^{-B_t}dt \\ &= e^{-B_t}B_t dt - \frac{1}{2}X_t e^{-B_t}dt. \end{aligned}$$

Hence, $y_t = X_t e^{-B_t}$ is differentiable and solves the ODE

$$y_t' = -\frac{1}{2}y_t + g_t,$$

where $g_t = e^{-B_t}B_t$. To solve this, multiply by $e^{t/2}$:

$$(y_t e^{t/2})' = g_t e^{t/2}.$$

Hence,

$$y_t e^{t/2} = y_0 + \int_0^t e^{s/2} g_s ds,$$

$$y_t = e^{-t/2} + e^{-t/2}\int_0^t e^{s/2}e^{-B_s}B_s ds.$$

Finally,

$$X_t = e^{B_t - t/2} + e^{B_t - t/2} \int_0^t e^{-B_s + s/2} B_s ds.$$

Of course, you can also use the general formula (5.31) from the book for solutions of linear SDEs to derive this solution.

Problem 5.8: Let $X_t = \mu(t) + \sigma(t)B_t$, for some deterministic functions $\mu(t)$ and $\sigma(t)$, $\sigma(t) \neq 0$ for $t > 0$, and $S_t = e^{X_t}$.

(a) Find the conditional expectation $E(S_t|S_s)$ for $s < t$.
(b) Give the relation between functions μ and σ so that the process S_t is a martingale.
(c) Assume that the functions μ and σ are differentiable. Derive the SDE for S_t.

Solution.

(a) $\mathrm{E}(S_t|S_s) = \mathrm{E}(e^{X_t}|S_s) = \mathrm{E}(e^{\mu(t)+\sigma(t)B_t}|S_s) = E(e^{\mu(t)+\sigma(t)B_t}|X_s)$,

since $S_s = \log X_s$ is a one-to-one transformation, so that the given S_s and X_s result in the same information. $B_s = \frac{X_s - \mu(s)}{\sigma(s)}$. Therefore, conditioning on X_s is the same as on B_s. Hence, using the independence of increments of Brownian motion,

$$\begin{aligned}\mathrm{E}(S_t|S_s) &= \mathrm{E}(S_t|B_s) = \mathrm{E}(e^{\mu(t)+\sigma(t)B_t}|B_s)\\ &= e^{\mu(t)+\sigma(t)B_s}\mathrm{E}(e^{\sigma(t)(B_t - B_s)}|B_s)\\ &= e^{\mu(t)+\sigma(t)B_s}e^{\sigma^2(t)(t-s)/2}.\end{aligned}$$

Next, by replacing B_s with X_s, we can write it as a function of S_s,

$$\begin{aligned}E(S_t|S_s) &= e^{\mu(t)+\sigma(t)(X_s - \mu(s))/\sigma(s)}e^{\sigma^2(t)(t-s)/2}\\ &= (S_s)^{\sigma(t)/\sigma(s)}e^{\mu(t)-\sigma(t)\mu(s)/\sigma(s)+\sigma^2(t)(t-s)/2}.\end{aligned}$$

Note here that if $\sigma(s) = 0$, then S_s (X_s) are deterministic and have no information about B_t for $t > s$. Therefore, $\mathrm{E}(S_t|S_s) = \mathrm{E}(S_t) = e^{\mu(t)+\sigma^2(t)t/2}$.

(b) Writing the martingale property

$$\mathrm{E}(S_t|\mathcal{F}_s) = e^{\mu(t)+\sigma(t)B_s+\sigma^2(t)(t-s)/2} = S_s = e^{\mu(s)+\sigma(s)B_s}$$

leads to

$$\mu(t) - \mu(s) + (\sigma(t) - \sigma(s))B_s + \sigma^2(t)(t-s)/2 = 0.$$

Since B_s is random, the only possibility is $\sigma(t) - \sigma(s) = 0$, which implies that σ is a constant, and $\mu(t) = \frac{\sigma^2}{2}t + \mu(0)$.

(c) $$dX_t = d(\mu(t) + \sigma(t)B_t) = \mu'(t)dt + \sigma'(t)B_t dt + \sigma(t)dB_t,$$

using integration by parts formula for the term $d(\sigma(t)B_t)$, since $d\sigma(t)dB_t = \sigma'(t)dtdB_t = 0$.

Use Itô's formula with $f(x) = e^x$, then $S_t = f(X_t)$.

$$\begin{aligned} dS_t &= df(X_t) = f'(X_t)dX_t + \frac{1}{2}f''(X_t)(dX_t)^2 \\ &= e^{X_t}dX_t + \frac{1}{2}e^{X_t}\sigma^2(t)dt. \end{aligned}$$

Since $(dX_t)^2 = (\sigma(t)dB_t)^2 = \sigma^2(t)dt$, all other terms with $dtdB_t$ and $(dt)^2$ are zero:

$$dS_t = \left(\mu'(t) + \sigma'(t)B_t + \frac{1}{2}\sigma^2(t)\right) S_t dt + \sigma(t)S_t dB_t.$$

Note that the drift term includes B_t also.

The same formula can be obtained by using Itô's formula for B_t with $f(x,t) = e^{\mu(t)+\sigma(t)x}$.

One can recover the condition for S_t to be a martingale by requiring the dt term to be zero.

Problem 5.9: Let $X_t = e^{B_t^3}$. Show that X is a diffusion process that does not have mean (and second moment).

Solution. It is clear that $X_0 = 1$ and $X_t > 0$ for all t. By Itô's formula, $X_t > 0$ satisfies the SDE

$$dX_t = \frac{1}{2}(6(\ln(X_t))^{1/3} + 9(\ln(X_t))^{4/3})X(t)dt + 3(\ln(X_t))^{2/3}X(t)dB_t.$$

The coefficients do not have linear growth. Further,

$$E(X_t) = E(e^{(\sqrt{t}B_1)^3}) = \int_{-\infty}^{\infty} e^{t^{3/2}x^3}\frac{1}{\sqrt{2\pi}}e^{-x^2}dx = \infty.$$

Problem 5.10: Find $d(\mathcal{E}(B)(t)^2)$.

Solution. Let $U(t) = \mathcal{E}(B)(t)$. Then,

$$\begin{aligned} dU(t) &= U(t)dB(t) \\ \text{and } dU^2(t) &= 2U(t)dU(t) + d[U,U](t) \\ &= 2U^2(t)dB(t) + U^2(t)dt = U^2(t)d(2B(t)+t). \end{aligned}$$

So, $U^2(t) = \mathcal{E}(2B(t)+t)$ and $d(\mathcal{E}(B)(t)^2) = \mathcal{E}(2B(t)+t)d(B(t)+t)$.

Problem 5.11: Let $X(t)$ satisfy $dX(t) = X^2(t)dt + X(t)dB(t)$, $X(0) = 1$. Show that $X(t)$ satisfies $X(t) = e^{\int_0^t (X(s)-1/2)ds + B(t)}$.

Solution. Note that $dX(t) = X(t)(X(t)dt + dB(t))$, and let $dY(t) = X(t)dt + dB(t)$, so $Y(t) = \int_0^t X(s)ds + B(t)$.

By definition of the stochastic exponential, $X(t) = \mathcal{E}(Y)(t) = e^{\int_0^t (X(s)-\frac{1}{2})ds + B(t)}$.

Problem 5.12: By definition, the stochastic logarithm satisfies $\mathcal{L}(\mathcal{E}(X)) = X$. Show that, provided $U(t) \neq 0$ for any t, $\mathcal{E}(\mathcal{L}(U))(t) = U(t)/U(0)$. In particular, if $U(0) = 1$, $\mathcal{E}(\mathcal{L}(U)) = U$.

Solution. The stochastic logarithm X of U (notation $X = \mathcal{L}(U)$) by definition satisfies the SDE $dX(t) = \frac{1}{U(t)}dU(t)$, $X(0) = 0$. Hence, $dU(t) = U(t)dX(t)$; therefore, $U/U(0)$ is the stochastic exponential of X, namely $U(t) = U(0)\mathcal{E}(X)(t)$. Thus, $\mathcal{E}(\mathcal{L}(U))(t) = U(t)/U(0)$.

Problem 5.13: Find the stochastic logarithm of $B^2(t) + 1$.

Solution. Let $U(t) = B^2(t) + 1$. Then, by Itô's formula, $dU(t) = 2B(t)dB(t) + dt$, $U(0) = 1$, and $d[U,U](t) = 4B^2(t)dt$.

Thus,

$$\mathcal{L}(U)(t) = \ln(B^2(t)+1) + 2\int_0^t \frac{B^2(t)dt}{(B^2(t)+1)^2}.$$

Problem 5.14: Let $\boldsymbol{B}(t)$ be a d-dimensional Brownian motion and $\mathbf{H}(t)$ a d-dimensional square-integrable adapted process. Show that

$$\mathcal{E}\left(\int_0^{\cdot} \boldsymbol{H}(s)d\boldsymbol{B}(s)\right)(t) = \exp\left(\int_0^t \boldsymbol{H}(s)d\boldsymbol{B}(s) - \frac{1}{2}\int_0^t |\boldsymbol{H}(s)|^2\, ds\right).$$

Here, $\boldsymbol{a}\cdot\boldsymbol{b}$ denotes the inner (scalar) product between vectors $\boldsymbol{a}$, $\boldsymbol{b}$.

Solution. Consider the scalar process

$$U(t) = \int_0^t \boldsymbol{H}(s)d\boldsymbol{B}(s) = \sum_{i=1}^d \int_0^t H_i(s)dB_i(s).$$

(Its stochastic differential can be written as $dU = \mathbf{H}\cdot d\mathbf{B}$.) Since all the Brownian motions are independent of each other, $[B_i, B_j] = 0$ for $i \neq j$, and using the rule $[\int HdX, \int KdY] = \int HKd[X,Y]$, we obtain that

$$\begin{aligned}[U,U](t) &= \sum_{i=1}^d \sum_{j=1}^d \left[\int H_i dB_i, \int H_j dB_j\right] \\ &= \sum_{i=1}^d \int_0^t H_i^2(s)ds = \int_0^t |\mathbf{H}(s)|^2 ds.\end{aligned}$$

Hence, by definition of stochastic exponential,

$$\begin{aligned}\mathcal{E}(U)(t) &= \exp\left(U(t) - U(0) - \frac{1}{2}[U,U](t)\right) \\ &= \exp\left(\int_0^t \boldsymbol{H}(s)d\boldsymbol{B}(s) - \frac{1}{2}\int_0^t |\mathbf{H}(s)|^2 ds\right).\end{aligned}$$

Problem 5.15: Give the forward and backward equations for the transition probability density function of Brownian motion $B(t)$ and verify that the function

$$p(y,t,x,s) = \frac{1}{\sqrt{2\pi}\sqrt{t-s}} e^{-\frac{(y-x)^2}{2(t-s)}}$$

is a solution. Introduce the variable $\tau = t - s$. Write the backward and forward equations in terms of τ for $p(\tau, x, y)$. Specify the solution of the backward equation for $t = 1$ and $y = 0$.

Solution. The backward equation is

$$\frac{\partial p}{\partial s} = -\frac{1}{2}\frac{\partial^2 p}{\partial x^2}.$$

The forward equation is

$$\frac{\partial p}{\partial t} = \frac{1}{2}\frac{\partial^2 p}{\partial y^2}.$$

In terms of τ, the forward equation is $\frac{\partial p}{\partial \tau} = \frac{1}{2}\frac{\partial^2 p}{\partial y^2}$, and the backward equation is the same as the forward equation because $\frac{d}{d\tau} = -\frac{d}{ds}$,

$\frac{\partial p}{\partial \tau} = \frac{1}{2}\frac{\partial^2 p}{\partial x^2}$. The solution to the backward equation for $t = 1$ and $y = 0$ is

$$p(1-s, x, 0) = \frac{1}{\sqrt{2\pi}\sqrt{1-s}} e^{-\frac{x^2}{2(1-s)}} \quad \text{for } s \in [0, 1).$$

Problem 5.16: Find the transition probability function $P(y, t, x, s)$ for Brownian motion with drift $B(t) + t$.

Solution. Let $X(t) = B(t) + t$.

By definition, its transition probability function is $P(y, t, x, s) = \mathrm{P}(X(t) \le y | X(s) = x)$. Hence,

$$\begin{aligned} &P(y, t, x, s) \\ &= \mathrm{P}(B(t) + t \le y | B(s) + s = x) \\ &= \mathrm{P}(B(t) - B(s) + t - s \le y - x | B(s) + s = x) \\ &= \mathrm{P}(B(t) - B(s) + t - s \le y - x) \text{ by independence of increments.} \\ &= \mathrm{P}(B(t) - B(s) \le y - x - t + s). \end{aligned}$$

Further, $B(t) - B(s)$ has $N(0, t-s)$ distribution, and thus $P(y, t, x, s) = \Phi\left(\frac{y-x-t+s}{\sqrt{t-s}}\right)$.

Problem 5.17: Let X_t solve the SDE $dX_t = -X_t dt + dB_t$, $X_0 = 0$.

(a) Give the transition probability function of X_t, i.e. $P(y, t, x, s) = \mathrm{P}(X_t \le y | X_s = x)$ and its density function.
(b) Show that the transition probability density solves the PDE

$$\frac{\partial u}{\partial s} + \frac{1}{2}\frac{\partial^2 u}{\partial x^2} - x\frac{\partial u}{\partial x} = 0.$$

Solution.

(a) Solving the SDE on the time interval $[s, t]$, we have

$$d(e^t X_t) = e^t dX_t + X_t e^t dt = e^t(-X_t dt + dB_t) + X_t e^t dt = e^t dB_t,$$

$$e^t X_t - e^s X_s = \int_s^t e^u dB_u,$$

$$X_t = X_s e^{-(t-s)} + e^{-t}\int_s^t e^u dB_u.$$

Since $\int_s^t e^u dB_u$ has $N(0, \int_s^t e^{2u}du)$ distribution, $e^{-t}\int_s^t e^u dB_u$ has $N(0, \frac{1-e^{-2(t-s)}}{2})$ distribution, independent of X_s. Therefore, the conditional distribution of X_t given $X_s = x$ is

$$N\left(xe^{-(t-s)}, \frac{1-e^{-2(t-s)}}{2}\right).$$

Hence,

$$\begin{aligned} P(y,t,x,s) &= \mathrm{P}(X_t \le y | X_s = x) \\ &= \mathrm{P}\left(N\left(xe^{-(t-s)}, \frac{1-e^{-2(t-s)}}{2}\right) \le y\right). \end{aligned}$$

This can be expressed using the standard normal cumulative probability function by taking away the mean and dividing it by the standard deviation:

$$P(y,t,x,s) = \Phi\Big(\sqrt{2}\frac{y - xe^{-(t-s)}}{\sqrt{1-e^{-2(t-s)}}}\Big).$$

Its density function is $\frac{\partial}{\partial y}P$:

$$p(y,t,x,s) = \frac{\partial P(y,t,x,s)}{\partial y} = \phi\Big(\sqrt{2}\frac{y - xe^{-(t-s)}}{\sqrt{1-e^{-2(t-s)}}}\Big)\frac{\sqrt{2}}{\sqrt{1-e^{-2(t-s)}}},$$

where $\phi(z) = \frac{1}{\sqrt{2\pi}}e^{-\frac{z^2}{2}}$ is the standard normal density.

(b) This is the backward PDE. $L = \frac{1}{2}\frac{\partial^2}{\partial x^2} - x\frac{\partial}{\partial x}$, $Lu + \frac{\partial u}{\partial s} = 0$.

We can obtain this PDE by differentiating $p(y,t,x,s)$ directly.

Direct calculations use $\phi'(z) = \phi(z)(-z)$, $\phi''(z) = \phi(z)(z^2-1)$ and somewhat tedious differentiations of composite functions.

Problem 5.18: Let $X(t)$ satisfy the following SDE for $0 \le t \le T$, $dX(t) = \sqrt{X(t)+1}dB(t)$, and $X(0) = 0$. Assuming that Itô integrals are martingales (this assumption will be verified later by using stopping times), find $\mathrm{E}X(t)$ and $\mathrm{E}(X^2(t))$. Let $m(u,t) = \mathrm{E}e^{uX(t)}$ be the moment-generating function of $X(t)$. Show that it satisfies the PDE

$$\frac{\partial m}{\partial t} = \frac{u^2}{2}\frac{\partial m}{\partial u} + \frac{u^2}{2}m.$$

Solution. From the stochastic differential $dX(t) = \sqrt{X(t)+1}dB(t)$, obtain the Itô integral:

$$X(t) = \int_0^t \sqrt{X(s)+1}dB(s).$$

Assume it is a martingale, by the zero mean property of Itô integrals $\mathrm{E}X(t) = 0$ and by the isometry property,

$$\mathrm{E}X^2(t) = \mathrm{E}\left(\int_0^t \sqrt{X(s)+1}dB(s)\right)^2 = \int_0^t \mathrm{E}\left(X(s)+1\right)ds = t.$$

To obtain the moment-generating function, first use Itô's formula with $f(x) = e^{ux}$, $(f'(x) = ue^{ux}$, $f''(x) = u^2e^{ux})$ and note $d[X,X](t) = (X(t)+1)dt$:

$$\begin{aligned} de^{uX(t)} &= ue^{uX(t)}dX(t) + \frac{1}{2}u^2e^{uX(t)}d[X,X](t) \\ &= ue^{uX(t)}\sqrt{X(t)+1}dB(t) + \frac{1}{2}u^2e^{uX(t)}(X(t)+1)dt. \end{aligned}$$

Hence, in integral form,

$$e^{uX(t)} = 1 + \int_0^t ue^{uX(s)}\sqrt{X(s)+1}dB(s) + \frac{1}{2}\int_0^t u^2e^{uX(s)}(X(s)+1)ds.$$

Take expectation and use the fact that Itô integral has zero mean:

$$m(t,u) = \mathrm{E}e^{uX(t)} = 1 + \frac{u^2}{2}\int_0^t \mathrm{E}e^{uX(s)}(X(s)+1)ds.$$

Note that we used Fubini's theorem to interchange the integral and expectation. Thus, $\frac{\partial m}{\partial t} = \frac{u^2}{2}\mathrm{E}\left(e^{uX(t)}X(t)\right) + \frac{u^2}{2}\mathrm{E}e^{uX(t)}$.

Finally, from the definition of the moment-generating function $m(t,u)$, $\frac{\partial m}{\partial u} = \mathrm{E}\left(e^{uX(t)}X(t)\right)$, and the desired PDE follows.

Problem 5.19: One can prove the following important generalization of Itô's formula (Theorem 4.16 in Klebaner's book): for an Itô process $\{X(t)\}_{t\in[0,T]}$ all values of which belong to an open interval $I \subseteq \mathbb{R}$ with probability 1 and a two times continuously differentiable function $f: I \to \mathbb{R}$, it holds that

$$df(X(t)) = f'(X(t))\,dX(t) + \frac{1}{2}f''(X(t))\,d[X,X](t) \quad \text{for } t \in [0,T].$$

Use this result to give a detailed proof of Theorem 5.3 in Klebaner's book.

Solution. Let $\{U(t)\}_{t\in[0,T]}$ be a strictly positive Itô process with probability 1. Then, we may apply the above-mentioned generalized Itô formula to the function $Y(t) = \log(U(t)) - \log(U(0))$ to conclude that

$$dY(t) = \frac{dU(t)}{U(t)} - \frac{1}{2}\frac{d[U](t)}{U(t)^2},$$

so that

$$U(t)\,d\Big(\log\Big(\frac{U(t)}{U(0)}\Big)+\frac{1}{2}\int_0^t\frac{d[U](r)}{U(r)^2}\Big)=U(t)\,d\Big(Y(t)+\frac{1}{2}\int_0^t\frac{d[U](r)}{U(r)^2}\Big)$$
$$=dU(t).$$

This means that the Itô process

$$\mathcal{L}(U(t))\equiv\log\Big(\frac{U(t)}{U(0)}\Big)+\frac{1}{2}\int_0^t\frac{d[U](r)}{U(r)^2}$$

has stochastic exponential $U(t)$ and therefore is the stochastic logarithm of $U(t)$. By multiplying both sides of the above equation by $1/U(t)$, we also see that $\mathcal{L}(U(t))$ obeys the equation

$$d\mathcal{L}(U(t))=\frac{1}{U(t)}\,dU(t),\quad \mathcal{L}(U(0))=0.$$

(Note that this SDE is not of diffusion type in general.)

Problem 5.20: The filtration $\{\mathcal{F}_t\}$ that features in the construction of the Itô integral process need not necessarily be the filtration $\{\mathcal{F}_t^B\}$ generated by B itself, but can more generally be as in Remark 3.1 in Klebaner's book. In particular, if $\{B_1(t)\}_{t\geq 0}$ and $\{B_2(t)\}_{t\geq 0}$ are independent Brownian motions, then we may employ the filtration $\{\mathcal{F}_t\}_{t\geq 0}$ given by $\mathcal{F}_t=\sigma(\mathcal{F}_t^{B_1},\mathcal{F}_t^{B_1})$ for $t\geq 0$ to be able to simultaneously consider the Itô integral process (and therefore the SDE as well) with respect to both B_1 and B_2.

The Nobel Prize-awarded Black–Scholes–Merton SDE

$$dX(t)=r\,X(t)\,dt+\sigma\,X(t)\,dB(t)\quad\text{for } t>0,\quad X(0)=x_0,$$

for future values $\{X(t)\}_{t>0}$ of a financial asset with an uncertain rate of return might be generalized to a model that can much more accurately model real-world financial assets, such as stock prices as follows: with the notation from the previous paragraph, consider the SDE (not in general of diffusion type)

$$dX(t)=r\,X(t)\,dt+\sigma(t)\,X(t)\,dB_1(t)\quad\text{for } t>0,\quad X(0)=x_0,$$

where the constant so-called volatility parameter $\sigma\in\mathbb{R}$ of the Black–Scholes–Merton SDE has been replaced with a random volatility process $\{\sigma(t)\}_{t\geq 0}$ that can model a market that features a time variable uncertainty for the rate of the return. Solve this more general SDE when the volatility process $\{\sigma(t)\}_{t\geq 0}$ is given by the SDE

$$d\sigma(t) = -\alpha\,\sigma(t)\,dt + \beta\,dB_2(t) \quad \text{for } t>0, \quad \sigma(0)=\sigma_0,$$

where $\alpha, \beta > 0$ are positive real constants (as is r).

Solution. Identifying X as a stochastic exponential we get

$$X(t) = x_0 \exp\left\{rt - \frac{1}{2}\int_0^t \sigma(s)^2\,ds + \int_0^t \sigma(s)\,dB_1(s)\right\} \quad \text{for } t\geq 0$$

(see Section 5.3 in Klebaner's book), where σ in turn is recognized as the solution to a Langevin-type SDE:

$$\sigma(t) = \exp\left\{-\int_0^t \alpha(s)\,ds\right\}\left(\sigma_0 + \beta\int_0^t \exp\left\{\int_0^s \alpha(r)\,dr\right\}dB_2(s)\right)$$
$$\text{for } t\geq 0$$

(see Example 5.6 and Section 5.3 in Klebaner's book).

Problem 5.21: Solve the SDE

$$dX(t) = \left(\sqrt{1+X(t)^2} + \frac{X(t)}{2}\right)dt + \sqrt{1+X(t)^2}\,dB(t)$$
$$\text{for } t>0, \quad X(0)=0.$$

Solution. First, note that all conditions of Theorem 5.4 in Klebaner's book are satisfied, so that it is clear that the SDE has a well-defined and unique solution. Now, by divine inspiration, we readily arrive at the idea to try the transformation $Y(t) = \sinh^{-1}(X(t))$. By an application of Itô's formula, Theorem 4.16 in Klebaner's book, we then get

$$\begin{aligned} dY(t) &= \frac{1}{\sqrt{1+X(t)^2}}\,dX(t) - \frac{X(t)}{2\,(1+X(t)^2)^{3/2}}\,d[X,X](t) \\ &= dt + \frac{X(t)}{2\sqrt{1+X(t)^2}}\,dt + dB(t) - \frac{X(t)}{2\sqrt{1+X(t)^2}}\,dt \\ &= dt + dB(t), \end{aligned}$$

with the obvious solution $Y(t) = t + B(t)$ [remembering that $Y(0)=0$]. Hence, the solution to the SDE must be $X(t) = \sinh(t+B(t))$. That this process X really solves the SDE is also easy to check by means of direct calculations using Itô's formula (Theorem 4.18 in Klebaner's book) together with the hyperbolic unit formula.

Problem 5.22: Show that

$$X(t) = \mathrm{e}^{-\alpha t}\Big(\frac{\sigma}{\sqrt{2\alpha}}\,(B(\mathrm{e}^{2\alpha t}) - B(1)) + x_0\Big) \quad \text{for } t \geq 0$$

is an Ornstein–Uhlenbeck process in the sense that it has got the same distributional properties (finite-dimensional distributions) as the solution

$$\{X(t)\}_{t\geq 0} = \Big\{\mathrm{e}^{-\alpha t}\Big(x_0 + \sigma \int_0^t \mathrm{e}^{\alpha r}\,dB(r)\Big)\Big\}_{t\geq 0}$$

to the Langevin SDE

$$dX(t) = -\alpha\, X(t)\,dt + \sigma\, dB(t) \quad \text{for } t > 0, \quad X(0) = x_0,$$

where $\alpha, \sigma > 0$ and $x_0 \in \mathbb{R}$ are constants.

Solution. As both the above X processes are Gaussian, they have the same finite-dimensional distributions if their mean and covariance functions agree. Here, we clearly have $\mathrm{E}\{X(t)\} = \mathrm{e}^{-\alpha t} x_0$ for $t \geq 0$ for both the X processes. Further, we have

$$\begin{aligned} \mathrm{Cov}\{X(s), X(t)\} &= \frac{\sigma^2}{2\alpha}\,\mathrm{e}^{-\alpha(s+t)}\,\mathrm{Cov}\left\{B(\mathrm{e}^{2\alpha s}) - B(1), B(\mathrm{e}^{2\alpha t}) - B(1)\right\} \\ &= \frac{\sigma^2}{2\alpha}\,\mathrm{e}^{-\alpha(s+t)}\left(\mathrm{e}^{2\alpha \min\{s,t\}} - 1 - 1 + 1\right) \\ &= \frac{\sigma^2}{2\alpha}\left(\mathrm{e}^{-\alpha|s-t|} - \mathrm{e}^{-\alpha(s+t)}\right) \quad \text{for } s, t \geq 0 \end{aligned}$$

for the first X process, while Theorem 4.11 in Klebaner's book shows that

$$\begin{aligned} \mathrm{Cov}\{X(s), X(t)\} &= \sigma^2\,\mathrm{e}^{-\alpha(s+t)} \int_0^{\min\{s,t\}} \mathrm{e}^{2\alpha r}\,dr \\ &= \frac{\sigma^2}{2\alpha}\left(\mathrm{e}^{-\alpha|s-t|} - \mathrm{e}^{-\alpha(s+t)}\right) \end{aligned}$$

for $s, t \geq 0$ for the second X process.

Problem 5.23: Use the expression for an Ornstein–Uhlenbeck process expressed in terms of B from the previous exercise to find the transition density function for the solution to the Langevin SDE (the Ornstein–Uhlenbeck process).

Solution. We have

$$X(t+s) = \mathrm{e}^{-\alpha(t+s)}\Big(\frac{\sigma}{\sqrt{2\alpha}}\,\big(B(\mathrm{e}^{2\alpha(t+s)}) - B(1)\big) + x_0\Big)$$
$$= \mathrm{e}^{-\alpha(t+s)}x_0 + \frac{\sigma}{\sqrt{2\alpha}}\,\mathrm{e}^{-\alpha(t+s)}$$
$$\times\big(\big(B(\mathrm{e}^{2\alpha(t+s)}) - B(\mathrm{e}^{2\alpha s})\big) + \big(B(\mathrm{e}^{2\alpha s}) - B(1)\big)\big)$$
$$= \frac{\sigma}{\sqrt{2\alpha}}\,\mathrm{e}^{-\alpha(t+s)}\big(B(\mathrm{e}^{2\alpha(t+s)}) - B(\mathrm{e}^{2\alpha s})\big) + \mathrm{e}^{-\alpha t}X(s),$$

where

$$\frac{\sigma}{\sqrt{2\alpha}}\,\mathrm{e}^{-\alpha(t+s)}\big(B(\mathrm{e}^{2\alpha(t+s)}) - B(\mathrm{e}^{2\alpha s})\big)$$

is an $\mathrm{N}\big(0, (\sigma^2/(2\alpha))\,(1-\mathrm{e}^{-2\alpha t})\big)$-distributed random variable independent of $\{X(r)\}_{r\le s}$.

It follows that $(X(t+s)\,|X(s)=x)$ is $\mathrm{N}\big(\mathrm{e}^{-\alpha t}x, (\sigma^2/(2\alpha))\,(1-\mathrm{e}^{-2\alpha t})\big)$-distributed, so that

$$p(y,t+s,x,s) = \frac{d}{dy}\,P(y,t+s,x,s)$$
$$= \frac{\sqrt{\alpha}}{\sqrt{\pi\,(1-\mathrm{e}^{-2\alpha t})}\,\sigma}\,\exp\Big\{-\frac{\alpha\,(y-x\,\mathrm{e}^{-\alpha t})^2}{\sigma^2\,(1-\mathrm{e}^{-2\alpha t})}\Big\}$$

for $t+s > s \ge 0$ and $x, y \in \mathbb{R}$.

Problem 5.24: Solve the following Stratonovich SDE $\partial U = U\partial B$, $U(0) = 1$, where $B(t)$ is Brownian motion.

Solution. Use the Stratonovich chain rule $\partial f(X(t)) = f'(X(t))\partial X$ to see that $U(t) = e^{B(t)}$ is the solution.

Problem 5.25: Solve the Stratanovich SDE

$$dX(t) = -\alpha\,dt + \sigma\,X(t)\,\partial B(t) \quad \text{for } t>0, \quad X(0)=x_0,$$

where $\alpha, \sigma > 0$ and $x_0 \in \mathbb{R}$ are constants.

Solution. By Theorem 5.20 in Klebaner's book, the above SDE is equivalent to the Itô SDE

$$dX(t) = \big(\tfrac{1}{2}\,\sigma^2\,X(t) - \alpha\big)\,dt + \sigma\,X(t)\,dB(t) \quad \text{for } t>0, \quad X(0)=x_0.$$

This in turn is a rather simple form of the linear SDE treated in Section 5.3 in Klebaner's book, with a solution given by

$$X(t) = U(t)\left(x_0 - \alpha \int_0^t \frac{ds}{U(s)}\right) \quad \text{where} \quad U(t) = \mathrm{e}^{\sigma B(t)},$$

which is to say that

$$X(t) = x_0\,\mathrm{e}^{\sigma B(t)} - \alpha\,\mathrm{e}^{\sigma B(t)} \int_0^t \mathrm{e}^{-\sigma B(s)}\,ds \quad \text{for } t \geq 0.$$

Problem 5.26: Show that the SDE $dX(t) = 3X^{1/3}(t)dt + 3X^{2/3}(t)dB(t)$ with initial condition $X(0) = 0$ does not have a unique solution. Explain which conditions fail in the existence and uniqueness theorems. Compare with Example 5.12 (Giranov's SDE).

Solution. The two solutions are $X(t) = 0$ and $X(t) = B^3(t)$. The coefficient $\mu(x) = 3x^{1/3}$ and $\sigma(x) = 3x^{2/3}$ are not Lipschitz on intervals including zero. Thus, the conditions of the uniqueness theorem (Theorem 5.4) fail. While $\sigma(x)$ is Hölder continuous of order 2/3, $\mu(x) = 3x^{1/3}$ is not Lipschitz, and thus the conditions of the Yamada–Watanabe uniqueness theorem (Theorem 5.5) also fail.

Problem 5.27: Let $f(x)$ satisfy the global Lipschitz condition:

$$|f(x) - f(y)| \leq C|x - y|, \quad \text{for all } x, y.$$

Show that f satisfies the linear growth condition.

Solution. Take $y = 0$. Then, $|f(x) - f(0)| \leq C|x|$. Hence,

$$|f(x)| = |f(x) - f(0) + f(0)| \leq |f(x) - f(0)| + |f(0)| \leq |f(0)| + C|x|.$$

Hence, for a suitable constant K, $|f(x)| \leq K(1 + |x|)$.

Problem 5.28: This exercise gives an idea for a proof of the uniqueness theorem under the Lipschitz condition.

Show the uniqueness of the solution of the SDE

$$dX_t = \mu(X_t)dt + \sigma(X_t)dB_t, \quad X_0 = x_0,$$

when the functions $\mu(x)$ and $\sigma(x)$ satisfy the Lipschitz condition, for any x, y,

$$|\mu(x) - \mu(y)| \leq C_1|x - y|, \quad |\sigma(x) - \sigma(y)| \leq C_2|x - y|.$$

You may assume without proof that expectation exists and Itô integrals have zero mean. In later chapters, this assumption will be proved by using stopping times, the so-called localization technique.

Solution. We assume that there are two solutions, X_t and Y_t, and show that for any t, $\mathrm{E}(X_t - Y_t)^2 = 0$, implying that they are the same. Let $Z_t = X_t - Y_t$. Then,

$$dZ_t = dX_t - dY_t = \big(\mu(X_t) - \mu(Y_t)\big)dt + \big(\sigma(X_t) - \sigma(Y_t)\big)dB_t,$$

and since $X_0 = Y_0$, $Z_0 = 0$. Consider dZ_t^2, by Itô's formula,

$$\begin{aligned} dZ_t^2 &= 2Z_t\big(\mu(X_t) - \mu(Y_t)\big)dt + \big(\sigma(X_t) - \sigma(Y_t)\big)^2 dt \\ &\quad + 2Z_t\big(\sigma(X_t) - \sigma(Y_t)\big)dB_t, \\ Z_t^2 &= 2\int_0^t Z_s\big(\mu(X_s) - \mu(Y_s)\big)ds + \int_0^t \big(\sigma(X_s) - \sigma(Y_s)\big)^2 ds \\ &\quad + 2\int_0^t Z_s\big(\sigma(X_s) - \sigma(Y_s)\big)dB_s. \end{aligned}$$

Now, take expectation, using that the expectation of the Itô integral is zero:

$$\mathrm{E}Z_t^2 = 2\mathrm{E}\int_0^t Z_s\big(\mu(X_s) - \mu(Y_s)\big)ds + \mathrm{E}\int_0^t \big(\sigma(X_s) - \sigma(Y_s)\big)^2 ds.$$

Use the Lipschitz condition next, with $C = \max(C_1, C_2)$:

$$\mathrm{E}Z_t^2 \leq 2C\mathrm{E}\int_0^t Z_s^2 ds + C^2\mathrm{E}\int_0^t Z_s^2 ds.$$

Interchanging the integral and expectation by Fubini, as the integrand is positive, we have with $C_3 = 2C + C^2$,

$$\mathrm{E}Z_t^2 \leq C_3 \textstyle\int_0^t \mathrm{E}Z_s^2 ds.$$

The Gronwall inequality now implies that $\mathrm{E}Z_t^2 = 0$. Thus, $\mathrm{P}(X_t = Y_t) = 1$.

Problem 5.29: For proving uniqueness of solutions under a weaker assumption than the Lipschitz condition on the coefficients, it is important to approximate the function $|x|$ by twice continuously differentiable functions. Let $\varepsilon \in (0, 1)$ and let

$$\psi_\varepsilon(u) = \frac{1}{u\log(1/\sqrt{\varepsilon})} 1_{[\varepsilon, \sqrt{\varepsilon}]}(u).$$

Define for $x \in \mathbb{R}$,

$$h_\varepsilon(x) = \int_0^{|x|} \int_0^y \psi_\varepsilon(u) du dy.$$

Show that for all x:

(a) $|x| \le h_\varepsilon(x) + \sqrt{\varepsilon}$,
(b) $0 \le |h'_\varepsilon(x)| \le 1$,
(c) $h''_\varepsilon(x) = \psi_\varepsilon(|x|) = \frac{1}{|x|\log(1/\sqrt{\varepsilon})} 1_{[\varepsilon,\sqrt{\varepsilon}]}(|x|)$.

In fact, we can take any function satisfying $\psi_\varepsilon(u) \ge 0$ zero outside $[\varepsilon, \sqrt{\varepsilon}]$, $\int_\varepsilon^{\sqrt{\varepsilon}} \psi_\varepsilon(u) du = 1$, and

$$\psi_\varepsilon(u) \le \frac{2}{u \log(1/\sqrt{\varepsilon})}.$$

Function $h_\varepsilon(x)$ is a smooth approximation of $|x|$, used in the proof of the Yamada–Watanabe theorem.

Solution. Obviously,

$$h'_\varepsilon(x) = \int_0^x \psi_\varepsilon(u) du = \begin{cases} 0 & 0 < x \le \varepsilon, \\ \dfrac{\log(x/\varepsilon)}{\log(1/\sqrt{\varepsilon})} & \varepsilon < x \le \sqrt{\varepsilon}, \\ 1 & x > \sqrt{\varepsilon}, \end{cases}$$

and hence $|h'_\varepsilon(x)| \le 1$. Moreover, since $h_\varepsilon(x)$ is symmetric around zero, $h'_\varepsilon(x) \ge 1_{[\sqrt{\varepsilon},\infty)}(x)$ for $x > 0$ and $h_\varepsilon(0) = 0$, then

$$|x| \le h_\varepsilon(x) + \sqrt{\varepsilon}.$$

Further, it follows directly that

$$h''_\varepsilon(x) = \psi_\varepsilon(|x|) = \frac{1}{|x| \log(1/\sqrt{\varepsilon})} 1_{[\varepsilon,\sqrt{\varepsilon}]}(|x|).$$

Problem 5.30: This exercise gives an idea for a proof of the Yamada–Watanabe uniqueness theorem. It uses the method of approximation of non-differentiable function $|x|$ by twice differentiable functions.

Show the uniqueness of the solution of the SDE (Feller diffusion)

$$dX_t = \mu X_t dt + \sigma \sqrt{X_t} dB_t, \quad X_0 = x_0 > 0.$$

You may assume without proof that expectation exists and Itô integrals have zero mean. In later chapters, this assumption will be proved by using stopping times, the so-called localization technique.

Solution. We assume that there are two solutions, X_t and Y_t, and show that for any t, $\mathrm{E}|X_t - Y_t| = 0$, implying that $\mathrm{P}(X_t = Y_t) = 1$, they are the same. Let $Z_t = X_t - Y_t$. Then,

$$dZ_t = dX_t - dY_t = \mu(X_t - Y_t)dt + \sigma\big(\sqrt{X_t} - \sqrt{Y_t}\big)dB_t,$$

with $Z_0 = 0$. Let $\varepsilon > 0$ and consider $dh_\varepsilon(Z_t)$, where h_ε is the function defined in the previous exercise. By Itô's formula,

$$dh_\varepsilon(Z_t) = h'_\varepsilon(Z_t)\mu Z_t dt + \frac{\sigma^2}{2}h''_\varepsilon(Z_t)\big(\sqrt{X_t} - \sqrt{Y_t}\big)^2 dt \\ + h'_\varepsilon(Z_t)\big(\sqrt{X_t} - \sqrt{Y_t}\big)dB_t.$$

Write this in integral form and take expectation. Since the expectation of the Itô integral is zero, we obtain

$$\mathrm{E}h_\varepsilon(Z_t) = \mu\mathrm{E}\int_0^t h'_\varepsilon(Z_s)Z_s ds + \frac{\sigma^2}{2}\mathrm{E}\int_0^t h''_\varepsilon(Z_s)\big(\sqrt{X_s} - \sqrt{Y_s}\big)^2 ds.$$

The function $\sqrt{x}$ is Hölder of order 1/2 (this can be seen by squaring both sides of the inequality):

$$|\sqrt{x} - \sqrt{y}| \le |x - y|^{1/2}.$$

Using the inequalities $|\mathrm{E}U| \le \mathrm{E}|U|$ and $|\int_0^t g(s)ds| \le \int_0^t |g(s)|ds$ in the above equation yields

$$\mathrm{E}h_\varepsilon(Z_t) \le |\mu|\int_0^t \mathrm{E}|h'_\varepsilon(Z_s)||Z_s|ds + \frac{\sigma^2}{2}\int_0^t \mathrm{E}h''_\varepsilon(Z_s)|Z_s|ds.$$

Now, we use the properties of the function h_ε and its derivatives given in the previous Problem 5.29 to have

$$h''_\varepsilon(Z_s)|Z_s| = \frac{1}{|Z_s|\log(1/\sqrt{\varepsilon})}1_{[\varepsilon,\sqrt{\varepsilon}]}(|Z_s|)|Z_s| \le \frac{1}{\log(1/\sqrt{\varepsilon})}.$$

Proceeding from the previous inequality, we have

$$\mathrm{E}|Z_t| \le \mathrm{E}h_\varepsilon(Z_t) + \sqrt{\varepsilon} \le |\mu|\int_0^t \mathrm{E}|Z_s|ds + \frac{\sigma^2 t}{2\log(1/\sqrt{\varepsilon})} + \sqrt{\varepsilon}.$$

The Gronwall inequality now gives

$$\mathrm{E}|Z_t| \le \left(\frac{\sigma^2 t}{2\log(1/\sqrt{\varepsilon})} + \sqrt{\varepsilon}\right)e^{|\mu|t}.$$

The left-hand side does not depend on ε, and taking $\varepsilon \to 0$ we obtain that $\mathrm{E}|Z_t| = 0$, implying $\mathrm{P}(X_t = Y_t) = 1$.

Chapter 6

Diffusion Processes

Martingales and Dynkin's Formula

Consider a general SDE of diffusion type,

$$dX(t) = \mu(X(t), t)\, dt + \sigma(X(t), t)\, dB(t) \quad \text{for } t \in [0, T],$$

with generator

$$L_t f(x, t) = (L_t f)(x, t) = \frac{\sigma(x, t)^2}{2} \frac{\partial^2 f(x, t)}{\partial x^2} + \mu(x, t) \frac{\partial f(x, t)}{\partial x}.$$

Itô's formula for $f(X(t), t)$ with $f \in C^{2,1}$ (when X solves the SDE) can be expressed with the help of the generator as

$$df(X(t), t) = \left(L_t f(X(t), t) + \frac{\partial f}{\partial t}(X(t), t) \right) dt$$
$$+ \left(\frac{\partial f}{\partial x}(X(t), t) \right) \sigma(X(t), t)\, dB(t).$$

And so, it follows that under suitable technical conditions (partial derivatives of f are at most of exponential growth, Equation (6.9*)) the following process is a martingale:

$$\left\{ f(X(t), t) - \int_0^t \left(L_s f + \frac{\partial f}{\partial s} \right)(X(s), s)\, ds \right\}_{t \in [0,T]}.$$

As an immediate corollary to the previous paragraph, it follows that (under technical conditions) $\{f(X(t), t)\}_{t \in [0,T]}$ is a martingale if f solves the backward equation

$$\left(L_t + \frac{\partial}{\partial t} \right) f(x, t) = 0.$$

Although simply a way of rewriting things, we already know this observation can be surprisingly useful.

Just by taking expectations of what we have found out above, it follows that if X is started deterministically $X(0) = x_0$, it holds for $f \in C^{2,1}$ (under technical conditions) that

$$\mathrm{E}(f(X(t),t)) = f(x_0,0) + \mathrm{E}\bigg(\int_0^t \Big(L_s f + \frac{\partial f}{\partial s}\Big)(X(s),s)\,ds\bigg) \quad \text{for } t \in [0,T].$$

By employing facts we learned in Chapter 7, this simple result can be extended to $t = \tau$, where τ is a bounded stopping time with $\tau \in [0,T]$. This extended version of the result is called Dynkin's formula (after the important contributor to the SDE theory, E.B. Dynkin) and is a very useful result with some arguably surprisingly sophisticated consequences.

Dynkin's formula can sometimes be extended to also apply to unbounded stopping times τ by first applying it to the bounded stopping time $\tau \wedge T$ and then sending $T \to \infty$ afterwards. This requires some integrability constraints on the items involved to be able to make use of the convergence theorems for the Lebesgue integral.

Calculation of Expectations and PDE

Consider a general SDE of diffusion type,

$$dX(t) = \mu(X(t),t)\,dt + \sigma(X(t),t)\,dB(t) \quad \text{for } t \in [0,T],$$

with generator

$$L_t f(x,t) = (L_t f)(x,t) = \frac{\sigma(x,t)^2}{2}\,\frac{\partial^2 f(x,t)}{\partial x^2} + \mu(x,t)\,\frac{\partial f(x,t)}{\partial x}.$$

Under technical conditions, a solution to the backward equation terminal boundary problem

$$\Big(L_t f + \frac{\partial f}{\partial t}\Big) f(x,t) = 0 \quad \text{for } t \in [0,T], \quad f(x,T) = g(x),$$

(if it exists) must be given by

$$f(x,t) = \mathrm{E}(g(X(T))|X(t) = x) \quad \text{for } t \in [0,T],$$

where X solves the SDE. (Just time shift the coefficients to get an SDE that starts at time 0 instead of time t if you want to.) As the backward equation ensures that $\{f(X(s),s)\}_{s\in[t,T]}$ is a martingale, we have

$$\mathrm{E}(g(X(T))|\mathcal{F}_t) = \mathrm{E}(f(X(T),T)|\mathcal{F}_t) = f(X(t),t) \quad \text{for } t \in [0,T].$$

Now, apply the Markov property to the left-hand side to obtain

$$\mathrm{E}(g(X(T))|X(t) = x) = f(x,t) \quad \text{for } t \in [0,T].$$

Conversely, $f(x,t) = \mathrm{E}(g(X(T))|X(t) = x)$ solves the backward equation terminal value problem. To see this, recall that for any f, we have

$$\begin{aligned}&\{f(X(s),s)\}_{s\in[t,T]}\\ &\quad = \left\{M(s) + f(X(t),t) + \int_t^s \left(L_u f + \frac{\partial f}{\partial u}\right)(X(u),u)\,du\right\}_{s\in[t,T]},\end{aligned}$$

where $\{M(s)\}_{s\in[t,T]}$ is a zero-mean martingale. Further, $\{E(g(X(T))|\mathcal{F}_s)\}_{s\in[t,T]}$ is a Doob–Lévy martingale. So, if we put $f(x,t) = E(g(X(T))|X(t) = x)$, that is,

$$f(X(t),t) = \mathrm{E}(g(X(T))|X(t)) = \mathrm{E}(g(X(T))|\mathcal{F}_t) \quad \text{for } t \in [0,T]$$

(by the Markov property), we have that

$$\begin{aligned}&\left\{\int_t^s \left(L_u f + \frac{\partial f}{\partial u}\right) f(X(u),u)\,du\right\}_{s\in[t,T]}\\ &\qquad = \{f(X(s),s) - M(s) - f(X(t),t)\}_{s\in[t,T]}\end{aligned}$$

is a zero-mean FV martingale. Hence, both the left- and right-hand sides are zero so that f satisfies the backward equation. See also Chapter 7.

There exist several generalisations of the first of the above-mentioned basic relation between the solution to the SDE and the corresponding backward equation, out of which we will mention two. (The proofs are similar to the proof we presented above but just notationally slightly more complicated.)

- Under technical conditions, the solution to the terminal boundary problem

  $$\left(L_t f + \frac{\partial f}{\partial t}\right) f(x,t) = \phi(x) \quad \text{for } t \in [0,T], \quad f(x,T) = g(x),$$

  (if it exists) must be given by

  $$f(x,t) = E\left(g(X(T)) - \int_t^T \phi(X(s))\,ds \,\middle|\, X(t) = x\right) \quad \text{for } t \in [0,T],$$

  where X solves the SDE.

- (FEYNMAN–KAC FORMULA) Under technical conditions, the solution to the terminal boundary problem

$$\left(L_t f + \frac{\partial f}{\partial t}\right) f(x,t) = r(x,t) f(x,t) \quad \text{for } t \in [0,T], \quad f(x,T) = g(x),$$

(if it exists) must be given by

$$f(x,t) = E\left(g(X(T)) \exp\left(-\int_t^T r(X(s),s)\,ds\right) \,\Big|\, X(t) = x\right)$$
$$\text{for } t \in [0,T],$$

where X solves the SDE.

Time Homogeneous Diffusions

Consider a time-homogeneous SDE of diffusion type

$$dX(t) = \mu(X(t))\,dt + \sigma(X(t))\,dB(t) \quad \text{for } t \in [0,T].$$

Due to (time) homogeneity, the generator now simplifies to

$$Lf(x) = (Lf)(x) = \frac{\sigma(x)^2}{2} f''(x) + \mu(x) f'(x).$$

Itô's formula for $f(X(t))$ with $f \in C^2$ (when X solves the SDE) can be expressed with the help of the generator as

$$df(X(t)) = Lf(X(t))\,dt + f'(X(t))\sigma(X(t))\,dB(t) \quad \text{for } t \in [0,T].$$

It follows that under suitable technical conditions, the following process is a martingale:

$$\left\{ f(X(t)) - \int_0^t Lf(X(s))\,ds \right\}_{t \in [0,T]}.$$

Weak solutions can be found by solving the corresponding martingale problem (to find an X that makes this process a martingale for a suitable class of functions $f \in C^2$).

The existence and uniqueness criteria for weak solutions that are strong Markov is as before with the t parameter of the coefficients $\mu(x,t)$ and $\sigma(x,t)$ removed. Further, it follows from applying time shifts that if there exists a unique weak solution for every starting value $X(0) = x$, then the transition CDF will be homogeneous:

$$P(y,t,x,s) = P(y,t-s,x,0) = P(t-s,x,y) \quad \text{for } t \in (s,T].$$

Of course, then again, the corresponding transition PDF $p(y,t,x,s) = p(t-s,x,y)$ will be homogeneous if it exists (which it usually does). Under appropriate technical conditions, it follows from what we did for the general (not necessarily time homogeneous) SDE that $p(t,x,y)$ satisfies the Kolmogorov backward equation

$$\frac{\partial p(t,x,y)}{\partial t} - Lp(t,x,y) = \frac{\partial p(t,x,y)}{\partial t} - \frac{\sigma(x)^2}{2}\frac{\partial^2 p(t,x,y)}{\partial x^2} - \mu(x)\frac{\partial p(t,x,y)}{\partial x} = 0.$$

Note that the time variable t now handles both the backward time variable s and the forward time variable t of $p(y,t,x,s)$ so that the sign of the time derivative changes. The corresponding Kolmogorov forward equation becomes

$$\frac{\partial p(t,x,y)}{\partial t} - L^\star p(t,x,y) = \frac{\partial p(t,x,y)}{\partial t} - \frac{\partial^2}{\partial y^2}\left(\frac{\sigma(y)^2}{2}p(t,x,y)\right) + \frac{\partial}{\partial y}(\mu(y)p(t,x,y) = 0.$$

There is a famous result that completely resolves the existence and uniqueness issues for a homogeneous SDE with zero drift due to Engelbert and Schmidt: the SDE

$$dX(t) = \sigma(X(t))\,dB(t) \quad \text{for } t \in [0,T], \quad X(0) = x_0,$$

has a non-exploding (see a later section for the explanation of this term) weak solution for every initial value x_0 if and only for every x, it holds that

$$\int_{-a}^{a} \frac{dy}{\sigma(x+y)^2} = \infty \quad \text{for } a > 0 \;\Rightarrow\; \sigma(x) = 0.$$

Moreover, the SDE has a unique (non-exploding) weak solution for every initial value x_0 if and only for every x, it holds that

$$\int_{-a}^{a} \frac{dy}{\sigma(x+y)^2} = \infty \quad \text{for } a > 0 \;\Leftrightarrow\; \sigma(x) = 0.$$

These results have no counterpart for non-homogeneous SDEs. However, by means of the application of Itô's formula, conclusions can also be drawn about homogeneous SDEs with non-zero drift. We will see how a result for SDEs with zero drift can be carried over to non-zero drift SDEs in the following section.

Exit Times from an Interval

Consider a time-homogeneous SDE of diffusion type,

$$dX(t) = \mu(X(t))\,dt + \sigma(X(t))\,dB(t) \quad \text{for } t \geq 0, \quad X(0) = x,$$

with generator

$$Lf(x) = (Lf)(x) = \frac{\sigma(x)^2}{2} f''(x) + \mu(x) f'(x).$$

We are interested in the stopping time

$$\tau = \inf\{t \geq 0 : X(t) \notin (a,b)\} \quad \text{for } a < x < b.$$

It holds that $v(x) = \mathrm{E}(\tau)$ satisfies the ODE

$$Lv(x) = -1 \quad \text{for } x \in (a,b), \quad v(a) = v(b) = 0.$$

The proof is by an application of Dynkin's formula to the stopping time $\tau \wedge t$ to obtain

$$\mathrm{E}(f(X(\tau \wedge t)) = f(x) + \mathrm{E}\bigg(\int_0^{\tau \wedge t} Lf(X(s))\,ds\bigg)$$

for "nice" functions $f : \mathbb{R} \to \mathbb{R}$. Now, take $f = v$, where v solves the above ODE and send $t \to \infty$ to obtain

$$0 = \mathrm{E}(v(X(\tau)) = v(x) - \mathrm{E}\bigg(\int_0^{\tau} ds\bigg) = v(x) - \mathrm{E}(\tau)$$

[since $X(\tau) = a$ or $X(\tau) = b$]. (The fact is that the values of $\{X(s)\}_{s \in [0, \tau \wedge t]}$ lie in $[a,b]$ makes a rigorous proof easier than it tends to be otherwise in these kinds of situations.) For example, for $X = B$, we readily get

$$v(x) = -\frac{1}{2}x^2 + \frac{a+b}{2}x - \frac{ab}{2} \quad \text{for } x \in (a,b).$$

For a more detailed study of exits from the interval (a,b), with τ as before, we consider the stopping times

$$T_a = \inf\{t \geq 0 : X(t) = a\} \quad \text{and} \quad T_b = \inf\{t \geq 0 : X(t) = b\}$$
$$\text{for } a < x < b.$$

Consider any solution S to the ODE $LS = 0$. Such a solution S is called a scale function for the SDE and is given by

$$S(x) = \int_{x_0}^{x} \exp\bigg(-\int_{x_0}^{y} \frac{2\mu(z)}{\sigma(z)^2}\,dz\bigg)dy + C$$

for some constants $C, x_0 \in \mathbb{R}$. By the application of Dynkin's formula to $\tau \wedge t$, we have

$$\mathrm{E}(S(X(\tau \wedge t))) = S(x) + \mathrm{E}\left(\int_0^{\tau\wedge t} LS(X(s))\,ds\right) = S(x).$$

Sending $t \to \infty$, we get

$$S(a)\,\mathrm{P}(T_a < T_b) + S(b)\,\mathrm{P}(T_b < T_a) = \mathrm{E}(S(X(\tau))) = S(x).$$

By rearrangement, this gives

$$\mathrm{P}(T_a < T_b) = \frac{S(x) - S(a)}{S(b) - S(a)} \quad \text{for } x \in (a, b).$$

For example, for $X = B$, we readily get

$$\mathrm{P}(T_a < T_b) = \frac{x - a}{b - a} \quad \text{for } x \in (a, b).$$

In particular,

$$\mathrm{P}(T_a < \infty) > 0 \quad \text{and} \quad \mathrm{P}(T_b < \infty) > 0.$$

As $LS = 0$, Itô's formula shows that the process $Y(t) = S(X(t))$ satisfies the SDE

$$dY(t) = S'(X(t))\sigma(X(t))\,dB(t) = S'(S^{-1}(Y(t)))\sigma(S^{-1}(Y(t))\,dB(t).$$

By applying our basic result about the existence and uniqueness of strong solutions to general (not necessarily homogeneous) SDEs to this SDE, it follows from careful inspection [that the diffusion coefficient $(S' \circ S^{-1})(\sigma \circ S^{-1})$ fits in with that result] that the original SDE has unique strong solution $X(t) = S^{-1}(Y(t))$ when μ is bounded and σ is globally Lipschitz and bounded away from zero. This is called Zvonkin's theorem. In the same fashion, one can apply the result of Engelbert–Schmidt to the SDE for Y and draw conclusions about weak solutions of the original SDE for X.

Explosion

Let $\{X(t)\}_{t \geq 0}$ solve the homogeneous SDE

$$dX(t) = \mu(X(t))\,dt + \sigma(X(t))\,dB(t) \quad \text{for } t \geq 0, \quad X(0) = x.$$

For simpler notation, we also assume that the values of X are not restricted to any finite or half-finite interval $(\alpha, \beta) \subseteq \mathbb{R}$. However, the things we do under the latter assumption can be extended in a more or less obvious way to the general case.

Let $\tau_n = \inf\{t \geq 0 : |X(t)| = n\}$. Clearly, the limit $\lim_{n\to\infty} \tau_n = \tau_\infty$ exists (although possibly $\tau_\infty = \infty$). We say that X explodes if $P(\tau_\infty < \infty) > 0$. Note that on the event $\{\tau_\infty < \infty\}$, we have $|X(\tau_\infty)| = \lim_{n\to\infty} |X(\tau_n)| = \infty$, which motivates the preceding language.

Clearly, under appropriate conditions on μ and σ (making no move of X impossible), X explodes started at any $x \in \mathbb{R}$ if X explodes started at a particular x.

Now, assume that σ is strictly positive and continuous and that μ is bounded on finite intervals (which are sufficient appropriate conditions in the previous paragraph). Given any $x_0 \in \mathbb{R}$, the diffusion X explodes when started at a particular $x \in \mathbb{R}$ if and only if it started at any $x \in \mathbb{R}$ if and only if at least one of the following two integrals are finite:

$$\int_{-\infty}^{x_0} \exp\biggl(-\int_{x_0}^{x} \frac{2\mu(y)}{\sigma(y)^2}\,dy\biggr)\biggl(\int_{x}^{x_0} \frac{1}{\sigma(y)^2}\exp\biggl(\int_{x_0}^{y} \frac{2\mu(z)}{\sigma(z)^2}\,dz\biggr)dy\biggr)dx$$

and

$$\int_{x_0}^{\infty} \exp\biggl(-\int_{x_0}^{x} \frac{2\mu(y)}{\sigma(y)^2}\,dy\biggr)\biggl(\int_{x_0}^{x} \frac{1}{\sigma(y)^2}\exp\biggl(\int_{x_0}^{y} \frac{2\mu(z)}{\sigma(z)^2}\,dz\biggr)dy\biggr)dx.$$

Due to the many integrals (primitive functions) involved and that it often is impossible to judge just by inspection if the above two integrals are finite or not, it can be a quite cumbersome task to check whether the above two integrals are finite or not.

The ODE $dx(t) = x(t)^2\,dt$, $x(0) = 1$, with solution $x(t) = 1/(1-t)$ for $t \in [0,1)$ explodes. No SDE with zero drift (and strictly positive continuous σ) explodes.

Recurrence and Transience

Let $\{X(t)\}_{t\geq 0}$ solve the homogeneous SDE

$$dX(t) = \mu(X(t))\,dt + \sigma(X(t))\,dB(t) \quad \text{for } t \geq 0, \quad X(0) = x.$$

For simpler notation, we also assume that the values of X are not restricted to any finite or half-finite interval $(\alpha, \beta) \subseteq \mathbb{R}$.

The starting point $x \in \mathbb{R}$ is called recurrent if $X(t_i) = x$ for a sequence of (usually random) times $0 < t_0 < t_1 < \cdots$, such that $t_n \to \infty$ as $n \to \infty$ with probability 1. If all starting points are recurrent, then X is called recurrent. [To require that X visits x infinitely many times is not the same thing because the (usually) infinite variation of X can give infinitely many visits in a bounded interval.]

The starting point $x \in \mathbb{R}$ is called transient if $|X(t)| \to \infty$ as $t \to \infty$ with probability 1. If all starting points are transient, then X is called transient.

We now make the same assumption that was used in the existence and uniqueness criteria for weak solutions that are strong Markov processes to the SDE: let σ be strictly positive and continuous and let both μ and σ have linear growth. Then, if there is one recurrent starting point X, then the diffusion is recurrent. Further, if there are no recurrent starting points, then the diffusion is transient. Moreover, given any $x_0 \in \mathbb{R}$, the diffusion is recurrent if and only if the following two integrals are both infinite:

$$\int_{-\infty}^{x_0} \exp\bigg(-\int_{x_0}^{x} \frac{2\mu(y)}{\sigma(y)^2}\,dy\bigg)\,dx \quad \text{and} \quad \int_{x_0}^{\infty} \exp\bigg(-\int_{x_0}^{x} \frac{2\mu(y)}{\sigma(y)^2}\,dy\bigg)\,dx.$$

Diffusion on an Interval

Consider a diffusion-type homogeneous SDE,

$$dX(t) = \mu(X(t))\,dt + \sigma(X(t))\,dB(t) \quad \text{for } t \geq 0, \quad X(0) = x,$$

such that the values of the solution $\{X(t)\}_{t\geq 0}$ are restricted to a finite or half-finite interval (α, β), $(\alpha, \beta]$, $[\alpha, \beta)$ or $[\alpha, \beta]$ in $\mathbb{R}$. The obvious way in which this can be accomplished is to have a μ that counteracts (by forcing the diffusion "inwards") any attempts from σ to take X outside the interval. We are interested in exits of X from the interval $(a, b) \subsetneq (\alpha, \beta)$, where $\alpha < a < x < b < \beta$.

Assume that $\alpha > -\infty$ (the case when $\beta < \infty$ is similarly handled), and let

$$L_1 = \int_{\alpha}^{b} \exp\bigg(-\int_{b}^{x} \frac{2\mu(y)}{\sigma(y)^2}\,dy\bigg)\,dx$$

and

$$L_2 = \int_{\alpha}^{b} \frac{1}{\sigma(x)^2}\bigg(\int_{\alpha}^{x} \exp\bigg(-\int_{b}^{y} \frac{2\mu(z)}{\sigma(z)^2}\,dz\bigg)dy\bigg)\exp\bigg(\int_{b}^{x} \frac{2\mu(y)}{\sigma(y)^2}\,dy\bigg)dx.$$

Under regularity and/or growth conditions on μ and σ, similar to those imposed before, it follows more or less from a judicious application of what we did in the last two sections that the following three statements hold:

- If $L_1 = \infty$, then $P(T_\alpha < T_b) = 0$.
- If $L_1 < \infty$ and $L_2 = \infty$, then $P(T_\alpha < T_b) > 0$ and

$$P((\min(T_\alpha, T_b) = \infty \text{ and } \lim_{t\to\infty} X(t) = \alpha) \bigcup (T_b < \infty)) = 1.$$

- If $L_1 < \infty$ and $L_2 < \infty$, then $P(T_\alpha < T_b) > 0$ and

$$P(\min(T_\alpha, T_b) < \infty) = 1 \quad \text{and} \quad P(T_\alpha < \infty) > 0.$$

Stationary Distributions

Let $\{X(t)\}_{t\geq 0}$ solve the homogeneous SDE

$$dX(t) = \mu(X(t))\,dt + \sigma(X(t))\,dB(t) \quad \text{for } t \geq 0, \quad X(0) = X_0.$$

Assume that X is a Markov process with transition CDF $P(t, x, y)$. For simpler notation, we also assume that the values of X are not restricted to any finite or half-finite interval $(\alpha, \beta) \subseteq \mathbb{R}$.

A CDF $\Pi : \mathbb{R} \to [0, 1]$ is called a stationary CDF or an invariant CDF for X if

$$\Pi(y) = \int_{-\infty}^{\infty} P(t, x, y)\,d\Pi(x) \quad \text{for } y \in \mathbb{R}.$$

This means that if $X(s)$ has CDF Π for an $s \geq 0$, then $X(t+s)$ also has CDF Π for $t > 0$.

Assume that the transition CDF $P(t, x, y)$ has a transition PDF $p(t, x, y) = \frac{\partial}{\partial y} P(t, x, y)$. A PDF $\pi : \mathbb{R} \to [0, \infty)$ is called a stationary PDF or an invariant PDF for X if

$$\pi(y) = \int_{-\infty}^{\infty} p(t, x, y)\pi(x)\,dx \quad \text{for } y \in \mathbb{R}.$$

This means that if $X(s)$ has PDF π for an $s \geq 0$, then $X(t+s)$ also has PDF π for $t > 0$. Making use of the formula

$$f_{X(t_1),\ldots,X(t_n)}(x_1, \ldots, x_n)$$
$$= f_{X(t_1)}(x_1) \prod_{i=2}^{n} p(t_i - t_{i-1}, x_{i-1}, x_i) \quad \text{for } x_1, \ldots, x_n \in \mathbb{R}$$

for the joint PDF of $(X(t_1),\ldots,X(t_n))$ for $0\le t_1<\cdots<t_n$, it follows that if X_0 has a stationary PDF π, then X is a stationary process, which is to say that

$$f_{X(t_1+h),\ldots,X(t_n+h)}(x_1,\ldots,x_n)=f_{X(t_1+h),\ldots,X(t_n+h)}(x_1,\ldots,x_n)\quad\text{for } h>0.$$

By applying the Kolmogorov forward equation to the definition of a stationary PDF, we obtain an ODE that π must satisfy:

$$\begin{aligned}0&=\int_{-\infty}^{\infty}\bigg[\bigg(\frac{1}{2}\frac{\partial^2}{\partial y^2}\sigma(y)^2-\frac{\partial}{\partial y}\mu(y)-\frac{\partial}{\partial t}\bigg)p(t,x,y)\bigg]\pi(x)\,dx\\&=\bigg(\frac{1}{2}\frac{\partial^2}{\partial y^2}\sigma(y)^2-\frac{\partial}{\partial y}\mu(y)\bigg)\pi(y)\\&=(L^*\pi)(y).\end{aligned}$$

Formally, a solution to this ODE is given by

$$\pi(x)=\frac{C}{\sigma(x)^2}\exp\bigg(\int_{x_0}^{x}\frac{2\mu(y)}{\sigma(y)^2}\,dy\bigg)\quad\text{for } x\in\mathbb{R}.$$

Here, $C>0$ and $x_0\in\mathbb{R}$ are constants that are determined by the normalisation $\int_{-\infty}^{\infty}\pi(x)\,dx=1$. [We really have only one "free" constant in the above solution to the second-order ODE before normalisation (as C and x_0 "interact".) The reason is that the second free constant in the general solution to the ODE disappears to even make normalisation possible.]

Under technical conditions, the SDE has the stationary PDF of the previous paragraph provided that the normalisation is possible and that the following two integrals are both infinite:

$$\int_{-\infty}^{x_0}\exp\bigg(-\int_{x_0}^{x}\frac{2\mu(y)}{\sigma(y)^2}\,dy\bigg)\,dx\quad\text{and}\quad\int_{x_0}^{\infty}\exp\bigg(-\int_{x_0}^{x}\frac{2\mu(y)}{\sigma(y)^2}\,dy\bigg)\,dx.$$

This of course is the necessary and sufficient criteria for the recurrence of a diffusion we have seen earlier.

If there exists a (non-probability) measure ν on $(\mathbb{R},\mathcal{B})$ with $\nu(\mathbb{R})=\infty$ such that

$$\nu(B)=\int_{-\infty}^{\infty}\bigg(\int_{y\in B}dP(t,x,y)\bigg)d\nu(x)=\int_{-\infty}^{\infty}P(t,x,B)\,d\nu(x)\quad\text{for } B\in\mathcal{B},$$

then ν is called an invariant measure for X. For example, while BM does not have a stationary CDF or PDF, it has the usual Euclidean measure of length $d\nu(x) = dx$ as an invariant measure. This follows from the fact that $p(t, x, y)$ for BM is a PDF both viewed as a function of y and as a function of x.

Multidimensional SDE

Let $\boldsymbol{\mu} : \mathbb{R}^n \times [0, T] \to \mathbb{R}^d$ and $\boldsymbol{\sigma} : \mathbb{R}^n \times [0, T] \to \mathbb{R}^{n\times d}$ be measurable functions and $\{\mathbf{B}(t)\}_{t\geq 0}$ an $\mathbb{R}^d$-valued BM. A multidimensional SDE is given in differential form by

$$d\mathbf{X}(t) = \boldsymbol{\mu}(\mathbf{X}(t), t)\, dt + \boldsymbol{\sigma}(\mathbf{X}(t), t)\, d\mathbf{B}(t) \quad \text{for } t \in [0, T], \quad X(0) = X_0.$$

A solution $\{\mathbf{X}(t)\}_{t\in[0,T]}$ to the SDE will be a kind of higher Itô integral process that was discussed at the end of Chapter 4. Many of the ideas we discussed for general (not time-homogeneous) SDEs in Chapter 5 have natural and similar extensions to multidimensional SDEs. Details of these are omitted here.

Problems

Generator

Problem 6.1: Find the generator for the Ornstein–Uhlenbeck process, write the backward equation and give its fundamental solution. Verify that it satisfies the forward equation.

Solution. The Ornstein–Uhlenbeck process solves the SDE $dX = -\alpha X(t)dt + \sigma dB(t)$, where $\alpha, \sigma > 0$. Thus, its generator is given by $L = \frac{\sigma^2}{2}\frac{\partial^2}{\partial x^2} - \alpha x\frac{\partial}{\partial x}$. The backward equation is $Lf = \frac{\partial f}{\partial t}$ (this is a time-homogeneous case, so the time enters as $t - s$ if compared to the general case, hence the derivative with respect to the backward time variable is multiplied by -1). That is,

$$\frac{\sigma^2}{2}\frac{\partial^2 f}{\partial x^2} - \alpha x\frac{\partial f}{\partial x} = \frac{\partial f}{\partial t}.$$

The fundamental solution is given by the transition probability density. While in the following we show how to find it by solving a PDE through the probability method, the transition probability function can be easily obtained from the solution to the SDE $X(t)$, as $\mathrm{P}(X(t) \leq y|X(0) = x)$. Since the solution to the linear SDE is known, $X(t) = xe^{-\alpha t} + \sigma e^{-\alpha t}\int_0^t e^{\alpha s}dB(s)$,

$$\begin{aligned} P(t,x,y) &= \mathrm{P}\left(xe^{-\alpha t} + e^{-\alpha t}\sigma \int_0^t e^{\alpha s}dB(s) \leq y\right) \\ &= \mathrm{P}\left(\sigma \int_0^t e^{\alpha s}dB(s) \leq ye^{\alpha t} - x\right) \\ &= \Phi\left(\frac{\sqrt{2\alpha}(ye^{\alpha t} - x)}{\sigma\sqrt{e^{2\alpha t} - 1}}\right), \end{aligned}$$

since $\sigma \int_0^t e^{\alpha s}dB(s)$ has $N(0, \frac{\sigma^2}{2\alpha}(e^{2\alpha t}-1))$ distribution. Thus, $p(t,x,y) = \frac{\partial}{\partial y}P(t,x,y) = \phi\left(\frac{\sqrt{2\alpha}(ye^{\alpha t}-x)}{\sigma\sqrt{e^{2\alpha t}-1}}\right)\frac{\sqrt{2\alpha}e^{\alpha t}}{\sigma\sqrt{e^{2\alpha t}-1}}$, with ϕ denoting the density of $N(0,1)$.

The fundamental solution can be found by solving the PDE by separation of variables. Let $f(x,t) = u(x)v(t)$, so $\frac{\partial f}{\partial t} = u(x)\frac{\partial v}{\partial t}$, $\frac{\partial f}{\partial x} = v(t)\frac{\partial u}{\partial x}$, and $\frac{\partial^2 f}{\partial x^2} = v(t)\frac{\partial^2 u}{\partial x^2}$. Substitute these into the backward equation and divide by $u(x)v(t)$ to get

$$\frac{1}{u(x)}\left(\frac{\sigma^2}{2}\frac{\partial^2 u(x)}{\partial x^2} - \alpha x\frac{\partial u(x)}{\partial x}\right) = -\frac{1}{v(t)}\frac{\partial v(t)}{\partial t}.$$

Since the left-hand side depends only on x and the right-hand side depends only on t, both sides must equal the same constant, say λ.

The equation for $v(t)$, $-\frac{1}{v(t)}\frac{\partial v(t)}{\partial t} = \lambda$, by integrating both sides and rearranging, one gets $v(t) = c_1e^{-\lambda t}$.

The equation for $u(t)$, $\frac{1}{u(x)}\left(\frac{\sigma^2}{2}\frac{\partial^2 u(x)}{\partial x^2} - \alpha x\frac{\partial u(x)}{\partial x}\right) = \lambda$.

It is a linear homogeneous ODE of order two, $\frac{\sigma^2}{2}u'' - \alpha x u' - \lambda u = 0$. Take $\lambda = \alpha$, so it is exact and can be rewritten as

$$\frac{\partial}{\partial x}\left(\frac{\sigma^2}{2\alpha}\frac{\partial u(x)}{\partial x} - xu(x)\right) = 0.$$

Now, it is left to solve a first-order linear ODE:

$$\frac{\sigma^2}{2\alpha}\frac{\partial u(x)}{\partial x} - xu(x) = c_2.$$

Multiply both sides by the integration factor $e^{-\frac{\sigma x^2}{\alpha}}$. Then, integrate and rearrange to obtain

$$u(x) = c_3 e^{\frac{\sigma x^2}{\alpha}} + e^{\frac{\sigma x^2}{\alpha}} \int^x c_2 e^{-\frac{\sigma y^2}{\alpha}} dy.$$

Problem 6.2: Let $X(t)$ be a stationary process. Show that the covariance function $\gamma(s,t) = \mathrm{Cov}(X(s), X(t))$ is a function of $|t-s|$ only. Deduce that for Gaussian processes, stationarity is equivalent to the requirements that the mean function is a constant and the covariance function is a function of $|t-s|$.
Hint: Take two-dimensional distributions.

Solution. Recall that the process X is stationary if its finite (k)-dimensional distributions do not change with the time shift, e.g. for two-dimensional distributions, for any s,t and any τ, for any $A \subset \mathbb{R}^2$,

$$\mathrm{P}((X(t), X(s)) \in A) = \mathrm{P}((X(t+\tau), X(s+\tau)) \in A).$$

In particular (by taking $s_1 = 0, t_1 = t-s, \tau = s$),

$$\mathrm{P}((X(t), X(s)) \in A) = \mathrm{P}((X(t-s), X(0)) \in A).$$

The last function is clearly a function of $t-s$ only. Therefore, $F(s,t,x,y) = \mathrm{P}(X(s) \leq x, X(t) \leq y))$ depends on s,t only through $t-s$. Consequently, all the moments as integrals with respect to this function depend on s,t only through $t-s$. Since the univariate distributions $k=1$ do not depend on t and are that of $X(0)$, the mean function is a constant in t. Hence,

$$\begin{aligned} \mathrm{Cov}(X(t), X(s)) &= \mathrm{E}X(t)X(s) - (\mathrm{E}X(0))^2 \\ &= \iint xydF(s,t,x,y) - (\mathrm{E}X(0))^2 \end{aligned}$$

is a function of $t-s$ only.

To show that a Gaussian X with $EX(t) = EX(0)$ and $\gamma(s,t) = \gamma(|s-t|)$ is stationary, use Problem 3.32 to conclude by inspection that the finite-dimensional distributions of $(X(t_1+\tau) - EX(0), \ldots, X(t_k+\tau) - EX(0))$ and $(X(t_1) - EX(0), \ldots, X(t_k) - EX(0))$ agree.

Problem 6.3: $X(t)$ is a diffusion with coefficients $\mu(x) = cx$ and $\sigma(x) = 1$. $X(0) = x$ with x deterministic. Give its generator and show that $X^2(t) - 2c\int_0^t X^2(s)ds - t$ is a martingale.

Solution. The generator of a process with the given coefficients is $L = \frac{1}{2}\frac{\partial^2}{\partial x^2} + cx\frac{\partial}{\partial x}$.

Take $f(x) = x^2$, then $Lf(x) = 1 + 2cx^2$, and by Theorem 6.3 (growth conditions on derivatives of f are satisfied), the process $M_f = f(X(t), t) - \int_0^t \left(L_u f + \frac{\partial f}{\partial t}\right)(X(u), u)\, du = X^2(t) - t - 2c\int_0^t X^2(s)ds$ is a martingale.

Problem 6.4: $X(t)$ is a diffusion with $\mu(x) = 2x$ and $\sigma^2(x) = 4x$. $X(0) = x$ with x deterministic. Give its generator L. Solve $Lf = 0$, and give a martingale M_f. Find the SDE for the process $Y(t) = \sqrt{X(t)}$, and give the generator of $Y(t)$.

Solution. The generator of $X(t)$ with the coefficients given above is $L = 2x\frac{\partial^2}{\partial x^2} + 2x\frac{\partial}{\partial x}$. Obtain a solution for $Lf = 0$ by integrating twice:

$$2x\frac{\partial^2 f}{\partial x^2} = -2x\frac{\partial f}{\partial x},$$
$$\int_{f_0}^{f} \frac{dg}{g} = -\int_{x_0}^{x} du,$$
$$\ln f - \ln f_0 = -x + x_0.$$

The ODE solution is $f(x) = f(0) + f'(0) - f'(0)e^{-x}$. Since $\frac{\partial f}{\partial t} = 0$, $f(x)$ solves the backward equation, and thus $M_f = e^{-X(t)}$ is a martingale.

Problem 6.5: The next few problems are more advanced. They are devoted to the question of what does a generator exactly generates. They are not used in subsequent problems and can be skipped. Let $X(t)$ be a time-homogeneous diffusion. Denote by P_t, for a fixed t, the operator defined by

$$P_t f(x) = \mathrm{E}(f(X(t))|X(0) = x).$$

Show the following:

(a) P_t is a linear operator acting on functions.
(b) The family of operators P_t, $t \geq 0$, has the property, for any t and s, $P_{t+s} = P_t P_s$, called the semigroup property. $P_0 = I$, is the identity operator. The family of operators P_t, $t \geq 0$, satisfying the above properties is called the semigroup of operators.

Solution.

(a) Since the expectation is linear, we have for constants a, b and functions f, g,

$$\begin{aligned}P_t(af+bg)(x) &= \mathrm{E}((af+bg)(X(t))|X(0)=x)\\&= a\mathrm{E}(f(X(t))|X(0)=x)+b\mathrm{E}(g(X(t))|X(0)=x)\\&= aP_tf(x)+bP_tg(x).\end{aligned}$$

(b) It is obtained by conditioning on $X(t)$ using the law of double expectation:

$$\begin{aligned}P_{t+s}f(x) &= \mathrm{E}(f(X(t+s))|X(0)=x)\\&= \mathrm{E}\big[\mathrm{E}(f(X(t+s))|X(t),X(0)=x)|X(0)=x\big].\end{aligned}$$

Now, since $X(t)$ has the Markov property,

$$\mathrm{E}(f(X(t+s))|X(t),X(0)=x)=\mathrm{E}(f(X(t+s))|X(t)).$$

Next, since $X(t)$ is time-homogeneous,

$$\mathrm{E}(f(X(t+s))|X(t))=\mathrm{E}(f(X(s))|X(0)=X(t))=P_sf(X(t)).$$

Denote $g(x) = P_sf(x)$. Thus, we have shown that $P_{t+s}f(x) = \mathrm{E}(g(X(t))|X(0)=x)$. But this, by definition, is $P_tg(x)$. Hence, we obtain

$$P_{t+s}f(x)=P_tg(x)=P_tP_sf(x),$$

and the semigroup property is proved.

We also have clearly $P_0f(x)=\mathrm{E}(f(X(0))|X(0)=x)=f(x)$, so that $P_0=I$.

Note that since $t+s=s+t$, we immediately obtain

$$P_{t+s}=P_tP_s=P_{s+t}=P_sP_t.$$

Problem 6.6: Let a function $\phi(t)$ satisfy the functional equation $\phi(t+s) = \phi(t)\phi(s)$, $\phi(0) = 1$. Assume that ϕ is differentiable with $\phi'(0) = a$. Show that $\phi(t) = e^{at}$. In fact, the conclusion holds under only the assumption that ϕ is measurable. See also Problem 3.28.

Solution. Consider $\phi'(t) = \lim_{\delta \to 0}(\phi(t+\delta) - \phi(t))/\delta$. Using the equation for ϕ,

$$\phi(t+\delta) - \phi(t) = \phi(t)\phi(\delta) - \phi(t)\phi(0) = \phi(t)(\phi(\delta) - \phi(0)).$$

Hence, we obtain

$$\phi'(t) = \phi(t) \lim_{\delta \to 0}(\phi(\delta) - \phi(0))/\delta = \phi(t)\phi'(0) = a\phi(t).$$

Solving this differential equation, $\phi(t) = e^{at}$.

Problem 6.7: Similarly to the above equation for functions, the operators P_t satisfy the semigroup property, where the composition of functions corresponds to multiplication. It can be shown that $P_t = e^{tL}$, where the operator L, called the generator of the semigroup (P_t), satisfies $L = \lim_{t \to 0} \frac{1}{t}(P_t - I)$, or with $Lf = \lim_{t \to 0} \frac{1}{t}(P_t f - f)$. Further, it can be shown that $P_t = e^{tL}$, where the operator e^{tL} is defined by the exponential power series formula

$$e^{tL} = \sum_{n=0}^{\infty} \frac{t^n L^n}{n!}.$$

(L is the generator because it generates the semigroup of operators P_t, $t \geq 0$, by the formula $P_t = e^{tL}$).

(a) Show that the "shift" process $X(t) = X(0) + t$ has generator $L = \frac{d}{dx}$ and that L indeed generates the "shift" process.
(b) Find the generator for the Brownian motion $B(t)$. Then, show that the operator $L = \frac{1}{2}\frac{d^2}{dx^2}$ generates Brownian motion $B(t)$.

Note that the domain of the operators L^n, $n = 1, 2, \ldots$, is progressively shrinking, requiring functions f to have more and more derivatives. e^L applies to infinitely differentiable functions, in C^∞.

Solution.

(a) The "shift" process $X(t) = X(0) + t$ is deterministic with $dX(t) = dt$. The corresponding semigroup of operators is the shift semigroup

$$P_t f(x) = \mathrm{E}(f(X(t))|X(0) = x) = \mathrm{E}(f(x+t)|X(0) = x) = f(x+t).$$

Computing the generator,

$$Lf(x) = \lim_{t \to 0} \frac{1}{t}(P_t f(x) - f(x)) = \lim_{t \to 0} \frac{1}{t}(f(x+t) - f(x)) = f'(x).$$

Hence, $L = \frac{d}{dx}$.

To show that L generates the operators P_t, consider $L^n = (\frac{d}{dx})^n = \frac{d^n}{dx^n}$. Then,

$$e^{tL} = \sum_{n=0}^{\infty} \frac{t^n L^n}{n!} = \sum_{n=0}^{\infty} \frac{t^n}{n!} \frac{d^n}{dx^n},$$

and for an infinitely differentiable function f, we have

$$e^{tL} f(x) = \sum_{n=0}^{\infty} \frac{t^n}{n!} \frac{d^n f}{dx^n}(x) = f(x+t).$$

We recognised the Taylor series expansion of $f(x+t)$ at the point x. Hence,

$$e^{tL} f(x) = P_t f(x) = \mathrm{E}(f(X(t))|X(0) = x) = \mathrm{E}f(x+t) = f(x+t).$$

(b) First, we compute the generator for Brownian motion using the formula $Lf(x) = \lim_{t \to 0} \frac{1}{t}(\mathrm{E}(f(B(t))|X_0 = x) - f(x))$ by using Itô's formula:

$$f(B(t)) = f(B(0)) + \int_0^t f'(B(s)dB(s) + \frac{1}{2}\int_0^t f''(B(s))ds.$$

Hence, by using $B(0) = x$ and taking expectations, we have noted that Itô integral has zero mean,

$$\mathrm{E}f(B(t)) = f(x) + \frac{1}{2}\int_0^t \mathrm{E}f''(B(s))ds.$$

Finally, using the continuity of $B(t)$ and the derivative of the integral, we obtain

$$L = \frac{1}{2}\frac{d^2}{dx^2}.$$

To show that $\frac{1}{2}\frac{d^2}{dx^2}$ generates the Brownian motion semigroup, consider $L^n = (\frac{1}{2}\frac{d^2}{dx^2})^n = \frac{1}{2^n}\frac{d^{2n}}{dx^{2n}}$. Hence,

$$e^{tL} f = \sum_{n=0}^{\infty} \frac{t^n L^n f}{n!} = \sum_{n=0}^{\infty} \frac{t^n}{2^n n!} \frac{d^{2n} f}{dx^{2n}}.$$

Unlike the previous example, where we recognised Taylor expansion, the presence of only even derivatives does not seem to simplify.

However, we derive the solution by looking at the operator P_t for Brownian motion:

$$P_tf(x) = \mathrm{E}(f(B(t)|B(0) = x) = \int f(y)p_t(x,y)dy,$$

$$P_tf(x) = \int f(y)\frac{1}{\sqrt{2\pi t}}e^{-(y-x)^2/(2t)}dy.$$

Now, change variable $y - x = u$ to have

$$P_tf(x) = \int f(x+u)\frac{1}{\sqrt{2\pi t}}e^{-u^2/(2t)}du.$$

Next, taking the Taylor expansion of f at x, $f(x+u) = \sum_{n=0}^{\infty}\frac{u^n}{n!}\frac{d^nf}{dx^n}(x)$, and interchanging summation with integration, we obtain

$$P_tf(x) = \sum_{n=0}^{\infty}\frac{1}{n!}\frac{d^nf}{dx^n}(x)\int u^n\frac{1}{\sqrt{2\pi t}}e^{-u^2/(2t)}du.$$

But $\int u^n\frac{1}{\sqrt{2\pi t}}e^{-u^2/(2t)}du$ is the nth moment of $N(0,t)$ distribution. In Problem 2.22, we have shown that the nth moment of $N(0,1)$ distribution is zero for n odd, and for even the $2n$th moment is given by $\frac{(2n)!}{2^nn!}$. Hence,

$$\int u^{2n}\frac{1}{\sqrt{2\pi t}}e^{-u^2/(2t)}du = t^n\frac{(2n)!}{2^nn!}.$$

Since only even terms are present in the series for P_tf, it follows that

$$P_tf(x) = \sum_{n=0}^{\infty}\frac{t^n}{2^nn!}\frac{d^{2n}f}{dx^{2n}}(x) = e^{tL}f(x).$$

Thus, $P_t = e^{tL}$.

Problem 6.8: Define the operators P_t, $t \geq 0$, acting on bounded continuous functions f, as follows. For a given continuous and bounded function $r(x)$ on $\mathbb{R}$,

$$P_tf(x) = \mathrm{E}_x\left(e^{\int_0^t r(B_u)du}f(B_t)\right).$$

(a) Show that the family (P_t) satisfies the semigroup property.
(b) Find the generator for this semigroup, $Lf = \lim_{t\to 0}\frac{1}{t}(P_tf - f)$, assuming that f has the first two derivatives, which are also bounded.

Solution.

(a) We show the semigroup property $P_{t+s} = P_t P_s$ by conditioning:

$$\begin{aligned}
P_{t+s}f(x) &= \mathrm{E}_x\big(e^{\int_0^{t+s} r(B_u)du} f(B_{t+s})\big)\\
&= \mathrm{E}_x\big(\mathrm{E}(e^{\int_0^{t+s} r(B_u)du} f(B_{t+s})|\mathcal{F}_t)\big)\\
&= \mathrm{E}_x\big(e^{\int_0^{t} r(B_u)du}\mathrm{E}(e^{\int_t^{t+s} r(B_u)du} f(B_{t+s})|\mathcal{F}_t)\big)\\
&= \mathrm{E}_x\big(e^{\int_0^{t} r(B_u)du}\mathrm{E}(e^{\int_0^{s} r(B_t+\hat{B}_u)du} f(B_t+\hat{B}_s)|B_t)\big),
\end{aligned}$$

where $\hat{B}_u = B_{u+t} - B_t$ is a new Brownian motion independent of B_t. Note that $\mathrm{E}(e^{\int_0^s r(B_t+\hat{B}_u)du} f(B_t+\hat{B}_s)|B_t) = g(B_t)$ for the function $g(x) = T_s f(x)$. Proceeding, we have

$$P_{t+s}f(x) = \mathrm{E}_x\big(e^{\int_0^t r(B_u)du} g(B_t)\big) = T_t g(x) = T_t T_s f(x).$$

(b) To calculate the generator of (P_t), we consider limits as $t \to 0$. Write

$$\begin{aligned}
P_t f(x) - f(x) &= \mathrm{E}_x\big(e^{\int_0^t r(B_u)du} f(B_t)\big) - f(x)\\
&= f(x)\mathrm{E}_x\big(e^{\int_0^t r(B_u)du} - 1\big)\\
&\quad + \mathrm{E}_x\big(e^{\int_0^t r(B_u)du}(f(B_t) - f(x))\big).
\end{aligned}$$

The limit after dividing by t of the first term on the right is the derivative at zero of the function $e^{\int_0^t r(B_u)du}$,

$$\begin{aligned}
\lim_{t\to 0}\frac{1}{t}\mathrm{E}_x\big(e^{\int_0^t r(B_u)du} - 1\big) &= \mathrm{E}_x \lim_{t\to 0}\frac{1}{t}\big(e^{\int_0^t r(B_u)du} - 1\big)\\
&= \frac{d}{dt}e^{\int_0^t r(B_u)du} = r(B_0) = r(x).
\end{aligned}$$

We use dominated convergence to exchange expectation and the limit; the inequality $|\int_0^t r(B_u)du| \le Ct$, where C is an upper bound for the function r. This together with $|e^x - 1| \le 2|x|$ for $|x| \le 1$ yields for small t's,

$$\left|e^{\int_0^t r(B_u)du} - 1\right| \le 2\left|\int_0^t r(B_u)du\right| \le 2Ct.$$

For the second term in the calculation of the generator, write

$$E_x\big((e^{\int_0^t r(B_u)du} - 1)(f(B_t) - f(x))\big) + \mathrm{E}_x(f(B_t) - f(x)).$$

As we have seen in Problem 6.7(b),

$$\lim_{t\to 0}\frac{1}{t}\mathrm{E}_x(f(B_t)-f(x))=\frac{1}{2}f''(x).$$

Further,

$$\lim_{t\to 0}\frac{1}{t}E_x\big(\big(e^{\int_0^t r(B_u)du}-1\big)(f(B_t)-f(x))\big)=0,$$

again by dominated convergence, because $\lim_{t\to 0}\frac{1}{t}E_x(e^{\int_0^t r(B_u)du}-1)$ exists and f is continuous so that $f(B_t)-f(x)\to 0$ as $t\to 0$.

Finally, putting it all together, the generator of (P_t) is given by

$$Lf(x)=\frac{1}{2}f''(x)+r(x)f(x),$$

defined for bounded twice continuously differentiable functions f.

Problem 6.9: Let $(P_t), t\geq 0$, be an operator semigroup with generator L.

(a) Show that $\frac{d}{dt}(P_tf)=LP_tf$.

(b) Apply this result to operators from the previous Problem 6.8 to obtain the forward Feynman–Kac formula: if

$$C(x,t)=\mathrm{E}_x\big(e^{\int_0^t r(B_u)du}f(B_t)\big),$$

then $C(x,t)$ solves the following PDE with the initial condition $f(x)$,

$$\frac{\partial C}{\partial t}(x,t)=\frac{1}{2}\frac{\partial^2 C}{\partial x^2}(x,t)+r(x)C(x,t),\quad C(x,0)=f(x).$$

Solution.

(a) Consider $\frac{1}{h}(P_{t+h}f-P_tf)$. By the semigroup property, $P_{t+h}f=P_hP_tf$, so that

$$\frac{1}{h}(P_{t+h}f-P_tf)=\frac{1}{h}(P_h-I)P_tf.$$

Taking limits as $h\to 0$ and taking into account that $L=\lim_{h\to 0}\frac{1}{h}(P_h-I)$, we obtain $\frac{d}{dt}(P_tf)=LP_tf$.

(b) Let $C(x,t)$ denote the function $C(x,t)=\mathrm{E}_x(e^{\int_0^t r(B_u)du}f(B_t))$. Then, by Problem 6.8, we can write it in terms of the linear operator P_t acting on the function f, $C(x,t)=P_tf(x)$. As it was shown, the family of operators (P_t) is a semigroup with the generator

$$Lf(x)=\frac{1}{2}f''(x)+r(x)f(x).$$

Writing the part (a) in terms of the function $C(x,t)$ by noting that $\frac{d}{dt}(P_t f) = \frac{\partial}{\partial t}C(x,t)$ and $LP_t f(x) = \frac{1}{2}\frac{\partial^2 C}{\partial x^2} + r(x)C(x,t)$, we obtain the desired PDE of Feynman–Kac type.

Backward Equation, Martingales

Problem 6.10: Show that for any u, $f(x,t) = \exp(ux - u^2t/2)$ solves the backward equation for Brownian motion. Take derivatives, first, second, etc., of $\exp(ux - u^2t/2)$ with respect to u, and set $u = 0$ to obtain that functions x, $x^2 - t$, $x^3 - 3tx$, $x^4 - 6tx^2 + 3t^2$, etc., also solve the backward equation. Deduce that $B^2(t) - t$, $B(t)^3 - 3tB(t)$, $B^4(t) - 6tB^2(t) + 3t^2$ are martingales.

Solution. Brownian motion has the generator (the differential operator L defined by $Lf = \frac{1}{2}\sigma^2\frac{\partial^2 f}{\partial x^2} + \mu\frac{\partial f}{\partial x}$) $L = \frac{1}{2}\frac{\partial^2}{\partial x^2}$. The backward equation is therefore

$$\frac{\partial f}{\partial t} + Lf = \frac{1}{2}\frac{\partial^2 f}{\partial x^2} + \frac{\partial f}{\partial t} = 0.$$

For $f(x,t) = \exp(ux - u^2t/2)$, $\frac{\partial f}{\partial t} = -\frac{u^2}{2}e^{ux-u^2t/2}$, and $Lf = \frac{1}{2}f_{xx}(x,t) = \frac{u^2}{2}e^{ux-u^2t/2}$, and it solves the backward equation.

Since f solves $Lf + \frac{\partial f}{\partial t} = 0$ for any fixed u, take the nth partial derivative with respect to u $\dfrac{\partial^n f}{\partial u^n}$, and interchange the order of differentiation to have that for any fixed u,

$$\frac{\partial^n}{\partial u^n}Lf + \frac{\partial^n}{\partial u^n}\frac{\partial f}{\partial t} = L\left(\frac{\partial^n f}{\partial u^n}\right) + \frac{\partial}{\partial t}\left(\frac{\partial^n f}{\partial u^n}\right) = 0,$$

so that $\frac{\partial^n f}{\partial u^n}$ solves the backward equation for any fixed u and in particular for $u = 0$.

Therefore, let $n = 1$ and $u = 0$ to obtain that

$$\left.\frac{\partial f}{\partial u}\right|_{u=0} = (x - ut)\exp(ux - u^2t/2)\big|_{u=0} = x$$

solves the backward equation.

Similarly, let $n = 2$ and $u = 0$ to obtain that

$$\left.\frac{\partial^2 f}{\partial u^2}\right|_{u=0} = (-t + (x - ut)^2)\exp(ux - u^2t/2)\big|_{u=0} = x^2 - t$$

solves the backward equation.

Let $n = 3$ and $u = 0$ to obtain that $x^3 - 3tx$ solves the backward equation, etc.

By Theorem 6.3 or Corollary 6.4, if $f(x,t)$ has derivatives that grow not faster than the exponential, then the following is a martingale:

$$f(B(t),t) - \int_0^t \left(Lf + \frac{\partial f}{\partial t}\right)(B(s),s)ds.$$

For $f(x,t) = e^{ux-u^2t/2}$,

$$\left|\frac{\partial f}{\partial t}\right| \le Ce^{u|x|}, \quad \left|\frac{\partial f}{\partial x}\right| \le Ce^{u|x|}, \quad \left|\frac{\partial^2 f}{\partial x^2}\right| \le Ce^{u|x|}.$$

Since f solves the backward equation, $Lf + \frac{\partial f}{\partial t} = 0$, $f(B(t),t)$ is a martingale. Thus, $f(B(t),t) = e^{uB(t)-u^2t/2}$ is a martingale for any u.

Since $\frac{\partial^n f}{\partial u^n}$ solves the backward equation for $u = 0$ and the conditions on the growth of derivatives are satisfied (all derivatives have a polynomial growth, e.g. $x^3 - 3tx \le C|x|^3 \le Ce^{|x|}$, we have that $B(t)$, $B^2(t) - t$, $B^3(t) - 3tB(t)$, etc., are martingales.

Problem 6.11: Find $f(x)$ such that $f(B(t)+t)$ is a martingale.

Solution. Let $X(t) = B(t) + t$. Then, $dX(t) = dB(t) + dt$, and $[X,X](t) = [B,B](t) = t$. By Itô's formula, $df(X(t)) = f'(X(t))dB(t) + f'(X(t))dt + \frac{1}{2}f''(X(t))dt$.

A necessary condition for $f(B(t)+t)$ to be a martingale is that the dt term is zero.

This gives $f'(x) + \frac{1}{2}f''(x) = 0$. Therefore, look for $f(x)$ which solves this.

For example, take $f(x) = e^{-2x}$ and check directly that $e^{-2(B(t)+t)}$ is a martingale ($e^{-2(B(t)+t)} = \mathcal{E}(2B)(t)$).

Problem 6.12: $X(t)$ is a diffusion with coefficients $\mu(x,t), \sigma(x,t)$. Find a differential equation for $f(x,t)$ such that $Y(t) = f(X(t),t)$ has a diffusion coefficient equal to 1.

Solution. By Itô's formula, $dY(t) = df(X(t),t) = \frac{\partial f}{\partial x}dX(t) + \frac{1}{2}\frac{\partial^2 f}{\partial x^2}d[X,X](t) + \frac{\partial f}{\partial t}dt$.

Since $dX(t) = \mu(x,t)dt + \sigma(x,t)dB(t)$, the diffusion coefficient of the process $Y(t) = f(X(t),t)$ (i.e. the coefficient of the term with $dB(t)$ in the expression for $dY(t)$) is given by $\sigma(X(t),t)\frac{\partial f}{\partial x}(X(t),t)$.

Thus, the PDE for f is $\sigma(x,t)\dfrac{\partial f}{\partial x}(x,t) = 1$.

Exit Times

Problem 6.13: Show that the mean exit time of a diffusion from an interval, which (by Theorem 6.16) satisfies the ODE $Lv = -1$ is given by

$$v(x) = -\int_a^x 2G(y)\int_a^y \frac{ds}{\sigma^2(s)G(s)}dy + \int_a^b 2G(y)\int_a^y \frac{ds}{\sigma^2(s)G(s)}dy\frac{\int_a^x G(s)ds}{\int_a^b G(s)ds},$$

where $G(x) = \exp\left(-\int_a^x \frac{2\mu(s)}{\sigma^2(s)}ds\right)$.

Solution. Recall Theorem 6.16, for a diffusion $X(t)$ with generator L, and $X(0) = x$, $a < x < b$, $\mathrm{E}_x(\tau) = v(x)$ satisfies the differential equation $Lv = -1$, with $v(a) = v(b) = 0$. This gives

$$\mu v' + \frac{\sigma^2}{2}v'' = -1.$$

Let $v' = y$ and rearrange

$$y' + \frac{2\mu}{\sigma^2}y + \frac{2}{\sigma^2} = 0.$$

Multiply by the integrating factor $u(x) = \exp\left(\int_a^x \frac{2\mu(s)}{\sigma^2(s)}ds\right)$ to have $(yu)' = -\frac{2}{\sigma^2}u$, and $y = -\frac{1}{u}\int\frac{2}{\sigma^2}u$. The result now follows.

Problem 6.14: Find the mean hitting time at 0 of the Ornstein–Uhlenbeck process, which starts at $x > 0$.

Solution. The Ornstein–Uhlenbeck process is given by $dX(t) = -\alpha X(t)dt + \sigma dB(t)$. We first consider the expected exit time of $X(t)$ from an interval $(0,b)$ and then take the limit $b \to \infty$. From Theorem

6.16 and Problem 6.13, with $\mu(x) = -\alpha x$ and $\sigma^2(x) = \sigma^2$, we have $G(x) = \exp(\int_0^x \frac{2\alpha s}{\sigma^2} ds) = \exp(\frac{\alpha}{\sigma^2} s^2)$ and

$$\begin{aligned}\mathrm{E}_x(T_0 \wedge T_b) = &-\frac{2}{\sigma^2}\int_0^x \exp\left(\frac{\alpha}{\sigma^2}y^2\right)\int_0^y \exp\left(-\frac{\alpha}{\sigma^2}s^2\right) dsdy \\ &+\frac{2}{\sigma^2}\int_0^b \exp\left(\frac{\alpha}{\sigma^2}y^2\right)\int_0^y \exp\left(-\frac{\alpha}{\sigma^2}s^2\right) dsdy \\ &\times \frac{\int_0^x \exp(\frac{\alpha}{\sigma^2}s^2)ds}{\int_0^b \exp(\frac{\alpha}{\sigma^2}s^2)ds}.\end{aligned}$$

Note that

$$\begin{aligned}&\lim_{b\to\infty} \frac{\int_0^b \exp(\frac{\alpha}{\sigma^2}y^2)\int_0^y \exp(-\frac{\alpha}{\sigma^2}s^2)dsdy}{\int_0^b \exp(\frac{\alpha}{\sigma^2}s^2)ds} \\ &= \lim_{b\to\infty} \frac{\exp(\frac{\alpha}{\sigma^2}b^2)\int_0^b \exp(-\frac{\alpha}{\sigma^2}s^2)ds}{\exp(\frac{\alpha}{\sigma^2}b^2)} = \int_0^\infty \exp\left(-\frac{\alpha}{\sigma^2}s^2\right) ds.\end{aligned}$$

Thus, letting $b \to \infty$ in $\mathrm{E}_x(T_0 \wedge T_b)$, we get

$$\begin{aligned}\mathrm{E}_x(T_0) &= -\frac{2}{\sigma^2}\int_0^x \exp\left(\frac{\alpha}{\sigma^2}y^2\right)\int_0^y \exp\left(-\frac{\alpha}{\sigma^2}s^2\right) dsdy \\ &\quad+\frac{2}{\sigma^2}\int_0^x \exp\left(\frac{\alpha}{\sigma^2}u^2\right) du\int_0^\infty \exp\left(-\frac{\alpha}{\sigma^2}s^2\right) ds \\ &= \frac{2}{\sigma^2}\int_0^x \exp\left(\frac{\alpha}{\sigma^2}y^2\right)\int_y^\infty \exp\left(-\frac{\alpha}{\sigma^2}s^2\right) dsdy \\ &= \frac{2\sqrt{\pi}}{\sigma\sqrt{\alpha}}\int_0^x \exp\left(\frac{\alpha}{\sigma^2}y^2\right)\Phi\left(-\frac{\sqrt{2\alpha}}{\sigma}y\right) dy,\end{aligned}$$

where Φ denotes the cumulative distribution function of standard normal.

Problem 6.15: Find $\mathrm{P}_x(T_b < T_a)$ for Brownian motion with drift when $\mu(x) = \mu$ and $\sigma^2(x) = \sigma^2$.

Solution. If $X(t)$ is a diffusion with generator L with continuous $\sigma(x) > 0$ on $[a, b]$ and $X(0) = x$, $a < x < b$, then $P_x(T_b < T_a) = \frac{S(x)-S(a)}{S(b)-S(a)}$, where $S(x) = \int^x \exp\left(-\int^u \frac{2\mu(y)}{\sigma^2(y)}dy\right) du$ is the scale function, Theorem 6.17.

$$\text{Hence, } P_x(T_b < T_a) = \frac{S(x) - S(a)}{S(b) - S(a)} = \frac{e^{-\frac{2\mu}{\sigma^2}a} - e^{-\frac{2\mu}{\sigma^2}x}}{e^{-\frac{2\mu}{\sigma^2}a} - e^{-\frac{2\mu}{\sigma^2}b}}.$$

Representation of Solutions

Problem 6.16: Give a probabilistic representation of the solution $f(x,t)$ of the PDE

$$\frac{1}{2}\frac{\partial^2 f}{\partial x^2} + \frac{\partial f}{\partial t} = 0, \quad 0 \le t \le T, \quad f(x,T) = x^2.$$

Solve this PDE using the solution of the corresponding stochastic differential equation.

Solution. $L = \frac{1}{2}\frac{\partial^2 f}{\partial x^2}$ is the generator of the process $X(t) = B(t)$ (the Brownian motion). If $f(x,t)$ solves the backward equation with $f(x,T) = g(x)$, then $f(x,t) = \mathrm{E}(g(X(T))|X(t) = x)$, Theorem 6.6. Hence, $f(x,t) = \mathrm{E}(B^2(T)|B(t) = x)$.

Using $B(T) = B(t) + B(T) - B(t)$, we obtain

$$\begin{aligned}
\mathrm{E}(B^2(T)|B(t) = x) &= \mathrm{E}(B^2(t)|B(t) = x) \\
&\quad - 2\mathrm{E}(B(t)(B(T) - B(t))|B(t) = x) \\
&\quad + \mathrm{E}((B(T) - B(t))^2|B(t) = x) \\
&= x^2 + 0 + (T - t) \\
&= f(x,t),
\end{aligned}$$

since $B(T) - B(t) \sim N(0, T - t)$ and using independence of increments.

Now, we can verify that $f(x,t) = x^2 + T - t$ solves the backward equation ($f_t = -1$, $f_{xx} = 2$) and satisfies $f(x,T) = x^2$.

Problem 6.17: Calculate $C(x,t) = \mathrm{E}(\int_t^T B_s^2 ds | B_t = x)$. Then, solve the partial differential equation $\frac{1}{2}\partial_x^2 f + \partial_t f = -x^2$ with the boundary condition $f(x,T) = 0$.

Solution. $B_s = B_s - B_t + x = \hat{B}_{s-t} + x$, for $\hat{B}$ independent of B_t. Hence,

$$C(x,t) = \mathrm{E}\left(\int_t^T (\hat{B}_{s-t} + x)^2 ds\right) = \int_0^{T-t} \mathrm{E}(\hat{B}_u + x)^2 du$$
$$= \int_0^{T-t} (u + x^2) du = x^2(T-t) + (T-t)^2/2.$$

Next, B has the generator $Lf = \frac{1}{2}\partial_x^2 f$. By Theorem 6.7, $f(x,t) = \mathrm{E}(\int_t^T \phi(B_s)ds | B_t = x)$ solves $Lf + \partial_t f = -\phi(x)$ with $f(x,T) = 0$.
Hence, $C(x,t)$ is the solution.

Problem 6.18: Let $C(x,t) = \mathrm{E}(e^{u\int_t^T B_s ds} | B_t = x)$, where $u \in \mathbb{R}$ is fixed.

(a) Give the partial differential equation for C in variables (x,t), including the boundary condition.
(b) Use probability calculations to derive the conditional distribution of $\int_t^T B_s ds$ given B_t. Hence, give the solution to the PDE in (a).

Solution.

(a) By the Feynman–Kac formula, $C(x,t)$ solves

$$\partial_t C + \frac{1}{2}\partial_x^2 C = -uxC, \quad C(x,T) = 1.$$

(b) $\int_t^T B_s ds = \int_t^T (B_t + \hat{B}_{s-t})ds = B_t(T-t) + \int_0^{T-t} \hat{B}_u du$. $\int_0^{T-t} \hat{B}_u du$ has $N(0,(T-t)^3/3)$ distribution. Hence, the conditional distribution of $\int_t^T B_s ds$ given B_t is $N(B_t(T-t),(T-t)^3/3)$. Hence, the moment-generating function is $e^{uB_t(T-t)+u^2(T-t)^3/6}$. Hence,

$$C(x,t) = e^{ux(T-t)+u^2(T-t)^3/6}.$$

Problem 6.19: Let $S_t = S_0 + \sigma B_t$ with $S_0, \sigma > 0$, where B_t represents a Brownian motion started at zero.

(a) Let $C(x,t) = \mathrm{E}(g(S_T)|S_t = x)$, for some function g and $0 \le t \le T$. Give the PDE that $C(x,t)$ solves.
(b) Solve the PDE when $g(x) = (x-K)^+$.
This gives the price of a call option in the Bachelier model (the function $x^+ = xI_{x>0}$).
(c) Show that the process $M_t = C(S_t,t)$ is a martingale.

Solution.

(a) $\frac{1}{2}\sigma^2\frac{\partial^2 f}{\partial x^2} + \frac{\partial f}{\partial t} = 0,\ \ 0 \le t \le T,\ \ f(x,T) = g(x).$

(b) $S_T = S_t + \sigma\hat{B}_{T-t}$,

$$\begin{aligned}
C(x,t) &= \mathrm{E}((S_T - K)^+ | S_t = x) = \mathrm{E}(\sigma\hat{B}_{T-t} + x - K)^+\\
&= \mathrm{E}(\sigma B_{T-t} + x - K)I_{B_{T-t}>(K-x)/\sigma}\\
&= \sigma\mathrm{E}B_{T-t}I_{B_{T-t}>(K-x)/\sigma} - (K-x)P(B_{T-t} > (K-x)/\sigma)\\
&= \frac{\sigma\sqrt{T-t}}{\sqrt{2\pi}}e^{-(K-x)^2/(2\sigma^2(T-t))} - (K-x)\Phi\left(\frac{x-K}{\sigma\sqrt{T-t}}\right)\\
&= (x-K)\Phi\left(\frac{x-K}{\sigma\sqrt{T-t}}\right) + \sigma\sqrt{T-t}\phi\left(\frac{x-K}{\sigma\sqrt{T-t}}\right),
\end{aligned}$$

where $\Phi(x)$ and $\phi(x)$ denote the cumulative probability function and the probability density function of the standard normal distribution, respectively.

We use for $X \sim N(0, v^2)$ the following result:

$$\begin{aligned}
\mathrm{E}X1_{X>a} &= \frac{1}{\sqrt{2\pi}v}\int_a^\infty xe^{-x^2/2v^2}dx\\
&= \frac{v}{\sqrt{2\pi}}\int_a^\infty e^{-x^2/2v^2}d(x^2/2v^2)\\
&= \frac{v}{\sqrt{2\pi}}\int_{a^2/2v^2}^\infty e^{-y}dy = \frac{v}{\sqrt{2\pi}}e^{-a^2/2v^2} = v\phi(a/v).
\end{aligned}$$

(c) Since $C(x,t)$ solves the backward PDE, $C(S_t,t)$ is a martingale (by using Ito's formula, provided $\partial C/\partial x$ grows not faster than the exponential). It is also possible but harder to verify directly from the explicit formula.

Alternatively, using the result from the following chapter, due to the process S_t being Markov (it is a Brownian motion), the conditional expectation of $g(S_T)$ given S_t is the same as the conditional expectation of $g(S_T)$ given $\mathcal{F}_t$:

$$M_t = C(S_t,t) = \mathrm{E}((S_T - K)^+|S_t) = \mathrm{E}((S_T - K)^+|\mathcal{F}_t).$$

But this of the form $\mathrm{E}(Y|\mathcal{F}_t)$, which is a martingale (Lévy martingale), $Y = (S_T - K)^+$.

Problem 6.20: Let $B(t)$ be a Brownian motion and g a bounded function possessing two derivatives, which are also bounded. Show that the function $f(x,t) = \mathrm{E}\big(g(B(t))|B(0) = x\big)$ solves the initial value problem

$$\frac{\partial f}{\partial t} = \frac{1}{2}\frac{\partial^2 f}{\partial x^2}, \quad f(x,0) = g(x).$$

Solution. Using Itô's formula for $g(B(t))$ and noting that $Lg = \frac{1}{2}g''$,

$$g(B(t)) = g(x) + \int_0^t Lg(B(s))ds + M(t),$$

where $M(t)$ is a local martingale. It is in fact a true martingale because of the boundedness of the second derivative of g. Taking expectations when started in x, E_x,

$$f(x,t) = \mathrm{E}_x\big(g(B(t))\big) = g(x) + \mathrm{E}_x\left(\int_0^t Lg(B(s))ds\right).$$

Interchanging expectation with the integral by using the Fubini theorem, using $Lg(x) = \frac{1}{2}g''(x)$, and taking partial derivative in t, we have

$$\frac{\partial}{\partial t}f(x,t) = \frac{1}{2}\mathrm{E}_x\big(g''(B(t))\big) = \frac{1}{2}\mathrm{E}\big(g''(x + \hat{B}(t))\big),$$

where $\hat{B}(t)$ is the standard Brownian motion started at zero.

Observe next that since g and its two derivatives are bounded, we can interchange the derivative and expectation by the Lebesgue-dominated convergence theorem — see Problem 2.24. Hence, we obtain

$$\frac{\partial}{\partial t}f(x,t) = \frac{1}{2}\mathrm{E}\big(g''(x + \hat{B}(t))\big) = \frac{1}{2}\frac{\partial^2}{\partial x^2}\mathrm{E}\big(g(x + \hat{B}(t))\big) = \frac{1}{2}\frac{\partial^2}{\partial x^2}f(x,t).$$

Clearly, $f(x,0) = g(x)$.

Problem 6.21: Let $X(t) = x + \mu t + \sigma B(t)$ be a diffusion with constant coefficients. Recall that its infinitesimal generator is $L = \frac{1}{2}\sigma^2\frac{\partial^2}{\partial x^2} + \mu\frac{\partial}{\partial x}$. Suppose that the function g and its two derivatives have at most exponential growth, namely $|g(x)| \leq Ce^{cx}$ for some positive constants c and C, and similarly for g' and g''. Show that the function $f(x,t) = \mathrm{E}_x(g(X(t))$ solves the initial value problem

$$\frac{\partial f}{\partial t} = Lf, \quad f(x,0) = g(x).$$

Solution. Note first that $X(t)$ is a Gaussian process. Then, $\mathrm{E}_x|g(X(t)| < \infty$, since exponential moments of any normal distribution exist. The same applies to the derivatives g' and g''. This also implies that the Itô integral appearing in Itô's formula is a martingale. Using Itô's formula for $g(X(t))$ and noting that $\frac{\partial g}{\partial t} = 0$,

$$g(X(t)) = g(x) + \int_0^t Lg(X(s))ds + M(t),$$

where $M(t)$ is a martingale. Taking expectations when started in x, E_x,

$$f(x,t) = \mathrm{E}_x g(X(t)) = g(x) + \mathrm{E}_x \int_0^t Lg(X(s))ds.$$

Interchanging expectation with the integral by the Fubini theorem, using the fact that $\int_0^t \mathrm{E}_x|Lg(X(s))|ds < \infty$, and taking partial derivative in t, we have

$$\frac{\partial}{\partial t} f(x,t) = \mathrm{E}_x\big(Lg(X(t))\big) = \frac{1}{2}\sigma^2 \mathrm{E}\big(g''(x + \mu t + \sigma B(t))\big) + \mu \mathrm{E}\big(g'(x + \mu t + \sigma B(t))\big).$$

Now, using Problem 2.24(b), we can interchange the derivatives and the expectation to have $\mathrm{E}g'(x + \mu t + \sigma B(t)) = (\mathrm{E}g(x + \mu t + \sigma B(t)))'$ and $\mathrm{E}g''(x + \mu t + \sigma B(t)) = (\mathrm{E}g(x + \mu t + \sigma B(t)))''$. This gives

$$\mathrm{E}_x Lg(X(t)) = L\mathrm{E}_x(g(X(t)) = Lf.$$

Clearly, $f(x,0) = g(x)$, and the proof is complete.

Explosion

Problem 6.22: Show that the solution of the following ordinary differential equation $dx(t) = cx^r(t)dt$, $c > 0$, $x(0) = x_0 > 0$, explodes (to plus infinity) if and only if $r > 1$. In a later question, we shall see that the diffusion obtained from this ODE by adding Brownian motion also explodes.

Solution. Obtain the solution of the ODE by separation of variables:

$$\int_{x_0}^{x} \frac{du}{cu^r} = \int_0^t ds = t.$$

For $r = 1$, $\ln x(t) = ct + \ln x_0$, $x(t) = x_0 e^{ct}$. For $r \neq 1$, $\frac{x^{1-r}}{1-r} = ct + \frac{x_0^{1-r}}{1-r}$, $x(t) = \big((1-r)ct + x_0^{1-r}\big)^{\frac{1}{1-r}}$. Thus, for $r \leq 1$, the function $x(t)$ does not explode at any value of t. For $r > 1$, the function $x(t)$ explodes to plus infinity at the time $t = \frac{x_0^{1-r}}{(r-1)c}$.

Problem 6.23: Show that the solution of the following ordinary differential equation $dx(t) = -cx^r(t)dt$, $c > 0$, $x(0) = x_0 > 0$, $r > 1$ does not explode. In a later question, we shall see that the diffusion obtained from this ODE by adding Brownian motion also does not explode.

Solution. We can see from the previous exercise that

$$x(t) = \frac{1}{\left((r-1)ct + x_0^{1-r}\right)^{\frac{1}{r-1}}}.$$

Since the denominator is always positive, there is no explosion.

Problem 6.24: Let X_t satisfy the SDE $dX_t = \mu(X_t)dt + dB_t$. Assume that $\mu(x) > 0$ is non-decreasing for $x > x_0$ for some x_0. Apply the Feller's test for explosion (Theorem 6.23) to show that X explodes if and only if

$$\int_{x_0}^{\infty} \frac{1}{\mu(x)} dx < \infty.$$

Solution. Feller's test for explosion states that $\mathrm{P}(\tau < \infty) > 0$, where τ is the explosion time, if and only if the following integral is finite:

$$\int_{x_0}^{\infty} e^{-2\int_{x_0}^{x} \mu(s)ds} \int_{x_0}^{x} e^{2\int_{x_0}^{y} \mu(s)ds} dydx < \infty.$$

Denote $g(x) = 2\int_{x_0}^{x} \mu(s)ds$. Then, in terms of g, the explosion occurs if and only if

$$\int_{x_0}^{\infty} e^{-g(x)} \int_{x_0}^{x} e^{g(y)} dydx < \infty.$$

We will show that this is equivalent to the integral test above.

First, suppose that $\int_{x_0}^{\infty} \frac{1}{\mu(y)} dy < \infty$. Write by swapping the order of integration using the Fubini theorem,

$$\begin{aligned}\int_{x_0}^{\infty} e^{-g(x)} \int_{x_0}^{x} e^{g(y)} dydx &= \int_{x_0}^{\infty} e^{g(y)} \int_{y}^{\infty} e^{-g(x)} dxdy \\ &= \int_{x_0}^{\infty} e^{g(y)} \int_{y}^{\infty} e^{-g(x)} \frac{g'(x)}{g'(x)} dxdy.\end{aligned}$$

Since under our assumption $1/g'(x) = 1/(2\mu(x)) \leq 1/(2\mu(y))$ for $x \geq y$

and $\int_y^\infty e^{-g(x)}g'(x)dx = e^{-g(y)}$, we have

$$\int_{x_0}^\infty e^{-g(x)}\int_{x_0}^x e^{g(y)}dydx \le \int_{x_0}^\infty e^{g(y)}\frac{1}{2\mu(y)}\int_y^\infty e^{-g(x)}g'(x)dxdy$$
$$= \int_{x_0}^\infty e^{g(y)}\frac{1}{2\mu(y)}e^{-g(y)}dy$$
$$= \int_{x_0}^\infty \frac{1}{2\mu(y)}dy \; < \; \infty.$$

For the other direction, suppose that $\int_{x_0}^\infty \frac{1}{\mu(y)}dy = \infty$. Using that $1/g'(y) = 1/(2\mu(y)) \ge 1/(2\mu(x))$ for $y \le x$, a similar calculation gives

$$\int_{x_0}^\infty e^{-g(x)}\int_{x_0}^x e^{g(y)}dydx = \int_{x_0}^\infty e^{-g(x)}\int_{x_0}^x e^{g(y)}g'(y)/g'(y)dydx$$
$$\ge \int_{x_0}^\infty e^{-g(x)}\frac{1}{2\mu(x)}\int_{x_0}^x e^{g(y)}g'(y)dydx$$
$$= \int_{x_0}^\infty e^{-g(x)}\frac{1}{2\mu(x)}\left(e^{g(x)} - e^{g(x_0)}\right)dx$$
$$\ge C\int_{x_0}^\infty \frac{1}{\mu(x)}dx = \infty.$$

Examples:

$\mu(x) = x^\alpha$: explosion if $\alpha > 1$ and no explosion if $\alpha \le 1$.
$\mu(x) = x\log x$: no explosion.
$\mu(x) = x(\log x)^{1+\epsilon}, \epsilon > 0$: explosion.

Problem 6.25: Show that if a diffusion with constant diffusion coefficient explodes to $+\infty$, then all other diffusions with the same diffusion coefficient and larger increasing drift coefficient on $[x_0, \infty)$, for any x_0, explode as well.

Solution. For the SDE $dZ(t) = \mu(Z(t))dt + \sigma dB(t)$, we have

$$Z(t) = Z(0) + \int_0^t \mu(Z(s))ds + \sigma B(t)$$

$$= Z(0) + \lim_{\max_{1\le i\le n} t_i - t_{i-1}\downarrow 0}\sum_{i=1}^n \mu(Z(t_{i-1}))(t_i - t_{i-1}) + \sigma B(t),$$

where $0 = t_0 < \cdots < t_n = t$. It follows that if we replace μ with a larger and increasing function $\hat{\mu}$, then the solution $\hat{Z}(t)$ of the SDE

$$d\hat{Z}(t) = \hat{\mu}(\hat{Z}(t))dt + \sigma dB(t), \;\; \hat{Z}(0) = Z(0),$$

satisfies $\hat{Z}(t) \geq Z(t)$.

To see this, note that if $\hat{Z}(t_{i-1}) \geq Z(t_{i-1})$, then

$$\begin{aligned}\hat{Z}(t_i) &= \hat{Z}(t_{i-1}) + \hat{\mu}(\hat{Z}(t_{i-1}))(t_i - t_{i-1}) + \sigma(B(t_i) - B(t_{i-1}))\\ &\geq Z(t_{i-1}) + \hat{\mu}(Z(t_{i-1}))(t_i - t_{i-1}) + \sigma(B(t_i) - B(t_{i-1})) \geq Z(t_i).\end{aligned}$$

Hence, if a diffusion with a constant diffusion coefficient explodes to $+\infty$, then all other diffusions with the same diffusion coefficient and larger increasing drift coefficient explode to $+\infty$ as well. The test integral for explosion to $+\infty$ only depends on the values of drift and diffusion coefficient on $[x_0, \infty)$ for any x_0. Hence the conclusion.

Problem 6.26: Show that the following process, the logistic equation with additive noise, explodes towards minus infinity:

$$dX(t) = X(t)(1 - X(t))dt + \sigma dB(t).$$

Solution. Let $Y(t) = -X(t)$. Then,

$$\begin{aligned}dY(t) = -dX(t) &= -X(t)(1 - X(t))dt - \sigma dB(t)\\ &= Y(t)(Y(t) + 1))dt + \sigma d\hat{B}(t),\end{aligned}$$

with $\hat{B}(t) = -B(t)$.

As $x(x+1) \geq x^2$ for $x \geq x_0 = 0$, where the function to the left is increasing and

$$dX(t) = X^2(t)dt + \sigma dB(t)$$

explodes to $+\infty$ by Problem 6.24, using Problem 6.25, the logistic equation explodes to $-\infty$ (cf. Problem 6.27).

Problem 6.27: Investigate for explosions, the following process:

$$dX(t) = X^2(t)dt + \sigma X^\alpha(t)dB(t).$$

Solution. While it is possible to investigate the explosion of X using Feller's test for explosions (Theorem 6.23), it becomes very technical here. Therefore, we give a more conceptual solution that communicates some useful skills in similar problems.

For $\alpha < 1$, by Itô's formula, $Y(t) = X(t)^{1-\alpha}$ satisfies

$$\begin{aligned}dY(t) &= \left[(1-\alpha)Y(t)^{1+1/(1-\alpha)} - \frac{1}{2}\alpha(1-\alpha)\sigma^2 Y(t)^{-1}\right]dt\\ &\quad + (1-\alpha)\sigma dB(t).\end{aligned}$$

As $(1-\alpha)x^{1+1/(1-\alpha)} - \frac{1}{2}\alpha(1-\alpha)\sigma^2 x^{-1}$ is increasing and larger than $\frac{1}{2}(1-\alpha)x^{1+1/(1-\alpha)}$ for $x \geq x_0$, Problem 6.24 gives explosion to $+\infty$.

For $\alpha = 1$, take $Y(t) = \ln(X(t))$ with $dY(t) = (e^{Y(t)} - \frac{1}{2}\sigma^2)dt + \sigma dB(t)$ to conclude explosion to $+\infty$ using Problem 6.24 again.

For $\alpha > 1$, again consider $Y(t) = X(t)^{1-\alpha}$ with the SDE

$$dY(t) = \left[\frac{1}{2}\alpha(\alpha-1)\sigma^2 Y(t)^{-1} - (\alpha-1)Y(t)^{1-1/(\alpha-1)}\right] dt \\ - (\alpha-1)\sigma dB(t).$$

Now, $X(t)$ explodes to $+\infty$ if and only if $Y(t)$ explodes to 0^+. (Theorem 6.23 still applies with obvious modifications.) The SDE

$$dZ(t) = \frac{1}{2}\rho Z(t)^{-1}dt - (\alpha-1)\sigma dB(t)$$

has the test integral

$$\int_{0+}^{1} x^{-\rho/(\sigma^2(\alpha-1)^2)} \left(\int_x^1 \frac{y^{\rho/(\sigma^2(\alpha-1)^2)}}{\sigma^2(\alpha-1)^2)} dy\right) dx \\ = \begin{cases} \displaystyle\int_{0+}^{1} \frac{x^{-\rho/(\sigma^2(1-\alpha)^2)} - x}{\rho + \sigma^2(1-\alpha)^2} dx, & \rho \neq -\sigma^2(1-\alpha)^2 \\ \displaystyle\int_{0+}^{1} \frac{-x\ln(x)}{\sigma^2(1-\alpha)^2} dx, & \rho = -\sigma^2(1-\alpha)^2 \end{cases}$$

for explosion to 0^+, which is infinite iff $\rho \geq \sigma^2(1-\alpha)^2$. Hence, the SDE for $Y(t)$ when $\alpha = \frac{3}{2}$

$$dY(t) = \left(\frac{3}{8}\sigma^2 - \frac{1}{2}\right) Y(t)^{-1}dt - \frac{1}{2}\sigma dB(t)$$

explodes to 0^+ iff $2(\frac{3}{8}\sigma^2 - \frac{1}{2}) < \frac{1}{4}\sigma^2$, i.e. iff $\sigma^2 < 2$.

For $1 < \alpha < \frac{3}{2}$, the SDE for $Y(t)$ has a negative decreasing drift coefficient close to 0^+ and therefore explodes to 0^+ as the SDE with zero drift coefficient does. For $\alpha > \frac{3}{2}$, the SDE for $Y(t)$ has a greater drift coefficient close to 0^+ than an SDE for $Z(t)$ (with decreasing drift) that does not explode to 0^+. Hence, $Y(t)$ does not explode to 0^+ [as then $Z(t)$ would].

In conclusion, the SDE for $X(t)$ explodes to $+\infty$ iff either $\alpha < \frac{3}{2}$ or $\alpha = \frac{3}{2}$ and $\sigma^2 < 2$.

Recurrence

Problem 6.28: Show that the Brownian motion $B(t)$ is recurrent. Show that $B(t) + t$ is transient.

Solution. Check the conditions of Theorem 6.28. The diffusion corresponding to L is recurrent if and only if both I_1 and I_2 are infinite, and transient otherwise, i.e. when at least one of I_1 or I_2 is finite, $I_1 = \int_{-\infty}^{x_0} \exp\left(-\int_{x_0}^{u} \frac{2\mu(s)}{\sigma^2(s)} ds\right) du$ and $I_2 = \int_{x_0}^{\infty} \exp\left(-\int_{x_0}^{u} \frac{2\mu(s)}{\sigma^2(s)} ds\right) du$.

For the Brownian motion $B(t)$, take $x_0 = 0$, $\mu(x) = 0$ and $\sigma(x) = 1$. $I_1 = \int_{-\infty}^{0} 1du = \infty$, and $I_2 = \int_0^\infty 1du = \infty$. Hence, $B(t)$ is recurrent.

For the process $B(t) + t$, $\mu(x) = 1$ and $\sigma(x) = 1$. $I_2 = \int_0^\infty e^{-2u} du = \frac{1}{2} < \infty$. Thus, $B(t) + t$ is transient.

Problem 6.29: Show that the Ornstein–Uhlenbeck process is positively recurrent and has a stationary distribution. Show that the limiting distribution for the Ornstein–Uhlenbeck process exists and is given by its stationary distribution.

Solution. Recall the Ornstein–Uhlenbeck process

$$dX(t) = -\alpha X(t)dt + \sigma dB(t)$$

for some positive constants α and σ. Thus, we have here $\mu(x) = -\alpha x$ and $\sigma(x) = \sigma$.

To show that the process is recurrent, we use Theorem 6.28. Let $x_0 = 0$. We have

$$I_1 = \int_{-\infty}^{0} \exp\left(-\int_0^u \frac{2\mu(s)}{\sigma^2(s)} ds\right) du = \int_{-\infty}^{0} \exp\left(-\int_0^u \frac{-2\alpha s}{\sigma^2} ds\right) du$$
$$= \int_{-\infty}^{0} \exp\left(\frac{\alpha}{\sigma^2} u^2\right) du = \infty$$

and

$$I_2 = \int_0^\infty \exp\left(-\int_0^u \frac{2\mu(s)}{\sigma^2(s)} ds\right) du = \int_0^\infty \exp\left(\frac{\alpha}{\sigma^2} u^2\right) du = \infty,$$

thus the diffusion is recurrent.

Furthermore,

$$\int_{-\infty}^{\infty} \frac{1}{\sigma^2(x)} \exp\left(\int_0^u \frac{2\mu(s)}{\sigma^2(s)} ds\right) du = \frac{1}{\sigma^2} \int_{-\infty}^{\infty} \exp\left(\int_0^u \frac{-2\alpha s}{\sigma^2} ds\right) du$$
$$= \frac{1}{\sigma^2} \int_{-\infty}^{\infty} \exp\left(-\frac{\alpha}{\sigma^2} u^2\right) du < \infty$$

for positive α. Therefore, an invariant density exists and is given by

$$\pi(x) = \frac{C}{\sigma^2(x)} \exp\left(\int_0^x \frac{2\mu(s)}{\sigma^2(s)} ds\right) = \frac{C}{\sigma^2} \exp\left(-\frac{\alpha}{\sigma^2} x^2\right),$$

where C is a constant such that $\int \pi(x)dx = 1$. Now,

$$\int \pi(x)dx = \int_{-\infty}^{\infty} \frac{C}{\sigma^2} \exp\left(-\frac{\alpha}{\sigma^2} x^2\right) dx = \frac{C}{\sigma} \sqrt{\frac{\pi}{\alpha}} \exp\left(\frac{\alpha}{\sigma^2}\right),$$

thus $C = \sqrt{\frac{\alpha}{\pi}}\sigma$. This gives

$$\pi(x) = \frac{1}{\sigma} \sqrt{\frac{\alpha}{\pi}} \exp\left(-\frac{\alpha}{\sigma^2} x^2\right).$$

Being recurrent and having a stationary distribution, the diffusion X is thus positive recurrent.

Note also that the Ornstein–Uhlenbeck process has a solution

$$X(t) = e^{-\alpha t} X(0) + \sigma e^{-\alpha t} \int_0^t e^{\alpha s} dB(s),$$

which is normally distributed with mean $e^{-\alpha t}X(0)$ and variance

$$\sigma^2 e^{-2\alpha t} \int_0^t e^{2\alpha s} ds = \frac{\sigma^2}{2\alpha}(1 - e^{-2\alpha t}).$$

Letting $t \to \infty$, we have that the limiting distribution of X is $N(0, \frac{\sigma^2}{2\alpha})$, which has density $\pi(x)$.

Problem 6.30: Show that the square of the Bessel process

$$dX(t) = ndt + 2\sqrt{X(t)}dB(t),$$

where n is a positive integer, comes arbitrarily close to zero when $n = 2$, that is, $P(T_y < \infty) = 1$ for any small $y > 0$, but when $n \geq 3$, $P(T_y < \infty) < 1$.

Solution. Consider the process X on the interval (α, β). The scale function is

$$S(x) = \int_{x_1}^{x} \exp\left(-\int_{x_0}^{u} \frac{2\mu(s)}{\sigma^2(s)} ds\right) du = \int_{x_1}^{x} \exp\left(-\int_{x_0}^{u} \frac{n}{2s} ds\right) du$$
$$= \int_{x_1}^{x} \exp\left(-\frac{n}{2}(\log(u) - \log(x_0))\right) du = C\int_{x_1}^{x} u^{-\frac{n}{2}} du,$$

where $x_0, x_1 \in (\alpha, \beta)$.

The probabilities of exit from an interval $(a, b) \in (\alpha, \beta)$ are then given by

$$P_x(T_a < T_b) = \frac{S(b) - S(x)}{S(b) - S(a)}.$$

Take $a = y$ (for any small $y > 0$) and let $b \to \infty$, so

$$P_x(T_y < \infty) = \lim_{b\to\infty} P_x(T_y < T_b) = \lim_{b\to\infty} \frac{S(b) - S(x)}{S(b) - S(a)}.$$

For $n = 2$, the scale function is $S(x) = \log(x) - \log(x_0)$, and

$$P_x(T_y < \infty) = \lim_{b\to\infty} \frac{\log(b) - \log(x)}{\log(b) - \log(y)} = 1.$$

But when $n \geq 3$; $S(x) = \dfrac{x^{-\frac{n}{2}+1}}{-\frac{n}{2}+1} - \dfrac{x_1^{-\frac{n}{2}+1}}{-\frac{n}{2}+1}$, and

$$P_x(T_y < \infty) = \lim_{b\to\infty} \frac{b^{1-\frac{n}{2}} - x^{1-\frac{n}{2}}}{b^{1-\frac{n}{2}} - y^{1-\frac{n}{2}}} = \left(\frac{y}{x}\right)^{\frac{n}{2}-1} < 1.$$

Stationary Distribution

Problem 6.31: Show that the diffusion $dX(t) = -X^3(t)dt + dB(t)$ is positively recurrent and gives the form of its stationary distribution. Generalise to see that a diffusion $dX(t) = p(X(t))dt + dB(t)$, where $p(x)$ is the polynomial in x of odd degree, with a negative leading coefficient is positively recurrent and does not explode.

Solution. Calculations similar to those in Problem 6.29 show that both integrals $I_1 = I_2 = \infty$. The invariant density has the form

$$\pi(x) = Ce^{-\frac{1}{2}x^4}.$$

In the case of a more general polynomial of odd degree with a negative leading coefficient, that coefficient dominates in the integrals I_1 and I_2, making them divergent. It is interesting to note that when the coefficient is positive, the process explodes (except for unit degree polynomials for which it is transient, cf. Problems 6.24 and 6.29), whereas when it is negative, not only that it does not explode, it has a stationary distribution.

Problem 6.32:

(a) Let diffusion $X(t)$ have $\sigma(x) = 1$, $\mu(x) = +1$ for $x < 0$, $\mu(x) = -1$ for $x > 0$ and $\mu(0) = 0$. Show that $\pi(x) = e^{-2|x|}$ is a stationary distribution for X.

(b) Find π by solving the equation $L^*\pi = 0$, where L^* is the adjoint operator of L.

Solution.

(a) A formal solution is given in Problem 10.24, as it uses techniques from later chapters. The coefficient μ in question does not satisfy the smoothness assumption needed in the following at one point. But, ignore the fact that μ is discontinuous for now.

If the coefficients μ and σ are twice continuously differentiable, then X admits a stationary distribution if and only if μ and σ satisfy the following conditions: there exists x_0 such that

i. $\int_{-\infty}^{x_0} \exp\big(-\int_{x_0}^{u} \frac{2\mu(s)}{\sigma^2(s)}ds\big)du = \int_{x_0}^{\infty} \exp\big(-\int_{x_0}^{u} \frac{2\mu(s)}{\sigma^2(s)}ds\big)du = \infty$.

ii. $\int_{-\infty}^{\infty} \frac{1}{\sigma^2(u)} \exp\big(\int_{x_0}^{u} \frac{2\mu(s)}{\sigma^2(s)}ds\big)du < \infty$.

Take now $x_0 = 0$. Then, clearly $\int_0^u \frac{2\mu(s)}{\sigma^2(s)}ds = -2|u|$ so that

$$I_1 = \int_{-\infty}^{0} \exp\left(-\int_0^u \frac{2\mu(s)}{\sigma^2(s)}ds\right)du = \int_{-\infty}^{0} e^{2|u|}du = \infty,$$

$$I_2 = \int_0^{\infty} \exp\left(-\int_0^u \frac{2\mu(s)}{\sigma^2(s)}ds\right)du = \int_0^{\infty} e^{2|u|}du = \infty.$$

Thus, condition (a) is satisfied. Further, condition (b) holds as

$$\int_0^{\infty} \frac{1}{\sigma^2(u)} \exp\left(\int_0^u \frac{2\mu(s)}{\sigma^2(s)}ds\right)du = \int_{-\infty}^{\infty} e^{-2|u|}du = 1.$$

Thus, a stationary distribution exists and has probability density function $\pi(x) = e^{-2|x|}$.

(b) If the stationary distribution of X exists, then it is given by solving $L^*\pi = 0$, that is,

$$L^*\pi(x) = \frac{1}{2}\frac{\partial^2}{\partial x^2}\pi(x) - \frac{\partial}{\partial x}\{-\text{sgn}(x)\pi(x)\} = 0.$$

This gives

$$\pi(x) = \begin{cases} Ce^{2x} & x < 0, \\ Ce^{-2x} & x > 0, \end{cases}$$

where C is a constant such that $\int \pi(x)dx = 1$. This yields $C = 1$ and $\pi(x) = e^{-2|x|}$.

Problem 6.33: Let diffusion on (α, β) be such that the transition probability density $p(t, x, y)$ is symmetric in x and y, $p(t, x, y) = p(t, y, x)$ for all x, y, and t. Show that if (α, β) is a finite interval, then the uniform distribution is invariant for the process $X(t)$.

Solution. From Equation (6.67*), if the stationary (invariant) distribution has a density, $\pi(x)$, and $p(t, x, y)$ is the transition probability density, then π satisfies

$$\pi(y) = \int p(t, x, y)\pi(x)dx.$$

Given the transition probability is symmetric, one has

$$\int_\alpha^\beta p(t, x, y)dy = \int_\alpha^\beta p(t, y, x)dy = 1,$$

and any $\pi(x) = C$ satisfies

$$C = \int_\alpha^\beta Cp(t, y, x)dy.$$

Since the invariant density $\pi(x) = C$ is constant for all $x \in (\alpha, \beta)$, the invariant distribution on (α, β) is uniform.

Problem 6.34: Investigate for absorption at zero the following process (used as a model for interest rates, the square root model of Cox, Ingersoll, and Ross):

$$dX(t) = b(a - X(t))dt + \sigma\sqrt{X(t)}dB(t),$$

where parameters b, a, and σ are constants.

Solution. While conditions for absorption can be checked directly, as shown in the following, this problem can be treated more conceptually, as suggested in Problem 6.27.

By Itô's formula, $Y(t) = 2\sqrt{X(t)}$ satisfies the SDE

$$\left[(2ab - \sigma^2/2)Y(t)^{-1} - \frac{1}{2}bY(t)\right]dt + \sigma dB(t).$$

From Problem 6.27, there is no explosion to 0^+ for $2ab \geq \sigma^2$.

0 is a natural boundary.

When there is an explosion, $2ab \geq \sigma^2$, then an inspection of the SDE for $X(t)$ (together with the Yamada–Watanabe Theorem 5.5) shows that 0 is absorbing for $ab \leq 0$, i.e. 0 is an absorbing boundary if $ab \leq 0$ and $2ab < \sigma^2$. $X(t)$ leaves 0 immediately after a visit there when $ab > 0$, i.e. 0 is a regular boundary if $ab \leq 0$ and $2ab < \sigma^2$.

Next, we give direct calculations for the original SDE.

Recall Remark 6.6 that, with $L_1 = \int_\alpha^\beta \exp(-\int_\beta^u \frac{2\mu(s)}{\sigma^2(s)}ds)du$, $L_2 = \int_\alpha^\beta \frac{1}{\sigma^2(y)}(\int_\alpha^y \exp(-\int_\beta^x \frac{2\mu(s)}{\sigma^2(s)}ds)dx)\exp(\int_\beta^y \frac{2\mu(s)}{\sigma^2(s)}ds)dy$, and $L_3 = \int_\alpha^\beta \frac{1}{\sigma^2(y)}\exp(\int_{x_0}^y \frac{2\mu(s)}{\sigma^2(s)}ds)dy$, the boundary point α is called

(a) natural if $L_1 = \infty$;
(b) attracting if $L_1 < \infty$, $L_2 = \infty$;
(c) absorbing if $L_1 < \infty$, $L_2 < \infty$, $L_3 = \infty$;
(d) regular if $L_1 < \infty$, $L_2 < \infty$, $L_3 < \infty$.

Here, $\mu(s) = b(a - s)$ and $\sigma^2(s) = \sigma^2 s$. First, note that

$$\int_\beta^u \frac{2\mu(s)}{\sigma^2(s)}ds = \frac{2b}{\sigma^2}\int_\beta^u \left(\frac{a}{s} - 1\right)ds = \frac{2b}{\sigma^2}(a\ln u - u - a\ln\beta + \beta).$$

Take $\alpha = 0$. We investigate the convergence of L_1, L_2, and L_3. First, we have

$$\begin{aligned} L_1 &= \int_0^\beta \exp\Big(-\frac{2b}{\sigma^2}(a\ln u - u - a\ln\beta + \beta)\Big)du \\ &= \exp\left(2b(a\ln\beta - \beta)/\sigma^2\right)\int_0^\beta u^{-\frac{2ab}{\sigma^2}}\exp\left(2bu/\sigma^2\right)du, \end{aligned}$$

which is finite if and only if $\frac{2ab}{\sigma^2} < 1$. Now, suppose that this holds. Then, we have

$$L_2 = \frac{1}{\sigma^2}\int_0^\beta \frac{1}{y}\int_0^y \exp\left(-\frac{2b}{\sigma^2}(a\ln x - x)\right)dx \exp\left(\frac{2b}{\sigma^2}(a\ln y - y)\right)dy$$
$$= \frac{1}{\sigma^2}\int_0^\beta y^{\frac{2ab}{\sigma^2}-1}\exp\big(-2by/\sigma^2\big)\int_0^y x^{-\frac{2ab}{\sigma^2}}\exp\big(2bx/\sigma^2\big)dxdy.$$

To see whether this integral converges, we inspect the behaviour of the integrand (of the outer integral) around zero. By L'Hopital's rule,

$$\lim_{y\to 0}\frac{\int_0^y x^{-\frac{2ab}{\sigma^2}}\exp\big(2bx/\sigma^2\big)dx}{y^{1-\frac{2ab}{\sigma^2}}\exp\big(2by/\sigma^2\big)}$$
$$= \lim_{y\to 0}\frac{y^{-\frac{2ab}{\sigma^2}}\exp\big(2by/\sigma^2\big)}{(1-\frac{2ab}{\sigma^2})y^{-\frac{2ab}{\sigma^2}}\exp\big(2by/\sigma^2\big)+\frac{2b}{\sigma^2}y^{1-\frac{2ab}{\sigma^2}}\exp\big(2by/\sigma^2\big)}$$
$$= \lim_{y\to 0}\frac{\sigma^2}{\sigma^2-2ab+2by}.$$

Thus,

$$L_2 \sim \int_0^\beta \frac{1}{\sigma^2-2ab+2by}dy = \int_{\sigma^2-2ab}^{\sigma^2-2ab+2b\beta}\frac{1}{2b}\frac{1}{v}dv,$$

which is finite if and only if $\sigma^2 > 2ab$. Next,

$$L_3 = \int_0^\beta \frac{1}{\sigma^2 y}\exp\left(\frac{2b}{\sigma^2}(a\ln y - u - a\ln x_0 + x_0)\right)dy$$
$$= \frac{1}{\sigma^2 y}\exp\big(-2b(a\ln x_0 - x_0)/\sigma^2\big)\int_0^\beta y^{\frac{2ab}{\sigma^2}-1}\exp\big(-2by/\sigma^2\big)dy,$$

which is finite if and only if $\frac{2ab}{\sigma^2} > 0$, that is, $ab > 0$.

In summary, the boundary point zero is

- natural if $2ab \geq \sigma^2$;
- regular if $2ab < \sigma^2$ and $ab > 0$;
- absorbing if $2ab < \sigma^2$ and $ab \leq 0$.

Problem 6.35: The Chan–Koralyi–Longstaff–Sanders (CKLS) SDE is given by

$$dX(t) = (\alpha + \beta X(t))\,dt + \sigma X(t)^{\gamma}\,dB(t) \quad \text{for } t > 0, \quad X(0) = x_0,$$

where $\alpha, \sigma, \gamma, x_0 > 0$ and $\beta \in \mathbb{R}$ are constants. This SDE is used in contemporary mathematical finance research as a model for interest rates deseasonalised electricity prices, etc., and is famous for being very hard to draw inferences from and very hard to simulate when $\gamma > 1$. Determine the stationary distribution for this SDE when it exists.

Solution. First, note the fact that $\alpha, x_0 > 0$ ensures that the solution is strictly positive when it exists. From Equation (6.69*) in Klebaner's book, we further see that the stationary probability density function must be given by

$$\pi(x) = \frac{1}{C\,x^{2\gamma}} \exp\left\{ \int_1^x \frac{2\,(\alpha + \beta\,y)}{\sigma^2\,y^{2\gamma}}\,dy \right\} \quad \text{for } x > 0,$$

whenever this function can be normalised to become a density, that is, whenever

$$C = \int_0^{\infty} \frac{1}{x^{2\gamma}} \exp\left\{ \int_1^x \frac{2\,(\alpha + \beta\,y)}{\sigma^2\,y^{2\gamma}}\,dy \right\} dx < \infty.$$

The issue is whether C is finite or not in turn clearly boils down to checking the integrability properties of the function

$$f(x) = \frac{1}{x^{2\gamma}} \exp\left\{ \int_1^x \frac{2\,(\alpha + \beta\,y)}{\sigma^2\,y^{2\gamma}}\,dy \right\},$$

as $x \downarrow 0$ and as $x \uparrow \infty$. Now, as $x \downarrow 0$, we see that

$$f(x) \sim \begin{cases} C_1\,x^{-2\gamma} & \text{for } \gamma \in (0, 1/2), \\ C_2\,x^{2\alpha/\sigma^2 - 1} & \text{for } \gamma = 1/2, \\ C_3\,x^{-2\gamma} \exp\{-(2\alpha/(\sigma^2(2\gamma - 1)))\,x^{-(2\gamma-1)}\} & \text{for } \gamma > 1/2, \end{cases}$$

where $C_1, C_2, C_3 > 0$ are constants. This is to say that we always have the integrability required as $x \downarrow 0$. When $x \uparrow \infty$, we further see that $f(x)$

is asymptotically equivalent as follows:

$$f(x) \sim \begin{cases} C_4\, x^{-2\gamma} & \text{for } \gamma > 1, \\ C_5\, x^{-2+2\beta/\sigma^2} & \text{for } \gamma = 1, \\ C_6\, x^{-2\gamma} \exp\{(\beta/(\sigma^2(1-\gamma)))x^{2-2\gamma}\} & \text{for } \gamma \in (1/2,1), \\ C_7\, x^{2\alpha/\sigma^2-1} \exp\{(2\beta/\sigma^2)x\} & \text{for } \gamma = 1/2, \\ C_8\, x^{-2\gamma} \exp\{(\beta/(\sigma^2(1-\gamma)))x^{2-2\gamma} & \\ \quad +(2\alpha/(\sigma^2(1-2\gamma)))x^{1-2\gamma}\} & \text{for } \gamma \in (0,1/2), \end{cases}$$

where $C_4, \ldots, C_8 > 0$ are constants. This is to say that we have the integrability required when

$$\gamma > 1 \quad \text{and} \quad \gamma = 1,\, 2\beta < \sigma^2 \quad \text{and}$$
$$\gamma \in (1/2,1),\, \beta \leq 0 \quad \text{and} \quad \gamma \in (0,1/2],\, \beta < 0.$$

To verify that $\pi(x)$ from above really is the stationary probability density, we must check that I_1 and I_2 in Theorem 6.28 (with $x_0 = 1$ say) are both infinite. The integrand of these integrals is proportional to $(x^{2\gamma} f(x))^{-1}$ and from the above established asymptotics, we see that $I_1 = \infty$ iff $\gamma > 1/2$ or $\gamma = 1/2$ and $\alpha \geq \sigma^2/2$. Further, for $\gamma \geq 1/2$, $I_2 = \infty$ iff $\gamma > 1$, $\gamma = 1$ and $2\beta \leq \sigma^2$, $\gamma \in (1/2,1)$ and $\beta \leq 0$ or $\gamma = 1/2$ and $\beta < 0$. Putting this together with our findings about $\pi(x)$, we conclude that the stationary distribution exists iff $\gamma > 1$, $\gamma = 1$ and $2\beta < \sigma^2$, $\gamma \in (1/2,1)$ and $\beta \leq 0$ or $\gamma = 1/2$ and $\beta < 0$ and $2\alpha \geq \sigma^2$.

Also, this problem can be treated more conceptually in the fashion of Problem 6.27.

When the stationary probability density exists, it is given by Equation (6.69*). Note that for a diffusion on $(0,\infty)$ with constant diffusion coefficient σ^2, if the criteria for that existence is satisfied for a certain drift coefficient μ, then it is also satisfied for all other drift coefficients that are smaller close to $+\infty$ and larger close to 0^+. And if the criteria are not satisfied for a certain drift coefficient μ, then it is also not satisfied for all other drift coefficients that are larger close to $+\infty$ and smaller close to 0^+. In particular, $\mu(x) = A/x$ for x close to $+\infty$ and $\mu(x) = B/x$ for x close to 0^+ means that a stationary density exists iff $A < -\sigma^2/2$ and $B \geq \sigma^2/2$.

For the CKLS diffusion $X(t)$, for $\gamma \neq 1$, take $Y(t) = X(t)^{1-\gamma}$ so that by Itô's formula,

$$dY(t) = (\gamma - 1)\left[\frac{1}{2}\gamma\sigma^2 Y(t)^{-1} - \alpha Y(t)^{1+1/(\gamma-1)} - \beta Y(t)\right]dt$$
$$- (\gamma - 1)\sigma dB(t).$$

For $\gamma > 1$, the drift coefficient is way more negative at $+\infty$ and way larger at 0^+ than required for the existence of stationary density, so the latter exists.

For $\gamma \in (0, 1)$, we rewrite the SDE for $Y(t)$ as

$$dY(t) = (1 - \gamma)\left[\alpha Y(t)^{1-1/(1-\gamma)} + \beta Y(t) - \frac{1}{2}\gamma\sigma^2 Y(t)^{-1}\right]dt$$
$$+ (1 - \gamma)\sigma dB(t).$$

For $\gamma \in (1/2, 1)$, we see that we must have $\beta \leq 0$ to pass the requirement for stationary density at $+\infty$ and in that case, the $Y(t)^{1-1/(1-\gamma)}$ term is dominated by the $Y(t)^{-1}$ term, which in turn has a factor less than $-\frac{1}{2}(1-\gamma)^2\sigma^2$ as required. At 0^+, the term $Y(t)^{1-1/(1-\gamma)}$ with positive factor dominates the term $Y(t)^{-1}$ so that the existence criteria for stationary density are satisfied there. And so stationary density exists iff $\beta \leq 0$.

For $\gamma = 1/2$, we see that $\beta \leq 0$ is necessary for the existence criteria to hold at $+\infty$. Clearly, $\beta < 0$ easily fixes that criteria, while $\beta = 0$ implies $(1 - \gamma)(\alpha - \frac{1}{2}\gamma\sigma^2) < -\frac{1}{2}(1 - \gamma)^2\sigma^2$, i.e. $\alpha < 0$, which is not allowed. At 0^+, we instead must have $(1-\gamma)(\alpha - \frac{1}{2}\gamma\sigma^2) \geq \frac{1}{2}(1-\gamma)^2\sigma^2$, i.e. $\alpha \geq \sigma^2/2$.

For $\gamma \in (0, 1/2)$, the term $Y(t)^{-1}$ with negative factor dominates the term $Y(t)^{1-1/(1-\gamma)}$ at 0^+ so that we cannot have a stationary density.

For $\gamma = 1$, take $Y(t) = \ln(X(t))$ to obtain by Itô's formula,

$$dY(t) = \left[\alpha e^{-Y(t)} + \left(\beta - \frac{1}{2}\sigma^2\right)\right]dt + \sigma dB(t).$$

Note that now $Y(t)$ is a diffusion on $(-\infty, \infty)$ unlike the case when $\gamma \neq 1$. We readily see that $2\beta < \sigma^2$ is what is required for the existence of a stationary density at $+\infty$, while nothing is required for existence at $-\infty$.

In conclusion, a stationary density exists iff $\gamma > 1$, $\gamma = 1$ and $2\beta < \sigma^2$, $\gamma \in (1/2, 1)$ and $\beta \leq 0$ or $\gamma = 1/2$ and $\beta < 0$ and $2\alpha \geq \sigma^2$.

Problem 6.36: Given some constants $\mu, \sigma \in \mathbb{R}$, consider the SDE

$$dX(t) = \mu dt + \sigma X(t) dB(t) \quad \text{for } t \in (0, T], X(0) = X_0.$$

(a) Show that the unique strong solution to this SDE is given by

$$X(t) = \mathrm{e}^{\sigma B(t) - \sigma^2 t/2} \left(X_0 + \mu \int_0^t \mathrm{e}^{-\sigma B(s) + \sigma^2 s/2}\, ds \right) \quad \text{for } t \in [0, T].$$

(b) Solve the SDE numerically for $\mu = \sigma = 1$, $T = 10$ and $X_0 = 0$ by the Euler method. Plot a sample path of the numerical P solution and compare it with the analytic solution.

Solution.

(a) According to Theorem 5.4 in Klebaner's book, the SDE has a unique strong solution. By application of Itô's formula, Theorem 4.17 in Klebaner's book with

$$f(y, z) = y\, z, \quad Y(t) = \mathrm{e}^{\sigma B(t) - \sigma^2 t/2} \quad \text{and}$$

$$Z(t) = X_0 + \mu \int_0^t \mathrm{e}^{-\sigma B(t) + \sigma^2 s/2}\, ds,$$

(cf. Example 4.25 in Klebaner's book), we further see that $Y(t)Z(t)$ is the solution as

$$\begin{aligned} d(Y(t)Z(t)) &= Y(t)\, dZ(t) + Z(t)\, dY(t) + dY(t) dZ(t) \\ &= \mu\, dt + Y(t)Z(t)\, \sigma\, dB(t) + (\mu\, dt)\, (\sigma\, dB(t)) \\ &= \mu\, dt + Y(t)Z(t)\, \sigma\, dB(t). \end{aligned}$$

(b) The exact solution is plotted by using the formula, first simulating the Brownian motion $B(t)$ at a number of points and then using the formula for $X(t)$. The simulated solution is obtained directly from the SDE by using the same Brownian motion as for the exact solution and then by iterations (Euler scheme).

```
In[25]:= Clear[steps, mu, sigma, T, X0, dB, Xnumeric, B, IntB, Xanalytic]; steps = 10000; mu = 1; sigma = 1;
         T = 10; X0 = 0; dB = Table[Random[NormalDistribution[0, Sqrt[T/steps]]], {i, 1, steps}];
In[35]:= For[i = 1; Xnumeric = {X0}, i ≤ steps, i++,
          AppendTo[Xnumeric, Xnumeric[[i]] + mu * T/steps + sigma * Xnumeric[[i]] * dB[[i]]]];
         ListPlot[Xnumeric, PlotJoined → True, PlotRange → {-0.51, 20.1},
          Ticks → {{{1000, ""}, {2000, "2"}, {3000, ""}, {4000, "4"}, {5000, ""},
             {6000, "6"}, {7000, ""}, {8000, "8"}, {9000, ""}, {10000, "10"}}, Automatic}]
```

Out[35]=

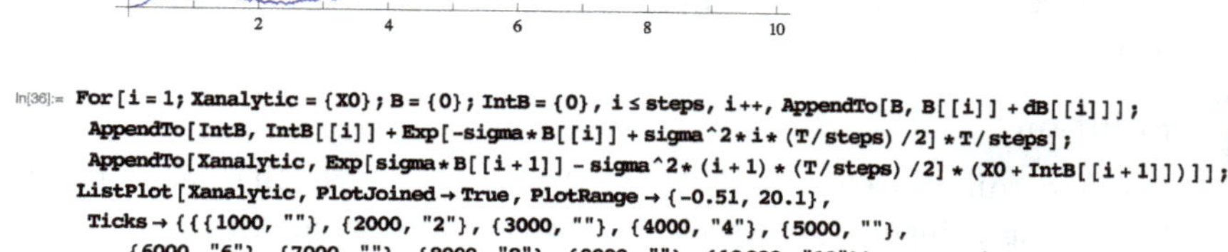

```
In[36]:= For[i = 1; Xanalytic = {X0}; B = {0}; IntB = {0}, i ≤ steps, i++, AppendTo[B, B[[i]] + dB[[i]]];
          AppendTo[IntB, IntB[[i]] + Exp[-sigma * B[[i]] + sigma^2 * i * (T/steps) / 2] * T/steps];
          AppendTo[Xanalytic, Exp[sigma * B[[i + 1]] - sigma^2 * (i + 1) * (T/steps) / 2] * (X0 + IntB[[i + 1]])]];
         ListPlot[Xanalytic, PlotJoined → True, PlotRange → {-0.51, 20.1},
          Ticks → {{{1000, ""}, {2000, "2"}, {3000, ""}, {4000, "4"}, {5000, ""},
             {6000, "6"}, {7000, ""}, {8000, "8"}, {9000, ""}, {10000, "10"}}, Automatic}]
```

Out[36]=

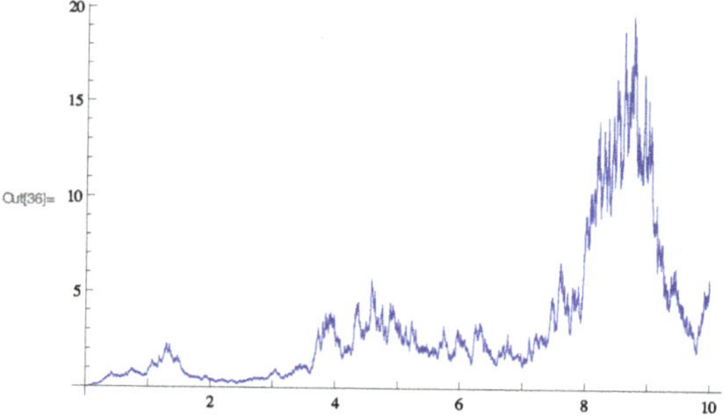

Problem 6.37: This question concerns the Engelbert–Schmidt condition on the existence and uniqueness of weak solutions. Consider the SDE $dX(t) = \sigma(X(t))dB(t)$. Let S and N be sets of singularities:

$$S = \left\{x \in \mathbb{R} : \int_{x-\varepsilon}^{x+\varepsilon} \frac{dy}{\sigma^2(y)} = \infty \text{ for any } \varepsilon > 0\right\}, \quad N = \{x \in \mathbb{R} : \sigma(x) = 0\}.$$

(a) State the Engelbert–Schmidt condition on the existence and uniqueness of weak solutions, Theorem 6.13, in terms of sets S and N.

(b) Show that if $\sigma(x)$ is continuous, the SDE has a weak solution. Moreover, if $\sigma(x) > 0$, then the weak solution is unique.

Solution.

(a) By the theorem of Engelbert–Schmidt, Theorem 6.13, a weak solution starting from any x exists if and only if the divergence of the integral $\int_{x-\varepsilon}^{x+\varepsilon} dy/\sigma^2(y) = \infty$ for any ε implies $\sigma(x) = 0$. In other words, the existence of a weak solution is the condition $S \subset N$. The weak solution is unique if the divergence of the integral is equivalent to $\sigma(x) = 0$. In other words, the weak solution is unique if $S = N$.

(b) If a continuous $\sigma^2(x) > 0$, then $\sigma^2(y) > \frac{1}{2}\sigma^2(x) > 0$ for $y \in (x - \varepsilon, x + \varepsilon)$, for some $\varepsilon > 0$, so that the above integral is finite, and $S = \emptyset$. As $\emptyset$ is a subset of any set, the weak solution exists. If $N = \emptyset$ as well, then $N = S$, and the solution is unique.

Chapter 7

Martingales

The concept of martingale, submartingale and supermartingale $\{M(t)\}_{t\geq 0}$ ($\{M(t)\}_{t\in[0,T]}$) with respect to a filtration $\{\mathcal{F}_t\}_{t\geq 0}$ ($\{\mathcal{F}_t\}_{t\in[0,T]}$) has been defined in Chapter 2. The union of these three families of processes can conveniently be called smartingales. So far, we have seen only continuous smartingales, but now we will also discuss more general smartingales without this restriction. However, our smartingales are always assumed to be right-continuous with left limits (càdlàg): it can be shown that they can be modified to be càdlàg "without loss" if they were not càdlàg from the beginning.

A smartingale M is a martingale if and only if the mean process $\mathrm{E}(M(t))$ is a constant (does not depend on t). Here, we have seen the only if statement before. To prove the if statement, we assume that M is a submartingale (supermartingale) but not a martingale. This means that $\mathrm{E}(M(t)|\mathcal{F}_s) > (<)\, M(s)$ for some $0 \leq s < t$ on a non-null event, while there is non-strict inequality $[\geq (\leq)]$ on the remainder of the sample space. But then (by positivity of expectation), it follows that

$$\mathrm{E}(M(t)) = \mathrm{E}(\mathrm{E}(M(t)|\mathcal{F}_s)) > (<)\, \mathrm{E}(M(s)).$$

This establishes the if statement.

Concepts of Integrability

A different concept of integrability turns out to be of interest for smartingales: we say that a smartingale $M(t)$ is integrable [square integrable] if $\sup_t \mathrm{E}(|M(t)|) < \infty[\sup_t \mathrm{E}(M(t)^2) < \infty]$. Of course, in general, integrability [square integrable] of a smartingale is a stronger requirement than the requirement that each process value is integrable [square integrable].

A smartingale M is uniformly integrable if

$$\lim_{n\to\infty} \sup_t \mathrm{E}\big(I_{\{|M(t)|>n\}}|M(t)|\big) = 0.$$

Clearly, a uniformly integrable martingale is integrable as

$$\sup_t \mathrm{E}(|M(t)|) \leq n + \sup_t \mathrm{E}\big(I_{\{|M(t)|>n\}}|M(t)|\big).$$

Further, if $\mathrm{E}(\sup_t |M(t)|) < \infty$, then it follows readily from dominated convergence that M is uniformly integrable.

One common way to check that a smartingale (or any random process) M is uniformly integrable is by the Vallée–Poussin[1] theorem: if there exists a function on $G : [0,\infty) \to [0,\infty)$ such that

$$\lim_{x\to\infty} \frac{G(x)}{x} = \infty \quad \text{and} \quad \sup_t \mathrm{E}(G(|M(t)|)) < \infty,$$

then M is uniformly integrable. This result is virtually always used with $G(x) = x^r$ for $r > 1$. The proof is more or less as that of the Chebyshev inequality:

$$\begin{aligned}\sup_t \mathrm{E}\big(I_{\{|M(t)|>n\}}|M(t)|\big) &\leq \sup_t \mathrm{E}\bigg(\frac{G(|M(t)|)}{|M(t)|}|M(t)|\bigg)\\ &\quad\times \left(\sup_{x>n} \frac{x}{G(x)}\right) \to 0 \quad \text{as } n\to\infty.\end{aligned}$$

A Doob–Lévy martingale $M(t) = \mathrm{E}(Y|\mathcal{F}_t)$ is uniformly integrable. The proof of this fact is a bit too long to be cited here. (In fact, this statement also has a converse which has to do with the convergence theorem we discuss in the following section.)

Martingale Convergence

The fundamental (and quite hard to prove) smartingale convergence theorem states that an integrable smartingale $\{M(t)\}_{t\geq 0}$ converges almost surely $M(t) \to_{\text{a.s.}} Y$ as $t\to\infty$, where Y is integrable.

---

[1]Baron Charles–Jean Étienne Gustave Nicolas Le Vieux de la Vallée Poussin (1866–1962).

For a martingale, the integrability condition is equivalent to $\lim_{t\to\infty} \mathrm{E}(|M(t)|) < \infty$ (because $|M|$ is a submartingale and therefore has increasing expectation) as well as to either $\lim_{t\to\infty} \mathrm{E}(M(t)^+) < \infty$ and/or $\lim_{t\to\infty} \mathrm{E}(M(t)^-) < \infty$ [as M^+ and M^- are submartingales and $\mathrm{E}(|M|) = \mathrm{E}(M^+) + \mathrm{E}(M^-) = 2\,\mathrm{E}(M^+) - \mathrm{E}(M(0)) = 2\,\mathrm{E}(M^-) + \mathrm{E}(M(0))$].

For a submartingale, the convergence theorem applies if $\sup_{t\geq 0} \mathrm{E}(M(t)^+) < \infty$ because this together with the fact that

$$\mathrm{E}(M(0)^-) \geq \mathrm{E}(M(0)^+) + \mathrm{E}(M(t)^-) - \mathrm{E}(M(t)^+)$$

ensures that $\sup_{t\geq 0} \mathrm{E}(M(t)^-) < \infty$ also, so that M is integrable.

From the convergence theorem, it follows that uniformly integrable smartingales and square-integrable smartingales converge (as they are integrable) and that positive or negative martingales converge [as they have $\mathrm{E}(M^-) = 0$ or $\mathrm{E}(M^+) = 0$].

The issue whether $\lim_{t\to\infty} \mathrm{E}(M(t)) = \mathrm{E}(Y)$ or not in the convergence theorem turns out to have a more or less complete answer: a martingale (negative submartingale) $\{M(t)\}_{t\geq 0}$ converges both almost surely and in L^1 to a limit Y as $t \to \infty$ if and only if M is uniformly integrable, and in that case M is a Doob–Lévy martingale (negative submartingale) $M(t) = (\leq)\,\mathrm{E}(Y|\mathcal{F}_t)$ for $t \geq 0$. Further, the if statement (but not necessarily the only if statement) is true for all smartingales, that is, uniformly integrable smartingales converge almost surely and in L^1.

A simple example of a martingale that converges almost surely but not in L^1 is the stochastic exponential of BM $\{e^{B(t)-t/2}\}_{t\geq 0}$. As this is a positive martingale, it converges almost surely to a limit that must be 0 as $B(t)/t \to_{\text{a.s.}} 0$ by the law of the iterated logarithm, so that $B(t) - t/2 = t(B(t)/t - 1/2) \to_{\text{a.s.}} -\infty$. However, this martingale is a unit mean, so it cannot converge in L^1.

Optional Stopping

Throughout this section, $\{M(t)\}_{t\geq 0}$ denotes a martingale with respect to the filtration $\{\mathcal{F}_t\}_{t\geq 0}$ and τ a stopping time. We only give results in this section for martingales, but they are also available in modified form for general smartingales.

The stopped process $\{M(\tau \wedge t)\}_{t\geq 0}$ is also a martingale with respect to $\{\mathcal{F}_t\}_{t\geq 0}$. This result, which is not easy to prove, turns out to be very important.

Any measurable and adapted process $\{X(t)\}_{t\geq 0}$ has a so-called progressively measurable modification which can be used "without loss" (which is a famously difficult to prove result). And for the modified process, it holds that $X(\tau)$ is $\mathcal{F}_\tau$ measurable. So, we can assume that $M(\tau \wedge t)$ is $\mathcal{F}_{\tau\wedge t}$-adapted in the previous paragraph. Therefore, $M(\tau \wedge t)$ is a martingale also with respect to $\mathcal{F}_{\tau\wedge t}$ as it (by towering) must be a martingale to any smaller than $\mathcal{F}_t$ filtration that it is adapted to (and it is readily shown that $\mathcal{F}_{\tau_1} \subseteq \mathcal{F}_{\tau_2}$ for stopping times $\tau_1 \leq \tau_2$).

We have from above that $\mathrm{E}(M(\tau \wedge t)) = \mathrm{E}(M(0))$ for $t \geq 0$. Here, it is tempting to try to send $t \to \infty$ to obtain the so-called basic stopping equation

$$\mathrm{E}(M(\tau)) = \mathrm{E}(M(0)).$$

That this equation is not true in general is seen from, for example, taking M to be BM B and $\tau = \inf\{t > 0 : B(t) = 1\}$.

We now list three versions of the very important so-called optional stopping theorem that ensures that the basic stopping equation holds:

- The basic stopping equation holds if τ is bounded from above by a real constant.
- The basic stopping equation holds (for all stopping times) if M is uniformly integrable.
- The basic stopping equation holds if the following three conditions hold:

$$\mathrm{P}(\tau < \infty) = 1, \quad \mathrm{E}(|M(\tau)|) < \infty \quad \text{and} \\ \lim_{t\to\infty} \mathrm{E}\big(I_{\{\tau>t\}}|M(t)|\big) = 0.$$

If $M(t)$ is a continuous martingale with $M(0) = x$ and the stopping time $\tau = \inf\{t \geq 0 : M(t) \notin [a,b]\}$ is finite for $a < x < b$, then the third of the above criteria gives that

$$a\mathrm{P}(M(\tau) = a) + b\mathrm{P}(M(\tau) = b) = x,$$

from which in turn of course $\mathrm{P}(M(\tau) = a)$ and $\mathrm{P}(M(\tau) = b)$ can be calculated [as $\mathrm{P}(M(\tau) = a) = 1 - \mathrm{P}(M(\tau) = b)$].

There is a converse to the first version of the optional stopping theorem: if $\{X(t)\}_{t\geq 0}$ is an adapted process such that $X(\tau)$ is integrable and $\mathrm{E}(X(\tau)) = \mathrm{E}(X(0))$ for every stopping time that is bounded from above by a constant, then X is a martingale.

The optional stopping theorem has a companion called the optional sampling theorem: if M is uniformly integrable and $\tau_1 \leq \tau_2$ are stopping times, then

$$\mathrm{E}(M(\tau_2)|\mathcal{F}_{\tau_1}) = M(\tau_1).$$

Of course, the concept of a time discrete martingale $\{M_n\}_{n\geq 0}$ with respect to a time discrete filtration is defined by an obvious modification of the definition in continuous time, as is the definition of a stopping time τ. We conclude this section by citing a useful version of the optional stopping theorem that is available only for discrete-time martingales: the basic stopping equation $\mathrm{E}(M_\tau) = \mathrm{E}(M_0)$ holds if $\mathrm{E}(\tau) < \infty$, $\mathrm{E}(|M(\tau)|) < \infty$ and

$$\mathrm{E}(|M_{n+1} - M_n|\,|\mathcal{F}_n) \leq K \quad \text{for } n \geq 0 \text{ for some constant } K > 0.$$

Of course, a sufficient criterion for the last condition to hold is that $|M_{n+1} - M_n| \leq K$ for $n \geq 0$.

Localisation and Local Martingales

In a more detailed treatment of stochastic calculus than ours (have been so far), the concept of local martingale is indispensable: an adapted process $\{M(t)\}_{t\geq 0}$ is a local martingale if there exists a so-called localising increasing sequence of stopping times $\{\tau_n\}_{n=1}^{\infty}$ such that $\tau_n \rightarrow_{\text{a.s.}} \infty$ as $n \rightarrow \infty$ and $\{M(t \wedge \tau_n)\}_{t\geq 0}$ is a uniformly integrable martingale for all n. (Local submartingales and supermartingales are defined accordingly.)

Clearly, any martingale is a local martingale as we may take $\tau_n = n$ (and use that any martingale over a finite interval is uniformly integrable as it is a Doob–Lévy martingale).

One main reason that local martingales are important in stochastic calculus is that the Itô integral process $\{\int_0^t X\,dB\}_{t\geq 0}$ is always a local martingale but not always a martingale: take

$$\tau_n = n \wedge \inf\left\{t \geq 0 : \int_0^t X(r)^2\,dr \geq n\right\} \quad \text{for } n \geq 1.$$

Then, we have $\{I_{[0,\tau_n]}X\}_{t\in[0,T]} \in E_T$ for $T \geq 0$ as

$$\mathrm{E}\bigg(\int_0^T I_{[0,\tau_n]}(s)^2 X(s)^2\,ds\bigg) \leq \mathrm{E}\bigg(\int_0^\infty I_{[0,\tau_n]}(s)^2 X(s)^2\,ds\bigg) \leq n.$$

And so, it follows that

$$\left\{\int_0^t I_{[0,\tau_n]}X\,dB\right\}_{t\in[0,T]} = \left\{\int_0^{t\wedge\tau_n} X\,dB\right\}_{t\in[0,T]}$$

is a square-integrable martingale and hence a uniformly integrable martingale for $T \geq 0$. However, the value of this process doesn't change after time $t = n$, and hence $\{\int_0^t X\,dB\}_{t\geq 0}$ is a local martingale with the same localising sequence $\{\tau_n\}_{n=1}^{\infty}$.

Beware that there are examples of uniformly integrable local martingales that are not martingales (and hence of integrable local martingales that are not martingales). However, if a local martingale $\{M(t)\}_{t\geq 0}$ satisfies $\mathrm{E}(\sup_{r\in[0,t]} |M(r)|) < \infty$ for $t \geq 0$, then it is a martingale as we may use the dominated convergence theorem for conditional expectations to conclude that

$$M(s) \leftarrow_{\text{a.s.}} M(s\wedge\tau_n) = \mathrm{E}(M(t\wedge\tau_n)|\mathcal{F}_s) \rightarrow_{\text{a.s.}} \mathrm{E}(M(t)|\mathcal{F}_s)$$

as $n \to \infty$ for $0 \leq s \leq t$. And if in addition $\mathrm{E}(\sup_{r\geq 0} |M(r)|) < \infty$, it is also uniformly integrable.

A positive local martingale $\{M(t)\}_{t\geq 0}$ is a supermartingale. This follows immediately from Fatou's lemma for conditional expectations as

$$\begin{aligned}
\mathrm{E}(M(t)|\mathcal{F}_s) &= \mathrm{E}\Big(\liminf_{n\to\infty} M(t\wedge\tau_n) \,\Big|\, \mathcal{F}_s\Big) \\
&\leq \liminf_{n\to\infty} \mathrm{E}(M(t\wedge\tau_n)|\mathcal{F}_s) \\
&= \liminf_{n\to\infty} M(s\wedge\tau_n) \\
&= M(s) \quad \text{for } 0 \leq s \leq t.
\end{aligned}$$

And so, a positive local martingale is a martingale if it has a constant mean.

A little more advanced result states that a local martingale $\{M(t)\}_{t\geq 0}$ is a uniformly integrable martingale if and only if the family $\{M(\tau) : \tau \text{ finite stopping time}\}$ is uniformly integrable (with an obvious modification of the definition of uniform integrability). Any process that satisfies the latter condition is said to be of Dirichlet class (D).

Quadratic Variation of Martingales

All local martingales (as well as the more general semimartingale processes we meet in Chapter 8) have well-defined quadratic variations and covariations: let $\{M_1(t)\}_{t\in[0,T]}$ and $\{M_2(t)\}_{t\in[0,T]}$ be local martingales with respect to a common filtration. Then, there exists a unique FV process $\{[M_1,M_2](t)\}_{t\in[0,T]}$ such that

$$\{M_1(t)M_2(t) - [M_1,M_2](t) - M_1(0)M_2(0)\}_{t\in[0,T]} \quad \text{is a local martingale.}$$

Here, of course, as indicated by the notation, $[M_1, M_2]$ is the covariation process for M_1 and M_2. The proof is very complicated.

A local martingale $\{M(t)\}_{t\in[0,T]}$ is constant $M(t) = M(0)$ for $t \in [0,T]$ if and only if $\{[M](t)\}_{t\in[0,T]}$ is a null process if and only if $[M](T) = 0$. Here, the only if statements are trivial. The if statements can be proved using the preceding paragraph. But they follow more directly from the fact that $\mathrm{E}([M](T)) = 0$ implies that $\mathrm{E}\big(\sup_{t\in[0,T]}(M(t) - M(0))^2\big) = 0$ according to the Burholder–Davis–Gundy inequality given in the following section.

Let $\{M(t)\}_{t\in[0,T]}$ be a continuous FV local martingale with $\mathrm{P}(V_M(T) < \infty) = 1$. Then, $[M](T) = 0$ so that M is constant $M(t) = M(0)$ for $t \in [0,T]$. So, every non-constant continuous local martingale has infinite variation with a strictly positive probability.

Of course, non-continuous (local) martingales can be FV without being constant. One example is the Poisson process with the mean process subtracted.

Martingale Inequalities

Inequalities are an exceedingly important topic for smartingales: the very powerful inequalities that are available for them are crucially used repeatedly in stochastic calculus as well as elsewhere. We discuss two important basic inequalities here and then indicate how other important inequalities can be derived from them.

The first basic inequality is labelled Doob's inequality (or Doob's maximal inequality) after the historically most prominent single contributor to martingale theory, J.L. Doob: for a non-negative submartingale $\{Y(t)\}_{t\in[0,T]}$, we have

$$\mathrm{P}\left(\sup_{t\in[0,T]} Y(t) > \lambda\right) \le \frac{\mathrm{E}\big(I_{\{\sup_{t\in[0,T]} Y(t)>\lambda\}} Y(T)\big)}{\lambda} \le \frac{\mathrm{E}(Y(T))}{\lambda} \quad \text{for } \lambda > 0.$$

The proof is neither long nor very difficult when set up right from the beginning but still a bit too advanced to be included here. By application of the convex function $|\cdot|^p$ with $p \ge 1$ to a martingale $\{M(t)\}_{t\in[0,T]}$, Doob's inequality gives the so-called Doob–Kolmogorov inequality:

$$\mathrm{P}\Big(\sup_{t\in[0,T]} |M(t)| > \lambda\Big) \le \frac{\mathrm{E}\big(|M(T)|^p\big)}{\lambda^p} \quad \text{for } \lambda > 0.$$

By clever repeated use of Doob's inequality together with Hölder's inequality (in an argument that is a little bit too long to be cited here), one can establish that for a non-negative submartingale $\{Y(t)\}_{t\in[0,T]}$, it holds that

$$\mathrm{E}\Big(\Big(\sup_{t\in[0,T]} Y(t)\Big)^p\Big) \le \Big(\frac{p}{p-1}\Big)^p \mathrm{E}\big(Y(T)^p\big) \quad \text{for } p>1.$$

Also, this inequality is often referred to as Doob's (maximal) inequality, but it really is the first version of that inequality we listed that is the more powerful result.

Putting together Doob's maximal inequality and the Vallée–Poussin theorm, we see that for a positive submartingale $\{Y(t)\}_{t\in[0,T]}$ with $\mathrm{E}(Y(T)^p) < \infty$ for some $p > 1$,

$$\sup_{t\in[0,T]} \mathrm{E}(Y(t)^p) \le \mathrm{E}\Big(\Big(\sup_{t\in[0,T]} Y(t)\Big)^p\Big) < \infty$$

so that the submartingale is uniformly integrable.

The second basic inequality has a much more complicated proof than the first one. It is called the Burkholder–Davis–Gundy inequality: let $\{M(t)\}_{t\in[0,T]}$ be a local martingale with $M(0) = 0$. There exist universal constants $c_p, C_p > 0$ depending on $p \ge 1$ only (but neither on T or M) such that

$$\begin{aligned} c_p\mathrm{E}\big(([M](T))^{p/2}\big) = \mathrm{E}\Big(\sup_{t\in[0,T]} |M(t)|^p\Big) &\le \mathrm{E}\Big(\Big(\sup_{t\in[0,T]} |M(t)|\Big)^p\Big) \\ &\le C_p\mathrm{E}\big(([M](T))^{p/2}\big). \end{aligned}$$

If M is continuous, the result also holds for $0 < p < 1$. Further, the result also holds when T is replaced with a stopping time τ.

From what we have done in this section, we are now ready to make the important conclusion that a local martingale $\{M(t)\}_{t\in[0,T]}$ with $\mathrm{E}(\sqrt{[M](T)}) < \infty$ is a uniformly integrable martingale. And if $\mathrm{E}([M](T)) < \infty$, it is a square-integrable martingale.

As an Itô integral process $\{\int_0^t X\, dB\}_{t\in[0,T]}$ is a local martingale with quadratic variation $[\int X\, dB](T) = \int_0^T X(t)^2 dt$, we conclude from the previous paragraph that the condition

$$\mathrm{E}\left(\sqrt{\int_0^T X(t)^2 dt}\right) < \infty$$

ensures that the Itô integral process is a zero-mean martingale. (Of course, this condition is weaker than requiring that $X \in E_T$.)

Continuous Martingales — Change of Time

A random process $\{X(t)\}_{t\geq 0}$ with $X(0) = 0$ is BM if and only if it is a continuous (local) martingale with a quadratic variation process $[X](t) = t$ for $t \geq 0$. This result is called the Lévy's characterisation of BM after the famous probabilist P. Lévy. It is a crucial tool in many proofs, for example, regarding weak solutions to SDE (such as solutions to martingale problems being weak solutions to SDE) The proof can be done in a not-too-long direct calculation that has to be omitted here.

A close companion to the Lévy's characterisation of BM is the Dambis–Dubins–Schwarz theorem: let $\{M(t)\}_{t\geq 0}$ be a continuous local martingale with $M(0) = 0$ and $[M](t) \to_{\text{a.s.}} \infty$ as $t \to \infty$ and define

$$\tau_t = \inf\{s \geq 0 : [M](s) > t\} \quad \text{for } t \geq 0.$$

Then, $B(t) = M(\tau_t)$ is a BM. Further, $[M](t)$ is a stopping time with respect to the filtration $\mathcal{F}_{\tau_t}$ and $M(t) = B([M](t))$.

Formally, we prove the Dambis–Dubins–Schwarz theorem using the Lévy's characterisation together with the fact that $[B](t) = [M]([M]^{-1})(t) = t$. On a more rigorous level, there are technically complicated continuity issues that have to be addressed to make this argument solid.

Let $\sigma : \mathbb{R} \to (0, \infty)$ be bounded and continuous and define stopping times $\{\tau(t)\}_{t\geq 0}$ by

$$\tau(t) = \inf\left(s \geq 0 : \int_0^s \frac{ds}{\sigma(B(s))^2} \geq t\right) \quad \text{for } t \geq 0.$$

Then, we have $\tau'(t) = \sigma(B(\tau_t))^2$, which implies that $\tau(t)$ has linear growth in the t-variable (as σ is bounded) and hence that each $\tau(t)$ is bounded. Since any martingale over a finite time interval is uniformly integrable (as it is a Doob–Lévy martingale), we may therefore apply the optional sampling theorem to the martingale

$$\left\{f(B(t)) - f(0) - \int_0^t \frac{f''(B(r))}{2}\, dr\right\}_{t\geq 0}$$

to conclude that

$$\mathrm{E}\left(f(B(\tau(t))) - f(0) - \int_0^{\tau(t)} \frac{f''(B(r))}{2}\, dr \,\middle|\, \mathcal{F}_{\tau(s)}\right)$$
$$= f(B(\tau(s))) - f(0) - \int_0^{\tau(s)} \frac{f''(B(r))}{2}\, dr$$

for $0 \le s \le t$. And so,

$$\left\{ f(B(\tau(t))) - f(0) - \int_0^{\tau(t)} \frac{f''(B(r))}{2}\, dr \right\}_{t \ge 0}$$

is a martingale with respect to the filtration $\{\mathcal{F}_{\tau(t)}\}_{t \ge 0}$. Now, taking $X(t) = B(\tau(t))$, that martingale by elementary calculus is equal to

$$\begin{aligned} f(X(t)) - f(0) - \int_0^t \tau'(r) \frac{f''(B(\tau(r)))}{2}\, dr \\ = f(X(t)) - f(0) - \int_0^t \frac{\sigma(X(r))^2}{2} f''(X(r))\, dr \end{aligned}$$

for $t \ge 0$ so that $\{X(t)\}_{t \ge 0}$ solves the martingale problem for the SDE

$$dX(t) = \sigma(X(t))\, dB(t) \quad \text{for } t \ge 0, \quad X(0) = 0.$$

In other words, we have constructed a weak solution to that SDE. Alternatively, this can be established (without martingale problem techniques) using the Dambis–Dubins–Schwarz theorem together with a time change of the SDE for the BM $dX = dB$.

One may also time change a general time-homogeneous SDE with drift

$$dX(t) = \mu(X(t))\, dt + \sigma(X(t))\, dB(t).$$

Let $g : [0, \infty) \to (0, \infty)$ satisfy

$$\infty > G(t) = \int_0^t g(X(s))\, ds \uparrow \infty \quad \text{as } t \to \infty.$$

Then, there exists a BM W such that $Y(t) = X(G^{-1}(t))$ satisfies the SDE

$$dY(t) = \frac{\mu(Y(t))}{g(Y(t))}\, dt + \frac{\sigma(Y(t))}{\sqrt{g(Y(t))}}\, dW(t) \quad \text{for } t \ge 0, \quad Y(0) = X(0).$$

Problems

Uniform Integrability: Vallee–Poussin

Problem 7.1 (Vallee–Poussin theorem) (Theorem 7.7): If for some positive function $G(x)$ on $[0, \infty)$ such that $\lim_{x \to \infty} G(x)/x = \infty$ and $\sup_t \mathrm{E}(G(|X_t|)) < \infty$, then X_t is uniformly integrable, that is,

$$\lim_{n \to \infty} \sup_t \mathrm{E}\big(|X_t| I(|X_t| > n)\big) = 0.$$

(Examples of such functions include x^2 and $x \log x$.)

Solution. Let $\epsilon > 0$ be fixed. Let $M = \sup_t \mathrm{E}(G(|X_t|))$ and $a = M/\epsilon$. Choose n large enough so that for $x \geq n$, $G(x)/x \geq a$ (or $x \leq G(x)/a$). Then,

$$\begin{aligned}\mathrm{E}\big(|X_t|I(|X_t| > n)\big) &\leq \frac{1}{a}\mathrm{E}\big(G(|X_t|)I(|X_t| > n)\big)\\ &\leq \frac{1}{a}\mathrm{E}\big(G(|X_t|)\big) \leq \frac{1}{a}M = \epsilon,\end{aligned}$$

and the uniform integrability follows.

Note that G need not be increasing or convex and that the time domain can be either finite $[0, T]$ or infinite $[0, \infty)$.

Sub- and Supermartingales and, Smartingales

Problem 7.2: Let $Y = e^{B_T}$.

(a) Find $M_t = E(Y|\mathcal{F}_t)$ and show that it is a martingale.
(b) Show that M_t is square integrable.
(c) Find the quadratic variation of M, $[M, M]_t$.
(d) Show that $M_t^2 - [M, M]_t$ is a martingale.

Solution.

(a) $M_t = \mathrm{E}(Y|\mathcal{F}_t) = \mathrm{E}(e^{B_T}|\mathcal{F}_t) = \mathrm{E}(e^{B_T - B_t + B_t}|\mathcal{F}_t) = e^{B_t}\mathrm{E}(e^{B_T - B_t}|\mathcal{F}_t)$ $= e^{B_t}\mathrm{E}(e^{B_T - B_t}) = e^{B_t + (T-t)/2}$ because B_t is $\mathcal{F}_t$ measurable, and $B_T - B_t$ is independent of $\mathcal{F}_t$.

$M_t = e^{T/2}e^{B_t - t/2}$, which is a constant times the exponential martingale of Brownian motion. Alternatively, $dM_t = M_t dB_t$, $M_0 = e^{T/2}$.

(b) $M_t^2 = e^T e^{2B_t - t}$. $\mathrm{E}M_t^2 = e^{T+t} \leq e^{2T}$, second moments are bounded.

(c) $d[M, M]_t = (dM_t)^2 = M_t^2 (dB_t)^2 = M_t^2 dt$.
$[M, M]_t = \int_0^t M_s^2 ds = \int_0^t e^T e^{2B_s - s} ds$.

(d) By Itô's formula with function $f(x) = x^2$ for M_t, $dM_t^2 = 2M_t dM_t + d[M, M]_t$. Hence, $d(M_t^2 - [M, M]_t) = dM_t^2 - d[M, M]_t = 2M_t dM_t = 2M_t^2 dB_t$, by using $dM_t = M_t dB_t$. Since by (b), $\mathrm{E}M_t^2 < C$ is bounded, the Itô integral is a martingale, $M_t^2 - [M, M]_t = \int_0^t 2M_s^2 dB_s$.

Problem 7.3 (A family of bounded martingales): Let S_t be a Markov process. Let T and a be fixed. Denote $F(x, t) = \mathrm{P}(S_T \leq a | S_t = x)$.

Show that $F(S_t, t)$ is a martingale. Obtain the martingales when S_t is Brownian motion and geometric Brownian motion.

Solution. By the Markov property,

$$\mathrm{P}(S_T \le a|S_t) = \mathrm{P}(S_T \le a|\mathcal{F}_t) = \mathrm{E}\big(I(S_T \le a)|\mathcal{F}_t\big),$$

which is a martingale of Doob–Lévy type. Therefore, $F(S_t, t) = \mathrm{P}(S_T \le a|S_t)$ is a martingale.

For Brownian motion, we obtain

$$\begin{aligned} F(x,t) &= \mathrm{P}(B_T \le a|B_t = x) = \mathrm{P}(B_t + B_T - B_t \le a|B_t = x) \\ &= \mathrm{P}(N(0, T-t) \le a - x) = \Phi\left(\frac{a-x}{\sqrt{T-t}}\right). \end{aligned}$$

The martingale is given by $\Phi\left(\frac{a-B_t}{\sqrt{T-t}}\right)$.

For geometric Brownian motion $S_t = e^{B_t}$, we obtain for $a > 0$,

$$\begin{aligned} F(x,t) &= \mathrm{P}(e^{B_T} \le a|S_t = x) = \mathrm{P}(e^{B_t}e^{B_T - B_t} \le a|S_t = x) \\ &= \Phi\left(\frac{\ln a - \ln x}{\sqrt{T-t}}\right). \end{aligned}$$

The martingale is given by $\Phi\left(\frac{\ln a - \ln S_t}{\sqrt{T-t}}\right)$.

Problem 7.4: Let $M(t)$ be an $\mathcal{F}_t$-martingale and denote its natural filtration by $\mathcal{G}_t$. Show that $M(t)$ is a $\mathcal{G}_t$-martingale.

Solution. Recall that the natural filtration of a process $M(t)$ ($\mathcal{G}_t = \sigma(\{M(s), 0 \le s \le t\})$) is the σ-field generated by random variables $M(s)$, $s \le t$. Since $\mathcal{G}_t$ is the smallest filtration for which the process $M(t)$ is adapted, $\mathcal{G}_t \subseteq \mathcal{F}_t$.

By the smoothing property of the conditional expectation, if $\mathcal{G} \subset \mathcal{F}$, then $\mathrm{E}(\mathrm{E}(X|\mathcal{F})|\mathcal{G}) = \mathrm{E}(X|\mathcal{G})$, and thus

$$\mathrm{E}(M(t)|\mathcal{G}_s) = \mathrm{E}(\mathrm{E}(M(t)|\mathcal{F}_s)|\mathcal{G}_s) = \mathrm{E}(M(s)|\mathcal{G}_s) = M(s),$$

verifying that $M(t)$ is a $\mathcal{G}_t$-martingale.

Problem 7.5: Show that an adapted increasing integrable process is a submartingale.

Solution. Let $M(t)$ be an increasing integrable process, $\mathrm{E}|M(t)| < \infty$, and all of its realisations are non-decreasing functions of t. For any t and s such that $0 \le s \le t \le T$, $M(s) \le M(t)$. By the monotonicity property of conditional expectation, if $X \le Y$, then $\mathrm{E}(X|\mathcal{G}) \le \mathrm{E}(Y|\mathcal{G})$, and it follows that

$$\mathrm{E}(M(t)|\mathcal{F}_s) \ge \mathrm{E}(M(s)|\mathcal{F}_s) = M_s.$$

Problem 7.6: Show that if $X(t)$ is a submartingale and g is a non-decreasing convex function such that $\mathrm{E}|g(X(t))| < \infty$, then $g(X(t))$ is a submartingale.

Solution. Since $g(x)$ is convex, by Jensen's inequality for the conditional expectation,

$$\mathrm{E}(g(X(t))|\mathcal{F}_s) \ge g(\mathrm{E}(X(t)|\mathcal{F}_s)).$$

By the submartingale property,

$$\mathrm{E}(X(t)|\mathcal{F}_s) \ge X(s).$$

Since g is a non-decreasing function,

$$g(\mathrm{E}(X(t)|\mathcal{F}_s)) \ge g(X(s)),$$

and combined with the first inequality, the result follows:

$$\mathrm{E}(g(X(t))|\mathcal{F}_s) \ge g(X(s)).$$

Problem 7.7: Show the following:

(a) A non-negative local martingale $M_t, 0 \le t \le T$ is a supermartingale (Theorem 7.23).

(b) If $M_t, t \le T$ is a martingale and $U = \mathcal{E}(M)$ the stochastic exponential of M, then $\mathrm{E}(U_T) \le 1$. You may use that U is a local martingale (cf. Section 8.8).

(c) A supermartingale $M_t, 0 \le t \le T$ is a martingale if and only if $\mathrm{E}M_T = M_0$ (Theorem 7.3).

(d) A non-negative local martingale $M_t, 0 \le t \le T$ is a martingale if and only if $\mathrm{E}M_T = M_0$ (Theorem 7.24).

(e) A stochastic exponential $U_t, t \le T$ of a martingale is a martingale if and only if $\mathrm{E}(U_T) = 1$.

Solution.

(a) Let τ_n be a localising sequence and $s < t$. Then, by the martingale property of $M(t \wedge \tau_n)$, for all n,

$$\mathrm{E}(M(t \wedge \tau_n)|\mathcal{F}_s) = M(s \wedge \tau_n).$$

While we don't know that the limit as $n \to \infty$ of the left-hand side exists, the limit as $n \to \infty$ of the right-hand side is $M(s)$. Hence, taking the lower limit, we obtain

$$\begin{aligned}\liminf_{n\to\infty} \mathrm{E}(M(t \wedge \tau_n)|\mathcal{F}_s) &= \liminf_{n\to\infty} M(s \wedge \tau_n)\\ &= \lim_{n\to\infty} M(s \wedge \tau_n) = M(s).\end{aligned}$$

Now, use Fatou's lemma for conditional expectation:

$$\liminf_{n\to\infty} \mathrm{E}(M(t \wedge \tau_n)|\mathcal{F}_s) \geq \mathrm{E}(\liminf_{n\to\infty} M(t \wedge \tau_n)|\mathcal{F}_s) = \mathrm{E}(M(t)|\mathcal{F}_s).$$

Hence,

$$\mathrm{E}(M(t)|\mathcal{F}_s) \leq \liminf_{n\to\infty} \mathrm{E}(M(t \wedge \tau_n)|\mathcal{F}_s) = M(s),$$

and the supermartingale property is proved.

Taking $s = 0$, we obtain integrability, $\mathrm{E}M(t) \leq M(0)$.

(b) U is a non-negative local martingale, so it is a supermartingale by (a). A supermartingale has a non-increasing mean; therefore, $\mathrm{E}(U_T) \leq U_0 = 1$.

(c) Let M be a supermartingale and $\mathrm{E}M(T) = M_0$. If for some $s < t$ $\mathrm{E}(M(t)|\mathcal{F}_s) < M(s)$ on a set of positive probability, then by taking expectation we have $\mathrm{E}M(t) < \mathrm{E}M(s)$. Then, by the supermartingale property, the supermartingale has a non-increasing mean, $\mathrm{E}M(T) \leq \mathrm{E}M(t) < \mathrm{E}M(s) \leq M_0$. This is a contradiction. Hence, for any $s < t$, $\mathrm{E}(M(t)|\mathcal{F}_s) < M(s)$.

(d) This follows from (a) and (c).

(e) This follows immediately from (c) or (d).

Problem 7.8: Let $M_t, t \leq T$ be a martingale with compensator $\langle M, M \rangle_t = A_t$ and let $Z_t = e^{M_t - \frac{1}{2}A_t}$ for $t \leq T$. If A_t is bounded, then $\mathrm{E}(Z_T) = 1$ and Z_t is a martingale.

Solution. Z_t is a non-negative local martingale. We apply the Vallee–Poussin theorem with $G(x) = x^2$ to show the uniform integrability of Z_τ, for τ in the family of finite stopping times:

$$\mathrm{E}(Z_\tau^2) = \mathrm{E}(e^{2M_\tau - A_\tau}) = \mathrm{E}(e^{2M_\tau - 2A_\tau + A_\tau}) \leq C\mathrm{E}(e^{2M_\tau - 2A_\tau})$$

for some finite constant C since A_t is bounded. Now $e^{2M_t - 2A_t}$ is a non-negative local martingale. Thus, it is a supermartingale and has an expectation bounded above by 1. Therefore, $\mathrm{E}(Z_\tau^2)$ is bounded and Z_τ is uniformly integrable by the Vallee–Poussin theorem. Since the local martingale Z is in Class D, it is indeed a martingale (by Theorem 7.26).

Problem 7.9: Let $M_t^{(n)}, t \leq T$, be a sequence of martingales with bounded compensators $A_t^{(n)}$. Show that if $A_t^{(n)}$ converges in probability to t, then $M_t^{(n)}$ converges weakly (in distribution) to a Brownian motion at each t.

Solution. Let $Z_t^{(n)} = e^{i\lambda M_t^{(n)} + \frac{1}{2}\lambda^2 A_t^{(n)}}$. From the previous result, $Z_t^{(n)}$ is a martingale for each n and $\mathrm{E}(Z_t^{(n)}) = 1$:

$$\begin{aligned}
|\mathrm{E}(e^{i\lambda M_t^{(n)}}) - \mathrm{E}(e^{i\lambda B_t})| &= |\mathrm{E}(e^{i\lambda M_t^{(n)}}) - e^{-\frac{1}{2}\lambda^2 t}| \\
&= |\mathrm{E}(e^{i\lambda M_t^{(n)}}) - e^{-\frac{1}{2}\lambda^2 t}\mathrm{E}(Z_t^{(n)})| \\
&= |\mathrm{E}(e^{i\lambda M_t^{(n)}}) - e^{-\frac{1}{2}\lambda^2 t}\mathrm{E}(e^{i\lambda M_t^{(n)} + \frac{1}{2}\lambda^2 A_t^{(n)}})| \\
&= |\mathrm{E}(e^{i\lambda M_t^{(n)}}(1 - e^{-\frac{1}{2}\lambda^2 t + \frac{1}{2}\lambda^2 A_t^{(n)}}))| \\
&\leq \mathrm{E}(|e^{i\lambda M_t^{(n)}}||1 - e^{\frac{1}{2}\lambda^2 (A_t^{(n)} - t)}|)
\end{aligned}$$

by Jensen's inequality. Now, since $|e^{i\lambda M_t^{(n)}}| = 1$ and $A_t^{(n)}$ converges in probability to t, thus the above expectation converges to 0, which implies the convergence in distribution of $M_t^{(n)}$ to B_t at each fixed t.

Problem 7.10: Let $M(t)$ be an $\mathcal{F}_t$-martingale and pick σ-fields $\mathcal{G}_t \subset \mathcal{F}_t$. Show that if $M(t)$ is $\mathcal{G}_t$-measurable, then it is a $\mathcal{G}_t$-martingale.

Solution. $\mathrm{E}(M(t)|\mathcal{G}_s) = \mathrm{E}(\mathrm{E}(M(t)|\mathcal{F}_s)|\mathcal{G}_s) = \mathrm{E}(M(s)|\mathcal{G}_s) = M(s)$.

Problem 7.11: Let $M(t)$ be an $\mathcal{F}_t$-martingale and pick σ-fields $\mathcal{G}_t \subset \mathcal{F}_t$. Show that $\hat{M}(t) = \mathrm{E}(M(t)|\ \mathcal{G}_t)$ is a $\mathcal{G}_t$-martingale.

Solution. We show the integrability:

$$\begin{aligned}\mathrm{E}(\hat{M}(t)|\mathcal{G}_s) &= \mathrm{E}(\mathrm{E}(M(t)|\ \mathcal{G}_t)|\mathcal{G}_s) = \mathrm{E}(M(t)|\mathcal{G}_s)\\ &= \mathrm{E}(\mathrm{E}(M(t)|\ \mathcal{F}_t)|\mathcal{G}_s) = \mathrm{E}(M(s)|\mathcal{G}_s) = \hat{M}(s).\end{aligned}$$

Problem 7.12: Let $B(t)$ be an $\mathcal{F}_t$-Brownian motion, and $\mathcal{G}_t \subset \mathcal{F}_t$. By the previous problem, $\widehat{B}(t) = \mathrm{E}(B(t)|\ \mathcal{G}_t)$ is a $\mathcal{G}_t$-martingale. Give an example of $\mathcal{G}_t$ such that $\widehat{B}(t)$ is not a Brownian motion.

Solution. Let $\mathcal{G}_t$ be the trivial σ-field, then $\widehat{B}(t) = \mathrm{E}(B(t)|\ \mathcal{G}_t) = 0$ is not a Brownian motion.

For $\mathcal{G}_t = \mathcal{F}^B_{t\wedge 1}$, one gets $\hat{B}(t) = B(t \wedge 1)$, which is not a Brownian motion.

Problem 7.13: Let $B(t)$ be an $\mathcal{F}_t$-Brownian motion, and denote by $X(t) = \int_0^t \mathrm{sign}(B_s)dB_s$. Let $\mathcal{G}_t$ be the natural filtration of the process $X(t)$. Then, it holds that $\mathcal{G}_t \subset \mathcal{F}_t$. Show that $X(t)$ is $\mathcal{G}_t$-Brownian motion.

Solution. $X(t)$ is a $\mathcal{G}_t$-martingale, it is continuous and has quadratic variation t. It is Brownian motion by the Lévy theorem.

Optional Stopping

Problem 7.14 (Expected exit time of Brownian motion from (a, b)): Let $B(t)$ be a Brownian motion started at $x \in (a, b)$, and $\tau = \inf\{t : B(t) = a \text{ or } b\}$. By stopping the martingale $M(t) = B(t)^2 - t$, show that $\mathrm{E}_x(\tau) = (x - a)(b - x)$.

Solution. Let τ be the first time $B(t)$ hits a or b when $B(0) = x$, $a < x < b$. τ is finite by Theorem 3.13.

This fact can also be seen by stopping martingale $M(t) = B^2(t) - t$. $\mathrm{E}M(\tau \wedge t) = M(0) = x^2$, or $\mathrm{E}B^2(\tau \wedge t) - \mathrm{E}(\tau \wedge t) = x^2$. Hence,

$$\mathrm{E}(\tau \wedge t) = \mathrm{E}B^2(\tau \wedge t) - x^2.$$

By definition of τ as the first hitting time of a or b having started at $x \in (a, b)$, we have that $B^2(\tau \wedge t)$ satisfies (as a can be negative with $a^2 > b^2$) $a^2 \wedge b^2 \le B^2(\tau \wedge t) \le a^2 \vee b^2$, $E(\tau \wedge t) \le a^2 \vee b^2$ and $E\tau \le a^2 \vee b^2 < \infty$.

Hence, $P(\tau < \infty) = 1$.

By stopping the martingale $B(t)$ (see Example 7.6), one obtains

$$\mathrm{P}(B(\tau) = b) = (x - a)/(b - a),$$

and $\mathrm{P}(B(\tau) = a) = 1 - \mathrm{P}(B(\tau) = b)$.

Taking limits as $t \to \infty$, we obtain $\tau \wedge t \to \tau$, $B^2(\tau \wedge t) \to B^2(\tau)$, and by dominated convergence, we obtain $\mathrm{E}B^2(\tau) - \mathrm{E}(\tau) = x^2$.

But $\mathrm{E}B^2(\tau) = a^2\mathrm{P}(B(\tau) = a) + b^2\mathrm{P}(B(\tau) = b) = a^2\frac{b-x}{b-a} + b^2\frac{x-a}{b-a} = bx + ax - ab$.

Thus, $\mathrm{E}(\tau) = bx + ax - ab - x^2 = (x-a)(b-x)$.

Problem 7.15: Find the probability of $B(t) - t/2$ reaching a before it reaches b when started at x, $a < x < b$.
Hint: Use the exponential martingale $M(t) = e^{B(t)-t/2}$.

Solution. Let $\tau = \inf\{t : B(t) - t/2 = a \text{ or } b\}$. As in the previous exercise, obtain $\mathrm{E}M(\tau) = M(0) = e^x$, or

$$e^a\mathrm{P}\left(M(\tau) = e^a\right) + e^b\left(1 - \mathrm{P}\left(M(\tau) = e^a\right)\right) = e^x.$$

This gives

$$\mathrm{P}(B(\tau) - \tau/2 = a) = \mathrm{P}\left(M(\tau) = e^a\right) = \frac{e^b - e^x}{e^b - e^a}.$$

Problem 7.16: Let τ be the time of ruin when gambling each time one dollar on the outcome of a fair coin. Let X_t denote the fortune at time t. $X_0 = 1$ and the game stops when ruin occurs or is played three times, whichever comes first. By calculating $\mathrm{E}X_\tau$, verify the optional stopping theorem.

Solution. Denote by Y_i the outcome of toss i, $i = 1, 2, 3$. Then, $X_t = 1 + Y_1 + \cdots + Y_t$, $t \le 3$. Let τ_1 be the time of ruin, i.e. the first time n, when $X_n = 0$. Then, game stops at $\tau = \min(\tau_1, 3) = \tau_1 \wedge 3$:

$$X_\tau = X_1 1_{\{\tau=1\}} + X_2 1_{\{\tau=2\}} + X_3 1_{\{\tau=3\}}.$$

By the definition of τ,

$$X_1 1_{\{\tau=1\}} = X_1 1_{\{\tau_1=1\}} = 0, \quad X_2 1_{\{\tau=2\}} = X_2 1_{\{\tau_1=2\}} = 0,$$
$$X_3 1_{\{\tau=3\}} = X_3 1_{\{\tau_1=3\}} + X_3 1_{\{\tau_1>3\}} = 0 + X_3 1_{\{\tau_1>3\}}.$$

Hence, $\mathrm{E}X_\tau = \mathrm{E}X_3 1_{\{\tau_1>3\}}$.

Clearly, $\{\tau_1 > 3\}$ consists of the outcomes $(Y_1 = 1, Y_2 = 1, Y_3 = -1)$, $(Y_1 = 1, Y_2 = -1, Y_3 = 1)$ and $(Y_1 = 1, Y_2 = 1, Y_3 = 1)$.

The corresponding values of X_3 are $2, 2, 4$. Therefore,

$$\mathrm{E}X_3 1_{\{\tau_1>3\}} = 2(1/8) + 2(1/8) + 4(1/8) = 1.$$

This verifies the optional stopping theorem because $\mathrm{E}X_\tau = 1 = X_0$.

Problem 7.17: A simple proof of the optional stopping theorem in discrete time for bounded stopping times. Let M_n, $n = 0, 1, 2\ldots$, be a martingale and τ be a bounded stopping time. Prove the optional stopping theorem $\mathrm{E}M_\tau = \mathrm{E}M_0$.

Solution. Let K be the bound for τ, $\tau = 0, 1, 2\ldots, K$. We have

$$M_\tau = \sum_{n=0}^{K} M_n 1_{\{\tau=n\}}.$$

Note that in the sum, only one term is non-zero. Here, the trick is in showing that $\mathrm{E}M_\tau = \mathrm{E}M_K$ rather than $\mathrm{E}M_0$. Of course, for a martingale, $\mathrm{E}M_K = \mathrm{E}M_0$.

By the martingale property $\mathrm{E}(M_K|\mathcal{F}_n) = M_n$ and because $1_{\{\tau=n\}}$ is $\mathcal{F}_n$-measurable,

$$\mathrm{E}(M_K 1_{\{\tau=n\}}|\mathcal{F}_n) = 1_{\{\tau=n\}}\mathrm{E}(M_K|\mathcal{F}_n) = 1_{\{\tau=n\}}M_n.$$

Hence, we have

$$\begin{aligned}
\mathrm{E}M_\tau &= \mathrm{E}\left(\sum_{n=0}^{K} M_n 1_{\{\tau=n\}}\right) = \sum_{n=0}^{K} \mathrm{E}(M_n 1_{\{\tau=n\}}) \\
&= \sum_{n=0}^{K} \mathrm{E}\mathrm{E}(M_K 1_{\{\tau=n\}}|\mathcal{F}_n) = \sum_{n=0}^{K} \mathrm{E}(M_K 1_{\{\tau=n\}}) \\
&= \mathrm{E}\sum_{n=0}^{K} M_K 1_{\{\tau=n\}} = \mathrm{E}M_K \sum_{n=0}^{K} 1_{\{\tau=n\}} \\
&= \mathrm{E}M_K = \mathrm{E}M_0,
\end{aligned}$$

where we used that $\sum_{n=1}^{K} 1_{\{\tau=n\}} = 1$.

Problem 7.18: Give the probability of ruin when playing a game of chance against an infinitely rich opponent (with initial capital $b \to \infty$; cf. Problem 2.19).

Solution. When the game is fair, i.e. $p = q = 1/2$, the ruin probability of a player starting with initial capital x against an opponent with initial capital b is given by $u = \dfrac{b}{x+b}$, and when $b \to \infty$, $u \to 1$.

If $p \neq q$, the ruin probability is given by $u = \dfrac{(q/p)^{b+x} - (q/p)^x}{(q/p)^{b+x} - 1}$.

Take now $b \to \infty$. If $p < q$, then $u \to 1$, and if $q < p$, then $u \to \left(\dfrac{q}{p}\right)^x$.

Problem 7.19: Find the mistake in the following argument. Let $X_1, X_2, \ldots$ define a sequence of independent and identically distributed random variables with distribution $\mathrm{P}(X_1 = -1) = \mathrm{P}(X_1 = 1) = \frac{1}{2}$. Define the following process: $M_0 = 0$, and

$$\begin{aligned} M_{n+1} &= M_n + \frac{1}{2} X_{n+1}\big((1 - M_n)1_{M_n \geq 0} + (1 + M_n)1_{M_n < 0}\big) \\ &= M_n + \frac{1}{2} X_{n+1}\big(1 - |M_n|\big). \end{aligned}$$

Then, M is a bounded martingale. Let $T = \min\{n : M_n = \frac{1}{2}\}$. Applying Doob's optional stopping theorem, we get that

$$\begin{aligned} 0 = \mathrm{E}(M_T) &= \mathrm{E}(M_T 1_{T<\infty}) + \mathrm{E}(M_T 1_{T=\infty}) = \frac{1}{2}\mathrm{P}(T < \infty) + \mathrm{E}(M_\infty 1_{T=\infty}) \\ &= \frac{1}{2}\mathrm{P}(T < \infty) - \mathrm{P}(T = \infty) = \frac{1}{2}\mathrm{P}(T < \infty) - 1 + \mathrm{P}(T < \infty); \end{aligned}$$

that is,

$$\mathrm{P}(T < \infty) = \frac{2}{3}.$$

Hint: Show that $|M_n| \leq 1$ by induction and that this ensures martingale convergence (Theorem 7.13), then argue the limit must be in $\{-1, 1\}$.

Solution. The mistake lies in the fact that M_∞ may equal 1 on $\{T = \infty\}$. M_n may jump over $\frac{1}{2}$.

Problem 7.20 (Ruin probability in insurance): A discrete-time risk model for the surplus U_n of an insurance company at the end of year n, $n = 1, 2, \ldots$, is given by $U_n = U_0 + cn - \sum_{k=1}^{n} X_k$, where c is the total annual premium and X_k is the total (aggregate) claim in year k. The time of ruin T is the first time when the surplus becomes negative, $T = \min\{n : U_n < 0\}$, with $T = \infty$ if $U_n \geq 0$ for all n. Assume that $\{X_k,\ k = 1, 2, \ldots\}$ are i.i.d. random variables, and there exists a constant $R > 0$ such that $\mathrm{E}(e^{-R(c - X_1)}) = 1$. Show that for all n, $\mathrm{P}_x(T \leq n) \leq e^{-Rx}$, where $U_0 = x$, the initial funds, and the ruin probability $\mathrm{P}_x(T < \infty) \leq e^{-Rx}$.
Hint: Show that $M_n = e^{-RU_n}$ is a martingale, and use the optional stopping theorem.

Solution. First, note that $U_{n+1} = U_n + c - X_{n+1}$. We have

$$\begin{aligned} \mathrm{E}(M_{n+1}|M_n) &= \mathrm{E}(e^{-RU_{n+1}}|U_n) = \mathrm{E}(e^{-R(U_n+c-X_{n+1})}|U_n) \\ &= e^{-RU_n}\mathrm{E}(e^{-R(c-X_{n+1})}) = e^{-RU_n}\mathrm{E}(e^{-R(c-X_1)}) = e^{-RU_n} \end{aligned}$$

and

$$\begin{aligned} \mathrm{E}(|M_n|) &= \mathrm{E}(e^{-RU_n}) = \mathrm{E}(e^{-R(U_0+cn-\sum_{k=1}^n X_k)}) \\ &= e^{-RU_0}\prod_{k=1}^n \mathrm{E}(e^{-R(c-X_k)}) \\ &= e^{-RU_0}\left(\mathrm{E}(e^{-R(c-X_1)})\right)^n = e^{-RU_0}, \end{aligned}$$

thus $M_n = e^{-RU_n}$ is a martingale. Now, applying optional stopping theorem, we have $\mathrm{E}(e^{-RU_{T\wedge n}}) = \mathrm{E}(e^{-RU_0}) = e^{-Rx}$. Since $T \wedge n = TI(T \le n) + nI(T > n)$, we have

$$\begin{aligned} e^{-Rx} &= \mathrm{E}(e^{-RU_T}I(T \le n)) + \mathrm{E}(e^{-RU_n}I(T > n)) \\ &\ge \mathrm{E}(e^{-RU_T}I(T \le n)) \\ &\ge \mathrm{E}(I(T \le n)) \quad \text{since } U_T < 0 \\ &= \mathrm{P}(T \le n). \end{aligned}$$

Letting $n \to \infty$, we obtain $e^{-Rx} \ge \mathrm{P}(T \le \infty)$.

Problem 7.21 (Ruin probability in insurance continued): Find the bound on the ruin probability when the aggregate claims have $N(\mu, \sigma^2)$ distribution. Give the initial amount x required to keep the ruin probability below level α.

Solution. Continuing from the previous problem, here $X_1 \sim N(\mu, \sigma^2)$. The moment-generating function gives $\mathrm{E}(e^{RX_1}) = e^{R\mu + \frac{1}{2}R^2\sigma^2}$. Thus, the constant $R > 0$ for $\mathrm{E}(e^{-R(c-X_1)}) = 1$ is $R = \frac{2(c-\mu)}{\sigma^2}$, assuming $c \ge \mu$. Therefore, we have the ruin probability $\mathrm{P}(T < \infty) \le e^{-Rx} = e^{-\frac{2(c-\mu)}{\sigma^2}x}$. For this to be below level α, we solve $e^{-\frac{2(c-\mu)}{\sigma^2}x} < \alpha$, which gives $x > -\frac{\sigma^2 \ln \alpha}{2(c-\mu)} = \frac{\sigma^2 \ln \frac{1}{\alpha}}{2(c-\mu)}$.

Problem 7.22 Let M_t, $t \ge 0$ be a continuous martingale. Prove the following inequality:

$$P\left(\sup_{t \le T} M_t > a, [M, M]_T \le L\right) \le e^{-\frac{a^2}{2L}}.$$

Solution. Let $\tau = \inf\{t : M_t > a\}$. Then, $\{\tau \le T\} = \{\sup_{t\le T} M_t > a\}$. Take now the martingale exponential of uM, $Z_t = e^{uM_t - \frac{u^2}{2}[M,M]_t}$. It is always a local martingale, and it is positive, hence it is a super martingale with an expectation less than or equal to 1, see Problem 7.7. By optional stopping, $\mathrm{E}Z_\tau \le 1$. Denote by A the required set

$$A = \{\tau \le T, [M]_T \le L\},$$

and note that on A, $M_\tau = a$, and $[M]_\tau \le [M]_T \le L$.

Hence, we obtain

$$1 \ge \mathrm{E}e^{uM_\tau - \frac{u^2}{2}[M]_\tau} I_A \ge e^{ua - \frac{u^2}{2}L}\mathrm{P}(A).$$

Taking $u = a/L$, we obtain the desired inequality.

Change of Time, DDS

Problem 7.23: Lévy characterization theorem. Let $M(t)$ be a continuous martingale with quadratic variation t. Then, $M(t)$ is Brownian motion.

Solution. In the following proof, we use the fact that a stochastic integral of a bounded function with respect to M is a martingale. To have a bounded function, we use the complex exponential, namely the characteristic function of $M(t)$. Let $\phi(t) = \mathrm{E}e^{iuM(t)}$, $u \in \mathbb{R}$, $i = \sqrt{-1}$.

Letting $Y(t) = e^{iuM(t)}$, we obtain by Itô's formula,

$$Y(t) = 1 + iu\int_0^t Y(s)dM(s) - \frac{u^2}{2}\int_0^t Y(s)ds.$$

Since $|Y(s)| \le 1$, $\int_0^t Y(s)dM(s)$ is a martingale, so that by taking expectations,

$$\phi(t) = \mathrm{E}Y(t) = 1 - \frac{u^2}{2}\int_0^t \phi(s)ds.$$

We have used the Fubini theorem to take the expectation inside the integral. Differentiating in t, we obtain a differential equation for the exponential, so that $\phi(t) = e^{-u^2t/2}$. This shows that the marginal distribution of $M(t)$ is $N(0,t)$. Similarly, we obtain the distribution of increments $M(t_2) - M(t_1)$. Using Itô's formula and integrating between t_1 and t_2, we have for $t > t_1$,

$$Y(t) = Y(t_1) + iu\int_{t_1}^t Y(s)dM(s) - \frac{u^2}{2}\int_{t_1}^t Y(s)ds.$$

Taking the conditional expectation given $\mathcal{F}_{t_1}$ and using that $\mathrm{E}(\int_{t_1}^t Y(s)dM(s)|\mathcal{F}_{t_1}) = 0$ being a martingale difference, we obtain

$$\mathrm{E}(Y(t)|\mathcal{F}_{t_1}) = Y(t_1) - \frac{u^2}{2}\int_{t_1}^t \mathrm{E}(Y(s)|\mathcal{F}_{t_1})ds.$$

Solving this equation, we obtain $\mathrm{E}(Y(t)|\mathcal{F}_{t_1}) = Y(t_1)e^{-u^2(t-t_1)/2}$. Rearranging, we have $\mathrm{E}(e^{iu(M(t_2)-M(t_1))}|\mathcal{F}_{t_1}) = e^{-u^2(t-t_1)/2}$. This demonstrates independence of $M(t_2)-M(t_1)$ of the past, as well as $N(0, t_2-t_1)$ distribution.

Problem 7.24: Let $B(t)$ be a Brownian motion, $X(t) = \int_0^t \mathrm{sign}(B(s))\, dB(s)$. Show that X is also a Brownian motion.

Solution. Recall Lévy's characterisation theorem: if $X(t)$ is a continuous martingale with quadratic variation t, then it is a Brownian motion. $X(t)$ is an Itô integral, and since $\mathrm{E}\int_0^T \mathrm{sign}^2(B(s))ds = \mathrm{E}\int_0^T 1ds = T < \infty$, it is a martingale. Next,

$$\begin{aligned}[X,X](t) &= \left[\int \mathrm{sign}(B(s))dB(s), \int \mathrm{sign}(B(s))dB(s)\right](t)\\ &= \int_0^t \mathrm{sign}^2(B(s))d[B,B](s) = \int_0^t 1ds = t,\end{aligned}$$

hence $X(t)$ is a Brownian motion.

Problem 7.25: Show that $X(t) = \int_0^t \mathrm{sign}(B(s))dB(s)$ is uncorrelated with $B(t)$, but they are not independent.

Solution.

$$\begin{aligned}\mathrm{E}(X(t)B(t)) &= \mathrm{E}\left(\int_0^t \mathrm{sign}(B(s))dB(s)\int_0^t 1dB(s)\right)\\ &= \mathrm{E}\left(\int_0^t \mathrm{sign}(B(s))ds\right) = 0,\end{aligned}$$

since $\mathrm{E}\mathrm{sign}(B(s)) = (1)1/2 + (-1)1/2 = 0$, and $X(t)$, $B(t)$ are uncorrelated. Now, consider $\mathrm{E}(X(t)B(t)^2)$. Using $dB^2(t) = 2B(t)dB(t) + dt$, we have

$$\begin{aligned}\mathrm{E}(X(t)B(t)^2) &= \mathrm{E}\left(\int_0^t \mathrm{sign}(B(s))dB(s)\left(2\int_0^t B(s)dB(s) + t\right)\right)\\ &= 2\mathrm{E}\left(\int_0^t |B(s)|ds\right) + tE\left(\int_0^t \mathrm{sign}(B(s))dB(s)\right) > 0.\end{aligned}$$

If X and B were independent, we would have

$$\mathrm{E}(X(t)B^2(t)) = \mathrm{E}X(t)\mathrm{E}B^2(t) = 0.$$

Problem 7.26: Let $M(t) = \int_0^t e^s dB(s)$. Find $g(t)$ such that $M(g(t))$ is a Brownian motion.

Solution. Since $\int_0^T e^{2s} ds < \infty$ for any T, $M(t)$ is a martingale on any finite time interval. Its quadratic variation is

$$[M, M](t) = \int_0^t e^{2s} ds = \frac{1}{2}(e^{2t} - 1).$$

The inverse function is

$$g(t) = \frac{1}{2}\ln(2t + 1).$$

Hence, $M(g(t))$ is a continuous martingale with quadratic variation t, $[M, M](g(t)) = t$. Hence, by Lévy's theorem, $M(g(t))$ is a Brownian motion.

(Note that this example is a simple case of the DDS Theorem 7.37, which also allows a random time change and, in essence, is as follows. If $M(t)$ is a continuous martingale, $M(0) = 0$, and τ_t are increasing stopping times, then $N(t) = M(\tau_t)$ is also a martingale (with respect to the filtration $\mathcal{F}_{\tau_t}$). Its quadratic variation is $[N, N](t) = [M, M](\tau_t)$. Hence, by taking τ_t as the inverse of $[M, M](t)$, $\tau_t = [M, M]^{-1}(t)$, we have that $[N, N](t) = t$. Since $N(t)$ is a continuous martingale, by Lévy's theorem, it is a Brownian motion. $[M, M](t)$ is non-decreasing, but inverse exists if it is strictly increasing. In general, the generalised inverse is used as defined by $\tau_t = \inf\{s : [M, M](s) > t\}$. It is also assumed that $[M, M](t)$ is non-decreasing to ∞.

Moreover, $[M, M](t)$ is a stopping time with respect to the filtration $\mathcal{F}_{\tau_t}$, and the martingale M can be obtained from the Brownian motion N by the change of time $M(t) = N([M, M](t)))$.

Problem 7.27: Let $X(t) = tB_t - \int_0^t B_s ds$, $t \geq 0$. Find an increasing function $g(t)$ such that $X(g(t))$ is a Brownian motion $W(t)$. Find an increasing function $h(t)$ such that $X(t) = W(h(t))$, $t \geq 0$.

Solution. $dX_t = d(tB_t) - B_t dt = t dB_t$. Hence, $X_t = \int_0^t s dB_s$. $[X, X]_t = \int_0^t s^2 ds = t^3/3$. Its inverse function is $g(t) = (3t)^{1/3}$. By DDS, $X(g(t)) = \int_0^{g(t)} s dB_s = W(t)$ is a Brownian motion. $h(t) = [X, X]_t = t^3/3$.

Problem 7.28: Let $B(t)$ be a Brownian motion. Give an SDE for $e^{-\alpha t}B(e^{2\alpha t})$.

Solution. This exercise is based on the following idea. For an increasing function $f(t)$, the time-changed Brownian motion $B(f(t))$ and the Itô integral $X(t) = \int_0^t \sqrt{f'(s)}d\hat{B}(s)$ have the same quadratic variation $f(t)$. This is the essence theorem 7.38, where $f(t)$ can be also random. Therefore, writing this fact in differential notations, we have that

$$dB(f(t)) = \sqrt{f'(t)}d\hat{B}(t).$$

Understanding these equalities in distribution, we drop the hat from Brownian motion on the right-hand side. Hence, we can write

$$dB(e^{2\alpha t}) = \sqrt{2\alpha e^{2\alpha t}}dB(t) = \sqrt{2\alpha}e^{\alpha t}dB(t).$$

Using integration by parts and the above rule for $dB(e^{2\alpha t})$, we obtain

$$dX(t) = d(e^{-\alpha t}B(e^{2\alpha t})) = -\alpha e^{-\alpha t}B(e^{2\alpha t})dt + e^{-\alpha t}dB(e^{2\alpha t})$$

$$dX(t) = -\alpha X(t)dt + \sqrt{2\alpha}dB(t).$$

Problem 7.29: A process Z_t is called a square-root process if it satisfies the SDE

$$dZ_t = \mu(Z_t)dt + \sigma\sqrt{Z_t}dW_t, \quad Z_0 = x > 0,$$

for some function $\mu(x)$, constant $\sigma > 0$, and a Brownian motion W_t.

(a) Let X_t be the Ornstein–Uhlenbeck process solving $dX_t = -aX_t dt + v dB_t$. Show that $Z_t = X_t^2$ is a square-root process. Specify μ, σ, W.

(b) Let Y_t be a Bessel process solving SDE $dY_t = \frac{b}{Y_t}dt + v dB_t$. Show that $Z_t = Y_t^2$ is a square-root process. Specify μ, σ, W.

Solution.

(a)
$$\begin{aligned} dZ_t &= 2X_t dX_t + d[X,X]_t = (-2aX_t^2 + v^2)dt + 2vX_t dB_t \\ &= (v^2 - 2aZ_t)dt + 2v\,\text{sign}(X_t)\sqrt{Z_t}dB_t \\ &= (v^2 - 2aZ_t)dt + 2v\sqrt{Z_t}dW_t, \end{aligned}$$

where $dW_t = \text{sign}(X_t)dB_t$. $W_t = \int_0^t \text{sign}(X_s)dB_s$ is a continuous martingale with quadratic variation t, so it is a Brownian motion. $\mu(x) = (v^2 - 2ax)$, $\sigma = 2v$.

(b) $dZ_t = 2Y_t dY_t + d[Y,Y]_t = (2b+v^2)dt + 2vY_t dB_t$
$$= (2b+v^2)dt + 2v\text{sign}(Y_t)\sqrt{Z_t}dB_t = (2b+v^2)dt + 2v\sqrt{Z_t}dW_t,$$

where $dW_t = \text{sign}(Y_t)dB_t$ s a Brownian motion, $\mu(x) = (2b+v^2)$, $\sigma = 2v$.

Problem 7.30: Let $X(t)$ solve SDE $dX(t) = \mu X(t)dt + \sigma\sqrt{X(t)}dB(t)$, $X(0) = x > 0$, Feller diffusion. Let $G(t) = \int_0^t X(s)ds$, and $\tau_t = G^{-1}(t)$. Derive the SDE for $Y(t) = X(\tau_t)$.

Solution. Write the equation in the integral form and then change the variables:

$$X(t) = x + \mu\int_0^t X(s)ds + \sigma\int_0^t \sqrt{X(s)}dB(s), \quad \text{so that}$$
$$X(\tau_t) = x + \mu\int_0^{\tau_t} X(s)ds + \sigma\int_0^{\tau_t}\sqrt{X(s)}dB(s).$$

Now, perform the change of variables $s = \tau_u$ in the integrals, and note that with $u = G(s)$, we have $du = X(s)ds$, so $ds = du/X(\tau_u)$:

$$\int_0^{\tau_t} X(s)ds = \int_0^t X(\tau_u)/X(\tau_u)du = t.$$

The Itô integral has a quadratic variation:

$$\int_0^{\tau_t} X(s)ds = \int_0^t X(\tau_u)/X(\tau_u)du = t.$$

Therefore, the Itô integral is a Brownian motion $\hat{B}(t)$. Hence, it follows that $Y(t) = X(\tau_t)$ solves the SDE

$$Y(t) = X(\tau_t) = x + \mu t + \sigma\hat{B}(t).$$

Remark that the SDE holds until time T when $X(t)$ becomes zero. The next question is the opposite direction: How to obtain the SDE for $X(t)$ from that of $Y(t)$.

Problem 7.31: Let $Y(t) = x + \mu t + \sigma B(t)$, $x > 0$, and let $T = \inf\{t : Y(t) = 0\}$. Let $G(t) = \int_0^{t\wedge T} ds/Y(s)$, and $\tau_t = G^{-1}(t)$. Show that $X(t) = Y(\tau_t)$ solves the SDE $dX(t) = \mu X(t)dt + \sigma\sqrt{X}(t)dB(t)$, $X(0) = x > 0$.

Solution. Write the equation in the integral form and then change the variables:

$$X(t) = Y(\tau_t) = x + \mu\tau_t + \sigma B(\tau_t) = x + \mu \int_0^{\tau_t} ds + \sigma \int_0^{\tau_t} dB(s).$$

Now, perform the change of variables $s = \tau_u$ in the integrals, and note that with $u = G(s)$, we have $du = ds/Y(s)$, so $ds = Y(s)du = Y(\tau_u)du$. $\int_0^{\tau_t} ds = \int_0^t Y(\tau_u)du$. The Itô integral has quadratic variation $\int_0^{\tau_t} ds = \int_0^t Y(\tau_u)du$. Hence, there is a Brownian motion $\hat{B}$ so that the Itô integral is given by $\int_0^t \sqrt{Y}(\tau_u)d\hat{B}(u)$. Hence, we obtain

$$X(t) = x + \mu \int_0^t X(s)ds + \sigma \int_0^t \sqrt{X}(s)dB(s).$$

The following question uses similar ideas but is presented more formally.

Problem 7.32: Prove the change of time result in SDEs, Theorem 7.41: let $X(t)$ be a solution to the SDE

$$dX(t) = \mu(X(t))dt + \sigma(X(t))dB(t).$$

Let $g(x)$ be a positive function for which $G(t) = \int_0^t g(X(s))ds$ is finite for finite t and increases to infinity almost surely. Define $\tau_t = G^{(-1)}(t)$ and $Y(t) = X(\tau_t)$. Then, $Y(t)$ is a weak solution to the SDE

$$dY(t) = \frac{\mu(Y(t))}{g(Y(t))}dt + \frac{\sigma(Y(t))}{\sqrt{g(Y(t))}}dB(t), \quad \text{with} \quad Y(0) = X(0).$$

Solution. Under the appropriate conditions on the coefficients, for any twice continuously differentiable function f vanishing outside a finite interval,

$$\begin{aligned} M(t) := f(X(t)) &- \int_0^t \mu(X(s))f'(X(s))ds \\ &- \frac{1}{2}\int_0^t \sigma^2(X(s))f''(X(s))ds \end{aligned}$$

is a martingale. Since τ_t is increasing and is a stopping time for each t, by using the optional stopping theorem, one can show that the process $M(\tau_t)$ is also a martingale, that is,

$$f(X(\tau_t)) - \int_0^{\tau_t} \mu(X(s))f'(X(s))ds - \frac{1}{2}\int_0^{\tau_t} \sigma^2(X(s))f''(X(s))ds$$

is a martingale. Now, perform the change of variables of $s = \tau_u$ in the integrals, and note that with $u = G(s)$, we have $du = g(X(s))ds$. The martingale $M(\tau_t)$ becomes

$$f(X(\tau_t)) - \int_0^t \frac{\mu(X(\tau_u))f'(X(\tau_u))}{g(X(\tau_u))}du - \frac{1}{2}\int_0^t \frac{\sigma^2(X(\tau_u))f''(X(\tau_u))}{g(X(\tau_u))}du.$$

That is, for any twice continuously differentiable function f vanishing outside a finite interval,

$$f(Y(t)) - \int_0^t \frac{\mu(Y(u))}{g(Y(u))}f'(Y(u))du - \frac{1}{2}\int_0^t \frac{\sigma^2(Y(u))}{g(Y(u))}f''(Y(u))du$$

is a martingale. Thus, Y is a solution of the martingale problem for $L = \frac{\mu}{g}\frac{d}{dx} + \frac{1}{2}\frac{\sigma^2}{g}\frac{d^2}{dx^2}$.

Problem 7.33: Let $X(t) = e^{t/2}\cos(B_t)$, $0 \le t \le T$.

(a) Show that X_t is a martingale.
(b) Find an increasing process $g(t)$ such that $X(g(t))$ is a Brownian motion. (g can be given implicitly).
(c) Show directly by using stochastic calculus that $dX(t) = d\hat{B}(t)$.

Solution.

(a) $dX_t = \frac{1}{2}e^{t/2}\cos(B_t)dt + e^{t/2}(-\sin(B_t))dB_t - 1/2\cos(B_t)]dt = e^{t/2}(-\sin(B_t))dB_t$. Since $\int_0^t e^t \sin^2(B_t)dt < \infty$, X_t being an Itô integral is a martingale.
(b) By Itô's formula, $[X, X](t) = \int_0^t e^s \sin^2(B_s)ds = h(t)$. h is increasing, hence it has an inverse $g(t)$, defined by $\int_0^{g(t)} e^s \sin^2(B_s)ds = t$. By DDS, $X(g(t)$ is a Brownian motion.
(c) $X(g(t)) = e^{g(t)/2}\cos(B(g(t))$. Using integration by parts and the time change rule, we obtain with $Y_t = B(g(t))$, $dY_t = \sqrt{g'(t)}d\hat{B}(t)$ and $d[Y, Y]_t = g'(t)dt$. Hence,

$$\begin{aligned} dX(g(t)) &= \frac{1}{2}e^{g(t)/2}g'(t)\cos(Y_t) + e^{g(t)/2} \\ &\quad \times \left(-\sin(Y_t)dY_t - \frac{1}{2}\cos(Y_t)g'(t)dt\right) \\ &= -e^{g(t)/2}\sin(Y_t)\sqrt{g'(t)}d\hat{B}(t). \end{aligned}$$

Since $g'(t) = 1/h'(g(t) = (e^{g(t)}\sin^2(Y_t))^{-1}$, $dX(g(t)) = -\text{sign}(\sin(Y_t))d\hat{B}(t)$, which is also a Brownian motion.

Local Martingales, Weak Solutions

Problem 7.34: Let $\mathbf{B}_t = (B_t^1, B_t^2, B_t^3)$ be a three-dimensional Brownian motion started at $\mathbf{1}$, and L be its generator:

$$L = \frac{1}{2}\frac{\partial^2}{\partial x_1^2} + \frac{1}{2}\frac{\partial^2}{\partial x_2^2} + \frac{1}{2}\frac{\partial^2}{\partial x_3^2}.$$

Let $f(\mathbf{x}) = \frac{1}{||\mathbf{x}||}$ for $\mathbf{x} = (x_1, x_2, x_3) \neq \mathbf{0}$, where $||\mathbf{x}|| = \sqrt{x_1^2 + x_2^2 + x_3^2}$. Show that $M_t = f(\mathbf{B}_t)$ is a local martingale.
Hint: Use Itô's formula and localising sequence $\tau_n = \inf\{t : ||\mathbf{B}_t|| = 1/n\}$. You can use (verifying it for yourself) that $Lf = 0$.

Solution. By Itô's formula,

$$df(B_t^1, B_t^2, B_t^3) = \frac{\partial f}{\partial x_1}(B_t^1, B_t^2, B_t^3)dB_t^1 + \frac{\partial f}{\partial x_2}(B_t^1, B_t^2, B_t^3)dB_t^2 + \frac{\partial f}{\partial x_3}(B_t^1, B_t^2, B_t^3)dB_t^3 + Lf(B_t^1, B_t^2, B_t^3)dt.$$

$$\frac{\partial f}{\partial x_i} = -\frac{x_i}{(x_1^2 + x_2^2 + x_3^2)^{3/2}}, \quad i = 1, 2, 3, \; Lf(x_1, x_2, x_3) = 0.$$

$$dM_t = df(B_t^1, B_t^2, B_t^3) = -\sum_{i=1}^{3} \frac{B_t^i}{((B_t^1)^2 + (B_t^2)^2 + (B_t^3)^2)^{3/2}} dB_t^i = -\sum_{i=1}^{3} \frac{B_t^i}{||\mathbf{B}_t||^3} dB_t^i.$$

For $t \leq \tau_n$,

$$\left(\frac{B_t^i}{||\mathbf{B}_t||^3}\right)^2 \leq \frac{||\mathbf{B}_t||^2}{||\mathbf{B}_t||^6} = \frac{1}{||\mathbf{B}_t||^4} \leq n^4 < \infty.$$

Therefore,

$$M_{t\wedge\tau_n} = -\sum_{i=1}^{3} \int_0^{t\wedge\tau_n} \frac{B_t^i}{||\mathbf{B}_t||^3} dB_t^i$$

is a martingale.

Problem 7.35: Let $X(t)$ satisfy the SDE $dX(t) = \mu(t)dt + \sigma(t)dB(t)$ on $[0, T]$. Show that $X(t)$ is a local martingale if and only if $\mu(t) = 0$ a.e.

Solution.

$$X(t) = X(0) + A(t) + M(t) = X(0) + \int_0^t \mu(s)ds + \int_0^t \sigma(s)dB(s).$$

If $\mu(t) = 0$, then $X(t) = X(0) + M(t)$ with $M(t) = \int_0^t \sigma(s)dB(s)$. It is a local martingale by the martingale property of the Itô integral.

Let now $X(t)$ be a local martingale. Then,

$$A(t) = \int_0^t \mu(s)ds = X(t) - M(t)$$

is a local martingale as a difference of two local martingales. Since $A(t)$ is continuous, it is a continuous local martingale. It also has finite variation, being an integral. But if a continuous local martingale has finite variation over an interval, then it must be a constant over that interval (Corollary 7.30). Hence, $A(t)$ is a constant, and $\mu(t) = A'(t) = 0$.

Problem 7.36: $f(x, t)$ is differentiable in t and twice in x. It is known that $X(t) = f(B(t), t)$ is of finite variation. Show that f is a function of t alone.

Solution. By Itô's formula,

$$f(B(t), t) - \int_0^t \left(\frac{\partial f}{\partial t}(B(s), s) + \frac{1}{2}\frac{\partial^2 f}{\partial x^2}(B(s), s) \right) ds$$
$$= \int_0^t \frac{\partial f}{\partial x}(B(s), s)dB(s).$$

The right-hand side is a continuous local martingale, and the left-hand side is of finite variation. This can only happen when the local martingale is a constant.

Thus, $\frac{\partial f}{\partial x} = 0$, and f is a constant in x and hence a function of t alone.

Problem 7.37: Let $Y(t) = \int_0^t B(s)dB(s)$ and $W(t) = \int_0^t \text{sign}(B(s))dB(s)$. Show that $dY(t) = \sqrt{t + 2Y(t)}dW(t)$. Show the uniqueness of the weak solution of the above SDE.

Solution. By solving the Itô integral for $Y(t)$, $Y(t) = \frac{1}{2}B^2(t) - \frac{1}{2}t$. Therefore, $B^2(t) = 2Y(t) + t$, and $B(t) = \text{sign}(B(t))\sqrt{2Y(t)+t}$. Hence,

$$\begin{aligned} dY(t) &= B(s)dB(s) \\ &= \text{sign}(B(t))\sqrt{t+2Y(t)}dB(t) \\ &= \sqrt{t+2Y(t)}dW(t), \end{aligned}$$

where $dW(t) = \text{sign}(B(t))dB(t)$ and W is a Brownian motion.

The uniqueness of the weak solution can be shown following Theorem 5.11, that is, to show that there is K_T such that for all $x \in \mathbb{R}$, $|\mu(x,t)| + |\sigma(x,t)| \leq K_T(1+|x|)$. Here, we have $\mu = 0$ and $\sigma(y,t) = \sqrt{t+2y}$. Note that $t + 2Y(t)$ is always positive by construction. Since for $x \leq 1$, $\sqrt{x} \leq 1$ and for $x \geq 1$, $\sqrt{x} \leq x$, we have $|\mu(x,t)| + |\sigma(x,t)| = |\sqrt{t+2x}| \leq c(1+|t+2x|) \leq K_T(1+|x|)$ for some constants c and K_T.

Alternatively, we show that any solution of the SDE can be written as $(B^2(t)-t)/2$, implying weak uniqueness. Let $Y(t)$ be a solution and denote $X(t) = 2Y(t) + t$. Then, our SDE is $dY(t) = \sqrt{X(t)}dW(t)$. Hence,

$$dX(t) = 2dY(t) + dt = 2\sqrt{X(t)}dW(t) + dt.$$

Consider now $\sqrt{X(t)}$. By Itô's formula,

$$d\sqrt{X(t)} = \frac{1}{2\sqrt{X(t)}}dX(t) - \frac{1}{8(X(t))^{3/2}}d[X,X](t) = dW(t).$$

Hence, $\sqrt{X(t)} = W(t)$, and $X(t) = W^2(t)$, thus $Y(t) = (W^2(t) - t)/2$.

Problem 7.38: Solve the SDE $dX_t = \sqrt{(X_t^2+1)}dB_t$ by using change of time.

Solution. Let $G(t) = \int_0^t \frac{ds}{B_s^2+1}$. τ_t is its inverse, $\int_0^{\tau_t} \frac{ds}{B_s^2+1} = t$.

Then, $X_t = B(\tau_t)$ solves the SDE weakly, i.e. there is a Brownian motion $\hat{B}_t$ such that $dB(\tau_t) = \sqrt{(B(\tau_t)^2+1)}d\hat{B}_t$.

This is because by the DDS theorem, $dB(\tau_t) = \sqrt{\tau_t'}d\hat{B}_t$. SDE follows by the rule of differentiation of the inverse function τ_t:

$$\tau_t' = \frac{1}{G'(\tau_t)} = \frac{1}{1/(B^2(\tau_t)+1)} = B^2(\tau_t) + 1.$$

Martingale Convergence

Problem 7.39: Let $X_n \geq 1$ be a sequence of integrable random variables. Show that $M_n = \sum_{i=1}^n X_i$ converges almost surely to a limit, which might be infinite.

M_n converges in L^1 if and only if $\sum_{i=1}^\infty \mathrm{E}X_i < \infty$.

Solution. M_n is non-decreasing since $X_i \geq 0$. Hence, $W = \lim_{n\to\infty} M_n = \sum_{i=1}^\infty X_i$ exists but might be infinite.

Using the Fubini theorem to interchange expectation and the infinite series, due to the positivity of terms, $\mathrm{E}W = \sum_{i=1}^\infty \mathrm{E}X_i$. $\mathrm{E}|W - M_n| = \mathrm{E}(W - M_n) = \sum_{i=n}^\infty \mathrm{E}X_i$. Hence, the convergence of the series $\sum_{i=0}^\infty \mathrm{E}X_i$ is equivalent to convergence in L^1.

Problem 7.40: (Martingale differences.) Let X_n be a sequence of integrable random variables adapted to $\mathcal{F}_n$. Show that $M_n = \sum_{i=1}^n (X_i - \mathrm{E}(X_i|\mathcal{F}_{i-1}))$ is a martingale. ($(X_i - \mathrm{E}(X_i|\mathcal{F}_{i-1}))$ is called a martingale difference.) Further, if X_n is in L^2 and $V = \sum_{i=0}^\infty \mathrm{E}(X_i - \mathrm{E}(X_i|\mathcal{F}_{i-1}))^2 < \infty$, then M_n converges in L^2 and almost surely to a limit with zero mean and a variance V.

Solution. Since

$$\mathrm{E}(X_n - \mathrm{E}(X_n|\mathcal{F}_{n-1})|\mathcal{F}_{n-1}) = \mathrm{E}(X_n|\mathcal{F}_{n-1}) - \mathrm{E}(X_n|\mathcal{F}_{n-1}) = 0,$$
$$\mathrm{E}(M_n|\mathcal{F}_{n-1}) = M_{n-1} + \mathrm{E}(X_n - \mathrm{E}(X_n|\mathcal{F}_{n-1})|\mathcal{F}_{n-1}) = M_{n-1},$$

which is the martingale property of (M_n).

$\mathrm{E}M_n^2 = \sum_{i=1}^n \mathrm{E}(X_i - \mathrm{E}(X_i|\mathcal{F}_{i-1}))^2$, as $\mathrm{E}(X_i - \mathrm{E}(X_i|\mathcal{F}_{i-1}))(X_j - \mathrm{E}(X_j|\mathcal{F}_{j-1})) = 0$ for $i \neq j$ by the smoothing property of conditional expectation with $\mathcal{F}_{i\wedge j}$, so that

$$\sup_n \mathrm{E}M_n^2 = \sum_{i=0}^\infty \mathrm{E}(X_i - \mathrm{E}(X_i|\mathcal{F}_{i-1}))^2.$$

If $V < \infty$, then $\sup_n \mathrm{E}M_n^2 = V < \infty$, and M_n is a square-integrable martingale. Thus it converges almost surely by the Martingale convergence theorem.

Since $\mathrm{E}M_n^2$ converges, and $\mathrm{E}(M_{n+k}-M_n)^2 = \mathrm{E}M_{n+k}^2 - \mathrm{E}M_n^2$, M_n is a Cauchy sequence in L^2. Hence, by the completeness of L^2, it converges to an element of L^2.

$W = \lim_{n\to\infty} M_n, \mathrm{E}W^2 = \lim_{n\to\infty} \mathrm{E}M_n^2 = V, \mathrm{E}W = \lim_{n\to\infty} \mathrm{E}M_n = 0.$

Problem 7.41: The following few questions show techniques of using martingale convergence and convergence in spaces L^p, $p = 1, 2$, by using classical results of Harris (for L^2) and Kesten–Stigum (for L^1) in the Galton–Watson branching process. Z_n, $n = 0, 1, 2, \ldots$ is defined inductively by $Z_0 = 1$ and for $n > 0$,

$$Z_{n+1} = \sum_{i=1}^{Z_n} \xi_{i,n},$$

where $\xi_{i,n}$ are i.i.d. integer-valued, non-negative random variables distributed as ξ. Let $\mathrm{E}\xi = m$. Assume $m > 1$. Show that $W_n = Z_n/m^n$ is a martingale and that it converges to a r.v. W, with $\mathrm{E}W \le 1$.

Solution.

$$\mathrm{E}(Z_{n+1}|Z_n) = \mathrm{E}\left(\sum_{i=1}^{Z_n} \xi_{i,n}|Z_n\right) = mZ_n.$$

Hence, W_n is a martingale. As it is non-negative, it converges almost surely by Doob's martingale convergence theorem to a limit W, $W \ge 0$. By Fatou's lemma, $\mathrm{E}W \le \liminf \mathrm{E}W_n = 1$.

Problem 7.42: Show that if $\mathrm{E}\xi^2 < \infty$, then W_n converges to a r.v. W almost surely and in L^2. In fact, $\mathrm{E}\xi^2 < \infty$ is also necessary for convergence in L^2

Solution. We need to show that $\mathrm{E}(W - W_n)^2 \to 0$, as $n \to \infty$.

Since $W = \lim_{n\to\infty} W_n$ finite exists, we can write W as a telescoping series $W = 1 + \sum_{k=0}^{\infty}(W_{k+1} - W_k)$ and also write $W_n = 1 + \sum_{k=0}^{n-1}(W_{k+1} - W_k)$. Hence, $W - W_n = \sum_{k=n}^{\infty}(W_{k+1} - W_k)$. Thus,

$$\mathrm{E}(W - W_n)^2 = \mathrm{E}\left(\sum_{k=n}^{\infty}(W_{k+1} - W_k)\right)^2.$$

Therefore, L^2 convergence is equivalent to convergence of the series $\mathrm{E}(\sum_{n=0}^{\infty}(W_{n+1} - W_n))^2 < \infty$. Since the terms are uncorrelated, being martingale differences, $E(W_{n+1} - W_n))(W_{j+1} - W_j) = 0$ for $n \neq j$, we have

$$E\left(\sum_{n=0}^{\infty}(W_{n+1} - W_n)\right)^2 = E\sum_{n=0}^{\infty}(W_{n+1} - W_n)^2 = \sum_{n=0}^{\infty} E(W_{n+1} - W_n)^2$$

by interchanging expectation and summation (using the Fubini theorem). Since W_n is a martingale, $EW_{n+1}W_n = EE(W_{n+1}W_n)|Z_n) = EW_n^2$, and $E(W_{n+1} - W_n)^2 = EW_{n+1}^2 - 2EW_{n+1}W_n + EW_n^2 = EW_{n+1}^2 - EW_n^2$. Hence,

$$E(W - W_n)^2 = \sum_{k=n}^{\infty} EW_{k+1}^2 - EW_k^2,$$

and convergence in L^2 is equivalent to convergence of the series. If $\lim_{n\to\infty} EW_n^2$ exists, then $\sum_{n=0}^{\infty} EW_{n+1}^2 - EW_n^2 = \lim_{n\to\infty} EW_n^2 - 1 < \infty$. The limit $\lim_{n\to\infty} EW_n^2$ is obtained by direct calculations:

$$\begin{aligned} E(Z_{n+1}^2|Z_n) &= Var\left(\sum_{i=1}^{Z_n} \xi_{i,n}|Z_n\right) + \left(E\left(\sum_{i=1}^{Z_n} \xi_{i,n}|Z_n\right)\right)^2 \\ &= Z_n E\xi^2 + m^2 Z_n^2. \end{aligned}$$

Hence, $EW_{n+1}^2 = EW_n^2 + \frac{E\xi^2}{m^2} m^{-n}$. Iterations show that the limit exists.

Notes. L^p convergence is equivalent to convergence in probability plus convergence of moments. Connection to analysis: L^p, the collection of all random variables with finite p-th moment, is a complete normed space with the norm $||X||_p = (EX^p)^{\frac{1}{p}}$. Hence, Cauchy sequences converge to a limit in L^p. The above calculations show that W_n is a Cauchy sequence in L^2 (i.e. $||W_{n+k} - W_n||$ is arbitrarily small whenever n is large) if and only if EW_n^2 is a Cauchy sequence in $\mathbb{R}$, that is, if and only if EW_n^2 converges to a finite limit.

Problem 7.43: Show that if $E\xi \log \xi < \infty$ (known as $X \log X$ condition), then Z_n/m^n converges to a r.v. W almost surely and in L^1. ($0 \log 0$ is defined as 0.) In fact, $E\xi \log \xi < \infty$ is also necessary for convergence in L^1.

Solution. The proof uses the truncation technique, useful when moments don't exist, and the Borel–Cantelli lemma. The proof is from the book, *Branching Processes*, by Asmussen and Herring. Consider the truncated variable $\xi I(\xi \leq m^n)$. It is bounded (by m^n) and has all the moments. Let $p_j = \mathrm{P}(\xi = j)$, $j = 0, 1, 2, \ldots$. Introduce

$$V_{n+1} = \frac{1}{m^{n+1}} \sum_{i=1}^{Z_n} \xi_{i,n} I(\xi_{i,n} \leq m^n).$$

We have

$$\begin{aligned}
\sum_{n=1}^{\infty} \mathrm{P}(V_{n+1} \neq W_{n+1}) &= \sum_{n=1}^{\infty} \mathrm{EP}(V_{n+1} \neq W_{n+1}|Z_n) \\
&= \sum_{n=1}^{\infty} \mathrm{EP}(\cup_{i=1}^{Z_n}\{\xi_{i,n} > m^n)\}|Z_n) \\
&\leq \sum_{n=1}^{\infty} \mathrm{E}\left(\sum_{i=1}^{Z_n} \mathrm{P}(\{\xi_{i,n} > m^n\})|Z_n\right),
\end{aligned}$$

where the inequality is by the union events bound. Proceeding,

$$\begin{aligned}
&= \sum_{n=1}^{\infty} m^n \mathrm{P}(\xi > m^n) = \sum_{n=1}^{\infty} m^n \sum_{j>m^n} p_j \\
&= \sum_{j=1}^{\infty} p_j \sum_{n<\log j/\log m} m^n = C \sum_{j=1}^{\infty} j p_j.
\end{aligned}$$

The last sum is finite because $m < \infty$. Hence, by the Borel–Cantelli lemma, eventually $V_n = W_n$, $\mathrm{P}(V_n = W_n \textit{ eventually}) = 1$.

Recall that $W = 1 + \sum_{n=0}^{\infty}(W_{n+1} - W_n)$, and it is enough to establish L^1 convergence of the series. It converges if $\sum_{n=0}^{\infty}(V_{n+1} - W_n)$ converges in L^1. This in turn is shown to converge by considering

$$\sum_{n=0}^{\infty}(V_{n+1} - \mathrm{E}(V_{n+1}|Z_n)) + (\mathrm{E}(V_{n+1}|Z_n) - W_n).$$

We show that $\sum_{n=0}^{\infty}(V_{n+1} - \mathrm{E}(V_{n+1}|Z_n))$ converges in L^2 and almost surely. $\sum_{n=0}^{\infty}(\mathrm{E}(V_{n+1}|Z_n) - W_n)$ converges in L^1 and almost surely iff the $X \log X$ condition holds. For the first series, we use criteria for martingale differences, Problem 7.40. For the second, we use the criteria for L^1 convergence of a series of non-negative random variables, Problem 7.39.

Note that $\mathrm{E}(V_{n+1}|Z_n) = \frac{1}{m}W_n \mathrm{E}\xi I(\xi \le m^n)$. Consider

$$\mathrm{E}((V_{n+1} - \mathrm{E}(V_{n+1}|Z_n))^2|Z_n)$$
$$= \mathrm{E}\left(\frac{1}{m^{2(n+1)}}\sum_{i=1}^{Z_n}(\xi_{i,n}I(\xi_{i,n} \le m^n)\right) - \mathrm{E}\xi I(\xi \le m^n))^2,$$

by using the independence of $\xi_{i,n}$, $i = 1, 2, \ldots, Z_n$. Hence, using $\mathrm{E}Z_n = m^n$,

$$\mathrm{E}(V_{n+1} - \mathrm{E}(V_{n+1}|Z_n))^2 = \frac{1}{m^{n+2}}Var(\xi I(\xi \le m^n)) \le \frac{\sigma^2}{m^{n+2}}.$$

This shows that $\sum_{n=0}^{\infty}\mathrm{E}(V_{n+1} - \mathrm{E}(V_{n+1}|Z_n))^2 < \infty$.

Next, consider $W_n - \mathrm{E}(V_{n+1}|Z_n) = \frac{1}{m}W_n\mathrm{E}\xi I(\xi > m^n) \ge 0$, where we have used that $\mathrm{E}\xi = m$. Hence,

$$\sum_{n=1}^{\infty}\mathrm{E}(W_n - \mathrm{E}(V_{n+1}|Z_n)) = \frac{1}{m}\sum_{n=1}^{\infty}\mathrm{E}\xi I(\xi > m^n) = \frac{1}{m}\sum_{n=1}^{\infty}\sum_{j>m^n} jp_j$$
$$= \frac{1}{m}\sum_{j=1}^{\infty} jp_j \sum_{n<\log j/\log m} 1 = C\sum_{j=1}^{\infty}(j\log j)p_j.$$

This is finite iff the $X \log X$ condition holds. The result now follows.

Chapter 8

Semimartingales

A semimartingale X_t, $0 \leq t \leq T$, is a process consisting of a sum of a local martingale and a finite variation process. Semimartingales are the most general processes for which a stochastic integral $\int_0^T H_t dX_t$ is defined. The process H_t must be predictable, i.e. measurable with respect to the predictable σ-field, generated by the adapted left-continuous processes. A measurable function of a predictable process gives a predictable process.

Doob–Meyer decomposition states that in a submartingale (or a local submartingale), X_t can be represented as a sum of a local martingale M_t and a unique increasing predictable process A_t:

$$X(t) = X(0) + M(t) + A(t).$$

This decomposition is instrumental for the definition of compensators.

If $N(t)$ is an adapted process of integrable or locally integrable variation, then its compensator $A(t)$ is the unique predictable process such that $M(t) = N(t) - A(t)$ is a local martingale. Compensators are essential for the analysis of pure jump processes considered in the following chapter.

The predictable quadratic variation or the sharp bracket process, $\langle X, X\rangle(t)$, of a semimartingale X is the compensator of its quadratic variation, i.e. the unique predictable process that makes $[X, X](t) - \langle X, X\rangle(t)$ a local martingale. Clearly, if X is continuous, then so is its quadratic variation process $[X, X](t)$, which means that it is predictable. Therefore, for continuous semimartingales, the square bracket process and the sharp bracket process are the same, $\langle X, X\rangle(t) = [X, X](t)$.

For a continuous semimartingale, its local time at a point $a \in \mathbb{R}$ is defined by the Meyer–Tanaka formula,

$$|X(t) - a| = |X(0) - a| + \int_0^t \text{sign}(X(s) - a) dX(s) + L^a(t).$$

L_t^a is a continuous, non-decreasing, and adapted process. The occupation time formula holds for a measurable function g:

$$\int_0^t g(X(s))d[X,X](s) = \int_{-\infty}^{\infty} g(a)L^a(t)da.$$

For a continuous local martingale null at zero $M(t), 0 \le t \le T < \infty$, its stochastic exponential is given by $\mathcal{E}(M) = e^{M(t)-\frac{1}{2}[M,M](t)}$, and it is a continuous positive local martingale. The question of when it is a true martingale is important in a number of applications, not least in the change of measure. Novikov's condition is a well-known sufficient condition for that:

$$\mathrm{E}\big(e^{\frac{1}{2}[M,M](T)}\big) < \infty.$$

The Kazamaki condition is another sufficient condition. It states that if $e^{\frac{1}{2}M(t)}$ is a submartingale, then $\mathcal{E}(M)$ is a true martingale. The Kazamaki condition is implied by the Novikov condition.

Itô formula for semimartingales:

Let $X(t)$ be a semimartingale and f be a C^2 function. Then, $f(X(t))$ is a semimartingale, and Itô's formula holds:

$$f(X(t)) = f(X(0)) + \int_0^t f'(X(s-))dX(s) + \frac{1}{2}\int_0^t f''(X(s-))d[X,X](s)$$
$$+ \sum_{s\le t}\Big(f(X(s)) - f(X(s-)) - f'(X(s-))\Delta X(s)$$
$$- \frac{1}{2}f''(X(s-))(\Delta X(s))^2\Big).$$

The quadratic variation $[X,X]$ jumps at the points of jumps of X and its jumps $\Delta[X,X](s) = (\Delta X(s))^2$. Thus, the jump part of the integral $\int_0^t f''(X(s-))d[X,X](s)$ is given by $\sum_{s\le t} f''(X(s-))(\Delta X(s))^2$, leading to an equivalent form of the formula:

$$f(X(t)) = f(X(0)) + \int_0^t f'(X(s-))dX(s) + \frac{1}{2}\int_0^t f''(X(s-))d[X,X]^c(s)$$
$$+ \sum_{s\le t}(f(X(s)) - f(X(s-)) - f'(X(s-))\Delta X(s)),$$

where $[X,X]^c$ is the continuous component of the finite variation function $[X,X]$. Clearly, this formula simplifies when X_t is a continuous semimartingale, with $X(s-) = X(s)$ and $\Delta X(s) = 0$ for all s. The number of jumps is at most countable so that the sums above make sense.

Another variant of Itô formula is obtained when we combine the terms including $f'(X(s-)$. Due to $X_t - \sum_{s\leq t}\Delta X(s) = X_t^c$, the continuous part of X, we have

$$f(X(t)) = f(X(0)) + \int_0^t f'(X(s-))dX^c(s) + \frac{1}{2}\int_0^t f''(X(s-))d[X,X]^c(s) + \sum_{s\leq t}(f(X(s)) - f(X(s-))).$$

Another approach is to consider the random measure of jumps. The measure of jumps of X (on $\mathbb{R}^+ \times \mathbb{R}^0$, with $\mathbb{R}^0 = \mathbb{R}\setminus 0$), denoted $\mu(dt,dx)$, is defined as

$$\mu((0,t]\times\Gamma) = \sum_{0<s\leq t} I_\Gamma(\Delta X(s))$$

for a Borel set Γ that does not include zero (there are no jumps of size 0, if $\Delta X(t) = 0$, then t is a point of continuity of X).

Observe that the sum of jumps $\sum_{s\leq t}\Delta X_s$ can be written as an integral of the function x with respect to the measure of jumps

$$\sum_{s\leq t}\Delta X_s = \int_0^t\int_{\mathbb{R}} x\mu(ds,dx).$$

More generally, for a function $h(x)$,

$$\sum_{s\leq t} h(\Delta X_s) = \int_0^t\int_{\mathbb{R}} h(x)\mu(ds,dx).$$

Itô formula can be written by using the measure of jumps,

$$f(X(t)) - f(X(0)) = \int_0^t f'(X(s-))dX^c(s) + \frac{1}{2}\int_0^t f''(X(s-))d[X,X]^c(s) + \int_0^t\int_{\mathbb{R}}(f(X(s-)+x) - f(X(s-)))\mu(ds,dx).$$

The canonical decomposition involves the compensator $\nu(dt,dx)$ of the random measure of jumps, which is a unique predictable process such that $\mu((0,t]\times\Gamma) - \nu((0,t]\times\Gamma)$ is a local martingale measure. ν is called the dual compensator because it makes the following expression into a local martingale for any function h:

$$\int_0^t\int_{\mathbb{R}} h(x)(\mu(ds,dx) - \nu(ds,dx)).$$

The canonical decomposition of a semimartingale is

$$X(t) = X(0) + A(t) + X^{cm}(t) + (h(x) * (\mu - \nu))(t) + ((x - h(x)) * \mu)(t),$$
$$= X(0) + A(t) + X^{cm}(t) + \int_0^t \int_{|x|\le 1} xd(\mu - \nu) + \int_0^t \int_{|x|>1} xd\mu,$$

where A is a predictable process of finite variation, X^{cm} is a continuous martingale component of X, μ is the measure of jumps of X, and ν its compensator. The following three processes appearing in the canonical decomposition (A, C, ν) are called the *triplet of predictable characteristics* of the semimartingale X, $C = \langle X^{cm}, X^{cm}\rangle$. Predictable characteristics determine the semimartingale uniquely; furthermore, they are often used to show convergence of semimartingales to a limit (we don't pursue this direction here).

Let $M(t)$ be a martingale, $0 \le t \le T$, adapted to the filtration $\mathbb{F} = (\mathcal{F}_t)$ and $H(t)$ be a predictable process satisfying $\int_0^T H^2(s) d\langle M, M\rangle(s) < \infty$ with probability one. Then, $\int_0^t H(s)dM(s)$ is a local martingale. The predictable representation property means that the converse is also true. Let $\mathbb{F}^M = (\mathcal{F}_t^M)$ denote the natural filtration of M. A local martingale M has the predictable representation property if for any $\mathbb{F}^M$-local martingale X, there is a predictable process H such that

$$X(t) = X(0) + \int_0^t H(s)dM(s).$$

This property has an important application in option pricing, called hedging, when an integrable $\mathcal{F}_T$-measurable random variable, Y has representation

$$Y = \mathrm{E}Y + \int_0^T H(t)dM(t).$$

In this application, the random variable Y is the option payoff and the process $H(t)$ is called the hedge. Both Brownian motion and Poisson process have the predictable representation property.

For the canonical decomposition of semimartingales, first large jumps are taken out, and then the small jumps are compensated as follows. Consider

$$(x - h(x)) * \mu(t) = \int_0^t \int_{\mathbb{R}\setminus 0} (x - h(x))\mu(ds, dx) = \sum_{s\le t} (\Delta X_s - h(\Delta X_s)),$$

where $h(x)$ is a truncation function. This is a sum over "large" jumps with $|\Delta X(s)| > 1$ (since $x - h(x) = 0$ for $|x| \le 1$). Since the sum of squares of jumps is finite, the above sum has only finitely many terms, hence it is finite.

Problems

Problem 8.1: Let $\tau_1 < \tau_2$ be stopping times. Show that $I_{(\tau_1,\tau_2]}(t)$ is a simple predictable process.

Solution. Let $X(t) = I_{(\tau_1,\tau_2]}(t)$. Trivially, $X(t)$ is a simple process. If $X(t)$ is adapted and left-continuous, then it is predictable (see Definition 8.3). As a stopping time, $\{\tau_1 \leq t\} \in \mathcal{F}_t$ for any t. Thus, $\{\tau_1 < t\} = \bigcup_{n=1}^{\infty}\{\tau_1 \leq t - \frac{1}{n}\} \in \mathcal{F}_t$ as $\{\tau_1 \leq t - \frac{1}{n}\} \in \mathcal{F}_{t-\frac{1}{n}} \subseteq \mathcal{F}_t$. Similarly, $\{\tau_2 < t\} \in \mathcal{F}_t$ and thus $\{\tau_2 \geq t\} = \{\tau_2 < t\}^c \in \mathcal{F}_t$. Therefore, $\{t \in (\tau_1, \tau_2]\} = \{\tau_1 < t\} \cap \{\tau_2 \geq t\} \in \mathcal{F}_t$, which implies $\{X(t) = 1\} \in \mathcal{F}_t$ and $\{X(t) = 0\} \in \mathcal{F}_t$. So, $X(t)$ is adapted. Since $X(t)$ is also left-continuous, it is a predictable process.

Problem 8.2: Let $H(t)$ be a regular adapted process, not necessarily left-continuous. Show that for any $\delta > 0$, $H(t-\delta)$ is predictable.

Solution. Consider $H(t-\delta+\epsilon)$ for $\epsilon < \delta$, the left-continuous modification of $H(t)$. The process $H(t-\delta+\epsilon)$ is adapted and $\lim_{\epsilon\to 0} H(t-\delta+\epsilon) = H(t-\delta)$. Thus, as a limit of left-continuous adapted processes, $H(t-\delta)$ is predictable.

Problem 8.3: Show that a continuous process is locally integrable. Show that a continuous local martingale is locally square integrable.

Solution. Let $X(t)$ be a continuous process. If it is bounded by K, then it is integrable, and there is nothing to show. Therefore, assume that it is unbounded. Define a sequence of stopping times $\tau_n = \inf\{t : |X(t)| \geq n\}$. Then, $\tau_n \uparrow \infty$ as $n \to \infty$, and $\sup_{t\geq 0} \mathrm{E}(|X(t \wedge \tau_n)|) \leq \sup_{t\geq 0} \mathrm{E}(n) = n < \infty$, that is, $X(t)$ is locally integrable. Now, let $X(t)$ be a continuous local martingale. Then, we have $\sup_{t\geq 0} \mathrm{E}(X^2(t\wedge\tau_n)) \leq \sup_{t\geq 0} \mathrm{E}(n^2) = n^2 < \infty$. Thus, $X(t)$ is locally square integrable.

Problem 8.4: M is a local martingale and $\mathrm{E}\int_0^T H^2(s)d[M,M](s) < \infty$. Show that $\int_0^t H(s)dM(s)$ is a square-integrable martingale.

Solution. Let $X(t) = \int_0^t H(s)dM(s)$. Then, $X(t)$ is a local martingale with $X(0) = 0$ and $[X,X](t) = \int_0^t H^2(s)d[M,M](s)$. Since

$\sup_{0<t\leq T} \mathrm{E}\big([X,X](t)\big) = \mathrm{E}\big([X,X](T)\big) < \infty$, $X(t)$ is a square-integrable martingale by Theorem 7.35.

Problem 8.5 (A càdlàg function of a Brownian motion, which is not a semimartingale): Show that the process $X(t) = I_{[a,b]}(B(t))$, $-\infty \leq a \leq b \leq \infty$, is not a semimartingale (but it is predictable). (In a stochastic integral $\int H(t)dX(t)$, H must be predictable and X must be a semimartingale).

Solution. An example of a function $f(x)$ such that $f(B(t))$ is not a semimartingale is provided by the function $\mathrm{sign}(x) = I_{(0,\infty]}(x)$, or the more general $I_{[a,b]}(x)$. The level set of hitting a, i.e. the (random) set of all times t, such that $B(t) = a$ is the set of zeros of $B(t) - a$. $B(t) - a$ is also a Brownian motion, started at $B(0) - a$. By Theorem 3.28, the set of zeroes is an uncountable, closed set without isolated points. Since it has no isolated points, any zero of Brownian motion is a limit of other zeros. This implies that Brownian motion changes sign uncountably often. Hence, crossings of a (and b) occur uncountably often, and $X(t)$ changes values from 0 to 1 and vice versa uncountably many times; these discontinuities are not jumps because between any two points t and s such that $X(t) = X(s) = a$, there is another point u at which $X(u) = a$. Since a semimartingale has at most countably many discontinuities and all of them are jumps, $X(t)$ is not a semimartingale.

$X(t)$ is predictable, as a composition of a regular function $I_{[a,b]}(x)$ and $B(t)$, which is continuous and predictable.

Problem 8.6: Find the variance of $\int_0^1 N(t-)dM(t)$, where M is the compensated Poisson process $N(t) - t$.

Solution. First, note that $[M,M](t) = [N,N](t) = N(t)$.
The integral $\int_0^1 N(t-)dM(t)$ has zero mean. Therefore,

$$
\begin{aligned}
Var\left(\int_0^1 N(t-)dM(t)\right) &= \mathrm{E}\left(\left(\int_0^1 N(t-)dM(t)\right)^2\right) \\
&= \mathrm{E}\int_0^1 N^2(t-)d[M,M](t) \\
&= \mathrm{E}\int_0^1 N^2(t-)dN(t) = \mathrm{E}\sum_{\tau_i \leq 1} N^2(\tau_{i-1}) \\
&= \mathrm{E}\sum_{i=1}^{N(1)} (i-1)^2 = \mathrm{E}\sum_{k=0}^{N(1)-1} k^2,
\end{aligned}
$$

where τ_i denotes the time of the ith jump of N, and we used the facts that $N(\tau_i) = i$ and $\{\tau_i \le 1\} = \{i \le N(1)\}$. Using the formula $\sum_{k=0}^{n} k^2 = \frac{1}{6}(2(n+1)^3 - 3(n+1)^2 + (n+1))$ or $\sum_{k=0}^{n-1} k^2 = \frac{1}{6}(2n^3 - 3n^2 + n)$, we obtain

$$Var\left(\int_0^1 N(t-)dM(t)\right) = \mathrm{E}\left(\frac{1}{6}(2N^3(1) - 3N^2(1) + N(1))\right)$$
$$= \frac{1}{3}\mathrm{E}(N^3(1)) - \frac{1}{2}\mathrm{E}(N^2(1)) + \frac{1}{6}\mathrm{E}(N(1)).$$

$N(1)$ is a Poisson random variable with parameter 1 with moment-generating function $m(s) = \mathrm{E}(e^{sN(1)}) = e^{e^s - 1}$. Computing the first three moments: $\mathrm{E}(N(1)) = m'(0) = 1$, $\mathrm{E}(N^2(1)) = m''(0) = 2$ and $\mathrm{E}(N^3(1)) = m'''(0) = 5$, it follows that $Var(\int_0^1 N(t-)dM(t)) = 5/6$.

Problem 8.7: If S and T are stopping times, show the following:

(a) $S \wedge T$ and $S \vee T$ are stopping times.
(b) The events $\{S = T\}$, $\{S \le T\}$ and $\{S < T\}$ are in $\mathcal{F}_S$.
(c) $\mathcal{F}_S \cap \{S \le T\} \subset \mathcal{F}_T \cap \{S \le T\}$.

Solution.

(a) The sets $\{S \le t\}$ and $\{T \le t\}$ in $\mathcal{F}_t$ imply that $\{S > t\}$ and $\{T > t\}$ in $\mathcal{F}_t$. Therefore, $\{S \wedge T > t\} = \{\min(S,T) > t\} = \{S > t\} \cap \{T > t\} \in \mathcal{F}_t$, and thus its compliment $\{S \wedge T \le t\} \in \mathcal{F}_t$, i.e. $S \wedge T$, is a stopping time.

Also, $\{S \vee T \le t\} = \{\max(S,T) \le t\} = \{S \le t\} \cap \{T \le t\} \in \mathcal{F}_t$.

(b) $\mathcal{F}_S = \{A \in \mathcal{F} : A \cap \{S \le t\} \in \mathcal{F}_t \text{ for any } t\}$.

First, we note that $\{S \ge u\} = \left(\bigcap_{n=1}^{\infty}\{S > u - \frac{1}{n}\}\right)$ and $\{S = u\} = \{S \le u\} \cap \{S \ge u\} = \{S \le u\} \cap \left(\bigcap_{n=1}^{\infty}\{S > u - \frac{1}{n}\}\right)$. Therefore, $\{S \ge u\}$ and $\{S = u\}$ are in $\mathcal{F}_u$ for any u. The same is true if we replace S with T.

To show $\{S=T\} \in \mathcal{F}_S$, we show $\{S=T\} \cap \{S \le t\} \in \mathcal{F}_t$ for any t. But $\{S = T\} \cap \{S \le t\} = \bigcup_{u \in \mathbb{R}} \left(\{S = u\} \cap \{T = u\} \cap \{S \le t\}\right)$. For $u > t$, the above is $\emptyset$ and thus in $\mathcal{F}_t$; while for $u < t$, $\{S = u\} \cap \{T = u\} \in \mathcal{F}_u \subseteq \mathcal{F}_t$. Thus, $\{S = u\} \cap \{T = u\} \cap \{S \le t\} \in \mathcal{F}_t$ for any u and $\bigcup_{u \in \mathbb{Q}} \left(\{S = u\} \cap \{T = u\} \cap \{S \le t\}\right) \in \mathcal{F}_t$, which implies $\bigcup_{u \in \mathbb{R}} \left(\{S = u\} \cap \{T = u\} \cap \{S \le t\}\right) \in \mathcal{F}_t$. Hence, this shows $\{S = T\} \in \mathcal{F}_S$.

To show $\{S \le T\} \in \mathcal{F}_S$, we show $\{S \le T\} \cap \{S \le t\} \in \mathcal{F}_t$ for any t. But $\{S \le T\} = \bigcup_{u \in \mathbb{R}} \left(\{S = u\} \cap \{T \ge u\}\right)$. For $u > t$,

$\{S = u\} \cap \{T \geq u\} \cap \{S \leq t\} = \emptyset \in \mathcal{F}_t$. For $u \leq t$, $\{S = u\} \cap \{T \geq u\} \in \mathcal{F}_u \subseteq \mathcal{F}_t$, thus $\{S = u\} \cap \{T \geq u\} \cap \{S \leq t\} \in \mathcal{F}_t$. Therefore, $\bigcup_{u \in \mathbb{Q}} (\{S = u\} \cap \{T \geq u\} \cap \{S \leq t\}) \in \mathcal{F}_t$ and thus $\bigcup_{u \in \mathbb{R}} (\{S = u\} \cap \{T \geq u\} \cap \{S \leq t\}) \in \mathcal{F}_t$. That is, $\{S \leq T\} \cap \{S \leq t\} \in \mathcal{F}_t$ and hence $\{S \leq T\} \in \mathcal{F}_S$.

Lastly, we show $\{S < T\} \in \mathcal{F}_S$, i.e. $\{S < T\} \cap \{S \leq t\} \in \mathcal{F}_t$, for any t. Using a similar argument as above, we can show that $\{S = u\} \cap \{T > u\} \cap \{S \leq t\} \in \mathcal{F}_t$ for any u. Therefore, $\bigcup_{u \in \mathbb{Q}} (\{S = u\} \cap \{T > u\} \cap \{S \leq t\}) \in \mathcal{F}_t$, and thus $\{S < T\} \cap \{S \leq t\} = \bigcup_{u \in \mathbb{R}} (\{S = u\} \cap \{T > u\} \cap \{S \leq t\}) \in \mathcal{F}_t$.

(c) We show that if $A \in \mathcal{F}_S \cap \{S \leq T\}$, then $A \in \mathcal{F}_T \cap \{S \leq T\}$. Suppose that $A \in \mathcal{F}_S \cap \{S \leq T\}$, i.e. $A \cap \{S \leq t\} \cap \{S \leq T\} \in \mathcal{F}_t$. Then, $A \cap \{T \leq t\} \cap \{S \leq T\} = A \cap \{T \leq t\} \cap \{S \leq T\} \cap \{S \leq t\} \in \mathcal{F}_t$ due to the assumption and that $\{T \leq t\} \in \mathcal{F}_t$. Thus, $A \in \mathcal{F}_T \cap \{S \leq T\}$.

Problem 8.8: Let U be a positive random variable on a probability space $(\Omega, \mathcal{F}, \mathrm{P})$, and let $\mathcal{G}$ be a sub-σ-field of $\mathcal{F}$.

(a) Let $t \geq 0$. Show that $\mathcal{F}_t := \{A \in \mathcal{F} : \exists B \in \mathcal{G} \text{ such that } A \cap \{U > t\} = B \cap \{U > t\}\}$ is a σ-field.
(b) Show that $\mathcal{F}_t$ is a right-continuous filtration on $(\Omega, \mathcal{F}, \mathrm{P})$.
(c) Show that U is a stopping time for $\mathcal{F}_t$.
(d) What are $\mathcal{F}_0$, $\mathcal{F}_{U-}$ and $\mathcal{F}_U$ equal to?

Solution.

(a) To show $\mathcal{F}_t$ is a σ-field, we show (i) $\Omega \in \mathcal{F}_t$, (ii) $A \in \mathcal{F}_t$ implies $A^c \in \mathcal{F}_t$, and (iii) $A_i \in \mathcal{F}_t$ for $i = 1, 2, \ldots$ implies $\bigcup_{i=1}^{\infty} A_i \in \mathcal{F}_t$.

i. Trivial (take $B = \Omega$).

ii. If $A \cap \{U > t\} = B \cap \{U > t\}$ for some $B \in \mathcal{G}$.
Then $\{U > t\} \setminus (A \cap \{U > t\}) = \{U > t\} \setminus (B \cap \{U > t\})$, equivalently, $A^c \cap \{U > t\} = B^c \cap \{U > t\}$.
But $B^c \in \mathcal{G}$. So, $A^c \in \mathcal{G}$.

iii. Suppose that for $i = 1, 2, \ldots$, $A_i \cap \{U > t\} = B_i \cap \{U > t\}$ for some $B_i \in \mathcal{G}$. Then, we have $(\bigcup_{i=1}^{\infty} A_i) \cap \{U > t\} = \bigcup_{i=1}^{\infty} (A_i \cap \{U > t\}) = \bigcup_{i=1}^{\infty} (B_i \cap \{U > t\}) = (\bigcup_{i=1}^{\infty} B_i) \cap \{U > t\}$, and $\bigcup_{i=1}^{\infty} B_i \in \mathcal{G}$. Therefore, $\bigcup_{i=1}^{\infty} A_i \in \mathcal{F}_t$.

(b) $\mathcal{F}_t$ is a right-continuous filtration if $\mathcal{F}_{t_0} \subseteq \mathcal{F}_{t_1} \subseteq \cdots \subseteq \mathcal{F}_T = \mathcal{F}$ for $t_0 < t_1 < \cdots < T$ and $\bigcap_{s>t} \mathcal{F}_s = \mathcal{F}_t$. To show $\bigcap_{s>t} \mathcal{F}_s = \mathcal{F}_t$, it is sufficient to show that $\bigcap_n \mathcal{F}_{t+1/n} = \mathcal{F}_t$.

First, suppose that $A \in \mathcal{F}_t$. Then, there exists $B \in \mathcal{G}$ such that $A \cap \{U > t\} = B \cap \{U > t\}$. It follows that $A \cap \{U > t + 1/n\} = B \cap \{U > t + 1/n\}$ for any n. Thus, $A \in \mathcal{F}_{t+1/n}$ and therefore $\mathcal{F}_t \subseteq \mathcal{F}_{t+1/n}$ for any n, or $\mathcal{F}_t \subseteq \bigcap_n \mathcal{F}_{t+1/n}$.

Now, suppose that $A \in \bigcap_n \mathcal{F}_{t+1/n}$. Then, for each n, there exists $B^{(n)} \in \mathcal{G}$ such that $A \cap \{U > t + 1/n\} = B^{(n)} \cap \{U > t + 1/n\}$. Take $B = \bigcup_n B^{(n)}$, then $B \in \mathcal{G}$ and $A \cap \{U > t\} = B \cap \{U > t\}$. So, $A \in \mathcal{F}_t$ and $\bigcap_n \mathcal{F}_{t+1/n} \subseteq \mathcal{F}_t$.

Hence, $\bigcap_n \mathcal{F}_{t+1/n} = \mathcal{F}_t$.

(c) Let $A = \{U \leq t\}$. Then, $A \cap \{U > t\} = \emptyset$. Take $B = \emptyset$, and we have $A = \{U \leq t\} \in \mathcal{F}_t$. So, U is a stopping time.

(d) $\mathcal{F}_0 = \{A \in \mathcal{F} : \exists B \in \mathcal{G} \text{ s.t. } A \cap \{U > 0\} = B \cap \{U > 0\}\}$. Since U is positive, $\{U > 0\} = \Omega$. Therefore, $\mathcal{F}_0 = \mathcal{G}$.

$\mathcal{F}_{U-} = \sigma(\{A \cap \{U > t\} : A \in \mathcal{F}_t, t \geq 0\} \cup \mathcal{F}_0)$, which is the sigma algebra generated by $\mathcal{G} \cap \{U > t\}, t \geq 0$.

$\mathcal{F}_U = \{A \in \mathcal{F} : A \cap \{U \leq t\} \in \mathcal{F}_t \text{ for any } t\}$. But $A \cap \{U \leq t\} \cap \{U > t\} = \emptyset$. Thus, one can take $B = \emptyset$, and it follows that $A \cap \{U \leq t\} \in \mathcal{F}_t$ for any t and any $A \in \mathcal{F}$. Therefore, $\mathcal{F}_U = \mathcal{F}$.

Problem 8.9: Let $U_1, U_2, \ldots$ be (strictly) positive random variables on a probability space $(\Omega, \mathcal{F}, \mathrm{P})$, and $\mathcal{G}_1, \mathcal{G}_2, \ldots$ be sub-filtration of $\mathcal{F}$. Suppose that for all n, $U_1, U_2, \ldots, U_n$ are $\mathcal{G}_n$-measurable and denote by T_n the random variable $\sum_{i=1}^n U_i$. Set $\mathcal{F}_t = \bigcap_n \{A \in \mathcal{F} : \ \exists B_n \in \mathcal{G}_n \text{ such that } A \cap \{T_n > t\} = B_n \cap \{T_n > t\}\}$.

(a) Show that $\mathcal{F}_t$ is a right-continuous filtration on $(\Omega, \mathcal{F}, \mathrm{P})$.
(b) Show that for all n, T_n is a stopping time for $\mathcal{F}_t$.
(c) Suppose that $\lim_n T_n = \infty$ almost surely. Show that $\mathcal{F}_{T_n} = \mathcal{G}_{n+1}$ and $\mathcal{F}_{T_n -} = \mathcal{G}_n$.

Solution.

(a) Write $\mathcal{F}_t = \{A \in \mathcal{F} : \exists B_m \in \mathcal{G}_m \text{ such that } A \cap \{T_m > t\} = B_m \cap \{T_m > t\} \ \forall m\}$.

$\mathcal{F}_t$ is a right-continuous filtration if $\bigcap_{s>t} \mathcal{F}_s = \mathcal{F}_t$. It is sufficient to show that $\bigcap_n \mathcal{F}_{t+1/n} = \mathcal{F}_t$.

First, suppose that $A \in \mathcal{F}_t$, that is, for any m, $\exists\ B_m \in \mathcal{G}_m$ such that $A \cap \{T_m > t\} = B_m \cap \{T_m > t\}$. It also follows that $A \cap \{T_m > t + 1/n\} = B_m \cap \{T_m > t + 1/n\}$ for any n. Thus, $A \in \mathcal{F}_{t+1/n}$ for any n. Therefore, $A \in \bigcap_n \mathcal{F}_{t+1/n}$ and $\mathcal{F}_t \subseteq \bigcap_n \mathcal{F}_{t+1/n}$.

Now, suppose that $A \in \bigcap_n \mathcal{F}_{t+1/n}$. Then, for each n, $\exists\ B_m^{(n)} \in \mathcal{G}_{\Updownarrow}$ such that $A \cap \{T_m > t + 1/n\} = B_m \cap \{T_m > t + 1/n\}$. For each m, take $B_m = \bigcup_n B_m^{(n)}$, then $B_m \in \mathcal{G}_m$ and $A \cap \{T_m > t\} = B_m \cap \{T_m > t\}$. Thus, $A \in \mathcal{F}_t$. So, we established $\bigcap_n \mathcal{F}_{t+1/n} = \mathcal{F}_t$.

(b) To show that T_n is a stopping time, we show $A = \{T_n \le t\} \in \mathcal{F}_t$. For $m \le n$, by the definition of T_m, $\{T_n \le t\} \cap \{T_m > t\} = \emptyset$, so we can take $B_m = \emptyset$. For $m > n$, since $\{T_n \le t\} \in \mathcal{G}_m$, we can take $B_m = \{T_n \le t\}$. Thus, there exists $B_m \in \mathcal{G}_m$ such that $\{T_n \le t\} \cap \{T_m > t\} = B_m \cap \{T_m > t\}$ for all m. Therefore, $\{T_n \le t\} \in \mathcal{F}_t$.

(c) Recall that $\mathcal{F}_{T_n} = \{A \in \mathcal{F} : A \cap \{T_n \le t\} \in \mathcal{F}_t \text{ for any } t\}$ and $\mathcal{F}_{T_n-} = \sigma(\{A \cap \{T_n > t\} : t \ge 0, A \in \mathcal{F}_t\} \cup \mathcal{F}_0)$.

Suppose that $A \in \mathcal{F}_{T_n}$. Then, $A \cap \{T_n \le t\} \in \mathcal{F}_t$ for all t, that is, for any m and t, $\exists B_{m,t} \in \mathcal{G}_m$ such that $A \cap \{T_n \le t\} \cap \{T_m > t\} = B_{m,t} \cap \{T_m > t\}$. In particular, for all t, $A \cap \{T_n \le t < T_{n+1}\} = B_{n+1,t} \cap \{T_{n+1} > t\} \in \mathcal{G}_{n+1}$. Thus, $A = \bigcup_{t \in \mathbb{Q}} A \cap \{T_n \le t < T_{n+1}\} = \bigcup_{t \in \mathbb{Q}} B_{n+1,t} \cap \{T_{n+1} > t\} \in \mathcal{G}_{n+1}$. This shows that $\mathcal{F}_{T_n} \subseteq \mathcal{G}_{n+1}$. Now suppose that $A \in \mathcal{G}_{n+1}$. For $m \le n$, since $\{T_n \le t\} \cap \{T_m > t\} = \emptyset$, we can take $B_m = \emptyset$. For $m > n$, since $A \cap \{T_n \le t\} \in \mathcal{G}_{n+1} \subseteq \mathcal{G}_m$, take $B_m = A \cap \{T_n \le t\}$. Then, we have $A \cap \{T_n \le t\} \in \mathcal{F}_t$. Thus, $A \in \mathcal{F}_{T_n}$ and consequently $\mathcal{G}_{n+1} \subseteq \mathcal{F}_{T_n}$. Therefore, $\mathcal{F}_{T_n} = \mathcal{G}_{n+1}$.

Next, we show $\mathcal{F}_{T_n-} = \mathcal{G}_n$. Let $A \in \mathcal{F}_t$. Then, there exists $B_n \in \mathcal{G}_n$ such that $A \cap \{T_n > t\} = B_n \cap \{T_n > t\} \in \mathcal{G}_n$. Thus, $\mathcal{F}_{T_n-} \subseteq \mathcal{G}_n$. For the opposite inclusion, let $B_n \in \mathcal{G}_n$. Note that for $m < n$, $\{T_{n-1} \le t\} \cap \{T_m > t\} = \emptyset$, and for $m \ge n$, $B_n \cap \{T_{n-1} \le t\} \in \mathcal{G}_m$. Thus, $B_n \cap \{T_{n-1} \le t\} \in \mathcal{F}_t$. Since $B_n = \bigcup_{t \in \mathbb{Q}} B_n \cap \{T_{n-1} \le t < T_n\} = \bigcup_{t \in \mathbb{Q}} B_n \cap \{T_{n-1} \le t\} \cap \{T_n > t\}$, with $B_n \cap \{T_{n-1} \le t\} \in \mathcal{F}_t$, we have $B_n \in \mathcal{F}_{T_n-}$. Thus, $\mathcal{G}_n \subseteq \mathcal{F}_{T_n-}$.

Problem 8.10: Let $B(t)$ be a Brownian motion and $H(t)$ be a predictable process. Show that $M(t) = \int_0^t H(s)dB(s)$ is a Brownian motion if and only if $\text{Leb}(\{t : |H(t)| \ne 1\}) = 0$ almost surely.

Solution. If $\text{Leb}(\{t : |H(t)| \ne 1\}) = 0$, then $M(t) = \int_0^t H(s)dB(s)$ is a continuous local martingale, and $[M,M](t) = \int_0^t H^2(s)ds = \int_0^t ds = t$. Thus, by the Lévy characterisation of Brownian motion (Theorem 7.36), $M(t)$ is a Brownian motion.

Conversely, if $M(t)$ is a Brownian motion, then $[M,M](t) = \int_0^t H^2(s)ds = t$. Differentiating both sides, we get $H^2(s) = 1$ Lebesgue almost surely.

Problem 8.11: Let T be a stopping time. Show that the process $M(t) = 2B(t \wedge T) - B(t)$ obtained by reflecting $B(t)$ at time T is a Brownian motion.

Solution. We verify the Lévy's characterisation of Brownian motion, namely we show that $M(t)$ is a continuous martingale with quadratic variation t. First, note that $B(t) = \int_0^t dB(s)$ and $B(t \wedge T) = \int_0^t I_{[0,T]}$ $dB(s)$. Then, $2I_{[0,T]}(s) - 1$ is predictable and $M(t) = \int_0^t (2I_{[0,T]}(s) - 1)dB(s)$ is a continuous local martingale. Since $\mathrm{E}(\int_0^t (2I_{[0,T]}(s) - 1)^2 ds) \leq 9t < \infty$, $M(t)$ is indeed a martingale. Its quadratic variation is $[M,M](t) = \int_0^t (2I_{[0,T]}(s) - 1)^2 ds$.

For $t \leq T$, $[M,M](t) = \int_0^t (2I_{[0,T]}(s) - 1)^2 ds = \int_0^t (2-1)^2 ds = t$. For $t > T$, $[M,M](t) = \int_0^T (2I_{[0,T]}(s) - 1)^2 ds + \int_T^t (2I_{[0,T]}(s) - 1)^2 ds = \int_0^T (2-1)^2 ds + \int_T^t (0-1)^2 ds = t$. Thus, $[M,M](t) = t$ and $M(t)$ is a Brownian motion by Lévy characterisation.

Problem 8.12: Let B and N be respectively a Brownian motion and a Poisson process on the same space and adapted to the same filtration $\mathbb{F}$. Show that, for any fixed t, B_t and N_t are independent.
Hint: Use Ito's formula with function $e^{i(ux+vy)}$ and use the fact that the covariation of B and N is zero.

Solution. The first step is to show that, for any fixed $t \geq 0$, B_t and N_t are independent. We start by writing Ito's formula for e^{uB_t} and $e^{v(N_t - \lambda t)}$:

$$de^{iuB_t} = iue^{iuB_t} dB_t - \frac{1}{2}u^2 e^{iuB_t} dt$$

$$\begin{aligned} de^{iv(N_t - \lambda t)} &= (e^{iv} - 1)e^{iv(N_{t-} - \lambda t)} d(N_t - \lambda t) \\ &\quad + \lambda(e^{iv} - iv - 1)e^{iv(N_{t-} - \lambda t)} dt. \end{aligned}$$

Note that N is a purely discontinuous semimartingale and B is a continuous martingale so that their quadratic covariation is nil, and the integration by parts formula leads to

$$\begin{aligned} de^{iuB_t + iv(N_t - \lambda t)} &= e^{iv(N_t - \lambda t)}(iue^{iuB_t} dB_t - \frac{1}{2}u^2 e^{iuB_t} dt) \\ &\quad + e^{iuB_t}\big((e^{iv} - 1)e^{iv(N_{t-} - \lambda t)} d(N_t - \lambda t) \\ &\quad + \lambda(e^{iv} - iv - 1)e^{iv(N_t - \lambda t)}\big)dt. \end{aligned}$$

Since the integrals in dB_t and $d(N_t - \lambda t)$ are true martingales (the integrands are bounded),

$$\mathrm{E}(e^{iuB_t+iv(N_t-\lambda t)}) = -\frac{1}{2}u^2 \int_0^t \mathrm{E}(e^{iuB_s+iv(N_s-\lambda s)})ds + \lambda(e^{iv} - iv - 1) \times \int_0^t \mathrm{E}(e^{iuB_s+iv(N_s-\lambda s)})ds.$$

It follows that $\phi(t) = \mathrm{E}\big(e^{iuB_t+iv(N_t-\lambda t)}\big)$ solves the ordinary differential equation $y' = \big(-u^2/2 + \lambda(e^{iv} - iv - 1)\big)y$, the solution of which is (note that $\phi(0) = 1$)

$$\mathrm{E}(e^{iuB_t+iv(N_t-\lambda t)}) = e^{(-u^2/2+\lambda(e^{iv}-iv-1))t}$$

or equivalently,

$$\mathrm{E}(e^{iuB_t+ivN_t}) = e^{(-u^2/2+\lambda(e^{iv}-1))t} = e^{-u^2/2t}e^{\lambda(e^{iv}-1)t} = \mathrm{E}(e^{iuB_t})\mathrm{E}(e^{ivN_t}).$$

Problem 8.13: Let $B(t)$ and $N(t)$ be respectively a Brownian motion and a Poisson process in the same space. Denote by $\bar{N}(t) = N(t) - t$ the compensated Poisson process. Show that the following processes are martingales: $B(t)\bar{N}(t)$, $\mathcal{E}(B)(t)\bar{N}(t)$ and $\mathcal{E}(\bar{N})(t)B(t)$.

Solution.

(a) $B(t)\bar{N}(t)$

$d(B(t)\bar{N}(t)) = B(t)d\bar{N}(t) + \bar{N}(t)dB(t) + d[B, \bar{N}](t)$. Note that $[B, \bar{N}](t) = 0$ since $B(t)$ is continuous and $\bar{N}(t)$ is of finite variation (cf. (8.18)). Thus, $B(t)\bar{N}(t) = \int_0^t B(s)d\bar{N}(s) + \int_0^t \bar{N}(s)dB(s)$. These two integrals are local martingales. We then check if they are martingales:

$$\begin{aligned}
\mathrm{E}\int_0^t B^2(s)d[\bar{N},\bar{N}](s) &= \mathrm{E}\int_0^t B^2(s)dN(s) \\
&= \mathrm{E}\sum_{\tau_i\le t} B^2(\tau_i) \quad \text{(where } \tau_i \text{ are the jump times of } N) \\
&= \mathrm{E}\mathrm{E}\left(\sum_{\tau_i\le t} B^2(\tau_i)|\tau_i\right) = \mathrm{E}\sum_{\tau_i\le t}\mathrm{E}(B^2(\tau_i)|\tau_i) = \mathrm{E}\sum_{\tau_i\le t}\tau_i \\
&\le \mathrm{E}\sum_{\tau_i\le t} t = \mathrm{E}(tN(t)) = t^2 < \infty.
\end{aligned}$$

Further,

$$\mathrm{E}\int_0^t \bar{N}^2(s)d[B,B](s) = \mathrm{E}\int_0^t \bar{N}^2(s)ds = \int_0^t \mathrm{E}(\bar{N}^2(s))ds$$

$$= \int_0^t Var(\bar{N}(s))ds$$

$$= \int_0^t Var(N(s))ds = \int_0^t sds < \infty.$$

Hence, as a sum of two martingales, $B(t)\bar{N}(t)$ is a martingale.

Alternatively, we can use a sufficient condition for a local martingale to be a true martingale. By Theorem 7.35, if a local martingale $M(t)$ has finite second moments $\mathrm{E}(M(t))^2 < \infty$, then it is a true martingale (even uniformly integrable. Note that $\mathrm{E}(M(t))^2) \leq \mathrm{E}(M(T))^2)$.

We have shown above that $M(t) = B(t)\bar{N}(t)$ is a local martingale as a sum of two stochastic integrals. Now, by independence, $\mathrm{E}M^2(t) = \mathrm{E}B^2(t)\mathrm{E}(\bar{N}(t))^2 = t^2 \leq T^2$.

(b) $X(t) = \mathcal{E}(B)(t)\bar{N}(t)$

Let $M(t) = \mathcal{E}(B)(t)$. Then, $M(t) = e^{B(t)-\frac{1}{2}t}$ is a continuous martingale with $dM(t) = M(t)dB(t)$. Since $\bar{N}$ is of finite variation, $[M,\bar{N}](t) = 0$, and we have that X is a local martingale, $dX(t) = d(M(t)\bar{N}(t)) = M(t)d\bar{N}(t) + \bar{N}(t)dM(t)$. By independence,

$$\mathrm{E}X^2(t) = \mathrm{E}M^2(t)\mathrm{E}(\bar{N}(t))^2 = t\mathrm{E}e^{2B(t)-t} = te^t \leq Te^T < \infty,$$

hence by Theorem 7.35, it is a martingale.

In the following, we give direct calculations verifying sufficient conditions for stochastic integrals to be martingales:

$$M(t)\bar{N}(t) = \int_0^t M(s)d\bar{N}(s) + \int_0^t \bar{N}(s)M(s)dB(s)$$

$$= \int_0^t e^{B(s)-\frac{1}{2}s}d\bar{N}(s) + \int_0^t \bar{N}(s)e^{B(s)-\frac{1}{2}s}dB(s).$$

$$\begin{aligned}
\mathrm{E}\int_0^t e^{2B(s)-s}d[\bar{N},\bar{N}](s) &= \mathrm{E}\int_0^t e^{2B(s)-s}dN(s)\\
&= \mathrm{E}\sum_{\tau_i\le t} e^{2B(\tau_i)-\tau_i}\\
&= \mathrm{E}\mathrm{E}\left(\sum_{\tau_i\le t} e^{2B(\tau_i)-\tau_i}\Big|\tau_i\right)\\
&= \mathrm{E}\sum_{\tau_i\le t} e^{2\tau_i-\tau_i} = \mathrm{E}\sum_{i=1}^{N(t)} e^{\tau_i}\\
&\le \mathrm{E}\sum_{i=1}^{N(t)} e^{t} = e^t\mathrm{E}(N(t)) = te^t < \infty,
\end{aligned}$$

$$\begin{aligned}
\mathrm{E}\int_0^t \bar{N}^2(s)e^{2B(s)-s}ds &= \int_0^t \mathrm{E}(\bar{N}^2(s)e^{2B(s)-s})ds\\
&= \int_0^t \mathrm{E}(\bar{N}^2(s))\mathrm{E}(e^{2B(s)-s})ds\\
&= \int_0^t se^s ds < \infty.
\end{aligned}$$

Thus, as a sum of two martingales, $\mathcal{E}(B)(t)\bar{N}(t)$ is a martingale.

(c) $\mathcal{E}(\bar{N})(t)B(t)$

Let $M(t) = \mathcal{E}(\bar{N})(t)$. Then, $dM(t) = M(t-)d\bar{N}(t)$ and using Theorem 8.33, Equation (8.63*), we have

$$\begin{aligned}
M(t) &= e^{\bar{N}(t)-\bar{N}(0)-\frac{1}{2}[\bar{N},\bar{N}](t)}\prod_{s\le t}(1+\Delta\bar{N}(s))e^{-\Delta\bar{N}(s)+\frac{1}{2}(\Delta\bar{N}(s))^2}\\
&= e^{N(t)-t-\frac{1}{2}N(t)}\prod_{i=1}^{N(t)}(1+\Delta N(\tau_i))e^{-\Delta N(\tau_i)+\frac{1}{2}(\Delta N(\tau_i))^2}\\
&= e^{\frac{1}{2}N(t)-t}2^{N(t)}e^{\sum_{i=1}^{N(t)}-\Delta N(\tau_i)+\frac{1}{2}(\Delta N(\tau_i))}\\
&= e^{-t}2^{N(t)} = e^{N(t)\log 2-t},
\end{aligned}$$

where we have used that $(\Delta N(\tau_i))^2 = \Delta N(\tau_i) = 1$. Since $[M,B](t) = 0$, $M(t)B(t)$ is a local martingale, $M(t)B(t) = \int_0^t M(s)dB(s) + \int_0^t B(s)dM(s)$,

$$\mathrm{E}(M(t)B(t))^2 = \mathrm{E}(M(t))^2\mathrm{E}(B(t))^2 = t\mathrm{E}e^{N(t)2\log 2-2t} = te^t.$$

Here, we have used that the random variable $N(t)$ has $Pn(t)$ distribution, and its exponential moment is given by $\mathrm{E}e^{uN(t)} = e^{t(e^u-1)}$.

Alternatively, we show that the two local martingales are true martingales:

$$\begin{aligned}
\mathrm{E}\int_0^t e^{-2s}2^{2N(s)}ds &= \int_0^t e^{-2s}\mathrm{E}(2^{2N(s)})ds \\
&= \int_0^t e^{-2s}e^{s(e^2-1)}ds \\
&= \int_0^t e^{s(e^2-3)}ds < \infty
\end{aligned}$$

as an integral of continuous function,

$$\begin{aligned}
&\mathrm{E}\int_0^t B^2(s)e^{-2s}2^{2N(s)-2}dN(s) \\
&= \mathrm{E}\sum_{\tau_i \le t} B^2(\tau_i)e^{-2\tau_i}2^{2N(\tau_i)-2} \\
&= \mathrm{E}\sum_{\tau_i \le t} \tau_i e^{-2\tau_i}2^{2N(\tau_i)-2} \quad \text{by conditioning} \\
&\le \mathrm{E}\sum_{i=1}^{N(t)} t2^{2N(t)-2} = t2^{-2}\mathrm{E}\big(N(t)2^{2N(t)}\big) < \infty.
\end{aligned}$$

Finally, as a sum of two martingales $\mathcal{E}(\bar{N})(t)B(t)$ is a martingale.

Problem 8.14: Let $N(t)$ be a Poisson process with rate λ and $B(t)$ a Brownian motion. $X(t)$ solves the SDE $dX(t) = \mu X(t)dt + aX(t-)dN(t) + \sigma X(t)dB(t)$, with $X(0) = 1$. Find the condition for $X(t)$ to be a martingale.

Solution.

$$\begin{aligned}
dX(t) &= 1 + \mu X(t)dt + aX(t-)\lambda dt + aX(t-)(dN(t) - \lambda dt) \\
&\quad + \sigma X(t)dB(t).
\end{aligned}$$

Write $\bar{N}(t) = N(t) - \lambda t$ to be the compensated Poisson process, which is a martingale. Since $\int X(t-)dt = \int X(t)dt$, we have

$$X(t) = 1 + \int_0^t (\mu + a\lambda)X(s)ds + \int_0^t aX(s-)d\bar{N}(s) + \int_0^t \sigma X(s)dB(s).$$

A necessary condition for $X(t)$ to be a martingale is that it must have no drift, i.e. the first integral has to be zero, which gives $\mu = -a\lambda$. We show that this is also a sufficient condition. In this case,

$$X(t) = 1 + \int_0^t X(s-)d\big(a\bar{N}(s) + \sigma B(s)\big) = 1 + \int_0^t X(s-)dM(s),$$

where $M(t) = a\bar{N}(t) + \sigma B(t)$ is a martingale. Hence, $X(t)$ satisfies the stochastic exponential equation, and therefore it is the stochastic exponential of M,

$$X(t) = \mathcal{E}(M)(t) = \mathcal{E}(a\bar{N} + \sigma B)(t).$$

Use the formula for semimartingale exponential (a general case with jumps involving square bracket):

$$\mathcal{E}(M)(t) = e^{M(t)-M(0)-\frac{1}{2}[M,M](t)} \prod_{s\le t}(1 + \Delta M(s))e^{-\Delta M(s)+\frac{1}{2}(\Delta M(s))^2}.$$

Now, since

$$\begin{aligned}
[\bar{N}, \bar{N}] &= [N, N] = N,\\
[\bar{N}, B] &= [N, B] = 0,\\
\Delta M(s) = \Delta(aN(s)) &= a\Delta N(s) = a1_{s \text{ jump}},
\end{aligned}$$

we have

$$\begin{aligned}
&[M, M] = [a\bar{N} + \sigma B, a\bar{N} + \sigma B] = a^2[\bar{N}, \bar{N}] + \sigma^2[B, B] = a^2N + \sigma^2 t,\\
&\prod_{s\le t}(1 + \Delta M(s))e^{-\Delta M(s)+\frac{1}{2}(\Delta M(s))^2} = (1 + a)^{N(t)}e^{-aN(t)+\frac{1}{2}a^2N(t)}
\end{aligned}$$

and

$$\begin{aligned}
\mathcal{E}(M)(t) &= e^{\sigma B(t)+aN(t)-at-\frac{1}{2}(a^2N(t)+\sigma^2 t)-aN(t)+\frac{1}{2}a^2N(t)}(1 + a)^{N(t)}\\
&= e^{\sigma B(t)+N(t)\log(1+a)-(\frac{1}{2}\sigma^2+a)t}\\
&= e^{\sigma B(t)-\frac{1}{2}\sigma^2 t}e^{N(t)\log(1+a)-at}.
\end{aligned}$$

Since $e^{\sigma B(t)-\frac{1}{2}\sigma^2 t} = \mathcal{E}(\sigma B)(t)$ and $e^{N(t)\log(1+a)-at} = \mathcal{E}(a\bar{N})(t)$, both martingales independent, $X(t)$ is a martingale.

Calculations can be skipped by using Theorem 8.13, since $[a\bar{N}, \sigma B] = 0$,

$$X(t) = \mathcal{E}(a\bar{N} + \sigma B)(t) = \mathcal{E}(a\bar{N})(t)\mathcal{E}(\sigma B)(t).$$

Problem 8.15: Let M_t be a martingale, $U_t = \mathcal{E}(M)_t$ its martingale exponential, and τ a stopping time. Show that $U_{t\wedge\tau}$ is a martingale exponential of $M_{t\wedge\tau}$.

Solution. It is easy to see by a direct verification when $t \le \tau$ and $t > \tau$ that

$$U_{t\wedge\tau} = 1 + \int_0^{t\wedge\tau} U_s dM_s = 1 + \int_0^t U_s 1_{s\le\tau} dM_s.$$

Since under the integral $s \le \tau$ nothing will change if we replace s by $s \wedge \tau$ in U_s,

$$\int_0^t U_s 1_{s\le\tau} dM_s = \int_0^t U_{s\wedge\tau} 1_{s\le\tau} dM_s.$$

Hence,

$$U_{t\wedge\tau} = 1 + \int_0^t U_{s\wedge\tau} 1_{s\le\tau} dM_s = 1 + \int_0^t U_{s\wedge\tau} d\tilde{M}_s,$$

where $d\tilde{M}_s = 1_{s\le\tau} dM_s$. This shows that $U_{t\wedge\tau}$ is a martingale exponential of $\tilde{M}_t = \int_0^t 1_{s\le\tau} dM_s$. But when $\tau \ge t$, then $\tilde{M}_t = M_t$, and when $\tau < t$, then $\tilde{M}_t = M_\tau$. Therefore, $\tilde{M}_t = M_{t\wedge\tau}$. Finally, since a stopped martingale $M_{t\wedge\tau}$ is a martingale, $U_{t\wedge\tau}$ is a martingale exponential.

Problem 8.16: Find the predictable representation for $Y = B^5(1)$.

Solution. Let

$$M(t) = \mathrm{E}(Y|\mathcal{F}_t) = \mathrm{E}(B^5(1)|\mathcal{F}_t).$$

Then, $M(t)$ is a (Doob–Lévy) martingale (Theorem 2.35), and we aim to write $M(t) = \int_0^t H(s)dB(s)$ for some predictable process H. To find conditional expectation, we separate the $\mathcal{F}_t$-measurable part:

$$\begin{aligned} M(t) &= \mathrm{E}(((B(1) - B(t)) + B(t))^5|\mathcal{F}_t) \\ &= \mathrm{E}((B(1) - B(t))^5 + 5B(t)(B(1) - B(t))^4 \\ &\quad + 10B^2(t)(B(1) - B(t))^3 + 10B^3(t)(B(1) - B(t))^2 \\ &\quad + 5B^4(t)(B(1) - B(t)) + B^5(t)|\mathcal{F}_t) \\ &= 0 + 5B(t)(3(1-t)^2) + 0 + 10B^3(t)(1-t) + 0 + B^5(t) \\ &= B^5(t) + 10B^3(t)(1-t) + 15B(t)(1-t)^2. \end{aligned}$$

Having obtained the expression for $M(t)$ as a function $f(B(t), t)$ of $B(t)$ and t, with $f(x,t) = x^5 + 10x^3(1-t) + 15x(1-t)^2$, it remains to write it as an Itô integral. By Itô's formula,

$$\begin{aligned} dM(t) &= \frac{\partial f}{\partial x}(B(t),t)dB(t) + \frac{\partial f}{\partial t}(B(t),t)dt + \frac{1}{2}\frac{\partial^2 f}{\partial x^2}(B(t),t)dt \\ &= \big(5B^4(t) + 30(1-t)B^2(t) + 15(1-t)^2\big)dB(t). \end{aligned}$$

Thus, $M(t) = \int_0^t H(s)dB(s)$ with

$$H(s) = 5B^4(s) + 30(1-s)B^2(s) + 15(1-s)^2.$$

Hence, $Y = B^5(1) = M(1) = \int_0^1 H(s)dB(s)$.

Problem 8.17: Find the predictable representation for the martingale $e^{B(t)-t/2}$.

Solution. Since $M(t) = e^{B(t)-t/2} = f(B(t),t)$ with $f(x,t) = e^{x-t/2}$, apply Itô's formula to have

$$\begin{aligned} d(e^{B(t)-t/2}) &= e^{B(t)}d(e^{-t/2}) + e^{-t/2}d(e^{B(t)}) + d(e^{B(t)})d(e^{-t/2}) \\ &= e^{B(t)}\left(-\frac{1}{2}e^{-t/2}dt\right) + e^{-t/2}\left(e^{B(t)}dB(t) + \frac{1}{2}e^{B(t)}dt\right) + 0 \\ &= e^{B(t)-t/2}dB(t). \end{aligned}$$

Thus, $e^{B(t)-t/2} = 1 + \int_0^t e^{B(s)-s/2}dB(s)$, and $H(s) = e^{B(s)-s/2}$.

Note that one can skip calculations by noting that $M(t)$ is the stochastic exponential of $B(t)$, and as such it satisfies $M(t) = 1+\int_0^t M(s)dB(s)$. Hence, $H(s) = M(s)$.

Problem 8.18: Let $Y = \int_0^1 \text{sign}(B(s))dB(s)$. Show that there is no deterministic function $H(s)$, such that $Y = \int_0^1 H(s)dB(s)$. Show that Y has a normal distribution.

This is a counter-example to Theorem 8.35. (Theorem 8.35 states that for $Y \in \mathcal{F}_1$ with $\mathrm{E}|Y| < \infty$, there is a predictable process $H(t)$ such that $Y = \mathrm{E}Y + \int_0^1 H(t)dB(t)$. Moreover, if Y and B are jointly Gaussian, then $H(t)$ is deterministic, i.e. does not depend on $B(t)$.) This example demonstrates that the assumption that Y, B are jointly Gaussian in Theorem 8.35 is essential, and Y being Gaussian alone is not enough.

Solution. We show the assertion by contradiction. Suppose that there is a deterministic function $H(s)$ such that $Y = \int_0^1 H(s)dB(s)$.

Then, $\int_0^1 (\text{sign}(B(s)) - H(s))dB(s) = 0$. Since a constant has zero variance, $\text{E} \int_0^1 (\text{sign}(B(s)) - H(s))^2 ds = 0$. This implies after interchanging integrals by Fubini's theorem that (for almost all s)

$$\text{E}\big((\text{sign}(B(s)) - H(s))^2\big) = 0.$$

Using that $\text{E}(\text{sign}(B(s)) = 0$ and the assumption that $H(s)$ is deterministic, we obtain

$$\text{E}\big((\text{sign}(B(s)) - H(s))^2\big) = 1 + H^2(s) = 0.$$

This is a contradiction. Hence, no such H exists.

To see that Y has a normal distribution, let $M(t) = \int_0^t \text{sign}(B(s)) dB(s)$, so that $Y = M(1)$. Then, $M(t)$ is a continuous martingale with $[M, M](t) = \int_0^t (\text{sign}(B(s)))^2 ds = \int_0^t 1 ds = t$, hence it is a Brownian motion. Thus, $Y = M(1)$ has distribution $N(0, 1)$.

Problem 8.19: Let W_t and B_t be two independent Brownian motions, and $M_t = \int_0^t W_s dB_s$. Show that the martingale M_t does not have the predictable representation property.

Solution. If it did, then any martingale X_t can be written as the stochastic integral $X_t = \int_0^t H_s dM_s$ for some predictable process H. Then,

$$[X, M]_t = \left[\int_0^t H_s dM_s, \int_0^t 1 dM_s\right] = \int_0^t H_s d[M, M]_s.$$

Hence,

$$H_t = \frac{d[X, M]_t}{d[M, M]_t}.$$

But if we take $X_t = W_t$, or any stochastic integral of W, then we have

$$[W, M]_t = \left[\int_0^t dW_s, \int_0^t H_s W_s dB_s\right] = \int_0^t H_s W_s d[W, B]_s = 0$$

because $[W, B]_t = 0$. This implies $H_t = 0$, which is a contradiction.

Problem 8.20: Find the quadratic variation of $|B(t)|$.

Solution. By Tanaka's formula (Theorem 8.9), we have

$$|B(t)| = \int_0^t \text{sign}(B(s))dB(s) + L^0(t),$$

where $L^0(t)$ is the local time of $B(t)$ at 0. $L^0(t)$ is continuous and non-decreasing, thus it is of finite variation. Therefore, it has zero quadratic covariation with any other semimartingale (Corollary 8.5). Hence, we obtain

$$\begin{aligned}[|B|,|B|](t) &= \left[\int_0^{\cdot} \operatorname{sign}(B(s))dB(s), \int_0^{\cdot} \operatorname{sign}(B(s))dB(s)\right](t)\\ &= \int_0^t (\operatorname{sign}(B(s)))^2 ds = \int_0^t ds = t.\end{aligned}$$

Problem 8.21 (Stochastic logarithm): Let U be a semimartingale such that $U(t)$ and $U(t-)$ are never zero. Show that there exists a unique semimartingale X with $X(0)=0$, ($X=\mathcal{L}(U)$) such that $dX(t)=\frac{dU(t)}{U(t-)}$, and

$$\begin{aligned}X(t) &= \ln\left|\frac{U(t)}{U(0)}\right| + \frac{1}{2}\int_0^t \frac{d\langle U^{cm},U^{cm}\rangle(s)}{U^2(s-)}\\ &\quad -\sum_{s\le t}\left(\ln\left|\frac{U(s)}{U(s-)}\right| + 1 - \frac{U(s)}{U(s-)}\right).\end{aligned}$$

Solution. Uniqueness of stochastic logarithm. If $U=\mathcal{E}(X)(t)=\mathcal{E}(Y)(t)$, then we must show that $X=Y$. But it follows immediately from the exponential equation $dU(t)=U(t-)dX(t)=U(t-)dY(t)$,

$$dX(t)=\frac{dU(t)}{U(t-)}, \quad \text{and} \quad dY(t)=\frac{dU(t)}{U(t-)}.$$

Let $f(x)=\ln|x|$. Applying Itô's formula for semimartingale (Equation (8.59*)) to f, we obtain

$$\begin{aligned}\ln|U(t)| &= \ln|U(0)| + \int_0^t \frac{1}{U(s-)}dU(s) - \frac{1}{2}\int_0^t \frac{1}{U^2(s-)}d\langle U^{cm},U^{cm}\rangle(s)\\ &\quad + \sum_{s\le t}\left(\ln|U(s)| - \ln|U(s-)| - \frac{U(s)-U(s-)}{U(s-)}\right).\end{aligned}$$

Therefore,

$$\begin{aligned}X(t) &= \int_0^t \frac{1}{U(s-)}dU(s)\\ &= \ln\left|\frac{U(t)}{U(0)}\right| + \frac{1}{2}\int_0^t \frac{d\langle U^{cm},U^{cm}\rangle(s)}{U^2(s-)}\\ &\quad -\sum_{s\le t}\left(\ln\left|\frac{U(s)}{U(s-)}\right| + 1 - \frac{U(s)}{U(s-)}\right).\end{aligned}$$

Problem 8.22: Let ξ be standard normal $N(0,1)$ random variable, and the random measure μ defined as $\mu(\omega, A) = I_A(\xi(\omega))$. Find the non-random measure ν such that for any bounded function h, $\mathrm{E}\int_{\mathbb{R}} h(x)\mu(dx) = \int_{\mathbb{R}} h(x)\nu(dx)$.

Solution. $\nu(A) = \mathrm{E}\mu(A) = \mathrm{E}I_A(\xi(\omega)) = \mathrm{P}(\xi \in A)$.

Indeed, $\int_{\mathbb{R}} h(x)\mu(dx) = h(\xi)$, and

$$\mathrm{E}\int_{\mathbb{R}} h(x)\mu(dx) = \mathrm{E}h(\xi) = \int_{\mathbb{R}} h(x)f(x)dx,$$

where $f(x)$ is the probability density of $N(0,1)$. Hence, $\nu(dx) = f(x)dx$, and $\nu(A) = \mathrm{P}(\xi \in A)$.

Problem 8.23: Let N_t be a Poisson process with rate λ. Give its representation as a random measure and then give its compensator.

Solution. $\mu((0,t] \times \Gamma) = \sum_{0<s\leq t} I_\Gamma(\Delta N_s)$. The compensator $\nu(dt, dx) = \lambda dt \delta_1(dx)$, where $\delta_1(\Gamma)$ is the Dirac measure at point 1.

We can prove it as follows. Clearly, ν is non-random, hence it is predictable. Next, consider for a continuous function $h(x)$:

$$\int_0^t \int_{\mathbb{R}} h(x)\mu(ds, dx) = \sum_{0<s\leq t} h(\Delta N_s) = N_t h(1)$$

because $\Delta N_s = 1$ when there is a jump, and up to time t there are N_t jumps. Further,

$$\int_0^t \int_{\mathbb{R}} h(x)\nu(ds, dx) = \int_0^t \int_{\mathbb{R}} h(x)\lambda ds \delta_1(dx) = \lambda t h(1).$$

Now, it is clear that

$$\int_0^t \int_{\mathbb{R}} h(x)\Big(\mu(ds, dx) - \nu(ds, dx)\Big) = (N_t - \lambda t)h(1),$$

which is a martingale.

Problem 8.24: Let N be a pure jump process with a compensator of its measure of jumps $\nu(dt, dx) = \lambda dt \delta_1(dx)$, where $\delta_1(\Gamma)$ is the Dirac measure at point 1. Show that N is a Poisson process with rate λ.

Solution. Since $\mu(\{t\}, \Gamma) = I_\Gamma(\Delta N_t)$, we have

$$\mathrm{E}I_\Gamma(\Delta N_t) = \mathrm{P}(\Delta N_t \in \Gamma) = \lambda dt \delta_1(\Gamma).$$

Hence, $\Delta N_t = 1$. Therefore, N_t equals to the number of jumps before t. Next,

$$\begin{aligned}
\mathrm{E}e^{iuN_t} &= \mathrm{E}\sum_{s\le t}(e^{iuN_s} - e^{iuN_{s-}}) = \sum_{\text{jump } s\le t} \mathrm{E}(e^{iu(N_{s-}+\Delta N_s)} - e^{iuN_{s-}}) \\
&= \sum_{\text{jump } s\le t} \mathrm{E}(e^{iu(N_{s-}+1)} - e^{iuN_{s-}}) = \sum_{\text{jump } s\le t} \mathrm{E}e^{iuN_{s-}}(e^{iu} - 1) \\
&= \mathrm{E}\int_0^t \int_{\mathbb{R}} (e^{iu(N_{s-}+x)} - e^{iuN_{s-}})\nu(ds, dx) \\
&= \mathrm{E}\int_0^t \int_{\mathbb{R}} (e^{iu(N_{s-}+x)} - e^{iuN_{s-}})\lambda ds\delta_1(dx) \\
&= \lambda(e^{iu} - 1)\int_0^t \mathrm{E}e^{iuN_{s-}}ds = \lambda(e^{iu} - 1)\int_0^t \mathrm{E}e^{iuN_s}ds,
\end{aligned}$$

where the last equality is due to integration with respect to ds. This gives a simple differential equation for $\mathrm{E}e^{iuN_t}$, and solving it we have

$$\mathrm{E}e^{iuN_t} = e^{\lambda t(e^{iu}-1)}.$$

But this is the characteristic function of the Poisson distribution with parameter λt, hence N_t has Poisson (λt) distribution.

Consider next the number of jumps between t_1 and t_2 given by

$$N((t_1, t_2]) := N(t_2) - N(t_1) = \sum_{t_1 < s \le t_2} = \int_{t_1}^{t_2} x\mu(ds, dx).$$

Repeating the argument above, we obtain that the characteristic function of the number of jumps in $(t_1, t_2]$ is given by $e^{\lambda(t_2-t_1)(e^{iu}-1)}$. This implies that $N((t_1, t_2])$ has Poisson $\lambda(t_2 - t_1)$ distribution; moreover,

$$e^{\lambda(t_2-t_1)(e^{iu}-1)}e^{\lambda t_1(e^{iu}-1)} = e^{\lambda t_2(e^{iu}-1)},$$

which means that the number of jumps over non-overlapping time intervals are independent. This verifies the defining properties of the Poisson process.

Problem 8.25: Let X be a semimartingale consisting only of jumps, with the measure of jumps $\mu(dt, dx)$. Show that its compensator ν determines μ.

Solution. Since X consists only of jumps, its value at any t is the sum of jumps up to t, $X_t = \sum_{s\leq t} \Delta X_s$. But it is also written as an integral of the measure of jumps, $X_t = \sum_{s\leq t} \Delta X_s = \int_0^t \int_{\mathbb{R}} x\mu(ds, dx)$.

Let t be fixed, and consider the characteristic function of X_t, $\mathrm{E}e^{iuX_t}$. Then, we have with $X_0 = 0$,

$$\begin{aligned}\mathrm{E}e^{iuX_t} &= \mathrm{E}\sum_{s\leq t}(e^{iuX_s} - e^{iuX_{s-}})\\ &= \mathrm{E}(e^{iu(X_{s-}+\Delta X_s)} - e^{iuX_{s-}})\\ &= \mathrm{E}\int_0^t \int_{\mathbb{R}} (e^{iu(X_{s-}+x)} - e^{iuX_{s-}})\mu(ds, dx)\\ &= \mathrm{E}\int_0^t \int_{\mathbb{R}} (e^{iu(X_{s-}+x)} - e^{iuX_{s-}})\nu(ds, dx),\end{aligned}$$

and integrate the function e^{iux}, with $u \in \mathbb{R}$. Then,

$$\mathrm{E}\int_0^t \int_{\mathbb{R}} e^{iux}\mu(ds, dx) = \mathrm{E}\int_0^t \int_{\mathbb{R}} e^{iux}\nu(ds, dx).$$

Problem 8.26: Let U_t be a locally square-integrable semimartingale, such that $U_t = U_0 + A_t + M_t$, where A_t is a predictable process of locally finite variation and M_t is a locally square-integrable local martingale, $A_0 = M_0 = 0$. Let $U_t^2 = U_0^2 + B_t + N_t$, where B_t is a predictable process and N_t is a local martingale. Show that

$$\langle M, M\rangle_t = B_t - 2\int_0^t U_{s-}dA_s - \sum_{s\leq t}(A_s - A_{s-})^2.$$

Solution. By the definition of the quadratic variation process (or integration by parts),

$$\begin{aligned}U_t^2 &= U_0^2 + 2\int_0^t U_{s-}dU_s + [U, U]_t\\ &= U_0^2 + 2\int_0^t U_{s-}dA_s + 2\int_0^t U_{s-}dM_s + [U, U]_t.\end{aligned}$$

Using the representation for U_t^2, we obtain that $[U,U]_t - B_t + 2\int_0^t U_{s-}dA_s$ is a local martingale. But $[U,U]_t = [A,A]_t + [M,M]_t$. Hence, $[M,M]_t = [U,U]_t - [A,A]_t$. Since $[A,A]_t = \sum_{s\leq t}(A_s - A_{s-})^2$,

$$[M,M]_t = [U,U]_t - \sum_{s\leq t}(A_s - A_{s-})^2.$$

Since $\langle M,M\rangle_t$ is the unique predictable process such that $[M,M]_t - \langle M,M\rangle_t$ is a local martingale, the result follows.

Chapter 9

Pure Jump Processes

A pure jump point process X is a process that changes only by jumps, jumping by size ξ_n at times T_n, $n = 1, 2, \ldots$. For such a process, $X(t)$ can be expressed as

$$X(t) = X(0) + \sum_{n=1}^{\infty} \xi_n I(T_n \leq t).$$

If $\xi_n = 1$, for all n, inter-arrival times $\tau_n = T_n - T_{n-1}$ are independent and identically distributed, and $X_0 = 0$, then the process X, often denoted by N, is called a renewal process. If moreover, the distribution of τ_1 is exponential, then X is a Poisson process.

The process X has finitely many jumps on finite time intervals if $\lim_{n\to\infty} T_n = \infty$. On the other hand, if $\lim_{n\to\infty} T_n < \infty$, then the process jumps infinitely many times in a finite time interval. This phenomenon is called an explosion.

Analysis of pure jump processes is often done by using their compensators. Firstly, the compensator determines its jump process uniquely. Secondly, the following important identity holds. For any bounded continuous function f,

$$\mathrm{E}\left(\int_0^t f(X_{s-})dX_s\right) = \mathrm{E}\left(\int_0^t f(X_{s-})dA_s\right),$$

where A is the compensator of the pure jump process X. It also allows to write the decomposition $X_t = A_t + M_t$, where M_t is a local martingale. This decomposition is useful because it represents the process as a sum of the "dynamics" A and "noise" M.

An important class of pure jump processes are Markov jump processes. In such processes, the times between the jumps are exponential, with the parameter depending on the current state. For such processes, analysis is often done by using their generators.

Heuristically, Markov jump processes can be described as follows. If the process is in state x, then it stays there for an exponential length of time with mean $1/\lambda(x)$ (i.e. with parameter $\lambda(x)$), after which it jumps from x to a new state. The distribution of the jump is denoted by $\pi(x, dy)$ and the jump itself by $\xi(x)$.

$\lambda(x)$ is always non-negative. If for some x, $\lambda(x) = 0$, then once the process enters state x, it stays there forever, making the "holding time" in x infinite,

$$\lambda(x) = 0 \Rightarrow \mathbb{P}(\forall s > 0,\ X(t+s) = x | X(t) = x) = 1.$$

Such a state x is called absorbing.

If $\lambda(x) = \infty$, then the process leaves x instantaneously. Here, we assume that $\lambda(x) < +\infty$.

Denote the mean $m(x) = \mathrm{E}\xi(x)$. Then, the compensator of X is given by

$$A(t) = \int_0^t \lambda(X(s))m(X(s))ds.$$

$M(t) = X(t) - A(t)$ is a local martingale. Its predictable quadratic variation, the sharp bracket is given by

$$\langle M, M \rangle (t) = \int_0^t \lambda(X(s))v(X(s))ds,$$

where $v(x) = \mathrm{E}\xi^2(x)$, the second moment of the jump $\xi(x)$.

This approach allows to write stochastic equations for such processes and analyse them like we would, for example diffusions.

Itô's formula for $f(X(t)$ can be written by using moments of the jumps of $f(X)$, $m_f(x) = \mathrm{E}(f(x + \xi(x)) - f(x))$. The compensator of $f(X(t))$ is given by

$$A^f(t) = \int_0^t \lambda(X(s))m_f(X(s))ds.$$

Using the second moments of the jumps of $f(X)$ $v_f(x) = \mathrm{E}(f(x+\xi(x)) - f(x))^2$, the sharp bracket of the local martingale $M^f(t) = f(X(t)) - A^f(t)$ is given by

$$\langle M, M \rangle (t) = \int_0^t \lambda(X(s))v_f(X(s))ds.$$

Markov population models fit this and can also have another useful representation as sums or differences of subordinated Poisson processes. In population models, the jumps have a size plus or minus 1, corresponding to the birth or death of a member. For example, if in a population of size x, $x \in \mathbb{N}$, each individual lives for an exponentially distributed lifetime with parameter $\lambda(x)$.

In a birth–death process, when the population size is x, a new particle is born at the rate $b(x)$ and a particle dies at the rate $d(x)$. This means that the process stays at x for an exponential time with parameter $\lambda(x) = b(x) + d(x)$ and then jumps to $x+1$ with probability $b(x)/\lambda(x)$ and to $x-1$ with the complimentary probability $d(x)/\lambda(x)$. It is a Markov process with the generator

$$Lf(x) = b(x)f(x+1) + d(x)f(x-1) - (b(x) + d(x))f(x).$$

This is developed further in Chapter 12 devoted to biological applications.

One of the questions in Markov jump processes is the question of explosion, i.e. a process having infinitely many jumps in a finite time. In the context of processes on integers, it corresponds to reaching infinity in a finite time. This question is important because if there is no explosion, then the forward equations hold and their solution is unique. Furthermore, one can represent the process as a stochastic equation with a jump martingale. A number of questions is devoted to developing sharp sufficient conditions for explosion and non-explosion.

Let $X(t)$ be a jump Markov process taking values in $\mathbb{R}$ without absorbing states. A necessary and sufficient condition for non-explosion is given in terms of the random series where T_n is the time of the nth jump,

$$\sum_{n=0}^{\infty} \frac{1}{\lambda(X(T_n))} = \infty. \qquad (*)$$

This result is useful because of the observation that for the above series to converge, it is necessary that the term $1/\lambda(X(T_n))$ converges to zero. This in turn implies that for the explosion to occur, we must have that $\lambda(X(T_n)) \to$

∞, and assuming λ is bounded on finite intervals, that $X(T_n) \to \infty$. In other words, when λ is bounded on finite intervals, the explosion can only occur at infinity.

When there is an absorbing state, which is typically zero in population processes, then the explosion can take place on the set of divergence to infinity. In this case, the following result on the convergence sets of submartingales is useful for checking this condition as well as in many other instances.

The following is useful in analysis. Let S be a submartingale, that is, an integrable process satisfying $\mathrm{E}(S_n|\mathcal{F}_{n-1}) \geq S_{n-1}$ for all $n = 1, 2, \ldots$. We recall Doob decomposition. We can clearly write a telescoping sum,

$$S_n = S_0 + \sum_{j=1}^{n}(S_j - S_{j-1}) = S_0 + \sum_{j=1}^{n}(S_j - \mathrm{E}(S_j|\mathcal{F}_{j-1}))$$

$$+ \sum_{j=1}^{n}(\mathrm{E}(S_j|\mathcal{F}_{j-1})) - S_{j-1}) = M_n + A_n,$$

where $M_n = S_0 + \sum_{j=1}^{n}(S_j - \mathrm{E}(S_j|\mathcal{F}_{j-1}))$ is a martingale and $A_n = \sum_{j=1}^{n}(\mathrm{E}(S_j|\mathcal{F}_{j-1})) - S_{j-1})$ is an increasing process by the submartingale property. This is the Doob decomposition of a submartingale S_n into a martingale and an increasing predictable process, $S_n = M_n + A_n$.

We prove results on convergence sets of a submartingale, Problems 9.17–9.19:

(a) If $S \geq 0$ or $\sup_n \mathrm{E}S_n^- < \infty$, then $\{A_\infty < \infty\} \subseteq \{S_n \text{ converges }\} \subseteq \{\sup S_n < \infty\}$.
(b) If S has bounded jumps, $|\Delta S_n| < C$, then $\{S_n \text{ converges }\} = \{\sup S_n < \infty\} \subseteq \{A_\infty < \infty\}$.

We finish by recalling the following convention. As one can only sum over a countable set, $\sum_{s \leq t}$ and similar sums (resp. products) can only be used when the summand is zero (resp. one) except for a countable number of terms. In this case, the sum (resp. product) is understood to mean the sum (resp. product) over the countable set on non-zero (resp. non-one) values. For example, for a Poisson process $N(t)$, $\{s \leq t : \Delta N(s) \neq 0\}$ is finite and

$$\sum_{s \leq t} \Delta N(s) \text{ stands for } \sum_{\substack{s \leq t \\ \Delta N(s) \neq 0}} \Delta N(s) = N(t)$$

and

$$\prod_{s\le t}(1+\Delta N(s)) \text{ stands for } \prod_{\substack{s\le t\\ \Delta N(s)\neq 0}}(1+\Delta N(s)) = 2^{N(t)}.$$

The chapter is divided into three sections. In particular, the third section, stopping times and filtrations, attempts to explain the intricacies of the concepts of stopping times and filtrations. Even in the simplest case of a one-jump process (often referred to as a toy model), a number of filtrations can be defined as offering a wealth of properties. It is hoped that this section will offer an insight into this often misunderstood area of stochastic processes. As an example, Problem 9.29 shows how a non-negative T, although a stopping time with respect to the natural filtration of the process $J_t = 1_{T\le t}$, is not a stopping time for the filtration generated by $T \wedge t$. The differences between these filtrations highlight the need for the so-called "usual conditions" often assumed and not always properly explained. In another example (see Problem 9.44), it is shown that for non-negative T with distribution function F, $(1-F(t))^{-1}1_{T>t}$ is a martingale.

Problems

Stochastic Calculus for Poisson Processes

Problem 9.1: Let $U \ge 0$ and denote $h(x) = \int_0^x \frac{dF(s)}{1-F(s-)}$, where $F(x)$ is the distribution function of U. Assume for simplicity that $F(0)=0$. Show that for $a>0$, $\mathrm{E}\big(h(U\wedge a)\big) = \int_0^a dF(x)$.

Solution.

$$\begin{aligned}
\mathrm{E}\big(h(U\wedge a)\big) &= \mathrm{E}\left(\int_0^{U\wedge a} \frac{dF(s)}{1-F(s-)}\right) = \mathrm{E}\left(\int_0^a I(U\ge s)\frac{dF(s)}{1-F(s-)}\right)\\
&= \int_0^a \mathrm{E}\big(I(U\ge s)\big)\frac{dF(s)}{1-F(s-)} = \int_0^a \mathrm{P}(U\ge s)\frac{dF(s)}{\mathrm{P}(U\ge s)}\\
&= \int_0^a dF(s).
\end{aligned}$$

Problem 9.2: Assume that the renewal process $N(t)$ has exponential inter-arrival distribution. Show that its compensator is λt, hence $N(t)$ is a Poisson process.

Solution. Let T_n be the arrival time in the renewal process. Then, $T_{n+1} - T_n$ are exponential with rate λ, with distribution function $F(x) = 1 - e^{-\lambda x}$ for $x > 0$. Thus, by Corollary 9.8, the compensator of $N(t)$ is

$$A(t) = -\sum_{n=1}^{\infty} \log(1 - F(t \wedge T_n - t \wedge T_{n-1}))$$

$$= \lambda \sum_{n=1}^{\infty} (t \wedge T_n - t \wedge T_{n-1}).$$

Note that for any t, the above sum has only finitely many non-zero terms, the last being $t \wedge T_{N(t)+1} - t \wedge T_{N(t)} = t - T_{N(t)}$, because by definition of $N(t)$, $T_{N(t)} \le t < T_{N(t)+1}$. Therefore, the sum simplifies to

$$A(t) = \lambda \sum_{n=1}^{N(t)+1} (t \wedge T_n - t \wedge T_{n-1})$$

$$= \lambda \sum_{n=1}^{N(t)} (t \wedge T_n - t \wedge T_{n-1}) + \lambda(t - T_{N(t)}) = \lambda t.$$

Recall Theorem 9.10, which states that a point process with a continuous deterministic compensator $A(t)$ has independent Poisson increments with parameter $A(t) - A(s)$, $0 \le s < t$. Since $A(t) = \lambda t$, $N(t)$ is a Poisson process with rate λ.

Problem 9.3: Show that when the distribution of inter-arrival times in a renewal process is geometric, then $N(t)$ has a binomial distribution.

Solution. Let T_n be the time of the nth arrival ($T_0 = 0$) so that $S_n = T_n - T_{n-1}$ is geometric with parameter p:

$$\mathrm{P}(S_n = t) = p(1-p)^{t-1}, \quad t = 1, 2, \ldots.$$

As the arrival times are integer-valued, $\Delta N(t) = 0$ for any non-integer t. Furthermore, $\mathrm{P}(N(k) = k) = \mathrm{P}(T_k \le k) = p^k$. Here, events occur on the "dot" and $N(k) = k$ means that an arrival occurred at every instant (between 1 and k).

Now, for $k \le n$,

$$\begin{aligned} &\mathrm{P}(N(n+1)=k)\\ &\quad= \mathrm{P}(T_k \le n+1) - \mathrm{P}(T_{k+1} \le n+1)\\ &\quad= \mathrm{P}(T_k \le n) + \mathrm{P}(T_k = n+1) - \mathrm{P}(T_{k+1} \le n) - \mathrm{P}(T_{k+1} = n+1)\\ &\quad= \mathrm{P}(N(n)=k) + \mathrm{P}(T_k = n+1) - \mathrm{P}(T_{k+1} = n+1); \end{aligned}$$

that is,

$$\mathrm{P}(N(n+1)=k) = \mathrm{P}(N(n)=k) + \mathrm{P}(T_k = n+1) - \mathrm{P}(T_{k+1} = n+1).$$

We use induction on n to prove that $N(n)$ is binomial (n,p); i.e.

$$\mathrm{P}(N(n)=k) = \binom{n}{k} p^k (1-p)^{n-k}, \quad n = k, k+1, \ldots. \qquad (*)$$

First, observe that, as demonstrated above, $\mathrm{P}(N(k) = k) = p^k$, and therefore (*) holds true for $n = k$.

Next, we show that if it holds true for n, it must hold true for $n+1$. Recall that T_k, as a sum of k independent geometric p random variables, is Pascal with parameters k and p. Therefore,

$$\begin{aligned} &\mathrm{P}(N(n+1)=k)\\ &\quad= \mathrm{P}(N(n)=k) + \mathrm{P}(T_k = n+1) - \mathrm{P}(T_{k+1} = n+1)\\ &\quad= \binom{n}{k} p^k (1-p)^{n-k} + \binom{n}{k-1} p^k (1-p)^{n+1-k} - \binom{n}{k} p^{k+1} (1-p)^{n-k}\\ &\quad= \binom{n}{k} p^k (1-p)^{n-k}(1-p) + \binom{n}{k-1} p^k (1-p)^{n+1-k}\\ &\quad= \left[\binom{n}{k} + \binom{n}{k-1}\right] p^k (1-p)^{n+1-k} = \binom{n+1}{k} p^k (1-p)^{n+1-k}, \end{aligned}$$

which completes the proof that $N(n)$ is binomial (n,p). It immediately follows that $N(t)$, which equals $N(\lfloor t \rfloor)$ (arrivals only occur on the dot), is binomial $(\lfloor t \rfloor, p)$.

Problem 9.4:

(a) Obtain the compensator of a renewal process $N(t)$ for which the inter-arrival times are geometric with parameter p.
(b) Deduce the distribution of $N(t)$.

Solution.

(a) The inter-arrival times $T_{n+1} - T_n$ are geometric with parameter p, with distribution function $F(t) = 1 - (1-p)^{\lfloor t \rfloor}$ for $t \geq 0$, where $\lfloor t \rfloor$ denotes the integer part of t. By Theorem 9.7, the compensator of $N(t)$ is

$$A(t) = \sum_{k=0}^{\infty} \int_0^{t\wedge T_{k+1} - t\wedge T_k} \frac{1}{1 - F_k(s-)} dF_k(s)$$
$$= \sum_{k=0}^{\infty} \sum_{s=1}^{\lfloor t\wedge T_{k+1} - t\wedge T_k \rfloor} \frac{1}{(1-p)^{s-1}} p(1-p)^{s-1} = \sum_{k=0}^{\infty} \sum_{s=1}^{\lfloor t\wedge T_{k+1} - t\wedge T_k \rfloor} p$$
$$= p \sum_{k=0}^{\infty} \lfloor t \wedge T_{k+1} - t \wedge T_k \rfloor = p\lfloor t \rfloor.$$

Hence, the compensator is a pure jump function that jumps by p at integer points $t = 1, 2, \ldots$.

(b) For the distribution of $N(t)$, we compute its moment-generating function (or characteristic function). For $u \in \mathbb{R}$, note that

$$e^{u\Delta N(s)} - 1 = (e^u - 1)\Delta N(s).$$

This is because $\Delta N(s)$ can take only two values, 0 and 1: when $\Delta N(s) = 0$ both expressions are zero, and when $\Delta N(s) = 1$ both expressions are $e^u - 1$. Using this, we write the telescoping sum over the jumps of $N(t)$:

$$e^{uN(t)} - e^{uN(0)}$$
$$= \sum_{s \leq t} (e^{uN(s)} - e^{uN(s-)}) = \sum_{s \leq t} e^{uN(s-)} (e^{u\Delta N(s)} - 1)$$
$$= \sum_{s \leq t} e^{uN(s-)} (e^u - 1)\Delta N(s) = (e^u - 1) \int_0^t e^{uN(s-)} dN(s).$$

Next, we compensate the integral to make a martingale,

$$e^{uN(t)} - 1 = (e^u - 1)\int_0^t e^{uN(s-)} d(N(s) - A(s)) + (e^u - 1)\int_0^t e^{uN(s-)} dA(s).$$

$N(s) - A(s)$ is a martingale with $N(0) = 0$. Hence, taking expectation, we have

$$\mathrm{E}(e^{uN(t)}) = 1 + (e^u - 1)\int_0^t \mathrm{E}(e^{uN(s-)}) dA(s).$$

Now, let $U(t) = \mathrm{E}(e^{uN(t)})$, then we obtain the equation

$$U(t) = 1 + \int_0^t U(s-)(e^u - 1) dA(s).$$

This equation is of the exponential type and letting $X(t) = (e^u - 1)$ $A(t)$, the solution (see (9.5)) is

$$U(t) = \mathcal{E}(X)(t) = e^{X(t)-X(0)} \prod_{s \le t} (1 + \Delta X(s)) e^{-\Delta X(s)} = e^{X^c(t)} \prod_{s \le t} (1 + \Delta X(s)),$$

where $X^c(t)$ is the continuous part of $X(t)$. Note that in this case, X is deterministic, $X^c = 0$ and $\Delta X(s) = (e^u - 1)\Delta A(s) = (e^u - 1)p$ for s integer. It follows that

$$\mathrm{E}(e^{uN(t)}) = \prod_{s \le t} (1 + (e^u - 1)\Delta A(s)) = \prod_{k=1}^{\lfloor t \rfloor} (1 - p + pe^u) = (1 - p + pe^u)^{\lfloor t \rfloor},$$

which is the moment-generation function of the binomial distribution with parameters $\lfloor t \rfloor$ and p.

Problem 9.5: (Itô's formula for a Poisson process.) Let N_t be a Poisson process and f be a differentiable function. Give Itô's formula for $f(N_t)$.

Solution. Using Itô's formula for semimartingales, we obtain, since the continuous component of N, $N^c = 0$,

$$f(N_t) = f(0) + \int_0^t f'(N_{s-})dN_s + \sum_{s\leq t} \big[f(N_s) - f(N_{s-}) - f'(N_{s-})\Delta N_s\big].$$

Since $\sum_{s\leq t} f'(N_{s-})\Delta N_s = \int_0^t f'(N_{s-})dN_s$, we have

$$f(N_t) = f(0) + \sum_{s\leq t} \big(f(N_s) - f(N_{s-})\big).$$

Since $N_s = N_{s-} + \Delta N_s$ and $\Delta N_s = 0$ or 1, we have

$$f(N_s) - f(N_{s-}) = f(N_{s-} + \Delta N_s) - f(N_{s-}) = 0 \quad \text{if } \Delta N_s = 0 \quad \text{and}$$
$$f(N_s) - f(N_{s-}) = f(N_{s-} + 1) - f(N_{s-}) = 0 \quad \text{if } \Delta N_s = 1.$$

Hence, we can write it as

$$\begin{aligned} f(N_t) &= f(0) + \sum_{s\leq t} \big(f(N_{s-} + 1) - f(N_{s-})\big)\Delta N_s \\ &= f(0) + \int_0^t \big(f(N_{s-} + 1) - f(N_{s-})\big)dN_s. \end{aligned}$$

Of course, we can start directly with a telescoping sum without appealing to Itô's formula for semimartingales. Letting $(T_n)_n$ be the sequence of arrival times, $\Delta N_t = 1$ if and only if $t \in \{T_n, n \in \mathbb{N}\}$. It follows that $N_t = \max\{n : T_n \leq t\}$ and

$$\begin{aligned} f(N_t) &= f(0) + \sum_{n\geq 1: T_n \leq t} \big(f(n) - f(n-1)\big) \\ &= f(0) + \sum_{s\leq t} \big(f(N_s) - f(N_{s-})\big) \\ &= f(0) + \int_0^t \big(f(N_{s-} + 1) - f(N_{s-})\big)dN_s. \end{aligned}$$

Problem 9.6: Let N be a Poisson process. Give Itô's formula for N_t^2.

Solution. Using the previous question with $f(x) = x^2$, we obtain

$$N_t^2 = \int_0^t ((N_{s-}+1)^2 - (N_{s-})^2)dN_s = \int_0^t (2N_{s-}+1)dN_s$$
$$= 2\int_0^t N_{s-}dN_s + N_t.$$

Problem 9.7 (Telegraphic signal process): Let N_t be a Poisson process with parameter λ and X_0 be a random variable independent of N taking values 0 or 1. Consider the following stochastic differential equation:

$$X_t = X_0 + \int_0^t (1 - 2X_{s-})dN_s.$$

Show that for any t, X_t can take only two values, 0 or 1. Find and solve the equation for $p(t) = \mathrm{P}(X_t = 1)$.

Solution. $\Delta X_t = \Delta \int_0^t (1-2X_{s-})dN_s = (1-2X_{t-})\Delta N_t$. Hence, if $\Delta N_t = 1$ and $X_{t-} = a$, then $\Delta X_t = 1 - 2a$ and $X_t = 1 - a$. Starting from $X_0 \in \{0, 1\}$ and evolving by jumps (only a finite number of jumps can occur on any finite interval), the process at the time any t must satisfy $X_t \in \{0, 1\}$. Hence, $\mathrm{P}(X_t = 1) = \mathrm{E}X_t$. Next, taking expectations in the defining SDE, we get

$$\mathrm{E}X_t = \mathrm{E}X_0 + \mathrm{E}\left(\int_0^t (1-2X_{s-})dN_s\right).$$

Since $M_t = N_t - \lambda t$ is a martingale, so is $\int_0^t (1 - 2X_{s-})dM_s$. In other words, the expectation of the integral with respect to N is the same as with respect to its compensator,

$$\mathrm{E}\left(\int_0^t X_{s-}dN_s\right) = \mathrm{E}\left(\int_0^t X_{s-}\lambda ds\right) = \int_0^t \mathrm{E}X_s \lambda ds.$$

We obtain, after rearrangement, that $p(t) = \mathrm{P}(X_t = 1)$ solves the integral equation

$$p(t) = p(0) + \lambda\int_0^t (1 - 2p(s))ds.$$

It follows that $p(t) = \mathrm{P}(X_t = 1)$ solves the differential equation

$$\frac{dp(t)}{dt} = 1 - 2p(t),$$

subject to the initial condition $p(0)$. This is the Fokker–Planck–Kolmogorov equation for the telegraphic signal process.

The solution of this equation is easily obtained:

$$p(t) = \frac{1}{2}\Big(1 - (1 - 2p(0))e^{-2t}\Big).$$

Problem 9.8: Let $X(t)$ be a Markov pure jump process. Suppose that the jumps $\xi(x)$ have finite moments of order k. Let $m_k(x) = \mathrm{E}\big(\xi(x)^k\big)$ and assume for all $\ell = 1, \dots, k$, $\lambda(x)\mathrm{E}\big(|\xi(x)|^\ell\big) \le C|x|^\ell$. Show that

$$\mathrm{E}\big(X^k(t)\big) = \mathrm{E}X^k(0) + \sum_{\ell=0}^{k-1}\binom{k}{\ell}\int_0^t \mathrm{E}\big(\lambda(X(s))m_{k-\ell}(X(s))X^\ell(s)\big)ds.$$

Solution. Apply Dynkin's formula to $f(x) = x^k$:

$$\mathrm{E}f(X(t)) = \mathrm{E}f(X(0)) + \mathrm{E}\int_0^t Lf(X(s))ds,$$

where the operator L here takes the form

$$\begin{aligned} Lf(x) &= \lambda(x)\mathrm{E}\big(f(x+\xi(x)) - f(x)\big) = \lambda(x)\mathrm{E}\big((x+\xi(x))^k - x^k\big) \\ &= \lambda(x)\mathrm{E}\left(\sum_{\ell=1}^{k}\binom{k}{\ell}x^\ell(\xi(x))^{k-\ell} - x^k\right) \\ &= \lambda(x)\mathrm{E}\left(\sum_{\ell=1}^{k-1}\binom{k}{\ell}x^\ell(\xi(x))^{k-\ell}\right) = \lambda(x)\sum_{\ell=1}^{k-1}\binom{k}{\ell}x^\ell m_{k-\ell}(x). \end{aligned}$$

Then, change the order of integration and summation to complete the proof.

Problem 9.9: Let τ_n, $n = 1, 2, \dots$, be a sequence of independent exponentially distributed random variables with parameter λ_n. Show that $\sum_{n=1}^\infty \tau_n < +\infty$ a.s. if and only if $\sum_{n=1}^\infty 1/\lambda_n < +\infty$.

Solution. Since $\mathrm{E}\tau_n = 1/\lambda_n$, the convergence of the series $\sum_{n=1}^\infty 1/\lambda_n$ implies, by the Fubini theorem, that

$$\mathrm{E}\left(\sum_{n=1}^\infty \tau_n\right) = \sum_{n=1}^\infty \frac{1}{\lambda_n} < +\infty.$$

Hence, the series $\sum_{n=1}^{\infty} \tau_n$ converges a.s.

The other direction is more challenging and follows from considering the Laplace transform of the partial sums. For $u > 0$,

$$\mathrm{E}\big(e^{-u\sum_{n=1}^{N}\tau_n}\big) = \mathrm{E}\prod_{n=1}^{N} e^{-u\tau_n} = \prod_{n=1}^{N} \mathrm{E}e^{-u\tau_n} = \prod_{n=1}^{N} \frac{1}{1+u/\lambda_n}$$
$$= \prod_{n=1}^{N}\left(1 - \frac{u}{u+\lambda_n}\right).$$

Here, we used the fact that the Laplace transform of a sum of (a finite number of) independent variables is the product of the Laplace transforms, as well as the form of the transform for exponential distribution. Since $\mathrm{P}(\tau_n \geq 0) = 1$, an application of the monotone convergence theorem shows that the extended random variable $\sum_n \tau_n$ is such that, for any $u > 0$,

$$\mathrm{E}\big(e^{-u\sum_n \tau_n}\big) = \prod_{n=1}^{\infty}\left(1 - \frac{u}{u+\lambda_n}\right).$$

Next, we use the following criterion for the convergence or divergence of products and sums for a sequence $a_n \in (0,1)$ that approaches 0:

$$\prod_{n=1}^{\infty}(1-a_n) = 0 \Leftrightarrow \sum_{n=1}^{\infty} a_n = +\infty.$$

Indeed, since $\ln(1-x) \sim -x$ in the neighbourhood of 0, the asymptotic behaviour (convergence or divergence) of $\ln\prod_{k=1}^{n}(1-a_k) = \sum_{k=1}^{n}\ln(1-a_k)$ is that of $-\sum_{k=1}^{n} a_k$.

Suppose $\lim_n \lambda_n = +\infty$. If $\sum_{n=1}^{\infty} 1/\lambda_n = +\infty$, then $\sum_{n=1}^{\infty} u/(u+\lambda_n) = +\infty$ and $\prod_{n=1}^{\infty}(1 - u/(u+\lambda_n)) = 0$. We conclude that $\mathrm{E}e^{-u\sum_n \tau_n} = 0$ or equivalently that $\sum_{n=1}^{\infty}\tau_n = +\infty$ (a.s.).

Finally, if λ_n does not converge to $+\infty$ so that there exists a subsequence $(\lambda_{n_k})_k$ that is bounded by say λ, then

$$0 \leq \mathrm{E}\big(e^{-u\sum_n \tau_n}\big) \leq \mathrm{E}\big(e^{-u\sum_k \tau_{n_k}}\big) = \prod_{k=1}^{\infty}\left(1 - \frac{u}{u+\lambda_{n_k}}\right)$$
$$\leq \prod_{k=1}^{\infty}\left(1 - \frac{u}{u+\lambda}\right) = 0,$$

and again we conclude that $\sum_{n=1}^{\infty}\tau_n = +\infty$ (a.s.).

Problem 9.10: Let $(N_t)_{t\geq 0}$ be a Poisson process with rate λ. N_t is a Markov process in continuous time. Obtain its (infinitesimal) generator.

Solution. We do direct calculations using the definition: for a bounded function f and $t > 0$,

$$Lf(x) = \lim_{\delta\downarrow 0} \frac{1}{\delta} \mathrm{E}[f(N_{t+\delta}) - f(N_t)|N_t = x].$$

Since the Poisson process has independent Poisson increments, we have

$$N_{t+\delta} = N_t + (N_{t+\delta} - N_t),$$

where the $N_{t+\delta} - N_t$ has a Poisson $\lambda\delta$ distribution and is independent of N_t. Hence,

$$\begin{aligned}
&\mathrm{E}[f(N_{t+\delta}) - f(N_t)|N_t = x] \\
&\quad = \mathrm{E}[f(x + N_{t+\delta} - N_t)] - f(x) = \sum_{n=0}^{\infty} f(x+n)e^{-\lambda\delta}\frac{(\lambda\delta)^n}{n!} - f(x).
\end{aligned}$$

Expanding the sum, we obtain

$$\begin{aligned}
\mathrm{E}[f(N_{t+\delta}) - f(N_t)|N_t = x] &= f(x)(e^{-\lambda\delta} - 1) + \lambda f(x+1)e^{-\lambda\delta}\delta \\
&\quad + \left(\sum_{n=2}^{\infty} f(x+n)e^{-\lambda\delta}\frac{\lambda^n\delta^{n-2}}{n!}\right)\delta^2.
\end{aligned}$$

Since, for $\delta < 1$,

$$\left|\sum_{n=2}^{\infty} f(x+n)e^{-\lambda\delta}\frac{\lambda^n\delta^{n-2}}{n!}\right| \leq \sup_x |f(x)| \sum_{n=2}^{\infty} e^{-\lambda\delta}\frac{\lambda^n}{n!} \leq \sup_x |f(x)|,$$

dividing by δ and taking limits, we see that the first term converges to $-\lambda f(x)$, the second term converges to $\lambda f(x+1)$, and the third term converges to zero. We conclude that

$$Lf(x) = \lambda\big(f(x+1) - f(x)\big).$$

Recall that the (infinitesimal) generator of a Markov jump process is

$$Lf(x) = \lambda(x)\big(\mathrm{E}[f(x + \xi(x)] - f(x)\big),$$

see Section 9.7 and Equation (9.45*) in the text. The Poisson process is a Markov jump process with a constant rate of jumps and jumps of size 1. Replacing $\lambda(x)$ with λ and $\xi(x)$ with 1, we recover the generator of $(N_t)_{t\geq 0}$.

Problem 9.11: For $t > 0$, obtain the joint law of T_1 and T_{N_t+1}; that is, obtain $\mathrm{P}(T_1 \leq u, T_{N_t+1} \leq v)$ for all u and v.

Solution. $\mathrm{P}(T_1 \leq u, T_{N_t+1} \leq v) = \mathrm{P}(T_1 \leq u, T_1 \leq v, N_t = 0) + \mathrm{P}(T_1 \leq u, T_2 \leq v, N_t = 1) + \sum_{n=2}^{\infty} \mathrm{P}(T_1 \leq u, T_{n+1} \leq v, N_t = n)$. We compute each term on the right separately.

First, $\mathrm{P}(T_1 \leq u, T_1 \leq v, N_t = 0) = \mathrm{P}(T_1 \leq u \wedge v, T_1 > t) = (e^{-\lambda t} - e^{-\lambda(u\wedge v)})1_{u\wedge v>t}$.

Then,

$$
\begin{aligned}
\mathrm{P}(T_1 \leq u, T_2 \leq v, N_t = 1) &= \mathrm{P}(T_1 \leq u, T_2 \leq v, T_1 \leq t < T_2)\\
&= \mathrm{P}(T_1 \leq t \wedge u, t < T_2 \leq v)\\
&= \int_0^{t\wedge u} \mathrm{P}(t - s < T_1 \leq v - s)\lambda e^{-\lambda s} ds 1_{v>t}\\
&= \int_0^{t\wedge u} (e^{-\lambda(t-s)} - e^{-\lambda(v-s)})\lambda e^{-\lambda s} ds 1_{v>t}\\
&= (e^{-\lambda t} - e^{-\lambda v}) \int_0^{t\wedge u} \lambda ds = (e^{-\lambda t} - e^{-\lambda v})\lambda(t\wedge u)1_{v>t}.
\end{aligned}
$$

Finally, for $n \geq 2$,

$$
\begin{aligned}
&\mathrm{P}(T_1 \leq u, T_{n+1} \leq v, N_t = n)\\
&= \mathrm{P}(T_1 \leq u, T_{n+1} \leq v, T_n \leq t < T_{n+1})\\
&= \int_0^{t\wedge u} \mathrm{P}(T_{n-1} \leq t - s < T_n \leq v - s)\lambda e^{-\lambda s} ds 1_{v>t}\\
&= \int_0^{t\wedge u} \int_0^{t-s} \mathrm{P}(t - s - r < T_1 \leq v - s - r)\\
&\quad \times \frac{\lambda^{n-1}}{(n-2)!} r^{n-2} e^{-\lambda r} dr \lambda e^{-\lambda s} ds 1_{v>t}\\
&= \int_0^{t\wedge u} \int_0^{t-s} \left(e^{-\lambda(t-s-r)} - e^{-\lambda(v-s-r)}\right)\\
&\quad \times \frac{\lambda^{n-1}}{(n-2)!} r^{n-2} e^{-\lambda r} dr \lambda e^{-\lambda s} ds 1_{v>t}
\end{aligned}
$$

$$= \left(e^{-\lambda t} - e^{-\lambda v}\right) \frac{\lambda^n}{(n-2)!} \int_0^{t\wedge u} \int_0^{t-s} r^{n-2} dr ds 1_{v>t}$$
$$= \left(e^{-\lambda t} - e^{-\lambda v}\right) \frac{\lambda^n}{(n-1)!} \int_0^{t\wedge u} (t-s)^{n-1} ds 1_{v>t}$$
$$= \left(e^{-\lambda t} - e^{-\lambda v}\right) \frac{\lambda^n}{n!} (t^n - (t - t\wedge u)^n) 1_{v>t}$$
$$= \left(e^{-\lambda t} - e^{-\lambda v}\right) \frac{\lambda^n}{n!} (t^n - ((t-u)^+)^n) 1_{v>t}$$

and

$$\sum_{n=2}^{\infty} \mathrm{P}(T_1 \le u, T_{n+1} \le v, N_t = n)$$
$$= \left(e^{-\lambda t} - e^{-\lambda v}\right) \sum_{n=2}^{\infty} \frac{\lambda^n}{n!} (t^n - ((t-u)^+)^n) 1_{v>t}$$
$$= \left(e^{-\lambda t} - e^{-\lambda v}\right) \left(e^{\lambda t} - \lambda t - e^{\lambda (t-u)^+} + \lambda (t-u)^+\right) 1_{v>t}$$
$$= \left(e^{-\lambda t} - e^{-\lambda v}\right) \left(e^{\lambda t} - e^{\lambda (t-u)^+} - \lambda (t\wedge u)\right) 1_{v>t}.$$

In summary,

$$\mathrm{P}(T_1 \le u, T_{N_t+1} \le v)$$
$$= (e^{-\lambda t} - e^{-\lambda (u\wedge v)}) 1_{u\wedge v>t} + (e^{-\lambda t} - e^{-\lambda v}) \lambda (t\wedge u) 1_{v>t}$$
$$+ \left(e^{-\lambda t} - e^{-\lambda v}\right) \left(e^{\lambda t} - e^{\lambda (t-u)^+} - \lambda (t\wedge u)\right) 1_{v>t}$$
$$= \left(e^{-\lambda t} - e^{-\lambda (u\wedge v)}\right) 1_{u\wedge v>t} + \left(e^{-\lambda t} - e^{-\lambda v}\right) \left(e^{\lambda t} - e^{\lambda (t-u)^+}\right) 1_{v>t}$$
$$= \left(1 - e^{-\lambda (v-t)} - e^{-\lambda u} + e^{-\lambda \left(u\vee v - (t-u)^+\right)}\right) 1_{v>t}.$$

Problem 9.12: For $t > 0$, obtain the law of the triple $(T_1, T_{N_t}, T_{N_t+1})$; that is, obtain $\mathrm{P}(T_1 \le u, T_{N_t} \le v, T_{N_t+1} \le w)$ for all (u, v, w).

Solution. As in the previous problem, we start with the case $N_t = 0$: $\mathrm{P}(T_1 \le u, T_{N_t} \le v, T_{N_t+1} \le w, N_t = 0) = \mathrm{P}(T_1 \le u \wedge w, T_1 > t) = \left(e^{-\lambda t} - e^{-\lambda (u\wedge w)}\right) 1_{u\wedge w>t}$.

Next,

$$
\begin{aligned}
&\mathrm{P}(T_1 \le u, T_{N_t} \le v, T_{N_t+1} \le w, N_t = 1) \\
&\quad = \mathrm{P}(T_1 \le u \wedge v, T_2 \le w, T_1 \le t < T_2) \\
&\quad = \mathrm{P}(T_1 \le t \wedge u \wedge v, t < T_2 \le w) \\
&\quad = \int_0^{t\wedge u\wedge v} \mathrm{P}(t-s < T_1 \le w-s)\lambda e^{-\lambda s} ds 1_{w>t} \\
&\quad = \int_0^{t\wedge u\wedge v} (e^{-\lambda(t-s)} - e^{-\lambda(w-s)})\lambda e^{-\lambda s} ds 1_{w>t} \\
&\quad = (e^{-\lambda t} - e^{-\lambda w}) \int_0^{t\wedge u\wedge v} \lambda ds = (e^{-\lambda t} - e^{-\lambda w})\lambda(t\wedge u\wedge v)1_{w>t}.
\end{aligned}
$$

Lastly, for $n \ge 2$,

$$
\begin{aligned}
&\mathrm{P}(T_1 \le u, T_n \le v, T_{n+1} \le w, N_t = n) \\
&\quad = \mathrm{P}(T_1 \le u, T_n \le v, T_{n+1} \le w, T_n \le t < T_{n+1}) \\
&\quad = \int_0^{t\wedge u\wedge v} \mathrm{P}(T_{n-1} \le t\wedge v - s, t-s < T_n \le w-s)\lambda e^{-\lambda s} ds 1_{w>t} \\
&\quad = \int_0^{t\wedge u\wedge v} \int_0^{t\wedge v-s} \left(e^{-\lambda(t-s-r)} - e^{-\lambda(w-s-r)}\right) \\
&\qquad \times \frac{\lambda^{n-1}}{(n-2)!} r^{n-2} e^{-\lambda r} dr \lambda e^{-\lambda s} ds 1_{w>t} \\
&\quad = \left(e^{-\lambda t} - e^{-\lambda w}\right) \frac{\lambda^n}{(n-2)!} \int_0^{t\wedge u\wedge v} \int_0^{t\wedge v-s} r^{n-2} dr ds 1_{w>t} \\
&\quad = \left(e^{-\lambda t} - e^{-\lambda w}\right) \frac{\lambda^n}{(n-1)!} \int_0^{t\wedge u\wedge v} (t\wedge v - s)^{n-1} ds 1_{w>t} \\
&\quad = \left(e^{-\lambda t} - e^{-\lambda w}\right) \frac{\lambda^n}{n!}((t\wedge v)^n - (t\wedge v - t\wedge u\wedge v)^n)1_{w>t} \\
&\quad = \left(e^{-\lambda t} - e^{-\lambda w}\right) \frac{\lambda^n}{n!}((t\wedge v)^n - ((t\wedge v - u)^+)^n)1_{w>t}
\end{aligned}
$$

and

$$\sum_{n=2}^{\infty} \mathrm{P}(T_1 \le u, T_n \le v, T_{n+1} \le w, N_t = n)$$

$$= \left(e^{-\lambda t} - e^{-\lambda w}\right) \sum_{n=2}^{\infty} \frac{\lambda^n}{n!}((t \wedge v)^n - ((t \wedge v - u)^+)^n) 1_{w>t}$$

$$= \left(e^{-\lambda t} - e^{-\lambda w}\right) \left(e^{\lambda(t\wedge v)} - \lambda(t \wedge v)\right.$$

$$\left. - e^{\lambda(t\wedge v - u)^+} + \lambda(t \wedge v - u)^+\right) 1_{w>t}$$

$$= \left(e^{-\lambda t} - e^{-\lambda w}\right) \left(e^{\lambda(t\wedge v)} - e^{\lambda(t\wedge v - u)^+} - \lambda(t \wedge u \wedge v)\right) 1_{w>t}.$$

In summary,

$$\mathrm{P}(T_1 \le u, T_{N_t} \le v, T_{N_t+1} \le w)$$

$$= \left(e^{-\lambda t} - e^{-\lambda(u\wedge w)}\right) 1_{u\wedge w>t} + (e^{-\lambda t} - e^{-\lambda w})\lambda(t \wedge u \wedge v) 1_{w>t}$$

$$+ (e^{-\lambda t} - e^{-\lambda w}) \left(e^{\lambda(t\wedge v)} - e^{\lambda(t\wedge v - u)^+} - \lambda(t \wedge u \wedge v)\right) 1_{w>t}$$

$$= \left(e^{-\lambda t} - e^{-\lambda(u\wedge w)}\right) 1_{u\wedge w>t} + \left(e^{-\lambda t} - e^{-\lambda w}\right)$$

$$\times \left(e^{\lambda(t\wedge v)} - e^{\lambda(t\wedge v - u)^+}\right) 1_{w>t}.$$

Problem 9.13: The remaining life in a Poisson or, more generally, a renewal process is defined as $X_t = T_{N_t+1} - t$.

(a) Argue that $(X_t)_{t\ge 0}$ is a Markov process.
(b) Let g be a bounded and measurable function. Obtain, for $x > 0$, $P_t g(x) = \mathrm{E}[g(X_t)|X_0 = x]$.
(c) Suppose g is continuous. Is $P_t g$ continuous?
(d) Suppose g is C^1. Obtain $\lim_{t\downarrow 0}(P_t g(x) - g(x))/t$.
(e) What conclusion can you draw?

See Problem 9.9 and Theorem 9.1 in the text.

Solution.

(a) Observe that the process $(X_t)_{t\ge 0}$ has jump times T_n, $n = 1, 2, \ldots$, the jump times of the Poisson process $(N_t)_{t\ge 0}$.

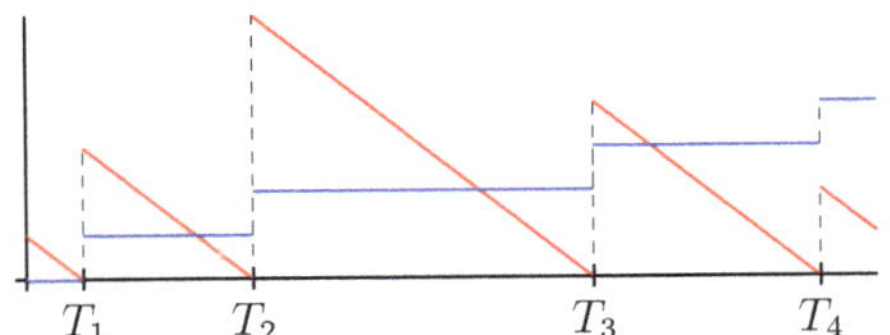

On the interval $[T_{n-1}, T_n)$, X_t decreases from a height of $T_n - T_{n-1}$ to 0. At T_n, X_t jumps to $T_{n+1} - T_n$. See the picture above where the path of $(X_t)_{t\geq 0}$ is drawn in red and the corresponding path of $(N_t)_{t\geq 0}$ is drawn in blue. The remaining life of a Poisson process can therefore be identified with a sequence of independent exponential random variables, $(T_n - T_{n-1})_{n\geq 1}$, and a piecewise linear function that jumps at T_{n-1} to $T_n - T_{n-1}$ and then decreases linearly to 0 on $[T_{n-1}, T_n)$.

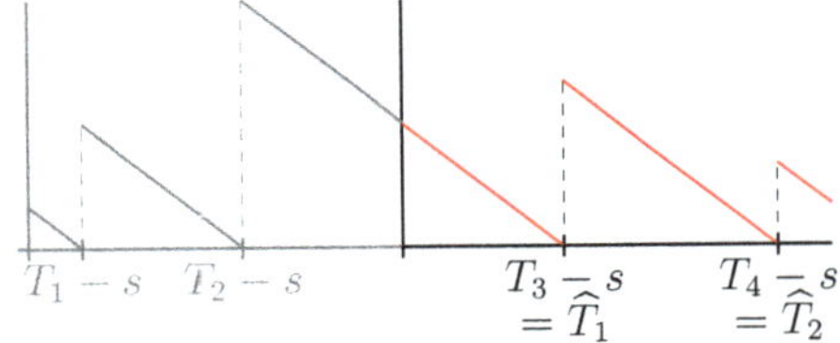

Since $T_{n+1} - T_n$ is exponential, so is $T_{n+1} - T_n - s$ conditional on $T_{n+1} - T_n > s$; this is the memoryless property of the exponential distribution. Restarting the clock at s therefore yields a path that is identical in law to the original. This is the essence of why $(X_t)_{t\geq 0}$ has the Markov property.

(b) We start by computing $\mathrm{E}[h(T_{N_t+1})]$ for h measurable and bounded:

$$
\begin{aligned}
\mathrm{E}[h(T_{N_t+1})] &= \sum_{n=0}^{\infty} \mathrm{E}[h(T_{n+1}), N_t = n] \\
&= \sum_{n=0}^{\infty} \mathrm{E}[h(T_{n+1}), T_n \leq t < T_{n+1}] \\
&= \mathrm{E}[h(T_1), T_1 > t] + \sum_{n=1}^{\infty} \int_0^t \mathrm{E}[h(s + T_1), T_1 > t - s] \\
&\quad \times \frac{\lambda^n}{(n-1)!} s^{n-1} e^{-\lambda s} ds
\end{aligned}
$$

$$
\begin{aligned}
&= \mathrm{E}[h(T_1), T_1 > t] + \lambda \int_0^t \mathrm{E}[h(s+T_1), T_1 > t-s]ds \\
&= \int_t^\infty h(r)e^{-\lambda r}dr + \lambda \int_0^t \int_{t-s}^\infty h(s+r)\lambda e^{-\lambda r}drds \\
&= \int_t^\infty h(r)e^{-\lambda r}dr + \int_0^t \int_t^\infty h(r)\lambda e^{-\lambda r}dr\lambda e^{\lambda s}ds \\
&= \int_t^\infty h(r)e^{-\lambda r}dr + (e^{\lambda t}-1) \int_t^\infty h(r)\lambda e^{-\lambda r}dr \\
&= e^{\lambda t}\mathrm{E}[h(T_1), T_1 > t].
\end{aligned}
$$

Therefore, $\mathrm{E}[h(T_{N_t+1})] = e^{\lambda t}\mathrm{E}[h(T_1), T_1 > t] = \mathrm{E}[h(T_1)|T_1 > t]$. In other words, the law of T_{N_t+1} is the conditional law of T_1 given $T_1 > t$, which is the exponential distribution shifted by t.

Alternatively, we can use Problem 9.11 (see also Problem 9.12) to reach the same conclusion. Indeed,

$$
\begin{aligned}
&\mathrm{P}(T_{N_t+1} \leq v) \\
&\quad = \lim_{u\uparrow+\infty} \mathrm{P}(T_1 \leq u, T_{N_t+1} \leq v) \\
&\quad = \lim_{u\uparrow+\infty} \left(1 - e^{-\lambda(v-t)} - e^{-\lambda u} + e^{-\lambda\left(u\vee v-(t-u)^+\right)}\right) 1_{v>t} \\
&\quad = \left(1 - e^{-\lambda(v-t)}\right) 1_{v>t}.
\end{aligned}
$$

Recall that $N_t = \sum_{n=1}^\infty 1_{T_n \leq t}$ and observe that X_t is a function of t and $\mathbb{T} = (T_n)_{n\geq 1}$, $X_t = \Phi_t(\mathbb{T})$. Furthermore, the conditional law of $\mathbb{T}$ given $\{T_1 = u\}$ is that of $\mathbb{T}_u = (u + T_{n-1})_{n\geq 1}$. It follows that the conditional law of X_t given $T_1 = u$ is that of $\Phi_t(\mathbb{T}_u) = u + T_{(1+N_{t-u})1_{u\leq t}} - t$. Note that the conditional law of N_t given $T_1 = u$ is that of $\sum_{n=1}^\infty 1_{u+T_{n-1}\leq t} = (1+N_{t-u})1_{u\leq t}$. It follows that for any bounded and measurable g, noting that $X_0 = T_1$,

$$
\begin{aligned}
&\mathrm{E}[g(X_t)|X_0 = x] \\
&\quad = \mathrm{E}\left[g\left(u + T_{(1+N_{t-x})1_{x\leq t}} - t\right)\right] \\
&\quad = 1_{x\leq t}\mathrm{E}\left[g\left(x + T_{1+N_{t-x}} - t\right)\right] + 1_{x>t}g(x-t) \\
&\quad = 1_{x\leq t}e^{\lambda(t-x)}\mathrm{E}[g(x-t+T_1), T_1 > t-x] + 1_{x>t}g(x-t) \\
&\quad = 1_{x\leq t}e^{\lambda(t-x)} \int_{t-x}^\infty g(x-t+s)\lambda e^{-\lambda s}ds + 1_{x>t}g(x-t)
\end{aligned}
$$

$$= 1_{x \le t} e^{\lambda(t-x)} \int_0^\infty g(s) \lambda e^{-\lambda(s-x+t)} ds + 1_{x>t} g(x-t)$$
$$= 1_{x \le t} \mathrm{E}[g(T_1)] + 1_{x>t} g(x-t).$$

(c) We see that even when g is continuous, $P_t g(x) = \mathrm{E}[g(X_t)|X_0 = x]$, as a function of x for t fixed, may not be continuous at t: in general,

$$P_t g(t) = \mathrm{E}[g(T_1)] \neq g(0) = \lim_{x \downarrow t} P_t g(x).$$

(d) For $x > 0$,

$$\lim_{t \downarrow 0} \frac{1}{t}(P_t g(x) - g(x)) = \lim_{t \downarrow 0} \frac{1}{t}\left(1_{x \le t} \mathrm{E}[g(T_1)] + 1_{x>t} g(x-t)\right)$$
$$= -g'(x).$$

(e) Although $\lim_{t \downarrow 0}(P_t g(x) - g(x))/t$ exists (and equals $-g'(x)$), it does not define a generator. Since, for g continuous and arbitrary, $P_t g$ is not continuous, and $(X_t)_{t \ge 0}$ is a Markov process that is not a Feller process. Recall that a Markov process is Feller if for any g continuous and vanishing at infinity, $P_t g$ must be continuous, vanish at infinity and satisfy $\lim_{t \downarrow 0} P_t g(x) = g(x)$.

Explosion for Markov Jump Processes

Problem 9.14: Let M_n be a non-negative sub-martingale in L^2 and $S_n = \max_{k \le n} M_k$. Show that

$$\mathrm{E}[S_n^2] \le 4\mathrm{E}[M_n^2].$$

Solution.

(a) Let, for $\lambda > 0$, $\tau_n = \tau_n(\lambda) = \inf\{k : M_k \ge \lambda\} \wedge n$ and $\Lambda_n = \Lambda_n(\lambda) = \{S_n \ge \lambda\}$. Then, τ_n is a stopping time, $M_{\tau_n} \ge \lambda$ on Λ_n and

$$\mathrm{E}[M_{\tau_n} 1_{\Lambda_n}] \ge \lambda \mathrm{P}(\Lambda_n).$$

On the other hand, $\forall k \le n-1$, $\Lambda_n \cap \{\tau_n \le k\} = \{\tau_n \le k\} \in \mathcal{F}_k$, and $\forall k \ge n$, $\Lambda_n \cap \{\tau_n \le k\} = \Lambda_n \in \mathcal{F}_n \subset \mathcal{F}_k$. In other words, $\Lambda_n \in \mathcal{F}_{\tau_n}$. It follows, by application of Doob's optional stopping theorem, that

$$\mathrm{E}[M_n 1_{\Lambda_n}] = \mathrm{E}[\mathrm{E}[M_n 1_{\Lambda_n} | F_{\tau_n}]] = \mathrm{E}[\mathrm{E}[M_n | F_{\tau_n}] 1_{\Lambda_n}] \ge \mathrm{E}[M_{\tau_n} 1_{\Lambda_n}].$$

In all, $\lambda \mathrm{P}(S_n \ge \lambda) = \lambda \mathrm{P}(\Lambda_n) \le \mathrm{E}[M_{\tau_n} 1_{\Lambda_n}] \le \mathrm{E}[M_n 1_{\Lambda_n}] = \mathrm{E}[M_n, S_n \ge \lambda]$.

(b) Using the above inequality, we get, for any $N > 0$,

$$\begin{aligned}\mathrm{E}[(S_n \wedge N)^2] &= 2\mathrm{E}\left[\int_0^{S_n \wedge N} \lambda d\lambda\right] = 2\int_0^N \lambda \mathrm{P}(S_n \geq \lambda) d\lambda \\ &\leq 2\int_0^N \mathrm{E}[M_n, S_n \geq \lambda] d\lambda = 2\mathrm{E}\left[M_n \int_0^{S_n \wedge N} d\lambda\right] \\ &= 2\mathrm{E}[M_n(S_n \wedge N)] \;\leq\; 2\sqrt{\mathrm{E}[M_n^2]}\sqrt{\mathrm{E}[(S_n \wedge N)^2]},\end{aligned}$$

from which we deduce that

$$\mathrm{E}[(S_n \wedge N)^2] \leq 4\mathrm{E}[M_n^2].$$

Letting $N \uparrow \infty$, we get the result by monotone convergence.

Problem 9.15: Let M_n be a martingale in L^2 and A_n be the predictable increasing process of the Doob decomposition of the sub-martingale M_n^2:

$$A_n = \sum_{k=1}^{n} \left(\mathrm{E}[M_k^2 | \mathcal{F}_{k-1}] - M_{k-1}^2\right).$$

Show that

$$\mathrm{E}\left[\left(\sup_n |M_n|\right)^2\right] \leq 4\mathrm{E}[A_\infty].$$

Solution. $|M_n|$ is a non-negative sub-martingale in L^2, and a direct application of Problem 9.14 yields

$$\mathrm{E}\left[\left(\max_{k \leq n} |M_k|\right)^2\right] \leq 4\mathrm{E}[M_n^2] = 4\mathrm{E}[A_n].$$

Letting $n \uparrow \infty$, we get the result by monotone convergence.

Problem 9.16: Let X be a discrete-time sub-martingale, A be the predictable increasing process of its Doob decomposition and τ be a stopping time. Show that the stopped process X^τ is a sub-martingale and find its Doob decomposition.

Solution.

(a) First, we show that X^τ is a sub-martingale. Indeed,

$$\mathrm{E}[|X_n^\tau|] = \sum_{k=0}^{n} \underbrace{\mathrm{E}[|X_k| 1_{\tau = k}]}_{\leq \mathrm{E}[|X_k|]} + \underbrace{\mathrm{E}[|X_n| 1_{\tau \geq n+1}]}_{\leq \mathrm{E}[|X_n|]} < +\infty$$

and, since for $k \leq n$, $X_k 1_{\tau=k}$ is $\mathcal{F}_k$-measurable and $\{\tau \geq n+1\} \in \mathcal{F}_n$,

$$\begin{aligned}\mathrm{E}[X^\tau_{n+1}|\mathcal{F}_n] &= \sum_{k=0}^{n} \mathrm{E}[X_k 1_{\tau=k}|\mathcal{F}_n] + \mathrm{E}[X_{n+1} 1_{\tau\geq n+1}|\mathcal{F}_n] \\ &= \sum_{k=0}^{n} X_k 1_{\tau=k} + \mathrm{E}[X_{n+1}|\mathcal{F}_n] 1_{\tau\geq n+1} \\ &\geq \sum_{k=0}^{n} X_k 1_{\tau=k} + X_n 1_{\tau\geq n+1} = X^\tau_n.\end{aligned}$$

Similarly, the martingale $M = X - A$ stopped at τ is again a martingale; simply replace $\geq$ with $=$ above.

(b) Next, we show that A^τ is predictable. Since $A_n\{\tau \geq n\}$ and, for $k \leq n-1$, $A_k 1_{\tau=k}$ are $\mathcal{F}_{n-1}$-measurable, so is

$$A^\tau_n = \sum_{k=0}^{n-1} A_k 1_{\tau=k} + A_n 1_{\tau\geq n}.$$

In other words, A^τ is predictable.

(c) Clearly, $X^\tau = M^\tau + A^\tau$ with A^τ, predictable and increasing and M^τ, martingale. In other words, $M^\tau + A^\tau$ is the Doob decomposition of the sub-martingale X^τ.

Problem 9.17: Let X_n be a non-negative sub-martingale and A_n be the predictable increasing process of its Doob decomposition. Show that $\lim_n X_n$ exists and is finite almost surely on $\{A_\infty < +\infty\}$.

Solution. Let, for $\lambda > 0$, $\tau = \tau(\lambda) = \inf\{n : A_{n+1} > \lambda\}$. Then, τ is a stopping time (recall that A is predictable), $A_{\tau\wedge n} \leq \lambda$ and

$$\mathrm{E}[X^\tau_n] = \mathrm{E}[M^\tau_n] + \mathrm{E}[A_{\tau\wedge n}] \leq \mathrm{E}[X_0] + \lambda.$$

It follows that the sub-martingale X^τ is bounded in L^1:

$$\sup_n \mathrm{E}[|X^\tau_n|] = \sup_n \mathrm{E}[X^\tau_n] \leq \mathrm{E}[X_0] + \lambda < +\infty$$

and therefore, by the martingale convergence theorem, almost surely convergent to a finite limit (in fact integrable).

On the other hand, on $\{A_\infty \leq \lambda\}$, $\tau = \tau(\lambda) = +\infty$ and $X_n = X_{\tau\wedge n}$ converges almost surely.

The proof is completed by observing that $\{A_\infty < +\infty\} = \bigcup_m \{A_\infty \leq m\}$.

Problem 9.18: Let X_n be a sub-martingale with bounded jumps ($|\Delta X_n| = |X_n - X_{n-1}| \le C$). Show that, almost surely, either $\lim_n X_n$ exists and is finite or $\limsup_n X_n = +\infty$; in particular, $\{X_n$ converges to a finite limit$\} = \{\sup_n X_n < +\infty\}$.

Solution. Let, for $\lambda > 0$, $\tau(\lambda) = \inf\{n : X_n > \lambda\}$. Then, $\tau(\lambda)$ is a stopping time. Furthermore, since $\tau(\lambda) \wedge n - 1 < \tau(\lambda)$, $X_{\tau(\lambda)\wedge n-1} \le \lambda$ and

$$X_{\tau(\lambda)\wedge n} \le X_0 1_{\tau(\lambda)\wedge n=0} + (X_{\tau(\lambda)\wedge n-1} + \Delta X_{\tau(\lambda)\wedge n})1_{\tau(\lambda)\wedge n\ge 1}$$
$$\le X_0 + \lambda + C.$$

It follows that $X_0 + \lambda + C - X^{\tau(\lambda)}$ is a positive super-martingale. As such, it converges (almost surely) to a finite limit, say $Y(\lambda)$. Equivalently, $\lim_n X_{\tau(\lambda)\wedge n}$ exists and is finite: $\lim_n X_{\tau(\lambda)\wedge n} = X_0 + \lambda + C - Y(\lambda)$.

Then, either $\limsup_n X_n = +\infty$ or $\limsup_n X_n < +\infty$ or equivalently that $\sup_n X_n < +\infty$. Furthermore, $\{\sup_n X_n < +\infty\} \subset \bigcup_m \{\tau(m) = +\infty\}$ and on $\{\tau(m) = +\infty\}$, $\lim_n X_n = \lim_n X_{\tau(m)\wedge n}$ exists, is finite and equals $X_0 + \lambda + C - Y(m)$.

Problem 9.19:

(a) Let $(X_n)_{n\ge 0}$ be a random walk; that is,

$$X_n = \sum_{k=1}^{n} \xi_k \qquad (X_0 = 0),$$

where $(\xi_n)_{n\ge 1}$ is a sequence of independent and identically distributed random variables.

Show that, if $\mathrm{P}(|\xi_1| < C) = 1$, for some $C > 0$, $\mathrm{E}[\xi_1] = 0$ and $\mathrm{var}(\xi_1) > 0$, then $\liminf_n X_n = -\infty$ and $\limsup_n X_n = +\infty$.

(b) Let $(\xi_n)_{n\ge 1}$ be a sequence of independent random variables with

$$\mathrm{P}(\xi_n = -n^2) = \frac{1}{1+n^2} \text{ and } \mathrm{P}(\xi_n = 1) = \frac{n^2}{1+n^2}.$$

Let $(X_n)_{n\ge 0}$ be the random walk

$$X_n = \sum_{k=1}^{n} (-1)^k \xi_k \qquad (X_0 = 0).$$

Describe the long-term behaviour of X_n.

Solution.

(a) Clearly, $(X_n)_{n\geq 0}$ is a martingale with bounded jumps. Therefore, Problem 9.18 applies to both X_n and $-X_n$. The result immediately follows as soon as we exclude the possibility that X_n converges to a finite random variable. However, for the latter to occur, we must have that ξ_n converges to 0, almost surely and therefore in distribution. This is of course not true as $(\xi_n)_{n\geq 1}$ is a sequence of independent and identically distributed random variables with strictly positive variance.

(b) Here, again $(X_n)_{n\geq 0}$ is a martingale. However, the jumps are not bounded and Problem 9.18 does not apply. In fact, by the first Borel–Cantelli lemma,

$$\sum_n \mathrm{P}(\xi_n = -n^2) = \sum_n \frac{1}{1+n^2} < +\infty,$$

for n large enough $\xi_n = 1$, and X_n oscillates with jumps of size ± 1. Therefore, it fails to converge and remains bounded ($-\infty < \inf_n X_n < \sup_n X_n < +\infty$).

Problem 9.20: Let X_n be a sub-martingale with bounded jumps ($|\Delta X_n| = |X_n - X_{n-1}| \leq C$) and A_n be the predictable increasing process of its Doob decomposition. Show that $A_\infty < +\infty$, whenever $\lim_n X_n$ exists and is finite.

Solution. Let, for $\lambda > 0$, $\tau = \tau(\lambda) = \inf\{n : |X_n| > \lambda\}$. Then, τ is a stopping time. Furthermore, since $\tau \wedge n - 1 < \tau$, $|X_{\tau\wedge n-1}| \leq \lambda$ and

$$|X_{\tau\wedge n}| \leq |X_0| 1_{\tau\wedge n=0} + (|X_{\tau\wedge n-1}| + |\Delta X_{\tau\wedge n}|) 1_{\tau\wedge n\geq 1} \leq |X_0| + \lambda + C.$$

It follows that the sub-martingale X^τ is bounded in L^1,

$$\sup_n \mathrm{E}[|X_n^\tau|] \leq \mathrm{E}[|X_0|] + \lambda + C < +\infty;$$

therefore, by the martingale convergence theorem, almost surely convergent to an integrable limit, say $Y(\lambda)$.

It is worth stressing here that τ may be infinite; therefore, the convergence of $X_{\tau\wedge n}$ is not guaranteed (of course, $\lim_n X_{\tau\wedge n} 1_{\tau<\infty} = X_\tau 1_{\tau<\infty}$).

Now, by the monotone and dominated convergence theorems,

$$\mathrm{E}[Y(\lambda)] = \mathrm{E}[\lim_n X_{\tau\wedge n}] = \lim_n \mathrm{E}[X_{\tau\wedge n}] = \mathrm{E}[X_0] + \lim_n \mathrm{E}[A_{\tau\wedge n}]$$
$$= \mathrm{E}[X_0] + \mathrm{E}[A_\tau]$$

and $\mathrm{E}[A_\tau]$ must be finite, leading to $\mathrm{P}(A_\tau < +\infty) = 1$.

On the other hand, $\{X_n$ converges to a finite limit$\} \subset \{\sup_n |X_n| < +\infty\} \subset \bigcup_m \{\tau(m) = \infty\}$ and therefore on $\{X_n$ converges to a finite limit$\}$, $\exists m$ such that $A_\infty = A_{\tau(m)}$. In all, almost surely on $\{X_n$ converges to a finite limit$\}$, $A_\infty < +\infty$.

Problem 9.21: Let X be a Markov pure jump process with state space a subset of non-negative reals and T_n be the time of the nth jump. Denote by $\mathcal{F}_t$ the natural filtration of X and let $\mathcal{G}_n = \mathcal{F}_{T_n}$. Assume that $\lambda(x)$ is bounded on finite intervals.

Suppose there exists a positive constant C and a positive increasing function $F(x)$ with $F(\infty) = \infty$, such that $S_n = F(X(T_n))$ satisfies for all n,

$$\mathrm{E}(S_{n+1}|\mathcal{G}_n) \le S_n + \frac{C}{\lambda(X(T_n))}. \qquad (*)$$

Show that the process X does not explode.

Solution. We start by observing that since λ is bounded on finite intervals,

$$\text{explosion} \Leftrightarrow \sum_n \frac{1}{\lambda(X(T_n))} < +\infty \Rightarrow \lambda(X(T_n)) \to +\infty$$
$$\Rightarrow X(T_n) \to +\infty.$$

In other words, explosion only occurs at infinity ($\lim_n X(T_n)) = +\infty$). Next, we show that this contradicts $(*)$.

Indeed, let $M_n = S_0 + \sum_{k=1}^n \left(S_k - \mathrm{E}(S_k|\mathcal{F}_{k-1})\right)$, and $A_n = \sum_{k=1}^n C/\lambda(X(T_{k-1}))$. Then, M_n is a martingale and A_n is a predictable and increasing process. Hence, $X_n = M_n + A_n$ is a sub-martingale. With the help of $(*)$, we show that $X_n \ge 0$:

$$S_k - \mathrm{E}(S_k|\mathcal{F}_{k-1}) + \frac{C}{\lambda(X(T_{k-1}))} \ge S_k - S_{k-1}.$$

Adding up these inequalities for $k = 1, \ldots, n$, we obtain

$$M_n - S_0 + A_n \ge S_n - S_0 \text{ or equivalently } X_n = M_n + A_n \ge S_n.$$

Since F is a positive function, $X_n = M_n + A_n \geq S_n > 0$. As a non-negative sub-martingale with A as the predictable increasing process in its Doob decomposition, it must converge to a finite limit on $\{A_\infty < +\infty\}$ — see Problem 9.17.

In conclusion, on $\{A_\infty < +\infty\}$, we must have that $\lim_n X(T_n) = +\infty$ and $\sup_n F(X(T_n)) = \sup_n S_n \leq \sup_n M_n < +\infty$. As $F(\infty) = +\infty$, these two facts cannot coexist and we deduce that $A_\infty = +\infty$ and explosion cannot occur.

Problem 9.22: Let X be a Markov pure jump process with state space a subset of non-negative reals and T_n be the time of the nth jump. Denote by $\mathcal{F}_t$ the natural filtration of X and let $\mathcal{G}_n = \mathcal{F}_{T_n}$. Assume that $\lambda(x)$ is bounded on finite intervals.

Suppose that there exist positive constants C and C_1 and a positive increasing function $F(x)$ with $F(\infty) < \infty$, such that $\inf_{x>C_1} \lambda(x) > 0$ — this condition is weaker than the condition $\lim_{x\to\infty} \lambda(x) = +\infty$, and such that $S_n = F(X(T_n))$ satisfies for all n,

$$\mathrm{E}(S_{n+1}|\mathcal{G}_n) \geq S_n + \frac{C}{\lambda(X(T_n))}. \qquad (*)$$

Show that the process explodes on the set $\{X(T_n) \to \infty\}$.

Solution. Multiplying both sides of $(*)$ by $1_{X(T_n)>C_1}$ and rearranging, we get

$$\mathrm{E}(S_{n+1}\ \mathcal{G}_n) \geq S_n + \frac{C}{\lambda(X(T_n))} 1_{X(T_n)>C_1} + \big(\mathrm{E}(S_{n+1}|\mathcal{G}_n) - S_n\big) 1_{X(T_n)\leq C_1}.$$

It follows that

$$W_n = S_n - \sum_{k=0}^{n-1} \left(\frac{C}{\lambda(X(T_k))} 1_{X(T_k)>C_1} + \big(\mathrm{E}(S_{k+1}|\mathcal{G}_k) - S_k\big) 1_{X(T_k)\leq C_1} \right)$$

is a sub-martingale. Furthermore, $|S_n - S_{n-1}| = S_n - S_{n-1} = F(X(T_n)) - F(X(T_{n-1})) \leq F(\infty) - F(0) \leq F(\infty)$ and

$$\begin{aligned} |W_n - W_{n-1}| &\leq |\Delta S_n| + \frac{C}{\lambda(X(T_{n-1}))} 1_{X(T_{n-1})>C_1} \\ &\quad + \big|\mathrm{E}(S_n|\mathcal{G}_{n-1}) - S_{n-1}\big| 1_{X(T_{n-1})\leq C_1} \\ &< 2F(\infty) + \frac{C}{\inf_{x>C_1} \lambda(x)} < +\infty. \end{aligned}$$

As a sub-martingale with bounded jumps, either $\limsup_n W_n = +\infty$ or $\lim_n W_n$ exists and is finite — see Problem 9.18.

Furthermore, $\{\lim_n X(T_n) = +\infty\} \subset \{\exists N \text{ s.t. } \forall n \geq N, X(T_n) > C_1\}$; therefore, on $\{\lim_n X(T_n) = +\infty\}$ and for $n \geq N$,

$$W_n = S_n - \sum_{k=0}^{N-1} \left(\frac{C}{\lambda(X(T_k))} 1_{X(T_k)>C_1} + (\mathrm{E}(S_{k+1}|\mathcal{G}_k) - S_k) 1_{X(T_k)\leq C_1} \right) - \sum_{k=N}^{n-1} \frac{C}{\lambda(X(T_k))}$$

and

$$\lim_n W_n = F(\infty) - \sum_{k=0}^{N-1} \left(\frac{C}{\lambda(X(T_k))} 1_{X(T_k)>C_1} + (\mathrm{E}(S_{k+1}|\mathcal{G}_k) - S_k) 1_{X(T_k)\leq C_1} \right) - \sum_{k=N}^{\infty} \frac{C}{\lambda(X(T_k))} < +\infty.$$

It follows that on $\{\lim_n X(T_n) = +\infty\}$, $\limsup_n W_n < +\infty$ and consequently that $\sum_n 1/\lambda(X(T_n)) < +\infty$ and the process explodes.

Problem 9.23: Let X be a Markov pure jump process with state space a subset of non-negative reals. Suppose that for all x sufficiently large, $x > C$ for some $C > 0$, the jumps are bounded below from zero $\xi(x) = \Delta X(x) > \delta$, for some $\delta > 0$. Assume that function $\lambda(x)$ is bounded on bounded intervals, it is monotone nondecreasing and also that there is a positive constant C_1, such that $\inf_{x>C_1} \lambda(x) > 0$ and that

$$\int_0^\infty \frac{dx}{\lambda(x)} < \infty.$$

Show that the process X explodes on the set $\{X(T_n) \to \infty\}$.

Solution. Note the following inequality for an increasing strictly positive function f: for any a and any $b > 0$,

$$\int_a^{a+b} \frac{dx}{f(x)} > \frac{b}{f(a+b)}.$$

Using it with $f = \lambda$ and $S_n = F(X(T_n))$ with $F(x) = \int_0^x \frac{dy}{f(y)}$, we obtain for all $X(T_n)$ sufficiently large,

$$\mathrm{E}(\Delta S_{n+1}|\mathcal{F}_n) \geq \mathrm{E}\left(\frac{\xi(X(T_n))}{\lambda(X(T_{n+1}))}|\mathcal{F}_n\right) \geq \delta \mathrm{E}\left(\frac{1}{\lambda(X(T_{n+1}))}|\mathcal{F}_n\right).$$

Repeating the argument is the previous question and replacing $1/\lambda(X(T_{n+1}))$ with its conditional expectation, $\mathrm{E}\left(\frac{1}{\lambda(X(T_{n+1}))}|\mathcal{F}_n\right)$, we obtain that

$$\sum_{n=0}^{\infty} \mathrm{E}\left(\frac{1}{\lambda(X(T_{n+1}))}|\mathcal{F}_n\right) < \infty,$$

on the set $\{X(T_n) \to \infty\}$. Now,

$$\sum_{k=0}^{n} \frac{1}{\lambda(X(T_{k+1}))} = \sum_{k=0}^{n} \left[\frac{1}{\lambda(X(T_{k+1}))} - \mathrm{E}\left(\frac{1}{\lambda(X(T_{k+1}))}|\mathcal{F}_k\right)\right] + \sum_{k=0}^{n} \mathrm{E}\left(\frac{1}{\lambda(X(T_{k+1}))}|\mathcal{F}_k\right),$$

which is a sum of a martingale and a predictable increasing process, resulting in a positive submartingale. It is easy to see that it has bounded jumps. Hence, by the convergence sets of submartingale result part (b), we have that the following sets are the same:

$$\left\{\sum_{n=0}^{\infty} \frac{1}{\lambda(X(T_{n+1}))} < \infty\right\} = \left\{\sum_{n=0}^{\infty} \mathrm{E}\left(\frac{1}{\lambda(X(T_{n+1}))}|\mathcal{F}_n\right) < \infty\right\}.$$

It follows now that $\sum_{n=0}^{\infty} \frac{1}{\lambda(X(T_{n+1}))}$ converges on the set $\{X(T_n) \to \infty\}$, which proves that the process explodes on this set.

Problem 9.24: Let X be a Markov pure jump process with state space a subset of the non-negative reals. Assume that the function $\lambda(x)$ is bounded on bounded intervals. Suppose that there exists a positive increasing function $f(x)$ such that $m(x)\lambda(x) \leq f(x)$ and that

$$\forall x > 0, \int_0^x \frac{dy}{f(y)} < \int_0^\infty \frac{dy}{f(y)} = +\infty. \qquad (*)$$

(a) Show that the process X does not explode.
(b) Suggest a setting for which

$$\int_0^\infty \frac{dy}{m(y)\lambda(y)} = +\infty$$

is a necessary condition.

Solution.

(a) Let $F(x) = \int_0^x \frac{dy}{f(y)}$ and $S_n = F(X(T_n))$. Since $f(x)$ is increasing, we have for $y > x > 0$,

$$F(y) - F(x) = \int_x^y \frac{dz}{f(z)} \leq \frac{y-x}{f(x)}.$$

Similarly, for $x > y > 0$,

$$F(y) - F(x) = -\int_y^x \frac{dz}{f(z)} \leq -\frac{x-y}{f(x)} = \frac{y-x}{f(x)}.$$

In all, for any $x, y > 0$, $F(y) - F(x) \leq (y-x)/f(x)$ and in particular,

$$F(X(T_{n+1})) - F(X(T_n)) \leq \frac{X(T_{n+1}) - X(T_n)}{f(X(T_n))},$$

and taking conditional expectations and recalling that $m(x)/f(x) \leq 1/\lambda(x)$,

$$\mathrm{E}(S_{n+1}|\mathcal{G}_n) \leq S_n + \frac{m(X(T_n))}{f(X(T_n))} \leq S_n + \frac{1}{\lambda(X(T_n))}.$$

The result now follows by application of Problem 9.21.

(b) Let λ and m be strictly positive functions such that $m\lambda$ is a non-decreasing function. Define the following increasing sequence: $x_0 = 0$ and $x_{n+1} = x_n + m(x_n)$, for $n \geq 0$, and let X be a jump process that jumps from x_n to x_{n+1} after an exponential waiting time with parameter $\lambda(x_n)$, $n \geq 0$. Then, since $m(x_n) = x_{n+1} - x_n$, we have

$$\begin{aligned}\sum_{n=0}^\infty \frac{1}{\lambda(X(T_n))} &= \sum_{n=0}^\infty \frac{1}{\lambda(x_n)} = \sum_{n=0}^\infty \frac{x_{n+1} - x_n}{\lambda(x_n)m(x_n)} \\ &\leq \sum_{n=0}^\infty \int_{x_n}^{x_{n+1}} \frac{dx}{m(x)\lambda(x)} = \int_0^\infty \frac{dx}{m(x)\lambda(x)}.\end{aligned}$$

It follows that non-explosion, that is the divergence of the series on the left, implies the divergence of the integral on the right.

This shows that $(*)$ cannot in general be weakened.

Problem 9.25: The jump rate λ alone does not determine explosion (or non-explosion). In other words, if $\Xi = \{X(T_n) : n \geq 1\}$ (note that set is random), the convergence or divergence of $\sum_{x \in \Xi} 1/\lambda(x)$ in general says nothing about the explosion of the jump process X. Consider the following example. Let the process X take values on the set of integers $\mathbb{N}$. Suppose that X jumps from any x to 1 or $4x - 1$ with probability $1/2$. Show that there is no explosion for any choice of the strictly positive rate function λ.

Solution. Since $\mathrm{P}(X(T_n) = 1) = 1/2$, $\sum_{n=1}^{\infty} \mathrm{P}(X(T_n) = 1) = \infty$ and the events $\{X(T_n) = 1\}$ are independent, we have by the second borel–Cantelli lemma that $\mathrm{P}(X(T_n) = 1 \text{ i.o.}) = 1$. Hence, there are infinitely many terms equal to $1/\lambda(1)$ in the series $\sum_{n=1}^{\infty} 1/\lambda(X(T_n))$. Therefore, the series diverges and there is no explosion.

Stopping Times and Filtrations

Problem 9.26: Let S and T be two stopping times with respect to a filtration $(\mathcal{F}_t)_{t \geq 0}$. Show that

(a) $\mathcal{F}_{T-} \subset \mathcal{F}_T$;
(b) T is $\mathcal{F}_{T-}$-measurable;
(c) $T \wedge t$ is $\mathcal{F}_t$-measurable;
(d) $\mathcal{F}_S \cap \{S \leq T\} \subset \mathcal{F}_T$ (and $\mathcal{F}_S \cap \{S \leq T\} \subset \mathcal{F}_T \cap \{S \leq T\}$) and $\mathcal{F}_S \cap \{S < T\} \subset \mathcal{F}_{T-}$. This proves Theorem 2.38 in the text.

Solution. Recall that for any stopping time T, $\mathcal{F}_T = \{A \in \mathcal{F}_\infty : \forall t, A \cap \{T \leq t\} \in \mathcal{F}_t\}$ and $\mathcal{F}_{T-} = \mathcal{F}_0 \vee \sigma(\mathcal{C}_T)$, where $\mathcal{C}_T = \{A \cap \{T > t\}, A \in \mathcal{F}_t, t \geq 0\}$.

(a) Clearly, for any $A \in \mathcal{F}_0$, $A \cap \{T \leq t\} \in \mathcal{F}_t$. It follows that $\mathcal{F}_0 \subset \mathcal{F}_T$. Also, for any fixed t, $A \in \mathcal{F}_t$ and $u > t$, $A \cap \{T > t\} \cap \{T \leq u\} \in \mathcal{F}_u$, showing that $\{A \cap \{T > t\}, A \in \mathcal{F}_t, t \geq 0\} \subset \mathcal{F}_T$. Note that $A \cap \{T > t\} \cap \{T \leq u\} = \emptyset$ if $u \leq t$. This completes the proof.
(b) This is trivial since $\{T > u\} \in \mathcal{C}_T \subset \mathcal{F}_{T-}$.
(c) For any $u < t$, $\{T \wedge t \leq u\} = \{T \leq u\}$, and for $u \geq t$, $\{T \wedge t \leq u\} = \Omega$. In both cases, $\{T \wedge t \leq u\} \in \mathcal{F}_t$. This shows that $T \wedge t$ is $\mathcal{F}_t$-measurable.

(d) Let $A \in \mathcal{F}_S$. $A \cap \{S \le T\} \cap \{T \le t\} = (A \cap \{S \le t\}) \cap \{T \le t\} \cap \{S \wedge t \le T \wedge t\} \in \mathcal{F}_t$ as the intersection of three $\mathcal{F}_t$-measurable sets. This shows that $A \cap \{S \le T\} \in \mathcal{F}_T$. Of course, for any $A \in \mathcal{F}_S$, $A \cap \{S \le T\} = (A \cap \{S \le T\}) \cap \{S \le T\} \in \mathcal{F}_T \cap \{S \le T\}$; that is, $\mathcal{F}_S \cap \{S \le T\} \subset \mathcal{F}_T \cap \{S \le T\}$.

For the second statement, we start by observing that $\{S < T\} = \bigcup_{t \in \mathbb{Q}^+} (\{S \le t\} \cap \{T > t\})$. Let $A \in \mathcal{F}_S$, $A \cap \{S \le t\} \cap \{T > t\} \in \mathcal{C}_T$. It follows that $A \cap \{S < T\} = \bigcup_{t \in \mathbb{Q}^+} (A \cap \{S \le t\} \cap \{T > t\}) \in \mathcal{F}_{T-}$.

Problem 9.27: Let $(N_t)_{t \ge 0}$ be a Poisson process, $(T_n)_n$ be its sequence of arrival times, and g be a bounded (or positive) measurable function.

(a) Obtain $\mathrm{E}[g(N_s)|\mathcal{F}_t]$.

(b) Deduce $\mathrm{P}(T_m > s|\mathcal{F}_t)$ and $\mathrm{P}(T_m > s, T_n > t)$.

Solution.

(a) A direct application of the Markov property yields

$$\mathrm{E}[g(N_s)|\mathcal{F}_t] = \begin{cases} g(N_s) & s \le t \\ \mathrm{E}[g(N_s)|N_t] & s > t \end{cases}$$

$$= \begin{cases} g(N_s) & s \le t, \\ \displaystyle\sum_{k=0}^{+\infty} g(k + N_t) e^{-\lambda(s-t)} \frac{(\lambda(s-t))^k}{k!} & s > t. \end{cases}$$

(b) Since $T_m > s$ if and only if $N_s < m$, we apply (a) to the function $g(k) = 1_{k<m}$:

$$\mathrm{P}(T_m > s|\mathcal{F}_t) = \mathrm{E}[g(N_s)|\mathcal{F}_t]$$

$$= \begin{cases} 1_{T_m > s} & s \le t \\ \displaystyle\sum_{k=0}^{+\infty} 1_{k+N_t<m} e^{-\lambda(s-t)} \frac{(\lambda(s-t))^k}{k!} & s > t \end{cases}$$

$$= \begin{cases} 1_{T_m > s} & s \le t, \\ \displaystyle 1_{T_m > t} e^{-\lambda(s-t)} \sum_{k=0}^{m-N_t-1} \frac{(\lambda(s-t))^k}{k!} & s > t. \end{cases}$$

Therefore, for $s > t$,

$$\begin{aligned}
\mathrm{P}(T_m > s, T_n > t) &= \mathrm{E}\Big[\mathrm{P}(T_m > s|\mathcal{F}_t)1_{T_n>t}\Big] \\
&= \mathrm{E}\left[1_{T_n>t}1_{T_m>t}e^{-\lambda(s-t)}\sum_{k=0}^{m-N_t-1}\frac{\big(\lambda(s-t)\big)^k}{k!}\right] \\
&= e^{-\lambda(s-t)}\mathrm{E}\left[1_{T_n>t}1_{T_m>t}\sum_{k=0}^{m-N_t-1}\frac{\big(\lambda(s-t)\big)^k}{k!}\right] \\
&= e^{-\lambda(s-t)}\mathrm{E}\left[1_{N_t<m\wedge n}\sum_{k=0}^{m-N_t-1}\frac{\big(\lambda(s-t)\big)^k}{k!}\right] \\
&= e^{-\lambda(s-t)}\mathrm{E}\left[\sum_{\ell=0}^{m\wedge n-1}1_{N_t=\ell}\sum_{k=0}^{m-\ell-1}\frac{\big(\lambda(s-t)\big)^k}{k!}\right] \\
&= e^{-\lambda s}\sum_{\ell=0}^{m\wedge n-1}\sum_{k=0}^{m-\ell-1}\frac{(\lambda t)^\ell\big(\lambda(s-t)\big)^k}{\ell!k!}.
\end{aligned}$$

For an arbitrary pair (s,t),

$$\mathrm{P}(T_m > s, T_n > t) = e^{-\lambda(s\vee t)}\sum_{\ell=0}^{m\wedge n-1}\sum_{k=0}^{m-\ell-1}\frac{(\lambda(s\wedge t))^\ell\big(\lambda|s-t|\big)^k}{\ell!k!}.$$

Problem 9.28: Let T be a non-negative random variable on a probability space $(\Omega, \mathcal{F}, \mathrm{P})$ and $(\mathcal{H}_t)_{t\geq 0}$ be the natural filtration of the process $T\wedge t$.

(a) Let $(\mathcal{G}_t)_{t\geq 0}$ be the smallest filtration that makes T a stopping time. Show that such a filtration exists.
(b) Show that $(\mathcal{G}_t)_{t\geq 0}$ is in fact the natural filtration of the process $J_t = 1_{T\leq t}$.
(c) Show that if $s < t$, then $\sigma(T\wedge s) \subset \sigma(T\wedge t)$. Deduce that $\mathcal{H}_t = \sigma(T\wedge t)$.
(d) Show that $\mathcal{G}_{t-} \subset \mathcal{H}_t \subset \mathcal{G}_t$.
(e) Show that $\mathcal{G}_T = \mathcal{G}_\infty = \sigma(T)$.
(f) Suppose that T has a continuous distribution and let $\mathcal{N}$ be the set of P-null sets: $\mathcal{N} = \{A \subset \Omega : \exists B \in \mathcal{F} \text{ such that } A \subset B \text{ and } \mathrm{P}(B) = 0\}$. Show that $\sigma(\mathcal{G}_t \cup \mathcal{N}) = \sigma(\mathcal{H}_t \cup \mathcal{N})$ and that in this case $\overline{\mathcal{G}}_{T-} = \sigma(T) \vee \sigma(\mathcal{N})$, where $\overline{\mathcal{G}}_t = \sigma(\mathcal{G}_t \cup \mathcal{N}) = \mathcal{G}_t \vee \sigma(\mathcal{N})$.

Solution.

(a) The constant filtration $\mathcal{F}_t = \mathcal{F}$ makes T a stopping time so that the family of filtrations that make T a stopping time is not empty. The smallest filtration that makes T a stopping time is simply the intersection of all filtrations in this family.

(b) Let $(\mathfrak{h}_t)_{t\geq 0}$ be a filtration. T is a stopping time with respect to $(\mathfrak{h}_t)_{t\geq 0} \Leftrightarrow \forall t \geq 0$, $\{T \leq t\} \in \mathfrak{h}_t \Leftrightarrow \forall t \geq 0$, $1_{T\leq t}$ is $\mathfrak{h}_t$-measurable $\Leftrightarrow (J_t)_{t\geq 0}$ is adapted to $(\mathfrak{h}_t)_{t\geq 0}$. Since $(\mathcal{G}_t)_{t\geq 0}$ is the intersection of all such filtrations $(\mathfrak{h}_t)_{t\geq 0}$ and so is the natural filtration of $(J_t)_{t\geq 0}$, the two must coincide.

(c) $\forall s < t$, $T\wedge s = (T\wedge t)\wedge s = g(T\wedge t)$, where $g(u) = u\wedge s$. Therefore, $\sigma(T\wedge s) \subset \sigma(T\wedge t)$ and $\sigma(T\wedge t) = \mathcal{H}_t$.

(d) $\forall s < t$, $\{T \leq s\} = \{T\wedge t \leq s\}$. Therefore, $\forall s < t$, $\mathcal{G}_s \subset \sigma(T\wedge t)$ and $\mathcal{G}_{t-} = \bigvee_{s<t} \mathcal{G}_s \subset \sigma(T\wedge t)$.

Conversely, $T1_{T\leq t} > u \Leftrightarrow u < T \leq t$. Therefore, $\{T1_{T\leq t} > u\} = \emptyset$, if $u \geq t$, and $\{T1_{T\leq t} > u\} = \{T > u\} \cap \{T \leq t\} \in \mathcal{G}_t$, if $u \leq t$. In total, $\{T1_{T\leq t} > u\} \in \mathcal{G}_t$ and $T\wedge t = T1_{T\leq t} + t1_{T>t}$ is $\mathcal{G}_t$-measurable; that is, $\sigma(T\wedge t) \subset \mathcal{G}_t$.

(e) Since $\mathcal{H}_t \subset \mathcal{G}_t \subset \mathcal{G}_\infty$, $T\wedge t$ is clearly $\mathcal{G}_\infty$-measurable and so is $T = \lim_{t\to\infty} T\wedge t$. It follows that $\sigma(T) \subset \mathcal{G}_\infty$-measurable. Conversely, since J_t is $\sigma(T)$-measurable, $\mathcal{G}_t \subset \sigma(T)$ and $\mathcal{G}_\infty = \bigvee_{t\geq 0} \mathcal{G}_t \subset \sigma(T)$. Next, we show that T is $\mathcal{G}_T$-measurable. Indeed, for any s and t, $\{T \leq t\} \cap \{T \leq s\} = \{T \leq s\wedge t\}$. Since T is a stopping time, $\{T \leq s\wedge t\} \in \mathcal{G}_{s\wedge t} \subset \mathcal{G}_s$ and $\{T \leq t\} \in \mathcal{G}_T$.

(f) Since, if T has a continuous distribution, $\{T = t\} \in \mathcal{N}$, and $\{T \leq t\} = \{T < t\} \cup \{T = t\} \in \mathcal{G}_{t-} \vee \sigma(\mathcal{N})$, we see that

$$\mathcal{G}_t \vee \sigma(\mathcal{N}) \subset \mathcal{G}_{t-} \vee \sigma(\mathcal{N}) \subset \mathcal{H}_t \vee \sigma(\mathcal{N}) \subset \mathcal{G}_t \vee \sigma(\mathcal{N}).$$

In other words, if all filtrations are completed, then $\mathcal{G}_t$ and $\mathcal{H}_t$ coincide: $\overline{\mathcal{G}}_t = \overline{\mathcal{H}}_t$.

Recall that $\overline{\mathcal{G}}_{T-} = \sigma(\{A \cap \{T > t\}, A \in \overline{\mathcal{G}}_t, t \geq 0\}) = \sigma(\{A \cap \{T > t\}, A \in \overline{\mathcal{H}}_t, t \geq 0\})$. However, $\overline{\mathcal{H}}_t = \sigma(T\wedge s, s \leq t) \vee \sigma(\mathcal{N}) = \sigma(T\wedge t) \vee \sigma(\mathcal{N})$ (for $s \leq t$, $T\wedge s = (T\wedge t)\wedge s$) and $A \in \sigma(T\wedge t) \cup \mathcal{N}$ if and only if there exists B such that $A\Delta\{T\wedge t \in B\} \in \mathcal{N}$. Therefore, $\overline{\mathcal{G}}_{T-} = \sigma(\{A \cap \{T > t\}, A \in \sigma(T\wedge t) \vee \mathcal{N}, t \geq 0\}) = \sigma(\{A \cap \{T > t\}, A \in \sigma(T\wedge t) \cup \mathcal{N}, t \geq 0\}) = \sigma(\{A \cap \{T > t\}, A \in \sigma(T\wedge t), t \geq 0\}) \vee \sigma(\mathcal{N}) = \sigma(T) \vee \sigma(\mathcal{N})$.

Problem 9.29: Let T be a non-negative random variable on a probability space $(\Omega, \mathcal{F}, \mathrm{P})$ and $(\mathcal{H}_t)_{t\geq 0}$ be the natural filtration of the process $T \wedge t$. We know (see Problem 9.28) that $\mathcal{H}_t = \sigma(T \wedge t)$.

(a) Let $\mathcal{C}_t = \{\{T \leq s\}, s < t\}$. Show that $\mathcal{H}_t = \sigma(\mathcal{C}_t)$.
(b) Express $\{T \geq t\}$ in terms of the elements of $\mathcal{C}_t$.
(c) Show that $\{T \geq t\}$ is an atom of $\mathcal{H}_t$; that is, if $A \in \mathcal{H}_t$ and $A \subset \{T \geq t\}$, then either $A = \emptyset$ or $A = \{T \geq t\}$.
(d) Let $\mathcal{H}_\infty = \bigvee_{t\geq 0} \mathcal{H}_t$. Show that $\mathcal{H}_\infty \cap \{T < t\} \subset \mathcal{H}_t$.
(e) Show that T is a stopping time for $\mathcal{H}$ if and only if it is constant.

Solution.

(a) For $s \geq t$, $\{T \wedge t \leq s\} = \Omega$ and for $s < t$, $\{T \wedge t \leq s\} = \{T \leq s\}$. Therefore, $\mathcal{H}_t = \sigma(T \wedge t) = \sigma(\{T \wedge t \leq s\}, s \geq 0) = \sigma(\{T \wedge t \leq s\}, s < t) = \sigma(\{T \leq s\}, s < t)$.
(b) $\{T \geq t\} = \{T < t\}^c = \left(\bigcup_{s<t}\{T \leq s\}\right)^c = \bigcap_{s<t}\{T \leq s\}^c = \bigcap_{A\in\mathcal{C}_t} A^c$.
(c) $A \in \mathcal{H}_t = \sigma(T \wedge t)$ if and only if $\exists B\mathcal{B}(\mathbb{R}_+)$ such that $A = \{T \wedge t \in B\}$. Therefore, if $A \subset \{T \geq t\}$, then

$$A = A \cap \{T \geq t\} = \{T \wedge t \in B\} \cap \{T \geq t\} = \begin{cases} \{T \geq t\} & t \in B, \\ \emptyset & t \notin B. \end{cases}$$

(d) $\mathcal{H}_\infty = \sigma(T \wedge s, s \geq 0)$ and since, for $s \leq t$, $T \wedge s = (T \wedge t) \wedge s$, $\mathcal{H}_\infty = \sigma(T \wedge s, s \geq t)$. Now, $\forall s \geq t, \forall a \geq 0$, $\{T \wedge s \leq a\} \cap \{T < t\} = \{T \leq a\} \cap \{T < t\} = \{T \wedge t \leq a\} \cap \{T < t\} \in \mathcal{H}_t$. It follows that $\forall s, a \geq 0$, $\{T \wedge s \leq a\} \cap \{T < t\} \in \mathcal{H}_t$; that is, $\mathcal{H}_\infty \cap \{T < t\} \subset \mathcal{H}_t$.
(e) Suppose T is a stopping time with respect to $\mathcal{H}$. Then, $\{T = t\} = \{T \leq t\} \cap \{T \geq t\} \in \mathcal{H}_t$. However, $\{T = t\}$ is a subset of the atom $\{T \geq t\}$. It follows that either $\{T = t\} = \emptyset$ or $\{T = t\} = \{T \geq t\}$, in which case $\{T > t\} = \emptyset$. Let $t^* = \inf\{t : \{T > t\} = \emptyset\}$. Then, $\{T > t^*\} = \bigcup_{t>t^*}\{T > t\} = \emptyset$ and for $t < t^*$, we must have $\{T = t\} = \emptyset$. In all, we have $T = t^*$ (everywhere).

Problem 9.30 (cont'd from Problem 9.29): Let τ be non-negative and measurable with respect to $\mathcal{H}_\infty$. τ is a stopping time with respect to $\mathcal{H}$ if and only if there exists $t^* \in [0, +\infty]$ such that

(a) $\tau > T$ on $\{T < t^*\}$;
(b) $\tau = t^*$ on $\{T \geq t^*\}$.

Solution. Suppose (a) and (b) hold true. If $t < t^*$, then $\{\tau \le t\} \cap \{T \ge t^*\} = \emptyset$ and $\{\tau \le t\} = \{\tau \le t\} \cap \{T < t^*\} \subset \{T < t\} \cap \{T < t^*\} = \{T < t\}$. Therefore, $\{\tau \le t\} = \{\tau \le t\} \cap \{T < t\} \in \mathcal{H}_\infty \cap \{T < t\} \subset \mathcal{H}_t$ (see Problem 9.29).

For $t \ge t^*$, $\{\tau \le t\} \cap \{T \ge t^*\} = \{T \ge t^*\} \in \mathcal{H}_t$ and $\{\tau \le t\} \cap \{T < t^*\} = \{\tau \le t\} \cap \{T < t^*\} \cap \{T < t\} \in \mathcal{H}_\infty \cap \{T < t\} \subset \mathcal{H}_t$. It follows that the union of these two sets belongs to $\mathcal{H}_t$; that is, $\{\tau \le t\} \in \mathcal{H}_t$.

In all, if (a) and (b) hold true, then τ must be a stopping time with respect to $\mathcal{H}_t$.

Conversely, suppose that τ is a stopping time with respect to $\mathcal{H}_t$. If $\tau > T$ (identically), then (a) and (b) hold true for $t^* = +\infty$. Suppose $\{\tau \le T\} \ne \emptyset$. Then, there exists t^* such that $\{\tau = t^*\} \cap \{\tau \le T\} \ne \emptyset$ (pick $\omega \in \{\tau \le T\}$ and let $t^* = \tau(\omega)$). Furthermore, $\{\tau = t^*\} \cap \{\tau \le T\} \in \mathcal{H}_{t^*}$; therefore, $\{\tau = t^*\} \cap \{\tau \le T\} = \{\tau = t^*\} \cap \{T \ge t^*\}$ is the atom $\{T \ge t^*\}$. In other words, on $\{T \ge t^*\}$, $\tau = t^*$.

If $T \ge t^*$ identically, then there is nothing else to prove and in this case, τ is constant (equal to t^*). Suppose $\{T < t^*\} \ne \emptyset$. We now prove by contradiction that $\{\tau \le T\} \cap \{T < t^*\} = \emptyset$.

If $\{\tau \le T\} \cap \{T < t^*\} \ne \emptyset$, then there exists $t < t^*$ such that $\{\tau = t\} \cap \{\tau \le T\} \cap \{T < t^*\} \ne \emptyset$. It follows that $\{\tau = t\} \cap \{T \ge t\}$ is a non-empty set in $\mathcal{H}_t$ and therefore that $\{\tau = t\} \cap \{T \ge t\} = \{T \ge t\}$ and that $\{T \ge t\} \subset \{\tau = t\}$. In all, $\{T \ge t^*\} \subset \{T \ge t\} \subset \{\tau = t\}$ and therefore on $\{T \ge t^*\}$, $\tau = t$. This contradicts the previous finding that $\tau = t^* \ne t$ on $\{T \ge t^*\}$.

Problem 9.31 (cont'd from Problem 9.30): Let $\mathcal{H}^+$ be the right continuous version of the filtration $\mathcal{H}$: $\mathcal{H}_t^+ = \mathcal{H}_{t+} = \bigcap_{u>t} \mathcal{H}_u$.

(a) Show that T is a stopping time for $\mathcal{H}^+$.
(b) Show that $\mathcal{H}_\infty \cap \{T \le t\} \subset \mathcal{H}_t^+$.
(c) Show that $\{T > t\}$ is an atom of $\mathcal{H}_t^+$; that is if $A \in \mathcal{H}_t^+$ and $A \subset \{T > t\}$, then either $A = \emptyset$ or $A = \{T > t\}$.
(d) Let Z be an integrable random variable ($\mathcal{H}_\infty$-measurable). Show that

$$\mathrm{E}[Z|\mathcal{H}_t^+] = Z1_{T \le t} + \frac{\mathrm{E}[Z1_{T>t}]}{\mathrm{P}(T > t)} 1_{T>t}.$$

Solution.

(a) For any $u > t$, $\{T \le t\} = \{T \wedge u \le t\} \in \sigma(T \wedge u) = \mathcal{H}_u$ (see Problem 9.28). Therefore, $\{T \le t\} \in \mathcal{H}_t^+$. It follows that T is stopping time with respect to $\mathcal{H}^+$.

(b) For $u \geq t$ and $a \geq 0$, $\{T \wedge u \leq a\} \cap \{T \leq t\} = \{T \leq a\} \cap \{T \leq t\} = \{T \wedge t \leq a\} \cap \{T \leq t\} \in \mathcal{H}_t \cap \{T \leq t\} \subset \mathcal{H}_t^+$. As in Problem 9.29, for $s < t$, $T \wedge s = (T \wedge t) \wedge s$ is $\mathcal{H}_t$-measurable and therefore $\mathcal{H}_t^+$-measurable. It follows that $\mathcal{H}_\infty \cap \{T \leq t\} \subset \mathcal{H}_t^+$.

(c) Suppose $A \in \mathcal{H}_t^+$ and $A \subset \{T > t\}$. Then, either $A = \emptyset$ or $\exists u > t$ such that $A \cap \{T \geq u\} \neq \emptyset$. Indeed, if $\forall u > t$, $A \cap \{T \geq u\} = \emptyset$, then

$$A = A \cap \{T > t\} = A \cap \left(\bigcup_{u>t} \{T \geq u\} \right) = \bigcup_{u>t} A \cap \{T \geq u\} = \emptyset.$$

Since $A \in \mathcal{H}_t^+ \subset \mathcal{H}_u$, we deduce that $A \cap \{T \geq u\}$ is in $\mathcal{H}_u$ and is a non-empty subset of $\{T \geq u\}$. It must be (see Problem 9.29) the whole of $\{T \geq u\}$. In other words, we must have $\{T \geq u\} \subset A \subset \{T > t\}$.

Since $\{T \geq v\}$ is a decreasing sequence of events, we let

$$t^* = \inf\{v > t : \{T \geq v\} \subset A\},$$

then $\{T > t^*\} \subset A \subset \{T > t\}$. Next, we show, by contradiction, that $t^* = t$.

Suppose $t^* > t$. Then, $A \in \mathcal{H}_{t^*}$ and if $\forall v \in (t, t^*)$, $A \cap \{T \geq v\} = \emptyset$, then

$$A = A \cap \{T > t\} = A \cap \left(\bigcup_{v \in (t,t^*)} \{T \geq v\} \right)$$

$$= \bigcup_{v \in (t,t^*)} A \cap \{T \geq v\} = \emptyset.$$

Therefore, $\exists v \in (t, t^*)$ such that $A \cap \{T \geq v\} \neq \emptyset$. Again, we deduce that $\{T \geq v\} \subset A$, which contradicts the fact that t^* is the smallest such v and leads to $A = \{T > t\}$.

(d) First, we show that $Z 1_{T \leq t}$ is $\mathcal{H}_t^+$-measurable. Indeed, for any $B \in \mathcal{B}(\mathbb{R})$, $\{Z \in B\} \cap \{T \leq t\} \in \mathcal{H}_\infty \cap \{T \leq t\} \subset \mathcal{H}_t^+$ and

$$\{Z 1_{T \leq t} \in B\} = \begin{cases} \{Z \in B\} \cap \{T \leq t\} & 0 \notin B, \\ (\{Z \in B\} \cap \{T \leq t\}) \cup \{T > t\} & 0 \in B. \end{cases}$$

In both cases, $\{Z 1_{T \leq t} \in B\} \in \mathcal{H}_t^+$. It follows that $\mathrm{E}[Z 1_{T>t} | \mathcal{H}_t^+] = Z 1_{T \leq t}$. On the other hand, for any $A \in \mathcal{H}_t^+$, $A \cap \{T > t\}$ either equals the empty set or the atom $\{T > t\}$. Therefore,

$$\begin{aligned}
&\mathrm{E}\big[Z1_{T>t}1_A\big]\\
&=\begin{cases}0 & A\cap\{T>t\}=\emptyset\\ \mathrm{E}\big[Z1_{T>t}\big] & A\cap\{T>t\}=\{T>t\}\end{cases}\\
&=\begin{cases}0 & A\cap\{T>t\}=\emptyset\\ (\mathrm{E}\big[Z1_{T>t}\big]/\mathrm{P}(T>t))\mathrm{P}(T>t) & A\cap\{T>t\}=\{T>t\}\end{cases}\\
&=\frac{\mathrm{E}\big[Z1_{T>t}\big]}{\mathrm{P}(T>t)}\mathrm{P}(A\cap\{T>t\})=\mathrm{E}\left[\frac{\mathrm{E}\big[Z1_{T>t}\big]}{\mathrm{P}(T>t)}1_{T>t}1_A\right].
\end{aligned}$$

We conclude that

$$\mathrm{E}\big[Z1_{T>t}|\mathcal{H}_t^+\big]=\frac{\mathrm{E}\big[Z1_{T>t}\big]}{\mathrm{P}(T>t)}1_{T>t}$$

and that

$$\mathrm{E}[Z|\mathcal{H}_t^+]=Z1_{T\le t}+\frac{\mathrm{E}\big[Z1_{T>t}\big]}{\mathrm{P}(T>t)}1_{T>t}.$$

Problem 9.32 (cont'd from Problem 9.31): Let τ be non-negative and measurable with respect to $\mathcal{H}_\infty$. τ is a stopping time with respect to $\mathcal{H}^+$ if and only if there exists $t^* \in [0,+\infty]$ such that

(a) $\tau \ge T$ on $\{T \le t^*\}$;
(b) $\tau = t^*$ on $\{T > t^*\}$.

Solution. The proof mimics that of Problem 9.30.

Suppose (a) and (b) hold true. If $t < t^*$, then $\{\tau \le t\} \cap \{T > t^*\} = \emptyset$ and $\{\tau \le t\} = \{\tau \le t\} \cap \{T \le t^*\} \subset \{T \le t\} \cap \{T \le t^*\} = \{T \le t\}$. Therefore, $\{\tau \le t\} = \{\tau \le t\} \cap \{T \le t\} \in \mathcal{H}_\infty \cap \{T \le t\} \subset \mathcal{H}_t^+$ (see Problem 9.31).

For $t \ge t^*$, $\{\tau \le t\} \cap \{T > t^*\} = \{T > t^*\} \in \mathcal{H}_t^+$ and $\{\tau \le t\} \cap \{T \le t^*\} = \{\tau \le t\} \cap \{T \le t^*\} \cap \{T \le t\} \in \mathcal{H}_\infty \cap \{T \le t\} \subset \mathcal{H}_t^+$. It follows that $\{\tau \le t\} \in \mathcal{H}_t^+$ and τ must be a stopping time with respect to $\mathcal{H}_t^+$.

Conversely, suppose that τ is a stopping time with respect to $\mathcal{H}_t^+$. If $\tau \ge T$ (identically), then (a) and (b) hold true for $t^* = +\infty$. Suppose $\{\tau < T\} \ne \emptyset$. Then, there exists t^* such that $\{\tau = t^*\} \cap \{\tau < T\} \ne \emptyset$. Furthermore, $\{\tau = t^*\} \cap \{\tau < T\} = \{\tau = t^*\} \cap \{T > t^*\} \in \mathcal{H}_{t^*}^+$; therefore, $\{\tau = t^*\} \cap \{\tau < T\}$ is the atom $\{T > t^*\}$. In other words, on $\{T > t^*\}$, $\tau = t^*$.

If $T > t^*$ identically, then there is nothing else to prove and in this case, τ is constant (equal to t^*). Suppose $\{T \le t^*\} \ne \emptyset$. We now prove by contradiction that $\{\tau < T\} \cap \{T \le t^*\} = \emptyset$.

If $\{\tau < T\} \cap \{T \le t^*\} \ne \emptyset$, then there exists $t < t^*$ such that $\{\tau = t\} \cap \{\tau < T\} \cap \{T \le t^*\} \ne \emptyset$. It follows that $\{\tau = t\} \cap \{T > t\}$ is a non-empty set in $\mathcal{H}_t^+$ and therefore that $\{\tau = t\} \cap \{T > t\} = \{T > t\}$ or that $\{T > t\} \subset \{\tau = t\}$. In all, $\{T > t^*\} \subset \{T > t\} \subset \{\tau = t\}$ and therefore on $\{T > t^*\}$, $\tau = t$. This contradicts the previous finding that $\tau = t^* \ne t$ on $\{T > t^*\}$.

Problem 9.33 (cont'd from Problem 9.32): Let τ be non-negative and measurable with respect to $\mathcal{H}_\infty$. If τ is a stopping time with respect to $\mathcal{H}^+$, then there exists $t^* \in [0, +\infty]$ such that $\tau \wedge T = t^* \wedge T$.

Solution. The proof immediately follows from Problem 9.32 applied to $\tau \wedge T$.

Problem 9.34 (cont'd from Problems 9.28–9.33):

(a) Show that $\mathcal{H}_t \subset \mathcal{G}_t \subset \mathcal{H}_t^+$.

(b) Let g be a right-continuous with a left-limit function. Show that $g(t \wedge T)$ is a predictable process with respect to all three filtrations, $\mathcal{H}$, $\mathcal{G}$, and $\mathcal{H}^+$.

(c) Let $F(t) = \mathrm{P}(T \le t)$ and suppose that $F(0) = 0$ and that $F(t) < 1$ for any $t > 0$. Define

$$\alpha(t) = \int_{(0,t]} \frac{dF(s)}{1 - F(s-)}.$$

Show that $A(t) = \alpha(t \wedge T)$ is the compensator of the increasing process $J(t) = 1_{T \le t}$ with respect to the filtration $\mathcal{H}^+$; i.e. A is $\mathcal{H}^+$-predictable and $J - A$ is an $\mathcal{H}^+$-local martingale (in fact an $\mathcal{H}^+$-martingale).

(d) Show that A is the compensator of the increasing process J with respect to the filtration $\mathcal{G}$.

(e) Show that for any $\varepsilon > 0$,

$$\mathrm{E}[J(t + \varepsilon) | \mathcal{H}_t^+] = J(t) + \frac{F(t + \varepsilon) - F(t)}{1 - F(t)} 1_{T > t}.$$

(f) Let

$$A^\varepsilon(t) = \frac{1}{\varepsilon} \int_0^t \big(\mathrm{E}[J(s + \varepsilon) | \mathcal{H}_s^+] - J(s)\big) ds.$$

Show that for any $\mathcal{H}^+$-stopping time τ,

$$\lim_{\varepsilon\downarrow 0} \mathrm{E}[A^\varepsilon(\tau)] = \mathrm{E}[J(\tau)].$$

(g) Show that $\lim_{\varepsilon\downarrow 0} A^\varepsilon(t) = A(t)$.

Solution.

(a) The first inclusion was established in Problem 9.28. Now, for $s \le t < u$, $\{T \le s\} = \{T \wedge u \le s\} \in \mathcal{H}_u$. We deduce that $\{T \le s\} \in \mathcal{H}_{t+}$ and therefore that $\mathcal{G}_t = \sigma(\{T \le s\}, s \le t) \subset \mathcal{H}_t^+$.

(b) Note that, as a continuous and adapted process, $t \wedge T$ is predictable. Now, $g(T)1_{t>T}$ is a left-continuous process. Furthermore, it is $\mathcal{H}$-adapted: for measurable B,

$$\{g(T)1_{t>T} \in B\} = \underbrace{(\{g(T) \in B\} \cap \{t > T\})}_{\in\mathcal{H}_\infty\cap\{t>T\}\subset\mathcal{H}_t\subset\mathcal{G}_t\subset\mathcal{H}_t^+} \cup \underbrace{\begin{cases} \emptyset & 0 \notin B, \\ \{t \le T\} & 0 \in B. \end{cases}}_{\in\mathcal{H}_t\subset\mathcal{G}_t\subset\mathcal{H}_t^+}$$

We deduce that $g(T)1_{t>T}$ is predictable with respect to all three filtrations. Similarly, $1_{t\le T}$ is predictable as a left-continuous adapted process. Finally, the non-random function g can be written as a limit of the left-continuous and (trivially) adapted process: $g(t) = \lim_{u\downarrow t} g(u-)$. We conclude that g is predictable and the same goes for the process $g(t \wedge T) = g(T)1_{t>T} + g(t)1_{t\le T}$.

(c) $A(t) = \alpha(t \wedge T)$ is clearly a non-decreasing predictable process. The only remaining point to prove is that $J - A$ is a martingale. From Problem 9.31, we see that

$$\mathrm{E}[\alpha(T)|\mathcal{H}_t^+] = \alpha(T)1_{T\le t} + \frac{\mathrm{E}\big[\alpha(T)1_{T>t}\big]}{\mathrm{P}(T > t)} 1_{T>t}.$$

On the other hand,

$$\begin{aligned}
\mathrm{E}\big[\alpha(T)1_{T>t}\big] &= \int_{(t,+\infty)} \left(\int_{(0,s]} \frac{dF(u)}{1 - F(u-)}\right) dF(s) \\
&= \int_{(0,t]} \left(\int_{(t,+\infty)} dF(s)\right) \frac{dF(u)}{1 - F(u-)} \\
&\quad + \int_{(t,+\infty)} \left(\int_{[u,+\infty)} dF(s)\right) \frac{dF(u)}{1 - F(u-)}
\end{aligned}$$

$$= \int_{(0,t]} (1 - F(t)) \frac{dF(u)}{1 - F(u-)}$$
$$+ \int_{(t,+\infty)} (1 - F(u-)) \frac{dF(u)}{1 - F(u-)}$$
$$= (1 - F(t))\alpha(t) + (1 - F(t)).$$

It follows that

$$\begin{aligned} \mathrm{E}[\alpha(T)|\mathcal{H}_t^+] &= \alpha(T)1_{T\le t} + \frac{\mathrm{E}\big[\alpha(T)1_{T>t}\big]}{\mathrm{P}(T>t)} 1_{T>t} \\ &= \alpha(T)1_{T\le t} + \frac{\big(1 - F(t)\big)\big(1+\alpha(t)\big)}{\mathrm{P}(T>t)} 1_{T>t} \\ &= \alpha(T)1_{T\le t} + \big(1+\alpha(t)\big)1_{T>t} = \alpha(t\wedge T) + 1_{T>t}, \end{aligned}$$

or equivalently that

$$J(t) - \alpha(t\wedge T) = 1 - 1_{T>t} - \alpha(t\wedge T) = 1 - \mathrm{E}[\alpha(T)|\mathcal{H}_t^+],$$

which is a martingale.

(d) Since A is $\mathcal{G}$-predictable, we need only prove that $J-A$ is martingale. For $s \le t$,

$$\begin{aligned} &\mathrm{E}[J(t) - A(t)|\mathcal{G}_s] \\ &\quad = \mathrm{E}[J(t) - \alpha(t\wedge T)|\mathcal{G}_s] = \mathrm{E}\big[\mathrm{E}[J(t) - \alpha(t\wedge T)|\mathcal{H}_s^+]|\mathcal{G}_s\big] \\ &\quad = \mathrm{E}[J(s) - \alpha(s\wedge T)|\mathcal{G}_s] = J(s) - \alpha(s\wedge T) = J(s) - A(s). \end{aligned}$$

(e) Since $J(t+\varepsilon)1_{T\le t} = 1_{T\le t} = J(t)$, it follows from Problem 9.31 that

$$\begin{aligned} \mathrm{E}[J(t+\varepsilon)|\mathcal{H}_t^+] &= J(t) + \frac{\mathrm{E}\big[J(t+\varepsilon)1_{T>t}\big]}{\mathrm{P}(T>t)} 1_{T>t} \\ &= J(t) + \frac{\mathrm{P}(t < T \le t+\varepsilon)}{\mathrm{P}(T>t)} 1_{T>t} \\ &= J(t) + \frac{F(t+\varepsilon) - F(t)}{1 - F(t)} 1_{T>t}. \end{aligned}$$

(f) Let τ be an $\mathcal{H}^+$-stopping time. Then, from Problem 9.31, we know that there exists a constant t^* such that $\tau\wedge T = t^*\wedge T$. Therefore, on the one hand,

$$\begin{aligned} &\mathrm{E}[J(\tau)] = \mathrm{P}(T\le\tau) \\ &\qquad = \mathrm{P}(T\le \tau\wedge T) = \mathrm{P}(T\le t^*\wedge T) = \mathrm{P}(T\le t^*) = F(t^*). \end{aligned}$$

On the other hand,

$$\begin{aligned}A^{\varepsilon}(t) &= \frac{1}{\varepsilon}\int_0^t (\mathrm{E}[J(s+\varepsilon)|\mathcal{H}_s^+] - J(s))ds \\ &= \frac{1}{\varepsilon}\int_0^t \frac{F(s+\varepsilon)-F(s)}{1-F(s)}1_{T>s}ds \\ &= \frac{1}{\varepsilon}\int_0^{t\wedge T} \frac{F(s+\varepsilon)-F(s)}{1-F(s)}ds\end{aligned}$$

and

$$\begin{aligned}\mathrm{E}[A^{\varepsilon}(\tau)] &= \frac{1}{\varepsilon}\mathrm{E}\left[\int_0^{\tau\wedge T} \frac{F(s+\varepsilon)-F(s)}{1-F(s)}ds\right] \\ &= \frac{1}{\varepsilon}\mathrm{E}\left[\int_0^{t^*\wedge T} \frac{F(s+\varepsilon)-F(s)}{1-F(s)}ds\right] \\ &= \frac{1}{\varepsilon}\int_0^{t^*} \frac{F(s+\varepsilon)-F(s)}{1-F(s)}\mathrm{P}(T>s)ds \\ &= \frac{1}{\varepsilon}\int_0^{t^*} (F(s+\varepsilon)-F(s))ds \\ &= \frac{1}{\varepsilon}\left(\int_{t^*}^{t^*+\varepsilon} F(s)ds - \int_0^{\varepsilon} F(s)ds\right) \underset{\varepsilon\downarrow 0}{\longrightarrow} F(t^*).\end{aligned}$$

The result immediately follows.

(g) Next, we use the Fubini theorem to give a new expression for A^{ε}:

$$\begin{aligned}A^{\varepsilon}(t) &= \frac{1}{\varepsilon}\int_0^{t\wedge T} \frac{F(s+\varepsilon)-F(s)}{1-F(s)}ds \\ &= \frac{1}{\varepsilon}\int_0^{t\wedge T} \frac{1}{1-F(s)}\left(\int_{(s,s+\varepsilon]} dF(u)\right) ds \\ &= \frac{1}{\varepsilon}\int_{(0,(t\wedge T)+\varepsilon]}\left(\int_{(u-\varepsilon)^+}^{(t\wedge T)\wedge u} \frac{ds}{1-F(s)}\right) dF(u).\end{aligned}$$

Therefore, A^{ε} is of the form

$$\frac{1}{\varepsilon}\int_{(0,t+\varepsilon]} (G(t\wedge u) - G((u-\varepsilon)^+))dF(u)$$

with $G(u) = \int_0^u g(s)ds$ and g is right-continuous with left limits. Note that we have written t instead of $t \wedge T$ as t is arbitrary.

Now, for $0 < \varepsilon < t$,

$$\begin{aligned}\frac{1}{\varepsilon}\int_{(0,\varepsilon]} G(u)dF(u) &= \frac{1}{\varepsilon}\left(F(\varepsilon)G(\varepsilon) - \int_0^{\varepsilon} F(u)g(u)du\right)\\ &\longrightarrow_{\varepsilon\downarrow 0} F(0)g(0) - F(0)g(0) = 0.\end{aligned}$$

For $\varepsilon < u < t$,

$$\frac{1}{\varepsilon}\int_{u-\varepsilon}^{u} g(s)ds = \frac{1}{\varepsilon}\big(G(u) - G(u-\varepsilon)\big) \underset{\varepsilon\downarrow 0}{\longrightarrow} g(u-) \quad \text{and}$$
$$\frac{1}{\varepsilon}\int_{u-\varepsilon}^{u} g(s)ds \le g(t).$$

It follows, by dominated convergence, that

$$\int_{(\varepsilon,t]}\left(\frac{1}{\varepsilon}\int_{u-\varepsilon}^{u} g(s)ds\right)dF(u) \underset{\varepsilon\downarrow 0}{\longrightarrow} \int_{(0,t]} g(u-)dF(u).$$

Finally, for $t < u < t+\varepsilon$,

$$\begin{aligned}&\int_{(t,t+\varepsilon]} \frac{1}{\varepsilon}\big(G(t) - G(u-\varepsilon)\big)dF(u)\\ &\quad\le \int_{(t,t+\varepsilon]} \frac{1}{\varepsilon}\big(G(u) - G(u-\varepsilon)\big)dF(u)\\ &\quad\le g(t+\varepsilon)\big(F(t+\varepsilon) - F(t)\big) \underset{\varepsilon\downarrow 0}{\longrightarrow} 0.\end{aligned}$$

Therefore,

$$\lim_{\varepsilon\downarrow 0} A^{\varepsilon}(t) = \int_{(0,T\wedge t]} g(u-)dF(u) = \int_{(0,T\wedge t]} \frac{dF(u)}{1-F(u-)} = A(t).$$

Problem 9.35: Let X be a three-state Markov chain, $X(t) = \xi 1_{T\le t}$, where the random variables ξ and T are independent and ξ takes values ± 1 with equal probability.

(a) For $s < t$, obtain the joint distribution of $X(s)$ and $X(t)$.
(b) Show that X is a martingale (with respect to its natural filtration) and deduce its compensator.

Solution. Note that since X is Markov, T must be an exponential random variable. This fact will not be used, and we will write F for the distribution function of T.

(a) $X(s)$ and $X(t)$ are both discrete (taking the three values $-1, 0, +1$). Their joint distribution can be described by the following table:

Probability	$X(t) = -1$	$X(t) = 0$	$X(t) = +1$
$X(s) = -1$	$F(s)/2$	0	0
$X(s) = 0$	$(F(t) - F(s))/2$	$1 - F(t)$	$(F(t) - F(s))/2$
$X(t) = +1$	0	0	$F(s)/2$

(b) Let $(\mathcal{F}_t)_{t\geq 0}$ be the natural filtration of X. To show that X is a martingale, we make the observation that, since it is a Markov process, $\mathrm{E}(X(t)|\mathcal{F}_s) = \mathrm{E}(X(t)|X(s))$ and the latter can easily be computed from (a) and found to be

$$\mathrm{E}(X(t)|X(s)) = \left\{\begin{array}{ll} -1 & \text{if } X(s) = -1 \\ 0 & \text{if } X(s) = 0 \\ +1 & \text{if } X(s) = +1 \end{array}\right\} = X(s).$$

X being a martingale, its compensator is nil. This is generalised in the following problem (Problem 9.36).

Problem 9.36 (cont'd from Problem 9.34): Let X be a one-jump point process, $X(t) = \xi 1_{T\leq t}$, where the random variables ξ and T are independent. For simplicity, assume that $\xi \neq 0$ (everywhere). Show that if $\mathrm{E}(\xi) = 0$, then X is a martingale (with respect to its natural filtration) and deduce its compensator.

Solution. Let $(\mathcal{F}_t)_{t\geq 0}$ be the natural filtration of X and let $\tau = \inf\{t : X(t) \neq 0\}$. Then, τ is a stopping time and $T = \tau$. It follows that T itself is a stopping time. We deduce that for $s < t$,

$$\begin{aligned} \mathrm{E}\big(X(t)\big|\mathcal{F}_s\big)1_{T\leq s} &= \mathrm{E}\big(\xi 1_{T\leq t}\big|\mathcal{F}_s\big)1_{T\leq s} = \mathrm{E}\big(\xi 1_{T\leq t}1_{T\leq s}\big|\mathcal{F}_s\big) \\ &= \mathrm{E}\big(\xi 1_{T\leq s}\big|\mathcal{F}_s\big) = \mathrm{E}\big(X(s)\big|\mathcal{F}_s\big) = X(s) = X(s)1_{T\leq s}. \end{aligned}$$

To compute $\mathrm{E}\big(X(t)\big|\mathcal{F}_s\big)1_{T>s}$, we have to look closer at the trace of $\mathcal{F}_s$ on $\{T > s\}$. Since $\mathcal{F}_s$ is generated by the events $\{X(r) \in B\}$, $r \leq s$ and $B \in \mathcal{B}(\mathbb{R})$, $\mathcal{F}_s \cap \{T > s\}$ is generated by the events $\{X(r) \in B\} \cap \{T > s\}$, $r \leq s$ and $B \in \mathcal{B}(\mathbb{R})$ (see Problem 2.38). However, $\{X(r) \in B\} \cap \{T > s\}$ can take only one of two values, depending on whether

$0 \in B$ or not, $\emptyset$ and $\{T > s\}$. In other words, $\mathcal{F}_s \cap \{T > s\} = \{\emptyset, \{T > s\}\}$ and, for any $A \in \mathcal{F}_s$,

$$\mathrm{E}(\xi 1_{T\le t} 1_{A\cap\{T>s\}}) = \begin{cases} 0 & A \cap \{T > s\} = \emptyset, \\ \mathrm{E}(\xi 1_{s<T\le t}) & A \cap \{T > s\} = \{T > s\}. \end{cases}$$

By the independence of ξ and T, we conclude that

$$\mathrm{E}(X(t)|\mathcal{F}_s) 1_{T>s} = 0 = X(s) 1_{T>s}.$$

Problem 9.37: Let X be the arrival process, that is a pure jump point process whose jumps are of size 1. Denote by T_n its jump times, $n = 1, 2, \ldots$, so that

$$X(t) = \sum_{n=1}^{\infty} 1_{T_n \le t}.$$

We assume that the process is regular (it has finitely many jumps on finite time intervals or equivalently that $\lim_{n\to\infty} T_n = \infty$). For convenience, we set $T_0 = 0$. Let, for $n = 0, 1, \ldots$,

$$F_n(s) = \mathrm{P}(T_{n+1} - T_n \le s | \mathcal{F}_{T_n}).$$

(a) Let $\mathcal{G}^{(n)}$ be the natural filtration of the process $J_t^{(n)} = 1_{T_n \le t}$. Show that $\mathcal{F}_t = \bigvee_n \mathcal{G}_t^{(n)}$.
(b) Show that $t \wedge T_{n+1} - t \wedge T_n$ is $\mathcal{G}_t^{(n)} \vee \mathcal{G}_t^{(n+1)}$-measurable.
(c) Show that

$$A(t) = \sum_{n=0}^{\infty} \int_0^{t\wedge T_{n+1} - t\wedge T_n} \frac{dF_n(s)}{1 - F_n(s-)}$$

is a predictable process.
(d) Use the fact that for each $\mathcal{F}$-stopping time τ and for any n, there exists an $\mathcal{F}_{T_n}$-measurable random variable ζ_n such that

$$\tau \wedge T_{n+1} = (T_n + \zeta_n) \wedge T_{n+1} \text{ on the set } \{T_n \le \tau\}$$

to deduce that A is the compensator of X. See Theorem 9.1 (and Problem 9.33).

Solution.

(a) We know that $\mathcal{G}^{(n)}$ is the smallest filtration that makes $T - n$ a stopping time (see Problem 9.28). Since $T_n = \inf\{t \ge 0 : X(t) = n\}$ is an $\mathcal{F}$-stopping time, $\mathcal{G}_t^{(n)} \subset \mathcal{F}_t$. It follows that $\bigvee_n \mathcal{G}_t^{(n)} \subset \mathcal{F}_t$.

Conversely, $X_N(t) = \sum_{n=1}^N 1_{T_n \le t}$ is $\bigvee_n \mathcal{G}_t^{(n)}$-measurable and so is $\lim_N X_N(t) = X(t)$.

(b) Let $\mathcal{H}^{(n)}$ be the natural filtration of the process $T_n \wedge t$. We know that $\mathcal{H}_t^{(n)} \subset \mathcal{G}_t^{(n)}$ (see Problem 9.28) and $\mathcal{H}_\infty^{(n)} \cap \{t > T_n\} \subset \mathcal{H}_t^{(n)}$ (see Problem 9.29). Therefore (see Problem 2.38),

$$\begin{aligned}
&(\mathcal{H}_\infty^{(n)} \vee \mathcal{H}_\infty^{(n+1)}) \cap \{t > T_{n+1}\} \\
&= \sigma_{\{t>T_{n+1}\}}\big((\mathcal{H}_\infty^{(n)} \cup \mathcal{H}_\infty^{(n+1)}) \cap \{t > T_{n+1}\}\big) \\
&= \sigma_{\{t>T_{n+1}\}}\big((\mathcal{H}_\infty^{(n)} \cap \{t > T_{n+1}\}) \cup (\mathcal{H}_\infty^{(n+1)} \cap \{t > T_{n+1}\})\big) \\
&= \sigma_{\{t>T_{n+1}\}}\big((\mathcal{H}_\infty^{(n)} \cap \{t > T_n\} \cap \{t > T_{n+1}\}) \\
&\quad \cup (\mathcal{H}_\infty^{(n+1)} \cap \{t > T_{n+1}\})\big) \\
&\subset \sigma_{\{t>T_{n+1}\}}\big((\mathcal{H}_t^{(n)} \cap \{t > T_{n+1}\}) \cup (\mathcal{H}_t^{(n+1)} \cap \{t > T_{n+1}\})\big) \\
&\subset \sigma_{\{t>T_{n+1}\}}\big((\mathcal{G}_t^{(n)} \cap \{t > T_{n+1}\}) \cup (\mathcal{G}_t^{(n+1)} \cap \{t > T_{n+1}\})\big) \\
&\subset \mathcal{G}_t^{(n)} \vee \mathcal{G}_t^{(n+1)}
\end{aligned}$$

and $\{(T_{n+1} - T_n)1_{t>T_{n+1}} \in B\} \cap \{t > T_{n+1}\} = \{(T_{n+1} - T_n) \in B\} \cap \{t > T_{n+1}\} \in (\mathcal{H}_\infty^{(n)} \vee \mathcal{H}_\infty^{(n+1)}) \cap \{t > T_{n+1}\} \subset \mathcal{G}_t^{(n)} \vee \mathcal{G}_t^{(n+1)}$.

Also, $\{(t - T_n)1_{T_n<t\le T_{n+1}} \in B\} \cap \{T_n < t \le T_{n+1}\} = \{(t - T_n) \in B\} \cap \{T_n < t \le T_{n+1}\} \in \mathcal{H}_\infty^{(n)} \cap \{t > T_n\} \cap \{t \le T_{n+1}\} \subset \mathcal{G}_t^{(n)} \vee \mathcal{G}_t^{(n+1)}$. It follows that

$$t \wedge T_{n+1} - t \wedge T_n = \begin{cases} 0 & t \le T_n \\ t - T_n & T_n < t \le T_{n+1} \\ T_{n+1} - T_n & t > T_{n+1} \end{cases}$$

is $\mathcal{G}_t^{(n)} \vee \mathcal{G}_t^{(n+1)}$-measurable.

(c) Combining (b) and (c), we deduce that $t \wedge T_{n+1} - t \wedge T_n$, as a continuous and $\mathcal{F}$-adapted process, is $\mathcal{F}$-predictable. Let

$$\alpha_n(t) = \int_{(0,t)} \frac{dF_n(s)}{1 - F_n(s-)}.$$

Then, α_n is left-continuous and $\alpha_n(t \wedge T_{n+1} - t \wedge T_n)$ is an $\mathcal{F}$-predictable process. It follows that

$$\int_0^{t\wedge T_{n+1}-t\wedge T_n} \frac{dF_n(s)}{1-F_n(s-)} = \int_{(0,t\wedge T_{n+1}-t\wedge T_n]} \frac{dF_n(s)}{1-F_n(s-)},$$

as a limit of $\mathcal{F}$-predictable processes, is itself $\mathcal{F}$-predictable and so is A (as the limit of sums of predictable processes).

(d) By definition, A is a compensator of X if it is predictable and $X - A$ is a local martingale. To see the latter, we show that $\mathrm{E}(X(\tau)) = \mathrm{E}(A(\tau))$ for any bounded stopping time τ (this is sufficient by Theorem 7.17). In fact, we show the stronger statement that, for any stopping time τ and for any n,

$$\mathrm{E}\left(\int_0^{\tau\wedge T_{n+1}-\tau\wedge T_n} \frac{dF_n(s)}{1-F_n(s-)}\right) = \mathrm{P}(T_{n+1} \le \tau).$$

To this end, we use Theorem 9.1 to rewrite $\tau\wedge T_{n+1} - \tau\wedge T_n$. Clearly, on $\{T_n > \tau\}$, $\tau\wedge T_{n+1} - \tau\wedge T_n = 0$ and, on $\{T_n \le \tau\}$,

$$\begin{aligned}(\tau\wedge T_{n+1} - \tau\wedge T_n)1_{T_n\le\tau} &= (T_n + \zeta_n)\wedge T_{n+1} - T_n \\ &= (T_{n+1} - T_n)\wedge \zeta_n.\end{aligned}$$

Consider both cases $T_n + \zeta_n > T_{n+1}$ and $T_n + \zeta_n \le T_{n+1}$ to see this. Therefore,

$$\begin{aligned}&\mathrm{E}\left(\int_0^{\tau\wedge T_{n+1}-\tau\wedge T_n} \frac{dF_n(s)}{1-F_n(s-)}\middle|\mathcal{F}_{T_n}\right) \\ &= \mathrm{E}\left(\int_0^{(T_{n+1}-T_n)\wedge\zeta_n} \frac{dF_n(s)}{1-F_n(s-)}1_{T_n\le\tau}\middle|\mathcal{F}_{T_n}\right) \\ &= \mathrm{E}\left(\int_0^{\zeta_n} 1_{s\le T_{n+1}-T_n}\frac{dF_n(s)}{1-F_n(s-)}\middle|\mathcal{F}_{T_n}\right)1_{T_n\le\tau} \\ &= \int_0^{\zeta_n} \mathrm{P}(T_{n+1}-T_n \ge s|\mathcal{F}_{T_n})\frac{dF_n(s)}{1-F_n(s-)}1_{T_n\le\tau} \\ &= \int_0^{\zeta_n} (1-F_n(s-))\frac{dF_n(s)}{1-F_n(s-)}1_{T_n\le\tau} = F_n(\zeta_n)1_{T_n\le\tau}.\end{aligned}$$

Here, we have used the facts that ζ_n is $\mathcal{F}_{T_n}$-measurable and $F_n(0) = \mathrm{P}(T_{n+1} = T_n|\mathcal{F}_{T_n}) = 0$.

On the other hand,

$$\begin{aligned}\mathrm{P}(T_{n+1} \le \tau | \mathcal{F}_{T_n}) &= \mathrm{P}(T_{n+1} - T_n \le \zeta_n | \mathcal{F}_{T_n}) 1_{T_n \le \tau} \\ &= F_n(\zeta_n) 1_{T_n \le \tau}.\end{aligned}$$

Taking expectations and summing over all $n \ge 0$, we complete the proof.

Problem 9.38: Obtain the compensator of a renewal process in terms of the distribution function F of its inter-arrival times.

Solution. This is a direct application of Problem 9.37. Here, $T_{n+1} - T_n$ is independent of $\mathcal{F}_{T_n}$ and, for all n, $F_n = F$. It follows that

$$A(t) = \sum_{n=0}^{\infty} \int_0^{t \wedge T_{n+1} - t \wedge T_n} \frac{dF(s)}{1 - F(s-)}.$$

Letting $G(t) = \int_0^t dF(s)/(1 - F(s-))$, we can write on $\{T_n < t \le T_{n+1}\}$,

$$A(t) = \sum_{k=0}^{n-1} G(T_{k+1} - T_k) + G(t - T_n).$$

Problem 9.39: Let $(N_t)_{t \ge 0}$ be a Poisson process with arrival times $(T_n)_n$ and completed natural filtration $(\mathcal{F}_t)_{t \ge 0}$, $(\mathcal{G}_t^{(n)})_{t \ge 0}$ be the completed natural filtration of the process $J_t^{(n)} = 1_{T_n \le t}$, $(\mathcal{H}_t^{(n)})_{t \ge 0}$ be the completed natural filtration of the process $T_n \wedge t$, $\mathcal{G}_t = \bigvee_n \mathcal{G}_t^{(n)}$ and $\mathcal{H}_t = \bigvee_n \mathcal{H}_t^{(n)}$. Show that $\mathcal{F}_t = \mathcal{G}_t = \mathcal{H}_t$.

Solution. The identity $\mathcal{G}_t = \mathcal{H}_t$ follows from Problem 9.28. Since $N_t = \sum_{n=0}^{\infty} 1_{T_n \le t} = \sum_{n=0}^{\infty} J_t^{(n)}$, we have that $\mathcal{F}_t \subset \mathcal{G}_t$. Conversely, $\{T_n \le t\} = \{N_t \ge n\}$. It follows that $\mathcal{G}_t^{(n)} \subset \mathcal{F}_t$ and therefore that $\mathcal{G}_t \subset \mathcal{F}_t$.

Problem 9.40: Let g be a bounded (or positive) measurable function.

(a) Show that

$$\mathrm{E}[g(T) | T \wedge t] = G(t, T \wedge t), \text{ where } G(t,s) = \begin{cases} g(s) & s < t, \\ \mathrm{E}[g(T) | T \ge t] & s \ge t. \end{cases}$$

(b) Deduce that, for $s < t$,

$$\mathrm{P}(T > t | T \wedge s) = \frac{1 - F(t)}{1 - F(s-)} 1_{T \ge s}.$$

Solution.

(a) Fix $s < t$. Since on $\{T \wedge t \leq s\}$, $T \wedge t < t$,

$$\mathrm{E}[G(t, T \wedge t)1_{T\wedge t\leq s}] = \mathrm{E}[g(T \wedge t)1_{T\leq s}] = \mathrm{E}[g(T)1_{T\leq s}].$$

Note that $T \wedge t \leq s \Leftrightarrow T \leq s$.

Now, fix $s \geq t$. Since $T \wedge t \leq s$,

$$\begin{aligned}
&\mathrm{E}[G(t, T \wedge t)1_{T\wedge t\leq s}] \\
&= \mathrm{E}[G(t, T \wedge t)] \\
&= \mathrm{E}[g(T \wedge t)1_{T\wedge t<t}] + \mathrm{E}[g(T)|T \geq t]\mathrm{P}(T \wedge t \geq t) \\
&= \mathrm{E}[g(T)1_{T<t}] + \mathrm{E}[g(T)1_{T\geq t}] = \mathrm{E}[g(T)] = \mathrm{E}[g(T)1_{T\wedge t\leq s}].
\end{aligned}$$

Since $\sigma(T \wedge t) = \sigma(\{T \wedge t \leq s\}, s \geq 0)$, $\forall A \in \sigma(T \wedge t)$, $\mathrm{E}[G(t, T \wedge t)1_A] = \mathrm{E}[g(T)1_A]$.

(b) Note that $T > s \Leftrightarrow T\wedge s > s$. Now, $\mathrm{P}(T > t|T\wedge s) = \mathrm{E}[1_{T>t}|T\wedge s] = G(s, T \wedge s)$, where

$$G(s, u) = \begin{cases} 1_{u>t} & u < s, \\ \mathrm{E}[1_{T>t}|T \geq s] & u \geq s. \end{cases}$$

Therefore, almost surely,

$$\begin{aligned}
\mathrm{P}(T > t|T \wedge s) &= 1_{T\wedge s>t}1_{T\wedge s<s} + \mathrm{P}(T > t|T \geq s)1_{T\wedge s\geq s} \\
&= \frac{1 - F(t)}{1 - F(s-)}1_{T\geq s}.
\end{aligned}$$

Problem 9.41: Suppose that T has a continuous distribution and let $(\mathcal{G}_t)_{t\geq 0}$ be the completed natural filtration of $J_t = 1_{T\leq t}$. Show that, for any integrable random variable X (measurable with respect to $\mathcal{G}_\infty$, which is of the form $h(T)$ — see Problem 9.28),

$$\mathrm{E}[X|\mathcal{G}_t]1_{T>t} = \mathrm{E}[X|T > t]1_{T>t}.$$

Solution. Since $\mathcal{G}_t = \sigma(T \wedge t)$, any $\mathcal{G}_t$-measurable random variable can be written as a measurable function of $T \wedge t$. Therefore, there exists a measurable function g such that $\mathrm{E}[X|\mathcal{G}_t] = g(T \wedge t)$. It follows that

$$\mathrm{E}[X1_{T>t}] = \mathrm{E}[\mathrm{E}[X|\mathcal{G}_t]1_{T>t}] = \mathrm{E}[g(T \wedge t)1_{T>t}] = g(t)\mathrm{P}(T > t),$$

from which we deduce that $g(t) = \mathrm{E}[X|T > t]$ and

$$\mathrm{E}[X|\mathcal{G}_t]1_{T>t} = g(t)1_{T>t} = \mathrm{E}[X|T > t]1_{T>t}.$$

Problem 9.42: Let F be the distribution function of T and H be its hazard function: $H(t) = -\ln(1 - F(t))$. We assume for simplicity that F is continuous and that, for each t, $F(t) < 1$.

(a) Show that, for $s \leq t$,

$$\int_s^\infty H(u \wedge t)dF(u) = H(s)(1 - F(s)) + F(t) - F(s).$$

(b) Show that $M_t = J_t - H(T\wedge t) = J_t - \int_0^t (1-J_s)\frac{dF(s)}{1-F(s)} = J_t - \int_0^{T\wedge t} \frac{dF(s)}{1-F(s)}$ is a martingale with respect to $(\mathcal{G}_t)_{t\geq 0}$ — see Theorem 9.5 of the text.

Solution.

(a) For a fixed t,

$$H(u \wedge t) = \begin{cases} H(u) & u < t \\ H(t) & u \geq t \end{cases} \quad \text{and} \quad dH(u \wedge t) = \frac{dF(u)}{1 - F(u)} 1_{u<t}.$$

By the integration by parts formula, for $s \leq t$,

$$\begin{aligned}
&\int_s^\infty H(u \wedge t)dF(u) \\
&= H(t) - H(s)F(s) - \int_s^\infty F(u)1_{u<t}\frac{dF(u)}{1 - F(u)} \\
&= H(t) - H(s)F(s) + \int_s^t \left(1 - \frac{1}{1 - F(u)}\right) dF(u) \\
&= H(t) - H(s)F(s) + F(t) - F(s) - \int_s^t \frac{dF(u)}{1 - F(u)} \\
&= H(t) - H(s)F(s) + F(t) - F(s) - H(t) + H(s) \\
&= H(s)(1 - F(s)) + F(t) - F(s).
\end{aligned}$$

(b) First, $\mathrm{E}[J_t|T \wedge s] = \mathrm{E}[1_{T\leq t}|T \wedge s] = G(s, T \wedge s)$, where

$$G(s, u) = \begin{cases} 1_{u\leq t} & u < s, \\ \mathrm{E}[1_{T\leq t}|T \geq s] & u \geq s. \end{cases}$$

Therefore,

$$\begin{aligned}\mathrm{E}[J_t|T\wedge s] &= 1_{T\wedge s\le t}1_{T\wedge s<s} + \frac{F(t)-F(s)}{1-F(s)}1_{T\wedge s\ge s}\\ &= 1_{T<s} + \frac{F(t)-F(s)}{1-F(s)}1_{T\ge s}.\end{aligned}$$

Second, $\mathrm{E}[H(T\wedge t)|T\wedge s] = \mathrm{E}[H(T\wedge t)|(T\wedge t)\wedge s] = G(s, T\wedge s)$, where

$$G(s,u) = \begin{cases} H(u) & u<s,\\ \mathrm{E}[H(T\wedge t)|T\wedge t\ge s] & u\ge s.\end{cases}$$

Therefore,

$$\begin{aligned}\mathrm{E}[H(T\wedge t)|T\wedge s] &= H(T\wedge s)1_{T\wedge s<s} + \mathrm{E}[H(T\wedge t)|T\ge s]1_{T\wedge s\ge s}\\ &= H(T)1_{T<s} + \frac{1}{1-F(s)}\int_s^\infty H(u\wedge t)dF(u)1_{T\ge s}\\ &= H(T)1_{T<s} + \left(H(s) + \frac{F(t)-F(s)}{1-F(s)}\right)1_{T\ge s}\\ &= H(T\wedge s) + \frac{F(t)-F(s)}{1-F(s)}1_{T\ge s}.\end{aligned}$$

In total, almost surely (recall that T has a continuous distribution),

$$\begin{aligned}\mathrm{E}[M_t|\mathcal{G}_s] &= \mathrm{E}[J_t - H(T\wedge t)|T\wedge s]\\ &= J_s + \frac{F(t)-F(s)}{1-F(s)}1_{T>s} - H(T\wedge s) - \frac{F(t)-F(s)}{1-F(s)}1_{T\ge s}\\ &= M_s.\end{aligned}$$

Furthermore,

$$\begin{aligned}H(T\wedge t) &= \int_0^{T\wedge t} dH(s) = \int_0^t 1_{T>s}\frac{dF(s)}{1-F(s)}\\ &= \int_0^t (1-J_s)\frac{dF(s)}{1-F(s)}.\end{aligned}$$

Problem 9.43: (cont'd from Problem 9.42) Obtain $[M,M]_t$ and $\langle M,M\rangle_t$ — see Theorem 9.3 in the text.

Solution. Clearly, $[M,M]_t = \sum_{s\leq t}(\Delta J_s)^2 = J_t$. Since $J_t - H(T\wedge t)$ is a martingale and $H(T\wedge t)$ is a non-decreasing continuous (and therefore predictable) process, we immediately deduce that $\langle M,M\rangle_t = H(T\wedge t)$.

Problem 9.44: Show that $\frac{1}{1-F(t)}1_{T>t}$ is a martingale with respect to its natural filtration (that is $(\mathcal{G}_t)_{t\geq 0}$). Note that, if $F(t) = 1$ so that $\mathrm{P}(T>t) = 0$, $\frac{1}{1-F(t)}1_{T>t}$ is taken to be 0.

Solution. For $s < t$,

$$\mathrm{E}\left[\frac{1}{1-F(t)}1_{T>t}|\mathcal{G}_s\right] = \frac{1}{1-F(t)}\mathrm{P}(T>t|T\wedge s) = \frac{1}{1-F(s)}1_{T>s}.$$

Problem 9.45: Obtain $\mathcal{E}(M)_t$ by solving the ordinary differential equation

$$X_t = 1 + \int_{(0,t]} X_{s-}dM_s$$

and also by computing

$$e^{M_t}\prod_{s\leq t}(1+\Delta M_s)e^{-\Delta M_s}.$$

Solution. First, for any measurable U,

$$\begin{aligned}\int_{(0,t]} U_s dM_s &= \int_{(0,t]} U_s dJ_s - \int_{(0,t]} U_s dH(T\wedge s)\\ &= \begin{cases} -\int_0^t U_s dH(s) & t<T\\ U_T - \int_0^T U_s dH(s) & t\geq T\end{cases}\\ &= U_T J_t - \int_0^{T\wedge t} U_s dH(s).\end{aligned}$$

Applied to X, this becomes

$$X_t = 1 + \int_{(0,t]} X_{s-}dM_s = 1 + X_{T-}J_t - \int_0^{T\wedge t} X_{s-}dH(s).$$

On $[0,T)$, the ODE simplifies to

$$X_t = 1 - \int_0^t X_{s-}\frac{dF(s)}{1-F(s)},$$

the solution of which is simply $X_t = 1 - F(t)$. In particular, $X_{T-} = 1 - F(T)$.

On $[T, +\infty)$, the ODE becomes

$$\begin{aligned} X_t &= 1 + X_{T-} - \int_0^T X_{s-} \frac{dF(s)}{1-F(s)} \\ &= 1 + (1 - F(T)) - \int_0^T (1-F(s)) \frac{dF(s)}{1-F(s)} \\ &= 2(1 - F(T)). \end{aligned}$$

In total,

$$X_t = 2(1 - F(T))J_t + (1 - F(t))(1 - J_t) = 2e^{-H(T)}J_t + e^{-H(t)}(1 - J_t).$$

On the other hand,

$$\begin{aligned} e^{M_t} \prod_{s \le t} (1 + \Delta M_s) e^{-\Delta M_s} &= e^{-H(T\wedge t)} e^{J_t} \prod_{s \le t} (1 + \Delta J_s) e^{-\Delta J_s} \\ &= \begin{cases} e^{-H(t)} & t < T \\ 2e^{-H(T)} & t \ge T \end{cases} \\ &= 2e^{-H(T)}J_t + e^{-H(t)}(1 - J_t). \end{aligned}$$

Problem 9.46: Suppose that T is exponential with parameter λ (and mean $1/\lambda$). Show that the following are martingales:

(a) $1_{T \le t} - \lambda(T \wedge t)$;
(b) $e^{\lambda t} 1_{T > t}$;
(c) $e^{-\lambda(T \wedge t)} + e^{-\lambda T} 1_{T \le t}$.

Solution. In this case, $H(t) = \lambda t$. The results follow from direct applications of 9.42, 9.44 and 9.45, respectively.

Problem 9.47: Suppose T has a continuous distribution. Show that any uniformly integrable $\mathcal{G}_t$-martingale X_t is of the form $G(t, T \wedge t)$ and can be written as $X_0 + \int_0^t k_s dM_s$, for some predictable process $(k_t)_{t \ge 0}$.

Solution. Any uniformly integrable martingale is a Doob–Lévy martingale; that is, it must be of the form $\mathrm{E}[X|\mathcal{G}_t]$, where X is a $\mathcal{G}_\infty$-measurable random variable. As a $\mathcal{G}_\infty$-measurable random variable, X must be of the form $g(T)$ for some measurable function g. It follows that the uniformly integrable martingale must be of the form

$$X_t = \mathrm{E}[g(T)|\mathcal{G}_t] = G(t, T\wedge t) = \begin{cases} g(T\wedge t) & T\wedge t < t \\ \mathrm{E}[g(T)|T\geq t] & T\wedge t \geq t \end{cases}$$

$$= g(T)1_{T<t} + \mathrm{E}[g(T)|T\geq t]1_{T\geq t}.$$

Furthermore, with

$$R(t) = \frac{1}{1-F(t)} \text{ and } U(t) = \mathrm{E}[g(T)1_{T\geq t}] = \int_t^\infty g(u)dF(u),$$

$X_s dJ_s = R(s)U(s)dJ_s$. This is a consequence of $J_{s-}\Delta J_s = 0$. Therefore,

$$\int_{(0,t]} (g(s) - X_s)dJ_s = (g(T) - R(T)U(T))J_t$$

and, since $J_s = 0$ for $s < T\wedge t \leq T$,

$$\begin{aligned}
&\int_{(0,t]} (g(s) - X_s)dH(T\wedge s) \\
&= \int_0^{T\wedge t} (g(s) - X_s)R(s)dF(s) \\
&= \int_0^{T\wedge t} \big(g(s) - g(T)J_s - (1-J_s)R(s)U(s)\big)R(s)dF(s) \\
&= \int_0^{T\wedge t} g(s)R(s)dF(s) - \int_0^{T\wedge t} U(s)\frac{dF(s)}{(1-F(s))^2} \\
&= \int_0^{T\wedge t} g(s)R(s)dF(s) \\
&\quad - \left(R(T\wedge t)U(T\wedge t) - U(0) + \int_0^{T\wedge t} R(s)g(s)dF(s)\right) \\
&= \mathrm{E}[g(T)] - R(T\wedge t)U(T\wedge t),
\end{aligned}$$

where we have used the integration by parts formula and the identity $U(0) = \mathrm{E}[g(T)]$. In summary,

$$\begin{aligned}
&\int_{(0,t]} (g(s) - X_s)dM_s \\
&= (g(T) - R(T)U(T))J_t - \mathrm{E}[g(T)] + R(T\wedge t)U(T\wedge t) \\
&= g(T)J_t + R(t)U(t)(1-J_t) - \mathrm{E}[g(T)] = X_t - \mathrm{E}[g(T)]
\end{aligned}$$

or equivalently,

$$X_t = \mathrm{E}[g(T)] + \int_{(0,t]} (g(s) - X_s)dM_s = X_0 + \int_{(0,t]} (g(s) - X_s)dM_s.$$

Problem 9.48:

(a) Let X be an integrable random variable, μ be its mean and F its distribution function. Show that $\mathrm{E}[|X-\mu|] = 2\int (x-\mu)1_{x\geq\mu}dF(x)$.
(b) Let N be a Poisson random variable with parameter (mean) λ. Show that $\mathrm{E}[|N-\lambda|] = 2\lambda\mathrm{P}(N = \lfloor\lambda\rfloor)$, where $\lfloor\lambda\rfloor$ is the integer part of λ.
(c) Let $(N_t)_{t\geq 0}$ be a Poisson process with rate λ. Obtain $\mathrm{E}[|N_t - \lambda t|]$.

Solution.

(a) Note that $\int (x-\mu)1_{x\geq\mu}dF(x) = \int (\mu - x)1_{x<\mu}dF(x)$,

$$\begin{aligned}
\mathrm{E}[|X-\mu|] &= \int |x-\mu|dF(x) \\
&= \int (\mu - x)1_{x<\mu}dF(x) + \int (x-\mu)1_{x\geq\mu}dF(x) \\
&= \mu\left(1 - \int 1_{x\geq\mu}dF(x)\right) - \left(\mu - \int x1_{x\geq\mu}dF(x)\right) \\
&\quad + \int (x-\mu)1_{x\geq\mu}dF(x) \\
&= 2\int (x-\mu)1_{x\geq\mu}dF(x).
\end{aligned}$$

(b) Applying (a) to the case of a Poisson random variable, we get

$$\begin{aligned}
\mathrm{E}[|N-\lambda|] &= 2\sum_{n\geq\lambda}(n-\lambda)e^{-\lambda}\frac{\lambda^n}{n!} = 2\sum_{n=\lfloor\lambda\rfloor+1}^{\infty}(n-\lambda)e^{-\lambda}\frac{\lambda^n}{n!} \\
&= 2\left(\sum_{n=\lfloor\lambda\rfloor+1}^{\infty} n\frac{\lambda^n}{n!} - \lambda\sum_{n=\lfloor\lambda\rfloor+1}^{\infty}\frac{\lambda^n}{n!}\right)e^{-\lambda} \\
&= 2\left(\lambda\sum_{n=\lfloor\lambda\rfloor}^{\infty}\frac{\lambda^n}{n!} - \lambda\sum_{n=\lfloor\lambda\rfloor+1}^{\infty}\frac{\lambda^n}{n!}\right)e^{-\lambda} \\
&= 2\lambda\frac{\lambda^{\lfloor\lambda\rfloor}}{\lfloor\lambda\rfloor!}e^{-\lambda} = 2\lambda\mathrm{P}(N = \lfloor\lambda\rfloor).
\end{aligned}$$

(c) Here, N_t is a Poisson random variable with parameter (mean) λt. Therefore,

$$\mathrm{E}[|N_t - \lambda t|] = 2\lambda t \frac{(\lambda t)^{\lfloor \lambda t \rfloor}}{\lfloor \lambda t \rfloor !} e^{-\lambda t} = 2\lambda t \mathrm{P}(N_t = \lfloor \lambda t \rfloor).$$

Problem 9.49:

(a) Let X be an integrable non-negative random variable, μ be its mean and F its distribution function. Obtain $\mathrm{E}[|X - \mu + \nu|]$ for $\nu \leq \mu$.
(b) Let N be a Poisson random variable with parameter (mean) λ. Obtain $\mathrm{E}[|N - \lambda + \nu|]$ for any ν.
(c) Let $(N_t)_{t\geq 0}$ be a Poisson process with rate λ and $(\mathcal{F}_t)_{t\geq 0}$ be its natural filtration. Obtain $\mathrm{E}[|N_t - \lambda t||\mathcal{F}_s]$ for $s < t$.

Solution.

(a) $\mathrm{E}[|X - \mu + \nu|]$

$$\begin{aligned}
&= \int (\mu - \nu - x) 1_{x<\mu-\nu} dF(x) + \int (x - \mu + \nu) 1_{x\geq\mu-\nu} dF(x) \\
&= (\mu - \nu)\left(1 - \int 1_{x\geq\mu-\nu} dF(x)\right) - \left(\mu - \int x 1_{x\geq\mu-\nu} dF(x)\right) \\
&\quad + \int (x - \mu + \nu) 1_{x\geq\mu-\nu} dF(x) \\
&= -\nu + 2\int (x - \mu + \nu) 1_{x\geq\mu-\nu} dF(x).
\end{aligned}$$

(b) First, we deal with the case $\nu > \lambda$. Since N is a non-negative random variable,

$$\mathrm{E}[|N - \lambda + \nu|] = \mathrm{E}[N - \lambda + \nu] = \nu.$$

If $\nu \leq \lambda$, then applying (a) to the case of a Poisson random variable, we get

$$\begin{aligned}
&\mathrm{E}[|N - \lambda + \nu|] \\
&= -\nu + 2 \sum_{n=\lfloor \lambda-\nu \rfloor+1}^{\infty} (n - \lambda + \nu) e^{-\lambda} \frac{\lambda^n}{n!} \\
&= -\nu + 2 \left(\sum_{n=\lfloor \lambda-\nu \rfloor+1}^{\infty} n \frac{\lambda^n}{n!} - (\lambda - \nu) \sum_{n=\lfloor \lambda-\nu \rfloor+1}^{\infty} \frac{\lambda^n}{n!} \right) e^{-\lambda}
\end{aligned}$$

$$= -\nu + 2\left(\lambda \sum_{n=\lfloor\lambda-\nu\rfloor}^{\infty} \frac{\lambda^n}{n!} - (\lambda - \nu) \sum_{n=\lfloor\lambda-\nu\rfloor+1}^{\infty} \frac{\lambda^n}{n!}\right) e^{-\lambda}$$

$$= -\nu + 2\left(\lambda \epsilon^{-\lambda} \frac{\lambda^{\lfloor\lambda-\nu\rfloor}}{\lfloor\lambda-\nu\rfloor!} + \nu \sum_{n=\lfloor\lambda-\nu\rfloor+1}^{\infty} e^{-\lambda}\frac{\lambda^n}{n!}\right)$$

$$= -\nu + 2\left(\lambda e^{-\lambda} \frac{\lambda^{\lfloor\lambda-\nu\rfloor}}{\lfloor\lambda-\nu\rfloor!} + \nu - \nu \sum_{n=0}^{\lfloor\lambda-\nu\rfloor} e^{-\lambda}\frac{\lambda^n}{n!}\right)$$

$$= \nu + 2\lambda \mathrm{P}(N = \lfloor\lambda-\nu\rfloor) - 2\nu \mathrm{P}(N \leq \lfloor\lambda-\nu\rfloor).$$

In summary, and noting that $\lfloor\lambda-\nu\rfloor \leq -1$ for $\lambda < \nu$,

$$\mathrm{E}[|N - \lambda + \nu|] = \nu + 2\lambda \mathrm{P}(N = \lfloor\lambda-\nu\rfloor) - 2\nu \mathrm{P}(N \leq \lfloor\lambda-\nu\rfloor).$$

(c) By the Markov property, $\mathrm{E}[|N_t - \lambda t||\mathcal{F}_s] = \mathrm{E}[|N_t - \lambda t||N_s]$, and the question reduces to computing $\mathrm{E}[|N_t - \lambda t||N_s = m]$ for any $m \in \mathbb{Z}^+$:

$$\begin{aligned}
\mathrm{E}[|N_t - \lambda t||N_s = m] &= \mathrm{E}[|N_t - N_s - \lambda(t-s) + (N_s - \lambda s)||N_s = m] \\
&= \mathrm{E}[|N_t - N_s - \lambda(t-s) + (m - \lambda s)|] \\
&= \mathrm{E}[|N_{t-s} - \lambda(t-s) + (m - \lambda s)|] \\
&= m - \lambda s + 2\lambda(t-s)\mathrm{P}(N_{t-s} = \lfloor\lambda t - m\rfloor) \\
&\quad - 2(m - \lambda s)\mathrm{P}(N_{t-s} \leq \lfloor\lambda t - m\rfloor).
\end{aligned}$$

Therefore,

$$\begin{aligned}
\mathrm{E}[|N_t - \lambda t||\mathcal{F}_s] &= N_s - \lambda s \\
&+ 2e^{-\lambda(t-s)}\left(\frac{(\lambda(t-s))^{\lfloor\lambda t - N_s\rfloor+1}}{\lfloor\lambda t - N_s\rfloor!} - (N_s - \lambda s)\sum_{n=0}^{\lfloor\lambda t - N_s\rfloor}\frac{(\lambda t)^n}{n!}\right) 1_{N_s \leq \lambda t}.
\end{aligned}$$

Problem 9.50: Let $(N_t)_{t\geq 0}$ be a Poisson process with rate λ and g be a continuous (or measurable non-negative) function on $\mathbb{R}^+ \times \mathbb{Z}^+$. Show that

$$\mathrm{E}\left[\int_0^t g(s, N_{s-})dN_s\right] = \lambda \int_0^t \mathrm{E}[g(s, N_s)]ds.$$

Deduce that $\int_0^t g(s, N_{s-})d(N_s - \lambda s)$ is a martingale.

Solution. Let $(T_n)_{n\geq 0}$ be the arrival times of the Poisson process $(N_t)_{t\geq 0}$ ($T_0 = 0$). Then,

$$\begin{aligned}
\mathrm{E}\left[\int_0^t g(s, N_{s-})dN_s\right] &= \mathrm{E}\left[\sum_{0<s\leq t} g(s, N_{s-})\Delta N_s\right] \\
&= \mathrm{E}\left[\sum_{n=1}^{\infty} g(T_n, n-1)1_{T_n\leq t}\right] \\
&= \sum_{n=1}^{\infty} \mathrm{E}\left[g(T_n, n-1)1_{T_n\leq t}\right] \\
&= \sum_{n=1}^{\infty}\int_0^t g(s, n-1)\frac{\lambda^n}{(n-1)!}s^{n-1}e^{-\lambda s}ds \\
&= \int_0^t \sum_{n=1}^{\infty} g(s, n-1)\frac{\lambda^n}{(n-1)!}s^{n-1}e^{-\lambda s}ds \\
&= \int_0^t \sum_{n=0}^{\infty} g(s, n)e^{-\lambda s}\frac{(\lambda s)^n}{n!}\lambda ds \\
&= \int_0^t \mathrm{E}[g(s, N_s)]\lambda ds.
\end{aligned}$$

Note that $\int_0^t g(s, N_{s-})ds = \int_0^t g(s, N_s)ds$. By the Markov property,

$$\begin{aligned}
\mathrm{E}\left[\int_0^t g(r, N_{r-})d(N_r - \lambda r)\,|\mathcal{F}_s\right] &= \int_0^s g(r, N_{r-})d(N_r - \lambda r) \\
&\quad + \mathrm{E}\left[\int_s^t g(r, N_{r-})d(N_r - \lambda r)\,|N_s\right].
\end{aligned}$$

Furthermore,

$$\begin{aligned}
&\mathrm{E}\left[\int_s^t g(r, N_{r-})d(N_r - \lambda r)\,|N_s = m\right] \\
&\quad = \mathrm{E}\left[\int_0^{t-s} g(r+s, N_{r-} + m)d(N_r - \lambda r)\right] = 0,
\end{aligned}$$

by direct application of the above applied to the function $(r, n) \hookrightarrow g(r+s, n+m)$. The result immediately follows.

Problem 9.51: Recall that if g is the difference of two convex functions on $[a, b]$, then for any $t \in (a, b)$, $g(t) = g(a) + \int_a^t g'_-(s)ds$.

Let $(N_t)_{t\geq 0}$ be a Poisson process with rate λ and f be a measurable function on $\mathbb{R}^+ \times \mathbb{R}$ such that, for each $x \in \mathbb{R}$, $t \hookrightarrow f(t, x)$ is the difference of two convex functions whose left derivative at t, for x fixed, is denoted $f'_-(t, x)$. Show that

$$f(t, N_t) = f(0, 0) + \sum_{s\leq t} (f(s, N_s) - f(s, N_{s-}))\Delta N_s + \int_0^t f'_-(s, N_s)ds$$

and that $f(t, N_t) - \int_0^t f(s, N_s)d(N_s - \lambda s) - \int_0^t f'_-(s, N_s)ds$ is a martingale.

Solution. Let $(T_n)_{n\geq 0}$ be the arrival times of the Poisson process $(N_t)_{t\geq 0}$ $(T_0 = 0)$. Then, for $t \in (T_{n-1}, T_n)$,

$$\begin{aligned} f(t, N_t) = f(t, n-1) &= f(T_{n-1}, n-1) + \int_{T_{n-1}}^t f'_-(s, n-1)ds \\ &= f(T_{n-1}, n-1) + \int_{T_{n-1}}^t f'_-(s, N_s)ds. \end{aligned}$$

Letting t approach T_n, we get ($t \hookrightarrow f(t, n-1)$ is continuous)

$$f(T_n, n-1) = f(T_{n-1}, n-1) + \int_{T_{n-1}}^{T_n} f'_-(s, N_s)ds.$$

It follows that for $t \in (T_n, T_{n+1})$,

$$\begin{aligned} f(t, N_t) &= f(T_n, n) + \int_{T_n}^t f'_-(s, N_s)ds \\ &= f(T_n, n) - f(T_n, n-1) + f(T_{n-1}, n-1) \\ &\quad + \int_{T_{n-1}}^t f'_-(s, N_s)ds. \end{aligned}$$

Adding and subtracting $f(T_{n-1}, n-2)$, we get

$$\begin{aligned} f(t, N_t) = &\sum_{k=n-1}^{n} (f(T_k, k) - f(T_k, k-1)) + f(T_{n-2}, n-2) \\ &+ \int_{T_{n-2}}^t f'_-(s, N_s)ds. \end{aligned}$$

Repeating the previous step for $k = n-2, n-3, \ldots, 1$, we conclude that

$$f(t, N_t) = f(0,0) + \sum_{k=1}^{n} \left(f(T_k, k) - f(T_k, k-1)\right) + \int_0^t f'_-(s, N_s)ds$$

$$= f(0,0) + \sum_{s \le t} \left(f(s, N_s) - f(s, N_{s-})\right)\Delta N_s + \int_0^t f'_-(s, N_s)ds.$$

Since N_s jumps at countably many points, $\int_0^t \left(f(s,N_s)-f(s,N_{s-})\right)ds = 0$ and

$$f(t, N_t) = f(0,0) + \int_0^t \left(f(s, N_s) - f(s, N_{s-})\right)d(N_s - \lambda s)$$

$$+ \int_0^t f'_-(s, N_s)ds$$

$$= f(0,0) - \int_0^t f(s, N_{s-})d(N_s - \lambda s) + \int_0^t f(s, N_s)d(N_s - \lambda s)$$

$$+ \int_0^t f'_-(s, N_s)ds.$$

Since $f(0,0) - \int_0^t f(s, N_{s-})d(N_s - \lambda s)$ is a martingale, then so is $f(t, N_t) - \int_0^t f(s, N_s)d(N_s - \lambda s) - \int_0^t f'_-(s, N_s)ds$.

Problem 9.52:

(a) Show that, for any $n \in \mathbb{Z}^+$,

$$\sum_{k=1}^{n} \frac{1}{(k-1)!} \int_{k-1}^{k} (k-r)r^{k-1}e^{-r}dr = \frac{n^n}{(n-1)!}e^{-n}.$$

(b) Let $(N_t)_{t \ge 0}$ be a Poisson process with rate λ and g be a continuous (or measurable non-negative) function on $\mathbb{R}^+ \times \mathbb{Z}^+$. Obtain $\mathrm{E}\left[\int_0^t g(s, N_s)dN_s\right]$. Deduce the expressions for $\mathrm{E}\left[\int_0^t |N_s - \lambda s|dN_s\right]$ and $\mathrm{E}[|N_t - \lambda t|]$.

Solution.

(a) First, we observe that $\int_\alpha^\beta (\kappa - r) r^{\kappa-1} e^{-r} dr = \beta^\kappa e^{-\beta} - \alpha^\kappa e^{-\alpha}$. Then, letting $\alpha = k-1$, $\beta = k$ and $\kappa = k$ and summing over k, we get

$$\begin{aligned}
&\sum_{k=1}^{n} \frac{1}{(k-1)!} \int_{k-1}^{k} (k-r) r^{k-1} e^{-r} dr \\
&= \sum_{k=1}^{n} \frac{1}{(k-1)!} \left(k^k e^{-k} - (k-1)^k e^{-(k-1)} \right) \\
&= \sum_{k=1}^{n} \frac{1}{(k-1)!} k^k e^{-k} - \sum_{k=2}^{n} \frac{1}{(k-1)!} (k-1)^k e^{-(k-1)} \\
&= \sum_{k=1}^{n} \frac{1}{(k-1)!} k^k e^{-k} - \sum_{k=2}^{n} \frac{1}{(k-2)!} (k-1)^{k-1} e^{-(k-1)} \\
&= \sum_{k=1}^{n} \frac{1}{(k-1)!} k^k e^{-k} - \sum_{k=1}^{n-1} \frac{1}{(k-1)!} k^k e^{-k} = \frac{1}{(n-1)!} n^n e^{-n}.
\end{aligned}$$

(b) Since whenever $\Delta N_s \neq 0$, $N_s = N_{s-1} + 1$, we deduce that

$$\begin{aligned}
\mathrm{E}\left[\int_0^t g(s, N_s) dN_s\right] &= \mathrm{E}\left[\int_0^t g(s, N_{s-} + 1) dN_s\right] \\
&= \lambda \int_0^t \mathrm{E}[g(s, N_s + 1)] ds.
\end{aligned}$$

It follows that

$$\mathrm{E}\left[\int_0^t |N_s - \lambda s| dN_s\right] = \lambda \int_0^t \mathrm{E}[|N_s + 1 - \lambda s|] ds$$

and

$$\begin{aligned}
&\mathrm{E}\left[\int_0^t |N_s - \lambda s| d(N_s - \lambda s)\right] \\
&= \lambda \int_0^t \mathrm{E}[|N_s + 1 - \lambda s| - |N_s - \lambda s|] ds.
\end{aligned}$$

However,

$$\begin{aligned}
&|N_s + 1 - \lambda s| - |N_s - \lambda s| + \operatorname{sign}(\lambda s - N_s)\\
&\quad = \begin{cases} 0 & N_s < \lambda s - 1, \\ 2(N_s - \lambda s + 1) & \lambda s - 1 \le N_s < \lambda s, \\ 0 & N_s \ge \lambda s, \end{cases}
\end{aligned}$$

and, almost everywhere in s (λs not an integer),

$$\begin{aligned}
&\mathrm{E}[|N_s + 1 - \lambda s| - |N_s - \lambda s| + \operatorname{sign}(\lambda s - N_s)]\\
&\quad = 2(\lfloor \lambda s \rfloor - \lambda s + 1)\mathrm{P}(N_s = \lfloor \lambda s \rfloor).
\end{aligned}$$

We conclude that

$$\begin{aligned}
\mathrm{E}[|N_t - \lambda t|] &= \mathrm{E}\left[\int_0^t |N_s - \lambda s| d(N_s - \lambda s)\right]\\
&\quad + \lambda \int_0^t \mathrm{E}[\operatorname{sign}(\lambda s - N_s)] ds\\
&= 2\lambda \int_0^t (\lfloor \lambda s \rfloor - \lambda s + 1)\mathrm{P}(N_s = \lfloor \lambda s \rfloor) ds\\
&= 2\sum_{k=1}^{\lfloor \lambda t \rfloor} \lambda \int_{(k-1)/\lambda}^{k/\lambda} (k - \lambda s) e^{-\lambda s} \frac{(\lambda s)^{k-1}}{(k-1)!} ds\\
&\quad + 2\lambda \int_{\lfloor \lambda t \rfloor/\lambda}^{t} (\lfloor \lambda t \rfloor - \lambda s + 1) e^{-\lambda s} \frac{(\lambda s)^{\lfloor \lambda t \rfloor}}{\lfloor \lambda t \rfloor !} ds\\
&= 2\sum_{k=1}^{\lfloor \lambda t \rfloor} \int_{k-1}^{k} (k - r) e^{-r} \frac{r^{k-1}}{(k-1)!} dr\\
&\quad + \frac{2}{\lfloor \lambda t \rfloor !} \int_{\lfloor \lambda t \rfloor}^{\lambda t} (\lfloor \lambda t \rfloor + 1 - r) r^{\lfloor \lambda t \rfloor} e^{-r} dr\\
&= 2\frac{\lfloor \lambda t \rfloor^{\lfloor \lambda t \rfloor}}{(\lfloor \lambda t \rfloor - 1)!} e^{-\lfloor \lambda t \rfloor}\\
&\quad + \frac{2}{\lfloor \lambda t \rfloor !}\left((\lambda t)^{\lfloor \lambda t \rfloor + 1} e^{-\lambda t} - \lfloor \lambda t \rfloor^{\lfloor \lambda t \rfloor + 1} e^{-\lfloor \lambda t \rfloor}\right)\\
&= 2\lambda t \frac{(\lambda t)^{\lfloor \lambda t \rfloor}}{\lfloor \lambda t \rfloor !} e^{-\lambda t} = 2\lambda t \mathrm{P}(N_t = \lfloor \lambda t \rfloor).
\end{aligned}$$

Problem 9.53: Let $(N_t)_{t\geq 0}$ be a Poisson process with rate λ and $(T_n)_{n\geq 0}$ the sequence of its arrival times. T_{N_t} is the time of the last arrival before t (0 if $N_t = 0$). Give an expression for $\mathrm{E}[T_{N_t}]$.

Solution. We start by observing that $T_{N_t} \leq t$ and that for $s \in [0,t]$, $T_{N_t} \leq s \Leftrightarrow N_t = N_s$. For $s \in [0,t]$,

$$\mathrm{P}(T_{N_t} \leq s) = \mathrm{P}(N_t - N_s = 0) = \mathrm{P}(N_{t-s} = 0) = e^{-\lambda(t-s)}.$$

Therefore,

$$\mathrm{E}[T_{N_t}] = \int_0^t \mathrm{P}(T_{N_t} > s)ds = \int_0^t \left(1 - e^{-\lambda(t-s)}\right)ds = t - \frac{1}{\lambda}\left(1 - e^{-\lambda t}\right).$$

Problem 9.54: Let $(N_t)_{t\geq 0}$ be a Poisson process with rate λ and $(T_n)_{n\geq 0}$ the sequence of its arrival times. T_{N_t} is the time of the last arrival before t (0 if $N_t = 0$) and $t - T_{N_t}$ is the time at t elapsed since the last arrival. For $s \leq t$, give an expression for $\mathrm{E}[t - T_{N_t}|N_s]$.

Solution. Recall that $T_{N_t} \leq t$ and that for $x \in [0,t]$, $t - T_{N_t} > x \Leftrightarrow N_t = N_{t-x}$. For $x \in [0,t]$ and $m \in \mathbb{Z}^+$,

$$\begin{aligned}
&\mathrm{P}(t - T_{N_t} > x, N_s = m)\\
&\quad = \mathrm{P}(N_t - N_{t-x} = 0, N_s = m)\\
&\quad = \begin{cases} \mathrm{P}(N_t - N_{t-x} = 0, N_s = m) & s \leq t - x \text{ (i.e. } x \leq t - s)\\ \mathrm{P}(N_t - N_{t-x} = 0, N_{t-x} = m) & s > t - x \text{ (i.e. } x > t - s)\end{cases}\\
&\quad = \mathrm{P}(N_t - N_{t-x} = 0, N_{s\wedge(t-x)} = m)\\
&\quad = \mathrm{P}(N_t - N_{t-x} = 0)\mathrm{P}(N_{s\wedge(t-x)} = m)\\
&\quad = e^{-\lambda x}e^{-\lambda(s\wedge(t-x))}\frac{\lambda^m(s\wedge(t-x))^m}{m!}.
\end{aligned}$$

Therefore,

$$\begin{aligned}
\mathrm{P}(t - T_{N_t} > x|N_s = m) &= \frac{e^{-\lambda x}e^{-\lambda(s\wedge(t-x))}\frac{\lambda^m(s\wedge(t-x))^m}{m!}}{e^{-\lambda s}\frac{\lambda^m s^m}{m!}}\\
&= e^{-\lambda(x\wedge(t-s))}\frac{(s\wedge(t-x))^m}{s^m}
\end{aligned}$$

and

$$\begin{aligned}
\mathrm{E}[t - T_{N_t}|N_s = m] &= \int_0^t \mathrm{P}(t - T_{N_t} > x|N_s = m)dx \\
&= \int_0^t e^{-\lambda(x\wedge(t-s))}\frac{(s\wedge(t-x))^m}{s^m}dx \\
&= \int_0^{t-s} e^{-\lambda x}dx + \int_{t-s}^t e^{-\lambda(t-s)}\frac{(t-x)^m}{s^m}dx \\
&= \frac{1}{\lambda}(1 - e^{-\lambda(t-s)}) + e^{-\lambda(t-s)}\frac{s}{m+1}.
\end{aligned}$$

We conclude that $\mathrm{E}[t - T_{N_t}|N_s] = \frac{1}{\lambda}(1 - e^{-\lambda(t-s)}) + e^{-\lambda(t-s)}\frac{s}{N_s+1}$. In particular, we see that $t - T_{N_t}$ is not independent of N_s. In fact, conditional on $\{N_s = m\}$, T_m has the law of $\max(U_1, \ldots, U_m)$ for a sequence of independent uniform over $[0, s]$ random variables. As such, for a large m, T_m is expected to be near s and the gap $t - T_{N_t}$ is expected to be reduced.

We can also recover the result of Problem 9.53 by letting $s = 0$ and $m = 0$ or alternatively taking expectation and observing that

$$\mathrm{E}\left[\frac{1}{N_s+1}\right] = \int_0^1 \mathrm{E}\left[u^{N_s}\right]du = \int_0^1 \left(e^{-\lambda s(1-u)}\right)du = \frac{1}{\lambda s}\left(1 - e^{-\lambda s}\right).$$

Warning: Let $(\mathcal{F}_t)_{t\geq 0}$ be the natural filtration of $(N_t)_{t\geq 0}$. One may be tempted to use the Markov property to reduce $\mathrm{E}[t - T_{N_t}|\mathcal{F}_s]$ to $\mathrm{E}[t - T_{N_t}|N_s]$. This would be wrong as, through the sequence $(T_n)_{n\geq 0}$, $t - T_{N_t}$ depends on information about the past (before s) of the process. In fact, $t - T_{N_t}$ is a Markov process that generates the same filtration as N_t and one can show that $\mathrm{E}[t - T_{N_t}|\mathcal{F}_s] = \mathrm{E}[t - T_{N_t}|s - T_{N_s}]$.

Problem 9.55: Let $(N_t)_{t\geq 0}$ be a Poisson process with arrival times $(T_n)_n$ and completed natural filtration $(\mathcal{F}_t)_{t\geq 0}$, and let S be a stopping time with respect to $(\mathcal{F}_t)_{t\geq 0}$. Use the optional stopping theorem 7.14 to show that if $\mathrm{P}(T_n \leq S < T_{n+1}) = 1$, then $\mathrm{P}(S = T_n) = 1$. Construct a stopping time S such that $\mathrm{P}(T_n \leq S \leq T_{n+1}) = 1$ and $\mathrm{P}(T_n < S < T_{n+1}) > 0$.

Solution. Applying the optional stopping theorem to the martingale $M_t = N_t - \lambda t$ and the bounded stopping times $T_n \wedge t$ and $S \wedge t$, we get that $\mathrm{E}[N_{T_n\wedge t} - \lambda(T_n \wedge t)] = \mathrm{E}[N_{S\wedge t} - \lambda(S \wedge t)]$ and equivalently $\mathrm{E}[\lambda(T_n \wedge t) - \lambda(S \wedge t)] = \mathrm{E}[N_{T_n\wedge t} - N_{S\wedge t}]$. However, $N_{T_n} = N_S$ and

$N_{T_n \wedge t} = N_{T_n} \wedge N_t = N_S \wedge N_t = N_{S \wedge t}$. Therefore, $\mathrm{E}[T_n \wedge t] = \mathrm{E}[S \wedge t]$ and, by monotone convergence, $\mathrm{E}[T_n] = \mathrm{E}[S]$. Since $T_n \leq S$, we deduce that almost surely, $S = T_n$.

The stopping time $S = (T_n \vee t) \wedge T_{n+1}$, for $t > 0$, satisfies the requirements. See Problem 9.57.

Problem 9.56: Let $(N_t)_{t\geq 0}$ be a Poisson process with arrival times $(T_n)_n$ and completed natural filtration $(\mathcal{F}_t)_{t\geq 0}$. Show that, for any given $t > 0$, $\{T_n \leq t < T_{n+1}\} \notin \mathcal{F}_{T_n}$ (or equivalently that $\{T_{n+1} > t\} \notin \mathcal{F}_{T_n}$).

Solution. We reason by contradiction and suppose that $\{T_n \leq t < T_{n+1}\} \in \mathcal{F}_{T_n}$. Then, for $s < t$, $\{T_n \leq t < T_{n+1}\} \cap \{T_n \leq s\} \in \mathcal{F}_s$ and

$$\mathrm{E}[N_t - \lambda t, T_n \leq s < t < T_{n+1}] = \mathrm{E}[N_s - \lambda s, T_n \leq s < t < T_{n+1}].$$

Since $N_t = N_s = n$ on $\{T_n \leq s < t < T_{n+1}\}$, we deduce that

$$(n - \lambda t)\mathrm{P}(T_n \leq s < t < T_{n+1}) = (n - \lambda s)\mathrm{P}(T_n \leq s < t < T_{n+1}),$$

which leads to the contradiction that $t\mathrm{P}(T_n \leq s < t < T_{n+1}) = s\mathrm{P}(T_n \leq s < t < T_{n+1})$. We conclude that $\{T_n \leq t < T_{n+1}\} \notin \mathcal{F}_{T_n}$.

Clearly, $\{T_n \leq t < T_{n+1}\} \in \mathcal{F}_{T_{n+1}-}$. Also, since $T_n < T_{n+1}$, $\mathcal{F}_{T_n} \subset \mathcal{F}_{T_{n+1}-}$. Therefore, we have just established that $\mathcal{F}_{T_n} \subsetneq \mathcal{F}_{T_{n+1}-}$.

Problem 9.57: Let $(N_t)_{t\geq 0}$ be a Poisson process with arrival times $(T_n)_n$ and completed natural filtration $(\mathcal{F}_t)_{t\geq 0}$. Show that, for any given $0 \leq t < u$, $\{(T_n \vee t) \wedge T_{n+1} \leq u < T_{n+1}\} \notin \mathcal{F}_{(T_n \vee t) \wedge T_{n+1}}$.

Solution. We follow the same approach as in Problem 9.56, reason by contradiction and suppose that $\{(T_n \vee t) \wedge T_{n+1} \leq u < T_{n+1}\} \in \mathcal{F}_{(T_n \vee t) \wedge T_{n+1}}$. Then, for $s \in (t, u)$, $\{(T_n \vee t) \wedge T_{n+1} \leq u < T_{n+1}\} \cap \{(T_n \vee t) \wedge T_{n+1} \leq s\} \in \mathcal{F}_s$. Since on this set, $T_n \leq s < u < T_{n+1}$ and $N_s = N_u = n$, we reach the contradiction that

$$\begin{aligned}
&(n - \lambda u)\mathrm{P}((T_n \vee t) \wedge T_{n+1} \leq s < u < T_{n+1}) \\
&= \mathrm{E}[N_u - \lambda u, (T_n \vee t) \wedge T_{n+1} \leq s < u < T_{n+1}] \\
&= \mathrm{E}[N_s - \lambda s, (T_n \vee t) \wedge T_{n+1} \leq s < u < T_{n+1} \\
&= (n - \lambda s)\mathrm{P}((T_n \vee t) \wedge T_{n+1} \leq s < u < T_{n+1}).
\end{aligned}$$

The result immediately follows:

In summary, although there are no stopping times strictly between T_n and T_{n+1} ($(T_n \vee t) \wedge T_{n+1}$ can equal either of the extremities T_n and T_{n+1}), the filtration does not remain constant and we have

$$\mathcal{F}_{T_n} \subsetneq \mathcal{F}_{(T_n \vee t)\wedge T_{n+1}} \subsetneq \mathcal{F}_{T_{n+1}}.$$

Chapter 10

Change of Probability Measure

Change of Measure for Random Variables

Let $(\Omega, \mathcal{F})$ be a measurable space on which two probability measures P and Q are defined. We make the following definitions:

- Q is absolutely continuous with respect to P $(\mathrm{Q} << \mathrm{P})$ if $\mathrm{P}(A) = 0 \Rightarrow \mathrm{Q}(A) = 0$ for $A \in \mathcal{F}$.
- P and Q are equivalent $(\mathrm{P} \sim \mathrm{Q})$ if $\mathrm{P} << \mathrm{Q}$ and $\mathrm{Q} << \mathrm{P}$.
- P and Q are singular $(\mathrm{P} \perp \mathrm{Q})$ if $\mathrm{P}(A) = 0$ and $\mathrm{Q}(A) = 1$ for *some* $A \in \mathcal{F}$.

If $\mathrm{Q} << \mathrm{P}$, then by the Radon–Nikodym theorem, there exists a (unique except for the values on a null event) random variable $\Lambda \geq 0$ with $\mathrm{E}(\Lambda) = 1$ such that

$$\mathrm{Q}(A) = \mathrm{E}_P(\Lambda I_A) = \int_A \Lambda \, d\mathrm{P} \quad \text{for } A \in \mathcal{F}.$$

For obvious reasons, one uses the notation $d\mathrm{Q}/d\mathrm{P}$ for Λ, and it is called the Radon–Nikodym derivative and also the density of Q with respect to P. Further, by the so-called Lebesgue decomposition (which can be established as an exercise on what we have done so far in this chapter), there exist two probability measures Q^c and Q^s on $(\Omega, \mathcal{F})$ such that $\mathrm{Q} = \mathrm{Q}^c + \mathrm{Q}^s$ with $\mathrm{Q}^c << \mathrm{P}$ and $\mathrm{Q}^s \perp \mathrm{P}$.

Conversely, given a probability space $(\Omega, \mathcal{F}, \mathrm{P})$ and a random variable $\Lambda \geq 0$ with $\mathrm{E}(\Lambda) = 1$, one may define a new probability Q on $(\Omega, \mathcal{F})$ by $\mathrm{Q}(A) = \mathrm{E}(\Lambda I_A)$ for $A \in \mathcal{F}$. In that case, we have $\mathrm{Q} << \mathrm{P}$ with $d\mathrm{Q}/d\mathrm{P} = \Lambda$.

There is an abstract version of Bayes' theorem from elementary probability theory called the general Bayes' formula: if $\mathrm{Q} \ll \mathrm{P}$ with $d\mathrm{Q}/d\mathrm{P} = \Lambda$ and $\mathcal{G} \subseteq \mathcal{F}$ is a σ-field, then we have

$$\mathrm{E}_Q(X|\mathcal{G}) = \frac{\mathrm{E}_P(X\Lambda|\mathcal{G})}{E_P(\Lambda|\mathcal{G})}$$

for random variables X with $\mathrm{E}_Q(|X|) < \infty$. The proof is a very useful exercise on what we have done so far in this chapter.

Let X be a continuously distributed (in the elementary sense) random variable with PDF $f : \mathbb{R} \to \mathbb{R}$ defined on a probability space $(\Omega, \mathcal{F}, P)$. For any other PDF $g : \mathbb{R} \to \mathbb{R}$ such that $g^{-1}(\{0\}) \supseteq f^{-1}(\{0\})$, we may define a new probability measure Q on $(\Omega, \mathcal{F})$ by

$$\mathrm{Q}(A) = \mathrm{E}_P\left(\frac{g(X)I_A}{f(X)}\right) \quad \text{for } A \in \mathcal{F}$$

so that $d\mathrm{Q}/d\mathrm{P} = g(X)/f(X)$ [where $g(x)/f(x) = 0$ when $f(x) = 0$]. It is easy to see that X will have PDF g when viewed as a random variable on $(\Omega, \mathcal{F}, \mathrm{Q})$. Note that the definitions of a random variable as a measurable function $X : \Omega \to \mathbb{R}$ do not involve any probability measure. The probability distribution of X will depend on what probability measure we select to use. If we change that measure, the probability distribution of (the one and same random variable), X will change in general.

Let X be a normal $N(\mu_1, \sigma_1^2)$-distributed random variable defined on a probability space $(\Omega, \mathcal{F}, \mathrm{P})$. Taking

$$\Lambda = \frac{\sigma_1}{\sigma_2} \exp\left(\frac{(X-\mu_1)^2}{2\sigma_1^2} - \frac{(X-\mu_2)^2}{2\sigma_2^2}\right),$$

it follows from what we did in the previous paragraph that X is normal $N(\mu_2, \sigma_2^2)$-distributed on the probability space $(\Omega, \mathcal{F}, \mathrm{Q})$ with probability measure $\mathrm{Q}(A) = \mathrm{E}_P(\Lambda I_A)$ for $A \in \mathcal{F}$ so that $\mathrm{Q} \ll \mathrm{P}$ with $d\mathrm{Q}/d\mathrm{P} = \Lambda$.

Change of Measure for Processes

Let $\{\Lambda(t)\}_{t\in[0,T]}$ be a positive martingale on a filtered probability space $(\Omega, \mathcal{F}, \{\mathcal{F}_t\}_{t\in[0,T]}, \mathrm{P})$ such that $E(\Lambda(T)) = 1$. Define a new probability Q on $(\Omega, \mathcal{F})$ by $\mathrm{Q}(A) = \mathrm{E}_P(\Lambda(T)I_A)$ for $A \in \mathcal{F}$. By application of the general Bayes' formula, it follows that

$$\mathrm{E}_Q(X|\mathcal{F}_t) = \mathrm{E}_P\left(\frac{\Lambda(T)}{\Lambda(t)}X \,\middle|\, \mathcal{F}_t\right) \quad \text{for } t \in [0, T]$$

whenever X is a random variable with $\mathrm{E}_Q(|X|) < \infty$. If in addition, X is $\mathcal{F}_t$-measurable for a $t \in [0,T]$, it further holds that

$$\mathrm{E}_Q(X|\mathcal{F}_s) = \mathrm{E}_P\left(\frac{\Lambda(t)}{\Lambda(s)}\, X \,\middle|\, \mathcal{F}_s\right) \quad \text{for } s \in [0,t].$$

Now, an adapted process $\{M(t)\}_{t\in[0,T]}$ is a Q-martingale if and only if $\{\Lambda(t)M(t)\}_{t\in[0,T]}$ is a P-martingale. In particular, $\{1/\Lambda(t)\}_{t\in[0,T]}$ is a Q-martingale.

As we have seen in the previous paragraph (and unsurprisingly one must say), conditional expectations change with the change of probability measure. However, quadratic variation and covariation do not. More specifically, if a sequence of random variables $\{X_n\}_{n=1}^{\infty}$ satisfy $X_n \to_P X$ and if $\mathrm{Q} << \mathrm{P}$, then it follows (from the so-called absolute continuity of the Lebesgue integral) that $X_n \to_Q X$. And so, if a quadratic variation or covariation is well defined in the sense of convergence in probability on the probability space $(\Omega, \mathcal{F}, \mathrm{P})$, then it is well defined on the probability space $(\Omega, \mathcal{F}, \mathrm{Q})$ as well and with the same value.

One of the two main results of this chapter is Girsanov's theorem for martingales: let $\{M_1(t)\}_{t\in[0,T]}$ and $\{X(t)\}_{t\in[0,T]}$ be continuous martingales on a filtered probability space $(\Omega, \mathcal{F}, \{\mathcal{F}_t\}_{t\in[0,T]}, \mathrm{P})$. Suppose that $\{\mathcal{E}(X)(t)\}_{t\in[0,T]}$ is a martingale and define a new probability measure Q by

$$\frac{d\mathrm{Q}}{d\mathrm{P}} = \mathcal{E}(X)(T) = e^{X(T)-\frac{1}{2}[X](T)}.$$

Then, the process

$$M_2(t) = M_1(t) - [M_1, X](t) \quad \text{for } t \in [0,T]$$

is a Q-martingale. The proof is by checking that M_2 can be written as a stochastic integral with respect to a martingale.

Recall the Kazamaki and Novikov conditions for checking whether stochastic exponentials of (local) continuous martingales are martingales from Chapter 8.

The other main result is Girsanov's theorem for change of drift in diffusions: let $\{X(t)\}_{t\in[0,T]}$ satisfy the (BM) SDE

$$dX(t) = \mu_1(X(t),t)\, dt + \sigma(X(t),t)\, dB(t) \quad \text{for } t \in [0,T],$$

where σ is assumed to be strictly positive for simplicity. Select a new drift coefficient $\mu_2 : \mathbb{R} \times [0,T] \to \mathbb{R}$ and set

$$H(t) = \frac{\mu_2(X(t),t) - \mu_1(X(t),t)}{\sigma(X(t),t)} \quad \text{for } t \in [0,T].$$

Suppose that $\{\mathcal{E}(\int H\,dB)(t)\}_{t\in[0,T]}$ is a martingale and define a new probability measure Q by

$$\frac{d\mathrm{Q}}{d\mathrm{P}} = \mathcal{E}(\textstyle\int H\,dB)(T) = \exp\biggl(\displaystyle\int_0^T H\,dB - \frac{1}{2}\int_0^T H(t)^2\,dt\biggr).$$

Then, the process

$$W(t) = B(t) - \int_0^t H(s)\,ds \quad \text{for } t \in [0,T]$$

is a BM on the filtered probability space $(\Omega, \mathcal{F}, \{\mathcal{F}_t^B\}_{t\in[0,T]}, \mathrm{Q})$ and X satisfies the SDE

$$dX(t) = \mu_2(X(t),t)\,dt + \sigma(X(t),t)\,dW(t) \quad \text{for } t \in [0,T]$$

(on that probability space). A direct proof that W is BM is by Lévy's characterisation of BM. It is quite easy when suitably set up from the beginning. However, by employing Girsanov's theorem for martingales, it is sufficient to prove that $[W](t) = t$, which in turn of course is immediate. (The fact that X satisfies the new SDE is by elementary algebra.)

Change of Measure for Point Processes

We state two theorems for change of measure for point processes. Although the second theorem strictly includes and generalises the first one, we state both because the second theorem is somewhat more complicated.

The first theorem shows how to change the rate for an ordinary (elementary) Poisson process: let $\{N(t)\}_{t\in[0,T]}$ be a Poisson process with rate $\lambda > 0$ on a probability space $(\Omega, \mathcal{F}, \mathrm{P})$. Define a new probability measure by

$$\frac{d\mathrm{Q}}{d\mathrm{P}} = e^{(\lambda-\mu)T - N(T)(\ln(\lambda)-\ln(\mu))}.$$

Then, $\{N(t)\}_{t\in[0,T]}$ is a Poisson process with rate $\mu > 0$ on the probability space $(\Omega, \mathcal{F}, \mathrm{Q})$. The proof can be carried out in a more or less elementary manner.

The second theorem shows that an ordinary Poisson process can acquire more or less any stochastic intensity by means of change of measure: let $\{N(t)\}_{t\in[0,T]}$ be a Poisson process with rate 1 on a filtered probability space $(\Omega, \mathcal{F}, \{\mathcal{F}_t\}_{t\in[0,T]}, \mathrm{P})$. Write $M(t) = N(t) - t$ for $t \in [0,T]$ and suppose

$\{\lambda(t)\}_{t\in[0,T]}$ is a predictable process such that $\{\mathcal{E}(\int(\lambda-1)\,dM)(t)\}_{t\in[0,T]}$ is a martingale. Define a new probability measure by

$$\frac{d\mathrm{Q}}{d\mathrm{P}}=\mathcal{E}(\textstyle\int(\lambda-1)\,dM)(T)=\exp\left(\displaystyle\int_0^T(1-\lambda(s))\,ds+\int_0^T\ln(\lambda)\,dN\right).$$

Then, $\{N(t)\}_{t\in[0,T]}$ is a point process with stochastic intensity $\{\lambda(t)\}_{t\in[0,T]}$ on the filtered probability space $(\Omega,\mathcal{F},\{\mathcal{F}_t\}_{t\in[0,T]},\mathrm{Q})$. Conversely, to each probability measure $\mathrm{Q}\ll\mathrm{P}$, there exists a predictable process $\{\lambda(t)\}_{t\in[0,T]}$ such that $\{N(t)\}_{t\in[0,T]}$ is a point process with stochastic intensity $\{\lambda(t)\}_{t\in[0,T]}$.

Likelihood Functions

Now, we see how Girsanov's theorem for change of drift in diffusions can be used to make statistical inferences (hypotheses testing and estimation) of the drift coefficient.

Again, let $\{X(t)\}_{t\in[0,T]}$ satisfy the (BM) SDE

$$dX(t)=\mu_1(X(t),t)\,dt+\sigma(X(t),t)\,dB(t)\quad\text{for }\ t\in[0,T],$$

where σ is strictly positive. Select another drift $\mu_2:\mathbb{R}\times[0,T]\to\mathbb{R}$ and set

$$H(t)=\frac{\mu_2(X(t),t)-\mu_1(X(t),t)}{\sigma(X(t),t)}\quad\text{for}\,t\in[0,T].$$

Suppose that $\{\mathcal{E}(\int H\,dB)(t)\}_{t\in[0,T]}$ is a martingale and define a new probability measure Q by the so-called likelihood ratio

$$\begin{aligned}\frac{d\mathrm{Q}}{d\mathrm{P}}&=\exp\left(\int_0^T H\,dB-\frac{1}{2}\int_0^T H(t)^2\,dt\right)\\&=\exp\left(\int_0^T\frac{\mu_2(X(t),t)-\mu_1(X(t),t)}{\sigma(X(t),t)^2}\,dX(t)\right.\\&\qquad\left.-\frac{1}{2}\int_0^T\frac{\mu_2(X(t),t)^2-\mu_1(X(t),t)^2}{\sigma(X(t),t)^2}\,dt\right).\end{aligned}$$

Recall that

$$W(t)=B(t)-\int_0^t H(s)\,ds\quad\text{for }\ t\in[0,T]$$

is a Q-BM and that X satisfies the SDE

$$dX(t)=\mu_2(X(t),t)\,dt+\sigma(X(t),t)\,dW(t)\quad\text{for }\ t\in[0,T].$$

With the framework in the previous paragraph, assume that we have made an observation of $\{X(t)\}_{t\in[0,T]}$ and want to make statistical inferences about the drift coefficient. In a hypotheses test of

$$H_0 : dX(t) = \mu_1(X(t),t)\,dt + \sigma(X(t),t)\,dB(t) \quad (= \text{the drift coefficient is } \mu_1)$$

against

$$H_1 : dX(t) = \mu_2(X(t),t)\,dt + \sigma(X(t),t)\,dW(t)$$
$$(= \text{the drift coefficient is } \mu_2),$$

we simply reject H_0 if $d\mathrm{Q}/d\mathrm{P}$ is sufficiently much larger than 1. Similarly, we can make the estimation of the drift by taking $\mu_1 = 0$ and maximising $d\mathrm{Q}/d\mathrm{P}$ with respect to μ_2, where $\mu_2 = \mu$, which gives the maximum, is the estimated drift. (Of course, typically, this requires a parametric choice of μ_2 to be carried out in practice.) These statistical procedures can be given a solid theoretical framework with appropriate central limit theorems, convergence rates, and optimality properties. However, the theory is rather complicated.

The diffusion coefficient $\sigma : \mathbb{R} \times [0,T] \to \mathbb{R}$ cannot be accessed by likelihood ratios. Here, one can instead employ quadratic variation and simply make use of the fact that

$$d[X](t) = \sigma(X(t),t)^2\,dt \quad \text{for } t \in [0,T].$$

One fits a suitable σ to this equation with the observation $\{X(t)\}_{t\in[0,T]}$ inserted, for example, using least squares. (Again, typically, this requires a parametric model for σ.)

Problems

Problem 10.1: In the following list of probability measures on the random variable X, find all pairs of equivalent measures:

(a)	$X \sim N(0,1)$,	(g)	$X \sim \text{Uniform}(0,2)$,
(b)	$X \sim \text{Cauchy}(0,1)$,	(h)	$X \sim \text{Binom}(1,0.5)$,
(c)	$X \sim LN(0,1)$,	(i)	$X \sim \text{Binom}(1,0.1)$,
(d)	$X \sim LN(1,2)$,	(j)	$X \sim \text{Binom}(2,0.5)$,
(e)	$X \sim \text{Exponential}(1)$,	(k)	$X \sim \text{Poisson}(1)$,
(f)	$X \sim \text{Uniform}(0,1)$,	(l)	$X \sim \text{Poisson}(2)$.

Solution. We can divide the choices into groups. By support, we mean sets that have positive measures, more precisely closure of the set of possible values:

- (a) and (b) are equivalent since the support is the real line. Note that (a) has mean zero but (b) has no mean.
- (c), (d), (e) are pair-wise equivalent since the support is the positive real line.
- (f): the support is the interval $[0, 1]$.
- (g): the support is the interval $[0, 2]$.
- (h) and (i) are equivalent since the support is $X = 0$ and $X = 1$.
- (j): the support is $X = 0$, $X = 1$ and $X = 2$.
- (k) and (l) are equivalent since the support is the non-negative integers.

Problem 10.2: Let P be $N(\mu_1, 1)$ and Q be $N(\mu_2, 1)$ on $\mathbb{R}$. Show that they are equivalent and that the Radon–Nikodym derivative $dQ/dP = \Lambda$ is given by $\Lambda(x) = e^{(\mu_2-\mu_1)x+\frac{1}{2}(\mu_1^2-\mu_2^2)}$. Give also dP/dQ.

Solution. Since P and Q are defined with the help of their densities with respect to the Lebesgue measure, for any (Borel) set A, $P(A) = 0$ if and only if $\text{Leb}(A) = 0$, if and only if $Q(A) = 0$, this shows that P and Q are equivalent. The Radon–Nikodym derivative

$$\Lambda(x) = \frac{dQ}{dP}(x) = \frac{\frac{1}{\sqrt{2\pi}}e^{-\frac{1}{2}(x-\mu_2)^2}}{\frac{1}{\sqrt{2\pi}}e^{-\frac{1}{2}(x-\mu_1)^2}} = e^{(\mu_2-\mu_1)x+\frac{1}{2}(\mu_1^2-\mu_2^2)},$$

and $\frac{dP}{dQ}(x) = \frac{1}{\Lambda(x)} = e^{(\mu_1-\mu_2)x+\frac{1}{2}(\mu_2^2-\mu_1^2)}$.

Problem 10.3: Show that if X has $N(\mu, 1)$ distribution under P, then there is an equivalent measure Q such that X has $N(0, 1)$ distribution under Q. Give the likelihood dQ/dP and also dP/dQ. Give the Q-distribution of $Y = X - \mu$.

Solution. Let $\Lambda(x) = e^{-\frac{1}{2}x^2+\frac{1}{2}(x-\mu)^2} = e^{-\mu x+\frac{1}{2}\mu^2}$. Define measure Q by $Q(A) = \int_A \Lambda(x)dP = E_P(I(A)\Lambda(X))$. Then, by the relation for expectation under different measures (Equation (10.4*)), for any random variable X,

$$E_Q X = E_P(\Lambda X),$$

we have for the moment generating function of X under Q,

$$\begin{aligned}\mathrm{E}_Q(e^{uX}) &= \mathrm{E}_P(e^{uX}\Lambda(X)) = \mathrm{E}_P(e^{(u-\mu)X+\frac{1}{2}\mu^2})\\ &= e^{\mu(u-\mu)+\frac{1}{2}(u-\mu)^2}e^{\frac{1}{2}\mu^2} = e^{\frac{1}{2}u^2},\end{aligned}$$

that is, X is $N(0,1)$ under Q. Since $0 < \Lambda(x) < \infty$ for all x, P and Q are equivalent with

$$\frac{d\mathrm{Q}}{d\mathrm{P}}(x) = \Lambda(x) = e^{-\mu x+\frac{1}{2}\mu^2}, \qquad \frac{d\mathrm{P}}{d\mathrm{Q}}(x) = \frac{1}{\Lambda}(x) = e^{\mu x-\frac{1}{2}\mu^2}.$$

The Q-distribution of $Y = X - \mu$ is $N(-\mu, 1)$.

Problem 10.4: Find $\mathrm{E}e^X 1_{X\in A}$ for $X \sim N(\mu, \sigma^2)$ by using the change of measure with $\Lambda = e^{-\mu-\sigma^2/2}e^X$, see Problem above 10.3.

Solution. $\mathrm{E}e^X = \mathrm{E}e^{N(\mu,\sigma^2)} = e^{\mu+\sigma^2/2}$. $\Lambda = e^X/\mathrm{E}e^X$. Hence, $\mathrm{E}\Lambda = 1$.

$$\begin{aligned}\mathrm{E}e^X 1_{X\in A} &= \mathrm{E}e^X \mathrm{E}(e^X/\mathrm{E}e^X)1_{X\in A} = e^{\mu+\sigma^2/2}\mathrm{E}\Lambda 1_{X\in A}\\ &= e^{\mu+\sigma^2/2}\mathrm{E}_Q 1_{X\in A} = e^{\mu+\sigma^2/2}Q(X\in A).\end{aligned}$$

Q distribution of X:

$$\begin{aligned}\mathrm{E}_Q e^{uX} &= \mathrm{E}\Lambda e^{uX} = \mathrm{E}e^{-\mu-\sigma^2/2}e^X e^{uX} = e^{-\mu-\sigma^2/2}\mathrm{E}e^{(u+1)X}\\ &= e^{-\mu-\sigma^2/2}\mathrm{MGF}_{N(\mu,\sigma^2)}(u+1) = e^{-\mu-\sigma^2/2}\\ &\quad \times e^{\mu(u+1)+\sigma^2(u+1)^2/2} = e^{(\mu+\sigma^2)u+\sigma^2u^2/2}.\end{aligned}$$

This is MGF of $N(\mu+\sigma^2, \sigma^2)$. Thus, the Q-distribution of X is $N(\mu+\sigma^2, \sigma^2)$.

Proceeding, $\mathrm{E}e^X 1_{X\in A} = e^{\mu+\sigma^2/2}Q(X \in A) = e^{\mu+\sigma^2/2}P(N(\mu + \sigma^2, \sigma^2) \in A)$.

Problem 10.5:

(a) Let X be a standard normal distributed random variable. Show how X can be made to have any given probability density function $f : \mathbb{R} \to [0, \infty)$ by means of a change of probability measure.

(b) Let X have probability density function $f : \mathbb{R} \to [0,\infty)$. Is it possible to make X have a standard normal distributed by a change of probability measure?

Solution.

(a) Clearly, X has probability density function f under the probability measure

$$\mathrm{Q}(A)=\int_A f(X)\,\sqrt{2\pi}\,\mathrm{e}^{X^2/2}\,d\mathrm{P}\quad\text{for } A\in\mathcal{F},$$

as this gives

$$\begin{aligned}\mathrm{Q}(X\in B)&=\mathrm{E}_Q I(X\in B)\\&=\mathrm{E}_P\big\{I_{\{X\in B\}}\,f(X)\,\sqrt{2\pi}\,\mathrm{e}^{X^2/2}\big\}\\&=\int_{\mathbb{R}} I_B(x)\,f(x)\,\sqrt{2\pi}\,\mathrm{e}^{x^2/2}\frac{1}{\sqrt{2\pi}}\,\mathrm{e}^{-x^2/2}\,dx\\&=\int_B f(x)\,dx\quad\text{for } B\subseteq\mathbb{R}.\end{aligned}$$

We have used that for a function h on $\mathbb{R}$,

$$\mathrm{E}h(X)=\int_\Omega h(X)dP=\int_{\mathbb{R}} h(x)dF(x),$$

see Theorem 2.6.

(b) If X has a strictly positive probability density function $f:\mathbb{R}\to(0,\infty)$, then X is standard normal distributed under the probability measure

$$\mathrm{Q}(A)=\int_A \frac{1}{\sqrt{2\pi}}\,\mathrm{e}^{-X^2/2}\,\frac{1}{f(X)}\,d\mathrm{P}\quad\text{for } A\in\mathcal{F},$$

as this gives

$$\begin{aligned}\mathrm{Q}(X\in B)&=\mathrm{E}_Q I_{\{X\in B\}}\\&=\mathrm{E}_P\bigg\{I_{\{X\in B\}}\,\frac{1}{\sqrt{2\pi}}\,\mathrm{e}^{-X^2/2}\,\frac{1}{f(X)}\bigg\}\\&=\int_{\mathbb{R}} I_B(x)\,\frac{1}{\sqrt{2\pi}}\,\mathrm{e}^{-x^2/2}\,\frac{1}{f(x)}\,f(x)\,dx\\&=\int_B \frac{1}{\sqrt{2\pi}}\,\mathrm{e}^{-x^2/2}\,dx\quad\text{for } B\subseteq\mathbb{R}.\end{aligned}$$

If f is not strictly positive, then it is not possible to make X standard normal distributed by means of this approach, as we then have

$$\begin{aligned}
\mathrm{Q}(\Omega) &= \mathrm{E}_Q 1 \\
&= \mathrm{E}_P\left\{ \frac{1}{\sqrt{2\pi}}\, \mathrm{e}^{-X^2/2}\, \frac{1}{f(X)} \right\} \\
&= \int_{\{x\in\mathbb{R}: f(x)>0\}} \frac{1}{\sqrt{2\pi}}\, \mathrm{e}^{-x^2/2}\, \frac{1}{f(x)}\, f(x)\, dx \\
&= \int_{\{x\in\mathbb{R}: f(x)>0\}} \frac{1}{\sqrt{2\pi}}\, \mathrm{e}^{-x^2/2}\, dx \ < 1,
\end{aligned}$$

so that Q is no longer a probability measure.

Problem 10.6: Let $X(t) = B(t) + \sin(t)$ for a P-Brownian motion $B(t)$. Let Q be an equivalent measure to P such $X(t)$ is a Q-Brownian motion. Give $\Lambda = d\mathrm{Q}/d\mathrm{P}$.

Solution. Since $\sin(t) = \int_0^t \cos(s)ds$, $X(t) = B(t) + \int_0^t \cos(s)ds$.

Theorem 10.16 states that if $\Lambda = \mathcal{E}(-\int_0^T H(s)dB(s))$ and $d\mathrm{Q}/d\mathrm{P} = \Lambda$, then the process $B(t) + \int_0^t H(s)ds$ is a Q-Brownian motion.

Hence, we obtain with $H(s) = \cos(s)$,

$$\Lambda = \frac{d\mathrm{Q}}{d\mathrm{P}} = \mathcal{E}\left(-\int_0^T \cos(s)dB(s)\right) = e^{-\int_0^T \cos(s)dB(s) - \frac{1}{2}\int_0^T \cos^2(s)ds}.$$

Problem 10.7: Find the measure change so that $X_t = e^{B_t}$ is a Q-martingale. What does B_t become under Q?

Solution. $dX_t = de^{B_t} = e^{B_t}dB_t + \frac{1}{2}e^{B_t}dt$.

Under Q, $B_t = W_t + \int_0^t H_s ds$ with W_t being a Q-Brownian motion.

Writing the above equation under Q, $dX_t = e^{B_t}dW_t + e^{B_t}H_t dt + \frac{1}{2}e^{B_t}dt$. For this to be a martingale, we need the coefficient of dt to be zero, hence $H_t = -\frac{1}{2}$.

So, $B_t = W_t - \frac{1}{2}t$. As we have seen before, $X_t = e^{B_t} = e^{W_t - \frac{1}{2}t}$ is a Q-martingale.

The change of measure $W_t = B_t + \frac{1}{2}t$ is a BM, $\Lambda = dQ/dP = e^{-\frac{1}{2}B_T - \frac{1}{8}T}$.

Problem 10.8: Find the measure change so that $X_t = e^{-t}B_t$ is a Q-martingale. What does B_t become under Q?

Solution. $dX_t = d(e^{-t}B_t) = -e^{-t}B_t dt + e^{-t}dB_t$.
Under Q, $B_t = W_t + \int_0^t H_s ds$ with W_t being Q-Brownian motion:

$$\begin{aligned}dX_t &= d(e^{-t}B_t) = -e^{-t}B_t dt + e^{-t}dB_t\\ &= -e^{-t}B_t dt + e^{-t}dW_t + e^{-t}H_t dt.\end{aligned}$$

Making drift zero, $H_t = B_t$.
$B_t = W_t + \int_0^t B_s ds$. $W_t = B_t - \int_0^t B_s ds$. $\Lambda = dQ/dP = e^{\int_0^T B_s dB_s - \frac{1}{2}\int_0^T B_s^2 ds}$.
We need to check that $e^{\int_0^t B_s dB_s - \frac{1}{2}\int_0^t B_s^2 ds}$ is a martingale. The Novikov condition fails for $T \geq 1$, but it can be shown that it is still a martingale, see Problem 10.26.

Problem 10.9: Let $\boldsymbol{B}$ be an n-dim Brownian motion and $\boldsymbol{H}$ an adapted regular process. Let $\boldsymbol{H}\cdot\boldsymbol{B}(t) = \sum_{i=1}^n \int_0^t H^i(s)dB^i(s)$ be a martingale. Show that the martingale exponential is given by $\exp(\boldsymbol{H}\cdot\boldsymbol{B}(t) - \frac{1}{2}\int_0^t |\boldsymbol{H}(s)|^2 ds)$. *Hint*: Show that the quadratic variation of $\boldsymbol{H}\cdot\boldsymbol{B}$ is given by $\int_0^T |\boldsymbol{H}(s)|^2 ds$, where $|\boldsymbol{H}(s)|^2$ denotes the length of vector $\boldsymbol{H}(s)$.

Solution. Let $X = \boldsymbol{H}\cdot\boldsymbol{B}$. We have $\mathcal{E}(X)(T) = e^{X(T)-X(0)-\frac{1}{2}[X,X](T)}$.
Consider

$$\begin{aligned}[X,X](T) &= [\boldsymbol{H}\cdot\boldsymbol{B}, \boldsymbol{H}\cdot\boldsymbol{B}](T) = \sum_{i=1}^n\sum_{j=1}^n [H^i\cdot B^i, H^j\cdot B^j](T)\\ &= \sum_{i=1}^n\sum_{j=1}^n \int_0^T H^i(s)H^j(s)d[B^i,B^j](s)\\ &= \sum_{i=1}^n \int_0^T \left(H^i(s)\right)^2 ds \quad (\text{since } [B^i,B^j]=0 \text{ for } i\neq j)\\ &= \int_0^T \sum_{i=1}^n \left(H^i(s)\right)^2 ds = \int_0^T |\boldsymbol{H}(s)|^2 ds.\end{aligned}$$

Thus,

$$\begin{aligned}\mathcal{E}(X)(T) = \mathcal{E}(\boldsymbol{H}\cdot\boldsymbol{B})(T) &= e^{\boldsymbol{H}\cdot\boldsymbol{B}(T) - \frac{1}{2}\int_0^T |\boldsymbol{H}(s)|^2 ds}\\ &= e^{\sum_{i=1}^n \int_0^T H^i(s)dB^i(s) - \frac{1}{2}\sum_{i=1}^n \int_0^T (H^i(s))^2 ds}.\end{aligned}$$

Problem 10.10: Let $W_\mu(t) = \mu t + B(t)$, where $B(t)$ is a P-Brownian motion. Show that

$$\mathrm{P}\Big(\max_{t\le T} W_\mu(t) \le y | W_\mu(T) = x\Big) = 1 - e^{\frac{-2y(y-x)}{T}}, \quad y \ge 0, x \le y,$$

and

$$\mathrm{P}\Big(\min_{t\le T} W_\mu(t) \ge y | W_\mu(T) = x\Big) = 1 - e^{\frac{-2y(y-x)}{T}}, \quad y \le 0, x \ge y.$$

Hint: Use the joint distributions (10.38*) and (3.16*).

Solution. By Girsanov's theorem, $W_\mu(t)$ is a Q-Brownian motion by the change of measure (see Equation 10.36*) $\Lambda = \frac{d\mathrm{P}}{d\mathrm{Q}} = e^{\mu W_\mu(T) - \frac{1}{2}\mu^2 T}$. Denote $W^* = W_\mu^*(T) = \max_{s\le T} W_\mu(s)$ and $W = W_\mu(T)$. Denote by $p(x,y)$ and $q(x,y)$ the joint density of random variables (W, W^*) under P and Q, respectively. The joint density of Brownian motion and its maximum is known and is given by Theorem 3.21:

$$q(x,y) = \sqrt{\frac{2}{\pi}} \frac{(2y-x)}{T^{3/2}} e^{-\frac{(2y-x)^2}{2T}} I(y \ge 0, x \le y).$$

For a bounded function of two variables h, we have by the change of measure for expectations,

$$\mathrm{E}_P h(W, W^*) = \mathrm{E}_Q \Lambda h(W, W^*) = \mathrm{E}_Q e^{\mu W - \frac{1}{2}\mu^2 T} h(W, W^*),$$

where we have used that $\Lambda = \frac{d\mathrm{P}}{d\mathrm{Q}} = e^{\mu W - \frac{1}{2}\mu^2 T}$.

Now, writing the expectation using joint densities, we obtain

$$\mathrm{E}_P h(W, W^*) = \iint h(x,y) p(x,y)\, dx dy,$$

$$\mathrm{E}_Q e^{\mu W - \frac{1}{2}\mu^2 T} h(W, W^*) = \iint h(x,y) e^{\mu x - \frac{1}{2}\mu^2 T} q(x,y)\, dx dy.$$

This implies the relation between the joint densities $p(x,y)$ and $q(x,y)$ (see also Equation (10.38*)):

$$p(x,y) = e^{\mu x - \frac{1}{2}\mu^2 T} q(x,y) = \sqrt{\frac{2}{\pi}} \frac{(2y-x)}{T^{3/2}} \times e^{\mu x - \frac{1}{2}\mu^2 T - \frac{(2y-x)^2}{2T}} I(y \ge 0, x \le y).$$

Since the P-density of W is that of $N(\mu T, T)$ distribution, $p_W(x) = \frac{1}{\sqrt{2\pi T}} e^{-\frac{(x-\mu T)^2}{2T}}$. Therefore, the conditional density

$$p_{W^*|W}(y|x) = \frac{p(x,y)}{p_W(x)} = \frac{2}{T}(2y-x)e^{-\frac{(2y^2-2xy)}{T}} I(y \geq 0, x \leq y).$$

Therefore, for $y \geq 0$ and $x \leq y$,

$$\begin{aligned}
\mathrm{P}(W^* \leq y | W = x) &= \int_{0 \vee x}^{y} p_{W^*|W}(z|x) dz \\
&= \int_{0 \vee x}^{y} \frac{2(2z-x)}{T} e^{-\frac{(2z^2-2xz)}{T}} dz = \int_{0}^{2y^2-2xy} \frac{1}{T} e^{-u/T} du \\
&= 1 - e^{\frac{-2y(y-x)}{T}}.
\end{aligned}$$

The second assertion

$$\mathrm{P}\Big(\min_{t \leq T} W_\mu(t) \geq y | W_\mu(T) = x\Big) = 1 - e^{\frac{-2y(y-x)}{T}}$$

for $y \leq 0$ and $x \geq y$ can be established similarly, or using the fact that $(-B(t)$ is also a Brownian motion, and $\min B(s) = -\max(-B(s))$.

Problem 10.11: Show that the SDE $dX_t = \cos(X_t)dt + dB_t$ with $X_0 = 0$, $0 \leq t \leq T$, has a unique solution. Give the equivalent probability measure Q, specifying $d\mathrm{Q}/d\mathrm{P}$, such that X_t is a Brownian motion under Q.

Solution. The coefficients of the SDE $\mu(x) = \cos x$, $\sigma(x) = 1$ are Lipschitz and satisfy the linear growth condition. The theorem on existence and uniqueness applies.

Let $H(t) = \cos(X_t)$, and

$$\frac{d\mathrm{Q}}{d\mathrm{P}} = \mathcal{E}\left(-\int H dB\right)_t = e^{-\int_0^t \cos(X_s)dB_s - \frac{1}{2}\int_0^t \cos^2(X_s)ds}.$$

Then, by Theorem 10.16, the process $B_t + \int_0^t H(s)ds = B_t + \int_0^t \cos(X_s)ds = X_t$ is a Q-BM.

Problem 10.12: Solve the SDE $dX_t = \mu(X_t)dt + dB_t$, $0 \leq t \leq T$, where the function μ is bounded and μ^2 is continuous.

Solution. We give a weak solution by a change of measure based on the Girsanov theorem for Brownian motion. If $B(t)$ is a P-Brownian motion and $M(t) = \int_0^t H_s dB(s)$ is a martingale such that its stochastic exponential is also a martingale, then with

$$\frac{dQ}{dP} = \Lambda = e^{M(T) - \frac{1}{2}[M,M](T)},$$

the process $W(t) = B(t) - \int_0^t H_s ds$ is a Q-Brownian motion.

Take $H_s = \mu(B(s))$, then $M(t) = \int_0^t \mu(B(s))dB(s)$ is a martingale by a property of Itô integral since $\mathrm{E}\int_0^T \mu^2(B(s))ds < CT < \infty$. Moreover, $\mathrm{E}e^{\frac{1}{2}[M,M]_t} \leq \mathrm{E}e^{\frac{Ct}{2}} < \infty$, and the Novikov condition holds. Hence, $\mathcal{E}(M)_t$ is a martingale. Therefore, $W(t) = B(t) - \int_0^t \mu(B(s))ds$ is a Q-Brownian motion, and we have that

$$B(t) = \int_0^t \mu(B(s))ds + W(t).$$

In other words, $B(t)$ under the new measure Q is the process that solves this SDE with another Q-Brownian motion $W(t)$.

Problem 10.13: Give a weak solution to the SDE $dX_t = -X_t dt + dB_t$, $X_0 = 0$, by using a change of measure.
Hint: You can use without proof that the martingale exponential Λ_T arising in the appropriate change of measure has expectation 1 for any T, $\mathrm{E}\Lambda_T = 1$.

Solution. Fix an arbitrary $T > 0$. Define $\Lambda_T = \mathcal{E}(-\int_0^{\cdot} B_s dB_s)_T$, and define a new probability measure $dQ/dP = \Lambda_T$.

By the Girsanov theorem, $dB_t + B_t dt = \tilde{B}_t$, is a Q-Brownian motion. Rewriting for B_t, $dB_t = -B_t dt + d\tilde{B}_t$, or in other words, $(V_t = B_t, \tilde{B}_t)$ is a weak solution to the SDE.

Note that another weak solution $(U_t, \hat{B}_t)$ can be obtained by using a change of time in the Itô integral.

Problem 10.14: Let B_t be a Brownian motion on $[0,1]$, and define the new measure Q by $dQ/dP = \Lambda = B_1^2$. Show that under the measure Q, B_t solves the SDE $dX_t = \frac{2X_t}{X_t^2+1-t}dt + dW_t$ with a Q-Brownian motion W.

Solution. We follow the ideas of Theorem 10.19 by finding a process q_t so that Λ_1 is the stochastic exponential of $\int_0^t q_s dB_s$ and then apply the Girsanov theorem.

Since $\Lambda = B_1^2 > 0$ a.s., Q is equivalent to P. Consider

$$\Lambda_t = \mathrm{E}(\Lambda|\mathcal{F}_t) = B_t^2 + 1 - t.$$

This is a martingale, which can be written as $\Lambda_t = 1 + 2\int_0^t B_s dB_s$. Define now

$$q_t = 2B_t/\Lambda_t = \frac{2B_t}{B_t^2 + 1 - t}.$$

Note that $q_t \Lambda_t = 2B_t$; therefore,

$$d\Lambda_t = 2B_t dB_t = q_t \Lambda_t dB_t = \Lambda_t(q_t dB_t) = \Lambda_t dM_t.$$

Hence, Λ_t is the stochastic exponential of M_t with $dM_t = q_t dB_t$:

$$\Lambda_t = \mathcal{E}(M)_t, \quad \Lambda_1 = \mathcal{E}(M)_1.$$

Hence, by the Girsanov theorem, with a Q Brownian motion W_t,

$$B_t = W_t + \int_0^t q_s ds = W_t + \int_0^t \frac{2B_s}{B_s^2 + 1 - s} ds,$$

which proves the result.

Problem 10.15: Let $X(t)$ and $Y(t)$ satisfy the stochastic differential equations with the same diffusion coefficient but different drift functions,

$$dX(t) = \mu_X(X(t), t)dt + \sigma(X(t), t)dB(t),$$
$$dY(t) = \mu_Y(Y(t), t)dt + \sigma(Y(t), t)dB(t),$$

$0 \le t \le T$, and $X_0 = Y_0$. Let $\mathrm{P}_X(A) = P(X \in A)$ and $\mathrm{P}_Y(A) = P(Y \in A)$, for a measurable set $A \subset C[0, T]$, be probability measures induced by these diffusions on the space of continuous functions $C[0, T]$. Show that P_X and P_Y are equivalent, with the Radon–Nikodym derivatives (the likelihoods) given by

$$\frac{d\mathrm{P}_Y}{d\mathrm{P}_X}(X_{[0,T]}) = e^{\int_0^T \frac{\mu_Y(X(t),t) - \mu_X(X(t),t)}{\sigma^2(X(t),t)} dX(t) - \frac{1}{2}\int_0^T \frac{\mu_Y^2(X(t),t) - \mu_X^2(X(t),t)}{\sigma^2(X(t),t)} dt}$$

and

$$\frac{d\mathrm{P}_X}{d\mathrm{P}_Y}(Y_{[0,T]}) = e^{-\int_0^T \frac{\mu_Y(Y(t),t) - \mu_X(Y(t),t)}{\sigma^2(Y(t),t)} dY(t) + \frac{1}{2}\int_0^T \frac{\mu_Y^2(Y(t),t) - \mu_X^2(Y(t),t)}{\sigma^2(Y(t),t)} dt}.$$

Solution. Let $H_t = \frac{\mu_X(X_t,t)-\mu_Y(X_t,t)}{\sigma(X_t,t)}$. Assume that $E\mathcal{E}(\int_0^T H_t dB_t) = 1$. Then, by the Girsanov theorem, with

$$\frac{dQ}{dP_X} = \Lambda = \mathcal{E}\left(-\int_0^T H_t dB_t\right) = e^{-\int_0^T H_t dB_t - \frac{1}{2}\int_0^T H_t^2 dt},$$

$Q \sim P$ and $\hat{B}_t = B_t + \int_0^t H_s ds$ is a Q-Brownian motion. This means that X_t satisfies the following SDE under Q:

$$dX(t) = \mu_Y(X(t),t)dt + \sigma(X(t),t)dB(t).$$

But this is precisely the SDE for Y_t. Hence, $Q = P_Y$. Therefore,

$$\frac{dP_Y}{dP_X} = e^{-\int_0^T H_t dB_t - \frac{1}{2}\int_0^T H_t^2 dt}.$$

It remains to express

$$dB_t = \frac{dX_t - \mu_X(X_t)dt}{\sigma(X_t)}$$

to obtain the desired form. dP_X/dP_Y is obtained by interchanging X and Y.

Problem 10.16: Let $N(t)$, $0 \le t \le T$ be a Poisson process with rate 1 under measure P. Define measure Q by $\frac{dQ}{dP} = e^{(1-\lambda)T - N(T)\ln\lambda}$. Prove that $N(t)$ is a Poisson process with rate λ under Q. (This is Theorem 10.22.)

Solution. Define the process $\Lambda(t) = e^{(1-\lambda)t - N(t)\ln\lambda} = e^{(1-\lambda)t}\lambda^{-N(t)}$. Then, it is easy to check that it is a martingale under P, with respect to the natural filtration of $N(t)$. Indeed, for $s < t$,

$$\begin{aligned} E_P\left(e^{(1-\lambda)t}\lambda^{-N(t)}|\mathcal{F}_s\right) &= e^{(1-\lambda)t}\lambda^{-N(s)}E_P\left(\lambda^{-(N(t)-N(s))}|\mathcal{F}_s\right) \\ &= e^{(1-\lambda)t}\lambda^{-N(s)}E_P\left(\lambda^{-(N(t)-N(s))}\right) \\ &= e^{(1-\lambda)t}\lambda^{-N(s)}e^{(\lambda-1)(t-s)} = \Lambda(s). \end{aligned}$$

Next, we have by the relation for conditional expectations under different measures (the generalized Bayes' formula, Theorem 10.10), we have

$$E_Q\left(e^{u(N(t)-N(s))}|\mathcal{F}_s\right) = E_P\left(e^{u(N(t)-N(s))}\frac{\Lambda(t)}{\Lambda(s)}\Big|\mathcal{F}_s\right).$$

$\Lambda(t)/\Lambda(s) = e^{(1-\lambda)(t-s)+\ln\lambda(N(t)-N(s))}$. Hence,

$$\begin{aligned}
\mathrm{E}_Q\big(e^{u(N(t)-N(s))}|\mathcal{F}_s\big) &= e^{(1-\lambda)(t-s)}\mathrm{E}_P\big(e^{(u+\ln\lambda)(N(t)-N(s))}|\mathcal{F}_s\big)\\
&= e^{(1-\lambda)(t-s)}\mathrm{E}_P\big(e^{(u+\ln\lambda)(N(t)-N(s))}\big)\\
&= e^{(1-\lambda)(t-s)}e^{(t-s)(e^{u+\ln\lambda}-1)} = e^{\lambda(t-s)(e^u-1)}.
\end{aligned}$$

The last line is by the independence property of the increments and a form of the moment-generating function of the Poisson distribution.

Hence, the moment-generating function under Q of the increment $N(t)-N(s)$ is that of the Poisson distribution with parameter $\lambda(t-s)$, and it is also independent of $\mathcal{F}_s$, see Problem 2.25. Thus, under Q, $N(t)$ is a Poisson process with rate λ.

Problem 10.17: Let $N(t)$ be a Poisson process with rate 1 and $\bar{N}(t) = N(t)-t$. Show that for an adapted, continuous, and bounded process $H(t)$, the process $M(t) = \int_0^t H(s)d\bar{N}(s)$ is a martingale for $0 \le t \le T$. Show that

$$\mathcal{E}(M)(t) = e^{-\int_0^t H(s)ds+\int_0^t \ln(1+H(s))dN(s)}.$$

Solution. Since $[\bar{N},\bar{N}](t) = N(t)$ and $H(t)$ is bounded, we get

$$E\int_0^T H^2(t)d[\bar{N},\bar{N}](t) = E\int_0^T H^2(t)dN(t) \le \sup_{t\le T} H(t)N(T) < \infty.$$

Thus, $M(t) = \int_0^t H(s)d\bar{N}(s)$ is a martingale for $0 \le t \le T$. Using the expression for martingale exponential, Equation (9.5*),

$$\begin{aligned}
\mathcal{E}(M)(t) &= e^{M(t)}\prod_{s\le t}(1+\Delta M(s))e^{-\Delta M(s)}\\
&= e^{M(t)+\sum_{s\le t}\log(1+\Delta M(s))-\Delta M(s)}.
\end{aligned}$$

The jumps of the stochastic integral $\int H(s)dX(s)$ occur at the points of jumps of X and $\Delta\int H(s)dX(s) = H(s)\Delta X(s)$ (property 1 of stochastic integral with respect to a semimartingale X, p. 219). Hence,

$$\Delta M(s) = H(s)\Delta\bar{N}(t) = H(s)\Delta N(s).$$

Thus,

$$\mathcal{E}(M)(t) = e^{M(t)+\sum_{s\le t}\log(1+H(s)\Delta N(s))-H(s)\Delta N(s)}.$$

Next, since $\Delta N(s) = 1$ or 0,

$$\sum_{s\le t} \log(1 + H(s)\Delta N(s)) = \int_0^t \log(1 + H(s))dN(s),$$

$$\sum_{s\le t} -H(s)\Delta N(s) = -\int_0^t H(s)dN(s).$$

Hence,

$$\begin{aligned}\mathcal{E}(M)(t) &= e^{\int_0^t H(s)d\bar{N}(s)+\int_0^t \log(1+H(s))dN(s)-\int_0^t H(s)dN(s)} \\ &= e^{-\int_0^t H(s)ds+\int_0^t \ln(1+H(s))dN(s)}.\end{aligned}$$

Problem 10.18 (Estimation of parameters): Find the likelihood corresponding to different values of μ of the process $X(t)$ given by $dX(t) = \mu X(t)dt + \sigma X(t)dB(t)$ on $[0, T]$. Give the maximum likelihood estimator of μ.

Solution. Consider two values of μ, μ_1 and μ_2. Let P correspond to the model with μ_1, $dX(t) = \mu_1 X(t)dt+\sigma X(t)dB(t)$ and Q correspond to the model with μ_2, $dX(t) = \mu_2 X(t)dt + \sigma X(t)dB(t)$. Then, the likelihood is given by

$$\begin{aligned}\Lambda(X)_T = \frac{d\mathrm{Q}}{d\mathrm{P}} &= \mathcal{E}\Big(-\int_0^{\cdot} \frac{\mu_1 X(t) - \mu_2 X(t)}{\sigma X(t)} dB(t)\Big)_T \\ &= \mathcal{E}\Big(-\int_0^{\cdot} \frac{\mu_1 - \mu_2}{\sigma} dB(t)\Big)_T \\ &= \exp\Big(\int_0^T \frac{\mu_2 - \mu_1}{\sigma} dB(t) - \frac{1}{2}\int_0^T \frac{(\mu_2 - \mu_1)^2}{\sigma^2} dt\Big).\end{aligned}$$

The stochastic integral is not observed directly, but only through the observed process $X(t)$, $0 \le t \le T$. To express this integral by using $X(t)$, substitute $dB(t) = \frac{dX(t)-\mu_1 X(t)dt}{\sigma X(t)}$ and simplify to obtain

$$\Lambda(X)_T = \exp\Big(\frac{\mu_2 - \mu_1}{\sigma^2}\int_0^T \frac{1}{X(t)} dX(t) + \frac{T}{2\sigma^2}(\mu_1^2 - \mu_2^2)\Big).$$

For the maximum-likelihood estimator, let P_μ be the measure of the model with μ, and let $\mu_1 = 0$ and $\mu_2 = \mu$. Then,

$$\Lambda(X)_T = \exp\Big(\frac{\mu}{\sigma^2}\int_0^T \frac{1}{X(t)} dX(t) - \frac{T}{2\sigma^2}\mu^2\Big).$$

The maximum-likelihood estimator $\widehat{\mu}$ is that value of μ which maximises $\Lambda(X)_T$. Since e^x is increasing, to find the maximum, take logarithm, differentiate with respect to μ, and set to zero to obtain

$$\widehat{\mu} = \frac{1}{T}\int_0^T \frac{dX(t)}{X(t)}.$$

In practice, the integral is replaced by the sum

$$\widehat{\mu} = \frac{1}{T}\sum_{i=0}^{N-1} \frac{X(t_{i+1}) - X(t_i)}{X(t_i)}, \quad t_i = \frac{iT}{N}.$$

It is often possible to simplify the stochastic integral $\int_0^T \frac{dX(t)}{X(t)}$. Let $U(t) = \int_0^t \frac{dX(s)}{X(s)}$. Then, $dU(t) = \frac{dX(t)}{X(t)}$, and $dX(t) = X(t)dU(t)$.

In other words, X is the stochastic exponential of U, and U is the stochastic logarithm of X. Applying Itô's formula to $\log X(t)$. We obtain $d(\log X(t)) = \frac{1}{X(t)}dX(t) - \frac{1}{2}\sigma^2 dt$ and thus $\int_0^T \frac{1}{X(t)}dX(t) = \log X(T) - \log X(0) + \frac{1}{2}\sigma^2 T$. This can be also obtained directly by using Theorem 5.3. Hence,

$$\widehat{\mu} = \frac{1}{T}\log\frac{X(T)}{X(0)} + \frac{1}{2}\sigma^2.$$

Problem 10.19: If N is a Poisson process with rate λ under P_λ and

$$\Lambda = \frac{d\mathrm{P}_\mu}{d\mathrm{P}_\lambda} = e^{(\lambda-\mu)T + N(T)(\ln\mu - \ln\lambda)},$$

then N is a Poisson process with rate μ under P_μ. Verify the martingale property of the likelihood occurring in the change of rate in a Poisson process, i.e. show that $\Lambda(t) = e^{(\lambda-\mu)t + N(t)(\ln\mu - \ln\lambda)}$ is a martingale under P_λ.

Solution.

$$\begin{aligned}
\mathrm{E}_\lambda\big(\Lambda(t)|\mathcal{F}_s\big) &= e^{(\lambda-\mu)t}\mathrm{E}_\lambda\big(e^{(N(t)-N(s)+N(s))(\ln\mu-\ln\lambda)}|\mathcal{F}_s\big) \\
&= e^{(\lambda-\mu)t}e^{N(s)(\ln\mu-\ln\lambda)}\mathrm{E}_\lambda\big(e^{(N(t)-N(s))(\ln\mu-\ln\lambda)}|\mathcal{F}_s\big) \\
&= e^{(\lambda-\mu)t}e^{N(s)(\ln\mu-\ln\lambda)}e^{(t-s)\lambda(e^{(\ln\mu-\ln\lambda)}-1)} \\
&= e^{(\lambda-\mu)s+N(s)(\ln\mu-\ln\lambda)} = \Lambda(s).
\end{aligned}$$

Problem 10.20: Let $dQ = \Lambda d\mathrm{P}$ on $\mathcal{F}_T$ and $\Lambda(t) = \mathrm{E}_P(\Lambda|\mathcal{F}_t)$ be a continuous process. For a P-martingale $M(t)$, find a finite variation process $A(t)$ such that $M'(t) = M(t) - A(t)$ is a Q-local martingale.

Solution. $M'(t)$ is a Q-local martingale if and only if $\Lambda(t)M'(t)$ is a P-local martingale (Corollary 10.11):

$$\begin{aligned} d(M'(t)\Lambda(t)) &= M'(t)d\Lambda(t) + \Lambda(t)dM'(t) + d[M',\Lambda](t) \\ &= M'(t)d\Lambda(t) + \Lambda(t)dM(t) - \Lambda(t)dA(t) \\ &\quad + d[M,\Lambda](t) - d[A,\Lambda](t). \end{aligned}$$

Since $[A,\Lambda] = 0$ for finite variation process A, $\Lambda(t)$ and $M(t)$ are also P-martingales. Thus, the condition for $M'(t)$ to be Q-local martingale is $d[M,\Lambda](t) - \Lambda(t)dA(t) = 0$, or $dA(t) = \frac{d[M,\Lambda](t)}{\Lambda(t)}$.

Problem 10.21: Let B and N be a Brownian motion and a unit rate Poisson process, independent, and defined on the same space $(\Omega, \mathcal{F}, \mathbb{F}, \mathrm{P})$, $0 \le t \le T$.

(a) Define $\Lambda(t) = e^{(\ln 2)N(t)-t}$ and $d\mathrm{Q}/d\mathrm{P} = \Lambda(T) = \Lambda$. Show that B remains a Brownian motion under Q, but N becomes a Poisson process with rate 2.

(b) Define $\Lambda_1(t) = e^{-B(T)-\frac{1}{2}T}$ and $d\mathrm{Q}_1/d\mathrm{P} = \Lambda_1(T) = \Lambda_1$. Show that N remains a Poisson process with rate 1 under Q_1, but B becomes a Brownian motion with drift $-t$.

Solution.

(a) First, we show that B remains a Brownian motion under Q. Since the quadratic variation of a process does not change under the change of measure (Corollary 10.14), $[B,B](t) = t$ is the same under P and Q. By the Lévy characterisation of Brownian motion, it remains to show that $B(t)$ is a Q-martingale, or equivalently, that $B(t)\Lambda(t)$ is a P-martingale. Writing both processes at time s plus their increments, we obtain

$$\begin{aligned} \mathrm{E}_P\big(B(t)\Lambda(t)|\mathcal{F}_s\big) &= \mathrm{E}_P\big(B(s)\Lambda(t)|\mathcal{F}_s\big) + \mathrm{E}_P\big((B(t)-B(s))\Lambda(t)|\mathcal{F}_s\big) \\ &= B(s)\mathrm{E}_P\big(\Lambda(t)|\mathcal{F}_s\big) + e^{(\ln 2)N(s)-t} \\ &\quad \times \mathrm{E}_P\big((B(t)-B(s))e^{(\ln 2)(N(t)-N(s))}|\mathcal{F}_s\big). \end{aligned}$$

The first term gives $B(s)\Lambda(s)$ since $\Lambda(t)$ is a martingale under P. For the second term, we use that the increments of both processes are independent of the past, implying that the conditioning can be dropped:

$$\mathrm{E}_P\big((B(t)-B(s))e^{(\ln 2)(N(t)-N(s))}|\mathcal{F}_s\big)$$
$$=\mathrm{E}_P\big((B(t)-B(s))e^{(\ln 2)(N(t)-N(s))}\big).$$

Finally, using the independence of the processes themselves, we obtain

$$\mathrm{E}_P\big((B(t)-B(s)e^{(\ln 2)(N(t)-N(s))}\big)$$
$$=\mathrm{E}_P(B(t)-B(s))\mathrm{E}_P\big(e^{(\ln 2)(N(t)-N(s))}\big)=0,$$

due to $\mathrm{E}_P(B(t)-B(s))=0$. This yields $\mathrm{E}_P(B(t)\Lambda(t)|\mathcal{F}_s)=B(s)\Lambda(s)$, as required.

The fact that N changes rate under Q is standard because Λ involves only the process N and not B. The proof is given in Problem 10.16.

(b) Let $u\in\mathbb{R}$, and for $s<t$, consider $\mathrm{E}_{Q_1}\big(e^{u(N(t)-N(s))}|\mathcal{F}_s\big)$. Hence,

$$\mathrm{E}_{Q_1}\big(e^{u(N(t)-N(s))}|\mathcal{F}_s\big)=\mathrm{E}_P\left(\frac{\Lambda_1(t)}{\Lambda_1(s)}e^{u(N(t)-N(s))}|\mathcal{F}_s\right)$$
$$=\mathrm{E}_P\left(e^{-(B(t)-B(s))-\frac{1}{2}(t-s)}e^{u(N(t)-N(s))}|\mathcal{F}_s\right).$$

The random variable in the conditional expectation is independent of $\mathcal{F}_s$. Hence, conditioning can be dropped, giving

$$=\mathrm{E}_P\big(e^{-(B(t)-B(s))-\frac{1}{2}(t-s)}e^{u(N(t)-N(s))}\big)$$
$$=E_P\big(e^{-(B(t)-B(s))-\frac{1}{2}(t-s)}\big)\mathrm{E}_P\big(e^{u(N(t)-N(s))}\big)$$
$$=\mathrm{E}_P\big(e^{u(N(t)-N(s))}\big),$$

since the increments of B and of N are independent, and the expectation of the first term is 1. Hence, the MGF under Q_1 of the increments of N have the same distribution as under P, and therefore the process N does not change its distribution. The process B changes its distribution according to the Girsanov theorem.

Problem 10.22: Let B and N be a Brownian motion and a unit rate Poisson process, independent, on the same space $(\Omega,\mathcal{F},\mathbb{F},\mathrm{P})$, $0\le t\le T$, and $X(t)=B(t)+N(t)$.

(a) Give an equivalent probability measure Q_1 such that $B(t) + t$ and $N(t) - t$ are Q_1-martingales. Deduce that $X(t)$ is a Q_1-martingale.
(b) Give an equivalent probability measure Q_2 such that $B(t) + 2t$ and $N(t) - 2t$ are Q_2-martingales. Deduce that $X(t)$ is a Q_2-martingale.
(c) Deduce that there are infinitely many equivalent probability measures Q such that $X(t) = B(t) + N(t)$ is a Q-martingale.

Solution.

(a) By the Girsanov theorem (Theorem 10.15), $B(t) + t$ is a Brownian motion (and thus a martingale) under Q_1 with $\Lambda(T) = \frac{dQ_1}{dP} = e^{-B(T)-\frac{1}{2}T}$. Moreover, $N(t) - t$ is also a Q_1-martingale:

$$\begin{aligned} E_P\big((N(t) - t)\Lambda(t)|\mathcal{F}_s\big) &= E_P\big((N(t) - t)e^{-B(t)-t/2}|\mathcal{F}_s\big) \\ &= E_P\big(N(t) - t|\mathcal{F}_s\big)E_P\big(e^{-B(t)-t/2}|\mathcal{F}_s\big) \\ &\quad \text{by independence} \\ &= (N(s) - s)e^{-B(s)-s/2} \quad \text{as P-martingales} \\ &= (N(s) - s)\Lambda(s). \end{aligned}$$

In fact, $N(t)$ remains a Poisson process with rate 1 under Q_1.
As a sum of two martingales,

$$(B(t) + t) + (N(t) - t) = B(t) + N(t) = X(t)$$

is a Q_1-martingale.

(b) Define measure Q_2 by

$$\Lambda(T) = \frac{dQ_2}{dP} = \Lambda_1(T)\Lambda_2(T) = e^{-2B(T)-2T}e^{-T+N(T)\ln 2}.$$

Then, by independence,

$$\begin{aligned} &E_P\big((B(t) + 2t)\Lambda(t)|\mathcal{F}_s\big) \\ &= E_P\big((B(t) + 2t)\Lambda_1(t)|\mathcal{F}_s\big)E_P\big(\Lambda_2(t)|\mathcal{F}_s\big) \\ &= (B(s) + 2s)\Lambda_1(s)\Lambda_2(s) = (B(s) + 2s)\Lambda(s) \end{aligned}$$

and

$$\begin{aligned} &E_P\big((N(t) - 2t)\Lambda(t)|\mathcal{F}_s\big) \\ &= E_P\big((N(t) - 2t)\Lambda_2(t)|\mathcal{F}_s\big)E_P\big(\Lambda_1(t)|\mathcal{F}_s\big) \\ &= (N(s) - 2s)\Lambda_2(s)\Lambda_1(s) = (N(s) - 2s)\Lambda(s), \end{aligned}$$

that is, $B(t)+2t$ and $N(t)-2t$ are Q_2-martingales. As a sum of two Q_2-martingales, $X(t)=B(t)+N(t)$ is a Q_2-martingale.

(c) For any $a>0$, define measure Q_a by

$$\frac{d\mathrm{Q}_a}{d\mathrm{P}} = e^{-aB(T)-\frac{1}{2}a^2T}e^{(1-a)T+N(T)\ln a}$$
$$= e^{N(T)\ln a - aB(T)+(1-a-\frac{1}{2}a^2)T}.$$

Then, $B(t)+at$ and $N(t)-at$ are Q_a-martingales and thus $X(t)=B(t)+N(t)$ is a Q_a-martingale as a sum of two martingales. Hence, there are infinitely many equivalent probability measures Q such that $X(t)=B(t)+N(t)$ is a Q-martingale.

Problem 10.23 (Feynman path integrals and the Weiner measure): This question gives heuristics for the Feynman path integrals. Consider the n-dimensional distribution of Brownian motion on $[0,1]$, $(B(t_1),B(t_2),\ldots,B(t_n))$ with $t_{i+1}-t_i=\delta$. Denote its measure on $\mathbb{R}^n$ by P_n. Give the density of P_n with respect to the Lebesgue measure L_n on $\mathbb{R}^n$ $(dL_n = dx_1\cdots dx_n)$. Obtain the formal limit as $\delta\to 0$, hence obtain a formal expression for the Weiner measure. Give an expression for the expectation of a random variable X on this space, and specify it for $X=e^{\int_0^1 V(B_t)dt}$.

Solution. Using the transition probability density function of Brownian motion, we obtain by letting $x_i=B(t_i)$,

$$p(\delta,x,y) = \frac{1}{\sqrt{2\pi}\sqrt{\delta}}e^{-\frac{1}{2}\frac{(y-x)^2}{\delta}},$$

$$d\mathrm{P}_n = C_n e^{-\frac{1}{2}\sum_{i=0}^{n-1}\frac{(x_{i+1}-x_i)^2}{\delta}}dL_n,$$

where $x_0=x$ and $C_n=(\frac{1}{\sqrt{2\pi}\sqrt{\delta}})^n$ is the normalizing constant. We can see P_n as the Wiener measure on cylinder sets, $S_n\subset C[0,1]$,

$$S_n=\{B(t),0\le t\le 1: B(t_1)\in A_1,\ldots,B(t_n)\in A_n\},$$

$W(S_n)=\mathrm{P}_n(A_1\times A_2\times\ldots\times A_n)$. Now if take $x_i=x(t_i)$ then as $\delta\to 0$,

$$\frac{(x_{i+1}-x_i)}{\delta} = \frac{(x(t_{i+1})-x(t_i))}{\delta} \approx x'(\hat{t}_i),$$

and

$$\sum_{i=0}^{n-1}\frac{(x_{i+1}-x_i)^2}{\delta} = \sum_{i=0}^{n-1}\frac{(x_{i+1}-x_i)^2}{\delta^2}\delta \approx \int_0^1 (x'(t))^2dt.$$

This gives the density of the Wiener measure with respect to L_∞ as

$$Ce^{-\frac{1}{2}\int_0^1 (x'(t))^2 dt}.$$

This is formal because L_∞ does not exist. But if x is seen as the limit of n-dim vectors, a function $x(t)$, and Dx as its infinitesimal measure, one can do formal calculations replacing dW by its formal expression:

$$dW = Ce^{-\frac{1}{2}\int_0^1 (x'(t))^2 dt} Dx.$$

This leads to the Feynman path integrals for considering the expectation of integrals of Brownian motion, $E\int_0^T V(B_t)dt$. One can see the integral $\int_0^T V(B_t)dt = X$ as a random variable on the Wiener space (with $\omega \in \Omega$ being a function $x(t)$ on $[0,T]$) and the integral as the usual expectation integral $\int_\Omega X(\omega)P(d\omega)$. In this way, with Ω denoting "the space of paths,"

$$\mathrm{E}\int_0^1 V(B_t)dt = C\int_\Omega V(x(t))e^{-\frac{1}{2}\int_0^1 (x'(t))^2 dt} Dx.$$

Replacing X with e^X, the Feynman path integral is

$$\mathrm{E}e^{\int_0^1 V(B_t)dt} = C\int_\Omega e^{\int_0^1 V(x(t))dt - \frac{1}{2}\int_0^1 (x'(t))^2 dt} Dx.$$

A rigorous treatment is by the Feynman–Kac formula. It states that a more general expression

$$f(x,t) = E\big(e^{\int_t^T V(B_t)dt} g(B(T)|B(t) = x\big)$$

solves the PDE

$$\frac{\partial f}{\partial t} + \frac{1}{2}\frac{\partial^2 f}{\partial x^2} + V(x)f(x,t) = 0,$$

with the boundary condition $f(x,T) = g(x)$.

Problem 10.24: Let diffusion $X(t)$ have $\sigma(x) = 1$, $\mu(x) = +1$ for $x < 0$, $\mu(x) = -1$ for $x > 0$ and $\mu(0) = 0$. Show that $\pi(x) = e^{-2|x|}$ is a stationary distribution for X.

Solution. We give a solution using comparison tricks from the solution to Problem 6.27.

Let $\mu_\varepsilon^-, \mu_\varepsilon^+ : \mathbb{R} \to \mathbb{R}$ be non-increasing twice continuously differentiable functions with $\mu_\varepsilon^-(x) \leq \mu(x) \leq \mu_\varepsilon^+(x)$ for all x and $\mu_\varepsilon^-(x) = \mu(x) = \mu_\varepsilon^+(x)$ for $x \notin [-\varepsilon, \varepsilon]$.

Let $dX_\varepsilon^-(t) = \mu_\varepsilon^-(X_\varepsilon^-(t))dy + dB(t)$ and $dX_\varepsilon^+(t) = \mu_\varepsilon^+(X_\varepsilon^+(t))dy + dB(t)$.

Note that $X_\varepsilon^-(t)$ and $X_\varepsilon^+(t)$ have stationary PDFs π_ε^- and π_ε^+, respectively, such that $\pi_\varepsilon^-(x), \pi_\varepsilon^+(x) \to \pi(x)$ as $\varepsilon \downarrow 0$ for all x making $\int_{-\infty}^x \pi_\varepsilon^-(y)dy, \int_{-\infty}^x \pi_\varepsilon^+(y)dy \to \int_{-\infty}^x \pi(y)dy$ by Scheffe's theorem.

Further, consider solutions $Y_\varepsilon^+(t)$, $Y(t)$ and $Y_\varepsilon^-(t)$ to the SDEs with non-decreasing drift coefficients $-\mu_\varepsilon^+(x) \leq -\mu(x) \leq -\mu_\varepsilon^-(x)$, respectively, and unit diffusion coefficient so that $Y_\varepsilon^+(t) \leq Y(t) \leq Y_\varepsilon^-(t)$ with $X_\varepsilon^+(t) =_D -Y_\varepsilon^+(t)$, $X(t) =_D -Y(t)$ and $X_\varepsilon^-(t) =_D -Y_\varepsilon^-(t)$. It follows that

$$\begin{aligned}\int_{-\infty}^x \pi(y)dy &= \lim_{\varepsilon\downarrow 0}\int_{-\infty}^x \pi_\varepsilon^-(y)dy = \lim_{\varepsilon\downarrow 0}\lim_{t\to\infty} \mathrm{P}(X_\varepsilon^-(t)\leq x)\\ &\geq \limsup_{t\to\infty} \mathrm{P}(X(t)\leq x), \int_{-\infty}^x \pi(y)dy\\ &= \lim_{\varepsilon\downarrow 0}\int_{-\infty}^x \pi_\varepsilon^+(y)dy = \lim_{\varepsilon\downarrow 0}\lim_{t\to\infty} P(X_\varepsilon^+(t)\leq x)\\ &\leq \liminf_{t\to\infty} P(X(t)\leq x).\end{aligned}$$

Hence, $\lim_{t\to\infty} \mathrm{P}(X(t)\leq x) = \int_{-\infty}^x \pi(y)dy$ and the transition density of $X(t)$ must converge pointwise to π.

An alternative solution, which is quite long but educational, is to compute the transition density of X and then show that it converges to π as $t\to\infty$. The diffusion $X(t)$ satisfies $dX(t) = -sgn(X(t))dt + dB(t)$ with $X(0) = x$. By the Girsanov theorem (Theorem 10.16), $W(t) := B(t) - \int_0^t sgn(X(s))ds$ is a Brownian motion under an equivalent measure Q defined through $\frac{dP}{dQ} = e^{-\int_0^T sgn(X(s))dW(s) - \frac{1}{2}T}$. Thus,

$$\begin{aligned}&\mathrm{E}[g(X(t)|X(0)=x]\\ &\quad= \mathrm{E}_Q[g(x+W(t))e^{-\int_0^T sgn(X(s))dW(s)-\frac{1}{2}T}|X(0)=x]\\ &\quad= \mathrm{E}[g(x+B(t))e^{-\int_0^t sgn(x+B(s))dB(s)-\frac{1}{2}t}],\end{aligned}$$

since $e^{-\int_0^t sgn(x+B(s))dB(s)-\frac{1}{2}t}$ is a martingale. By Tanaka's formula (Theorem 8.9),

$$\int_0^t sgn(x+B(s))dB(s) = |B(t)+x| - |x| - L_t^{-x},$$

where L_t^a denotes the local time of B at level a. Substituting this in the expectation above,

$$\begin{aligned}\mathrm{E}[g(X(t)|X(0)=x] &= e^{|x|-\frac{1}{2}t}\mathrm{E}[g(x+B(t))e^{-|B(t)+x|+L_t^{-x}}]\\ &= e^{|x|-\frac{1}{2}t}\left\{\mathrm{E}[g(x+B(t))e^{-|B(t)+x|}1_{T_{-x}>t}]\right.\\ &\quad\left.+\int_0^t \mathrm{E}[g(x+B(t))e^{-|B(t)+x|+L_t^{-x}}|T_{-x}=s]f_{T_{-x}}(s)ds\right\},\end{aligned}$$

where T_a is the first hitting time of B at level a. For the first term, as

$$\begin{aligned}&\mathrm{E}[g(x+B(t))e^{-|B(t)+x|}1_{T_{-x}>t}]\\ &= \begin{cases}\mathrm{E}[g(x+B(t))e^{-|x+B(t)|}1_{M_t<-x}], & x<0,\\ \mathrm{E}[g(x-B(t))e^{-|x-B(t)|}1_{M_t<x}], & x>0,\end{cases}\end{aligned}$$

where $M_t=\sup_{s\le t}B(s)$, and the joint density of B_t and M_t is

$$f_{B_t,M_t}(y,z)=\sqrt{\frac{2}{\pi}}\frac{2z-y}{t^{3/2}}e^{-\frac{(2z-y)^2}{2t}},\quad y\le z, z\ge 0,$$

it follows, after simplification, that

$$\begin{aligned}&\mathrm{E}[g(x+B(t))e^{-|B(t)+x|}1_{T_{-x}>t}]\\ &= \begin{cases}\displaystyle\int_{-\infty}^0 g(u)e^{u}\frac{1}{\sqrt{2\pi t}}\left(e^{-\frac{(u-x)^2}{2t}}-e^{-\frac{(u+x)^2}{2t}}\right)du, & x<0,\\ \displaystyle\int_0^{\infty} g(u)e^{-u}\frac{1}{\sqrt{2\pi t}}\left(e^{-\frac{(u-x)^2}{2t}}-e^{-\frac{(u+x)^2}{2t}}\right)du, & x>0.\end{cases}\end{aligned}$$

For the second term, by the strong Markov property of Brownian motion, for $s<t$,

$$\mathrm{E}[g(x+B(t))e^{-|B(t)+x|+L_t^{-x}}|T_{-x}=s]=\mathrm{E}[g(B(t-s))e^{-|B(t-s)|+L_{t-s}^0}].$$

Note that

$$f_{B_t,L_t^0}(y,v)=\frac{|y|+v}{\sqrt{2\pi t^3}}e^{-\frac{(|y|+v)^2}{2t}},\quad y\in\mathbb{R}, v\in[0,\infty)$$

and

$$f_{T_a}(s)=\frac{|a|}{\sqrt{2\pi s^3}}e^{-\frac{a^2}{2s}}, s\ge 0.$$

Thus,

$$\int_0^t \mathrm{E}[g(x+B(t))e^{-|B(t)+x|+L_t^{-x}}|T_{-x}=s]f_{T_{-x}}(s)ds$$
$$= \int_0^t \frac{|x|}{\sqrt{2\pi s^3}}e^{-\frac{x^2}{2s}} \int_{-\infty}^{\infty} g(y) \int_0^{\infty} e^{-|y|+v}$$
$$\times \frac{|y|+v}{\sqrt{2\pi(t-s)^3}}e^{-\frac{(|y|+v)^2}{2(t-s)}} dvdyds.$$

Since for any x and u, $\int_0^t f_{T_{|x|}}(s)f_{T_{|u|}}(t-s)ds = f_{T_{|x|+|u|}}(t)$, the integral above reduces to

$$\int_{-\infty}^{\infty} g(y) \int_0^{\infty} e^{-|y|+v} \frac{|x|+|y|+v}{\sqrt{2\pi t^3}} e^{-\frac{(|x|+|y|+v)^2}{2t}} dvdy$$
$$= \int_{-\infty}^{\infty} g(y) \left(\frac{1}{\sqrt{2\pi t}} e^{-|y|-\frac{(|x|+|y|)^2}{2t}} + e^{\frac{t}{2}-2|y|-|x|} \Phi\left(\frac{t-|x|-|y|}{\sqrt{t}}\right)\right) dy.$$

Putting both terms together, we have

$$\mathrm{E}[g(X(t)|X(0)=x]$$
$$= e^{|x|-\frac{1}{2}t} \left\{ 1_{x<0} \int_{-\infty}^{0} g(u)e^{u} \frac{1}{\sqrt{2\pi t}} \left(e^{-\frac{(u-x)^2}{2t}} - e^{-\frac{(u+x)^2}{2t}}\right) du \right.$$
$$+ 1_{x>0} \int_0^{\infty} g(u)e^{-u} \frac{1}{\sqrt{2\pi t}} \left(e^{-\frac{(u-x)^2}{2t}} - e^{-\frac{(u+x)^2}{2t}}\right) du$$
$$+ \int_{-\infty}^{\infty} g(y) \left(\frac{1}{\sqrt{2\pi t}} e^{-|y|-\frac{(|x|+|y|)^2}{2t}} \right.$$
$$\left. \left. + e^{\frac{t}{2}-2|y|-|x|} \Phi\left(\frac{t-|x|-|y|}{\sqrt{t}}\right)\right) dy \right\}.$$

That is, for $x<0$,

$$\mathrm{E}[g(X(t)|X(0)=x]$$
$$= \int_{-\infty}^{0} g(y)\left(\frac{1}{\sqrt{2\pi t}} e^{-\frac{(x-y+t)^2}{2t}} + e^{2y}\Phi\left(\frac{t+x+y}{\sqrt{t}}\right)\right)dy$$
$$+ \int_0^{\infty} g(y)\left(\frac{1}{\sqrt{2\pi t}} e^{-2x-\frac{(y-x+t)^2}{2t}} + e^{-2y}\Phi\left(\frac{t+x-y}{\sqrt{t}}\right)\right)dy;$$

for $x > 0$,

$$\begin{aligned} &\mathrm{E}[g(X(t)|X(0) = x] \\ &= \int_{-\infty}^{0} g(y)\Big(\frac{1}{\sqrt{2\pi t}}e^{2x-\frac{(x-y+t)^2}{2t}} + e^{2y}\Phi\Big(\frac{t-x+y}{\sqrt{t}}\Big)\Big)dy \\ &+ \int_{0}^{\infty} g(y)\Big(\frac{1}{\sqrt{2\pi t}}e^{-\frac{(y-x+t)^2}{2t}} + e^{-2y}\Phi\Big(\frac{t-x-y}{\sqrt{t}}\Big)\Big)dy. \end{aligned}$$

From these, we obtain the transition density $p(t, x, y)$ of X, which converges as $t \to \infty$ to $\pi(y) = e^{-2|y|}$.

Problem 10.25: Let $M_t = \int_0^t H_s B_s$ be a martingale, $U_t = \mathcal{E}(M)_t$ its martingale exponential, and τ a stopping time, such that $EU_{T\wedge\tau} = 1$. Let $dQ/dP = U_{T\wedge\tau}$. Find the semimartingale decomposition of B_t under Q.

Solution. $U_{t\wedge\tau}$ is the stochastic exponential of the stopped martingale $M_{t\wedge\tau}$, see Problem 8.15. $M_{t\wedge\tau} = \int_0^t H_s 1_{s<\tau} dB_s = \int_0^t H_s dB_{s\wedge\tau}$. It follows from the proof of the Girsanov theorem that the following process is a Q-Brownian motion stopped at τ:

$$B_{t\wedge\tau} - \int_0^{t\wedge\tau} H_s ds = \hat{B}_{t\wedge\tau}. \text{ Hence, } B_{t\wedge\tau} = \int_0^{t\wedge\tau} H_s ds + \hat{B}_{t\wedge\tau}.$$

Note that while $\hat{B}_t = B_t - \int_0^t H_s ds$ is defined for all t, we cannot claim that it is a Q-Brownian motion for times larger than τ.

Problem 10.26: Let $M_t = \int_0^t B_s dB_s$. Show that its martingale exponential $\mathcal{E}(M)_t$, $0 \le t \le T$ is a martingale for any T. However, sufficient conditions of Novikov and Kazamaki fail if $T \ge 2$.

Solution. First, we try the sufficient condition of Kazamaki. Note that $M_t = \frac{1}{2}B_t^2 - \frac{1}{2}t$. Hence, $\mathrm{E}e^{\frac{1}{2}M_T} = e^{-\frac{1}{4}T}Ee^{\frac{1}{4}B_T^2}$.

$$\mathrm{E}e^{\frac{1}{4}B_T^2} = \mathrm{E}e^{\frac{T}{4}B_1^2} = \frac{1}{\sqrt{2\pi}}\int e^{\frac{T}{4}x^2}e^{-\frac{x^2}{2}}dx = \infty,$$

for $T \ge 2$. Hence, Kazamaki's condition fails. Since Novikov's condition implies Kazamaki's (Equation (8.41*), p. 228), Novikov's condition also fails. Therefore, another method of proof is given, which is also useful for other situations.

Note that $U_t = \mathcal{E}(M)_t$ is a positive local martingale, hence it is a supermartingale, Theorem 7.23. Hence, it is a martingale if (and only if) $\mathrm{E}U_T = 1$, Theorem 7.24. Next, let $\tau_n = \inf\{t : U_t = n\}$. Then, τ_n is a stopping time. $\tau_{n+1} \geq \tau_n$; therefore, $a = \lim_{n\to\infty} \tau_n$ exists. By continuity of U_t, $U_a = \infty$. Since $U_t < \infty$ for any finite t, we conclude that $a = \infty$ a.s. Thus, τ_n is a localising sequence for U_t. Therefore for any n, $\mathrm{E}U_{T\wedge\tau_n} = 1$. Since $\tau_n \to \infty$, $U_T = \lim_{n\to\infty} U_{T\wedge\tau_n}$. Therefore, the question now is whether $\mathrm{E}U_T$ is the limit of expectations $\mathrm{E}U_{T\wedge\tau_n}$. A sufficient condition to interchange the limit with expectation is the uniform integrability of the sequence $U_{T\wedge\tau_n}$ (in n). A sufficient condition for uniform integrability is the de la Vallee–Poussin criterion, which states that if for an increasing convex function g, such that $\lim_{x\to\infty} g(x)/x = \infty$,

$$\sup_n \mathrm{E}g(X_n) < \infty.$$

We take the function $g(x) = x\log x$. Moreover, since $U_{T\wedge\tau_n} > 0$ and $\mathrm{E}U_{T\wedge\tau_n} = 1$, we use this random variable to change the measure $dQ/dP = U_{T\wedge\tau_n}$. $x\log x$ is convenient for calculations of the expectation of a stochastic exponential. Namely,

$$\begin{aligned}
\mathrm{E}g(U_{T\wedge\tau_n}) &= \mathrm{E}U_{T\wedge\tau_n}\log U_{T\wedge\tau_n} = \mathrm{E}_Q \log U_{T\wedge\tau_n} \\
&= \mathrm{E}_Q \log e^{\int_0^{T\wedge\tau_n} B_t dB_t - \frac{1}{2}\int_0^{T\wedge\tau_n} B_t^2 dt} \\
&= \mathrm{E}_Q\left(\int_0^{T\wedge\tau_n} B_t dB_t - \frac{1}{2}\int_0^{T\wedge\tau_n} B_t^2 dt\right).
\end{aligned}$$

By the previous questions, with $H_s = B_s$, we have $B_t = \int_0^t B_s ds + \hat{B}_t$, where for $t \leq \tau_n$, $\hat{B}_t$, $0 \leq t \leq T$ is a Q-Brownian motion. But this is an Ornstein–Uhlenbeck equation for B_t. Hence, by writing its solution, $B_t = e^t \int_0^t e^{-s} d\hat{B}_s$. Moreover,
$B_{t\wedge\tau_n} = \int_0^t e^{t-s} d\hat{B}_{t\wedge\tau_n}$. This shows that

$$\mathrm{E}B^2_{T\wedge\tau_n} = \mathrm{E}\int_0^{T\wedge\tau_n} e^{2(t-s)} ds \leq e^{2T}.$$

Proceeding from where we left and using that an Itô integral has zero mean, we obtain

$$\begin{aligned}
\mathrm{E}g(U_{T\wedge\tau_n}) &= \mathrm{E}_Q\left(\int_0^{T\wedge\tau_n} B_t d\hat{B}_t + \frac{1}{2}\int_0^{T\wedge\tau_n} B_t^2 dt\right) \\
&= \frac{1}{2}\mathrm{E}_Q \int_0^{T\wedge\tau_n} B_t^2 dt.
\end{aligned}$$

Using the bound on $\mathrm{E}_Q B_t^2$ again, we obtain

$$\mathrm{E}g(U_{T\wedge\tau_n}) \le \mathrm{E}_Q \int_0^{T\wedge\tau_n} B_t^2 dt \le \int_0^T \mathrm{E}_Q B_{t\wedge\tau_n}^2 dt \le \int_0^T e^{2t} dt < e^{2T}.$$

Since the right-hand side is independent of n, we obtain $\sup_n \mathrm{E}g(U_{T\wedge\tau_n}) < \infty$. This proves that $\mathrm{E}U_T = \lim_{n\to\infty} \mathrm{E}U_{T\wedge\tau_n} = 1$, and U_t, $0 \le t \le T$, is a martingale.

We also give a more elementary solution.

We prove that for all $n \in \mathbb{N}$ and some $T > 0$,

$$(\star)\ \ \mathrm{E}(\mathcal{E}(M)_{nT}) = 1.$$

By Novikov's condition, $(\star)$ holds for $n = 1$ for some $T > 0$.

Take this T and assume that $(\star)$ holds for $n = k$. Consider

$$\begin{aligned}\mathrm{E}(\mathcal{E}(M)_{(k+1)T}) &= \mathrm{E}\left(\frac{\mathcal{E}(M)_{(k+1)T}}{\mathcal{E}(M)_{kT}}\mathcal{E}(M)_{kT}\right)\\ &= \lim_N \mathrm{E}\left(\mathrm{E}\left(N \wedge \frac{\mathcal{E}(M)_{(k+1)T}}{\mathcal{E}(M)_{kT}}\,\middle|\, \mathcal{F}_{kT},\right) N \wedge \mathcal{E}(M)_{kT}\right)\end{aligned}$$

by monotone convergence because

$$\begin{aligned}&\mathrm{E}\left(N \wedge \frac{\mathcal{E}(M)_{(k+1)T}}{\mathcal{E}(M)_{kT}}\,\middle|\, \mathcal{F}_{kT}\right)\\ &= \mathrm{E}\left(N \wedge \exp\left(\int_0^T (\hat{B}_s + B_{kT})d\hat{B}_s - \frac{1}{2}\int_0^T (\hat{B}_s + B_{kT})^2 ds\right)\right) \nearrow 1,\end{aligned}$$

and

$$\mathrm{E}(N \wedge \mathcal{E}(M)_{kT}) \nearrow 1$$

so that $(\star)$ holds for $n = k + 1$.

Problem 10.27: Let $M_t = \int_0^t \sigma(B_s)dB_s$, where $\sigma(x)$ satisfies the linear growth condition, $\sigma^2(x) \le C(1 + x^2)$. Show that $\mathcal{E}(M)_t$, $0 \le t \le T$ is a martingale for any T. This generalises the previous question. As before, Kazamaki's condition fails, but M_t is a martingale.

Solution. The proof is similar; however, when doing the change of measure, we cannot solve explicitly for B_t in terms of $\hat{B}_t$, as we have done in the previous question. This complicates the derivation of the bound for expectations $Eg(U_{T\wedge\tau_n})$. To this end, we use different stopping times

$\tau_n = \inf\{t : U_t \vee B_t^2 \geq n\}$. Since we cannot perform direct calculations, we use the linear growth condition and the Gronwall inequality to bound $\int_0^t \mathrm{E}_Q B_{t\wedge\tau_n}^2 dt$.

$U_t = e^{M_t - \frac{1}{2}\langle M,M\rangle_t} = e^{M_t - \frac{1}{2}\int_0^t \sigma^2(B_s)ds}$ is a local martingale. Note that $\mathrm{E}(U_{\tau_n\wedge T}) = 1$. We can define an equivalent measure Q_n by $\frac{d\mathrm{Q}_n}{d\mathrm{P}} = U_{\tau_n\wedge T}$. Then, by the change of measure, we have

$$\hat{B}_{\tau_n\wedge t} = B_{\tau_n\wedge t} - \int_0^{\tau_n\wedge t} \sigma(B_s)ds,$$

with $\hat{B}$ a Brownian motion under measure Q_n, and

$$\begin{aligned}
\mathrm{E}_P(U_{\tau_n\wedge T}\log U_{\tau_n\wedge T}) &= \mathrm{E}_{Q_n}(\log U_{\tau_n\wedge T}) \\
&= \mathrm{E}_{Q_n}\left(M_{\tau_n\wedge T} - \frac{1}{2}\int_0^{\tau_n\wedge T} \sigma^2(B_s)ds\right) \\
&= \mathrm{E}_{Q_n}\left(\int_0^{\tau_n\wedge T} \sigma(B_s)d\hat{B}_s + \frac{1}{2}\int_0^{\tau_n\wedge T} \sigma^2(B_s)ds\right) \\
&= \frac{1}{2}\mathrm{E}_{Q_n}\left(\int_0^{\tau_n\wedge T} \sigma^2(B_s)ds\right).
\end{aligned}$$

Using the linear growth condition for σ^2,

$$\sigma^2(B_t) \leq C(1 + B_t^2),$$

and that $(a+b)^2 \leq 2(a^2+b^2)$, we have

$$B_{\tau_n\wedge t}^2 \leq 2\hat{B}_{\tau_n\wedge t}^2 + 2\left(\int_0^t I_{s\leq\tau_n}\sigma(B_s)ds\right)^2.$$

Next, using the Cauchy–Schwarz inequality for the integrals with $g(s) = 1$)

$$\left(\int_0^t f(s)ds\right)^2 \leq t\int_0^t f^2(s)ds,$$

we have

$$\left(\int_0^t I_{s\leq\tau_n}\sigma(B_s)ds\right)^2 \leq t\int_0^t I_{s\leq\tau_n}\sigma^2(B_s)ds.$$

Since $\mathrm{E}_{Q_n}\hat{B}_{\tau_n\wedge t}^2 \leq t$, this yields, by the Gronwall inequality, that $\mathrm{E}_{Q_n}\big(\int_0^{\tau_n\wedge T} \sigma^2(B_s)ds\big)$ is bounded by a constant depending only on t.

This in turn implies that $\sup_n \mathrm{E}(U_{\tau_n \wedge T} \log U_{\tau_n \wedge T}) < \infty$, and $U_{\tau_n \wedge T}$, $n \geq 1$ is uniformly integrable. It then follows that

$$\mathrm{E}(U_T) = \lim_{n \to \infty} \mathrm{E}(U_{\tau_n \wedge T}) = 1.$$

There is also an elementary solution for Problem 10.26, in which $\hat{B}_s + B_{kT}$ is replaced by $\sigma(\hat{B}_s + B_{kT})$.

Chapter 11

Applications in Finance

Stock and FX Options

Financial risk management and pricing of derivative contracts (interchangeably options or claims) are two sides of the same question. Mathematically, options are some functions of the underlying asset, such as stock or interest rate. Typically such contracts specify a payoff in the future, with the payoff depending on the realisation of the underlying asset. To value such contracts, a model for asset price is used, say S_t at time t with $0 \leq t \leq T$, where T is usually the time of expiration of the contract. Typical examples include widely traded call options with $X = (S_T - K)^+$ and put options $X = (K - S_T)^+$, which represent the option to buy or sell a stock at time T for K. An example of a path-dependent option is given by a contract that depends on S_T as well as on $\int_0^T S_t dt$ (average rate options). The arbitrage pricing theory allows to determine C_t, the price of claim X, at time $t < T$, as well as manage risk. An important assumption for this theory is that claim X is integrable, $\mathrm{E}|X| < \infty$.

Models for S_t include discrete time and discrete state space models as well as models of stochastic differential equations, with Brownian motion modelling the uncertainty in the movement of prices. The most general model for asset prices used with no arbitrage pricing is that of a semimartingale. The basic principle for obtaining a fair price for a contract is the principle of no arbitrage. Simply, there should be no free lunch, so to speak, which might be possible to obtain if the contract is not priced accordingly.

To define arbitrage mathematically, the concept of a self-financing portfolio is used. A portfolio is a pair of predictable processes (a_t, b_t) that denote the number of shares of stock and money in the savings account. The value of the savings account at time t is $b_t\beta_t$, so that b_t is the number of dollars rather than dollars themselves. The value of the portfolio at time t is $V_t = a_tS_t + b_t\beta_t$. A portfolio $V_t = a_tS_t + b_t\beta_t$ is *self-financing* if no new funds are withdrawn or injected into the portfolio, and only redistribution is allowed. The self-financing condition in continuous time is given by

$$dV_t = a_t dS_t + b_t d\beta_t, \quad or \quad V_t = V_0 + \int_0^t a_u dS_u + \int_0^t b_u d\beta_u,$$

with the value equal to the initial funds plus the gain from trade in stocks and savings accounts and similar equations in discrete time models. This portfolio is often called a *hedge* since it can be used to completely negate the risk of the option payoff. Arbitrage is a self-financing portfolio with $V_0 = 0$ and $V_T \geq 0$, with $\mathrm{P}(V_T > 0) > 0$. The price of X at time $t < T$, say C_t, is the value of a self-financing portfolio that replicates X, namely a self-financing portfolio with the terminal value $V_T = X$. It is easy to see that if the price of X at time t is not equal to V_t, then by trading the contract against the portfolio gives arbitrage. This statement applies to any model, Problem 11.3.

There is a direct link between arbitrage opportunities and equivalent martingale measures (EMMs). We define an EMM as a probability measure Q such that it is equivalent to the real-world probability measure P, and under Q, the discounted price of any *principal tradable asset* is a martingale.

The key concept is that, under the EMM, all self-financing portfolios (which only include tradable assets) are martingales, as long as they are integrable. This is easy to see, as the self-financing property implies that the discounted value of such portfolios is a stochastic integral with respect to the discounted stock price process, and under integrability conditions, stochastic integrals with respect to martingales are again martingales.

The first fundamental theorem states that a model does not have arbitrage opportunities (in the stock and the savings account) if there is an EMM Q such that the discounted price process of the tradable assets are martingales.

Sufficiency is easy, see Problem 11.4. The opposite direction, where no arbitrage implies the existence of an EMM is true in discrete time models, although the proof is much more difficult.

For continuous time models, the result is "mostly true". In order to achieve the opposite implication, the definition of arbitrage needs to be modified to a much more technical version, which is not considered here.

The second fundamental theorem of asset pricing answers the question whether it is always possible to replicate the payoff of any option by a self-financing portfolio. A market model is called *complete* if any option can be replicated by a self-financing portfolio. The theorem states that a market model is complete if and only if there is only one EMM.

Finding self-financing replicating portfolios relies on the predictable representation theorems, $X = \mathrm{E}_Q X + \int_0^T H_t dZ_t$, where Z is the discounted stock price martingale. The predictable process H_t can be found explicitly in many practical models, in particular in the Black–Scholes model and other models based on Brownian motion. Completeness in these models follows directly from the predictable representation property for Brownian filtration.

It is also possible to give the price C_t of X at time $t < T$ as an expectation of X under the EMM. Since we know that $C_t = V_t$ is the value of a self-financing replicating portfolio and that such a portfolio when discounted is a martingale under the EMM, we obtain the pricing formula as an expectation under the EMM, as follows: $\mathrm{E}_Q(V_T/\beta_T|\mathcal{F}_t) = V_t/\beta_t$. Finally, since the portfolio replicates the claim $V_T = X$, it gives

$$C_t = V_t = \mathrm{E}_Q\left(\frac{\beta_t}{\beta_T}C_T|\mathcal{F}_t\right) = E_Q\left(\frac{\beta_t}{\beta_T}X|\mathcal{F}_t\right).$$

Discrete Multi-Period Models

In these models, stocks take finitely many values and trading occurs at discrete times $t = 0, 1, 2, \ldots, T$. The risk-less savings account is given by $\beta_t = \beta^t$, where β is the simple interest over one period. A portfolio or trading strategy (a_t, b_t) is decided on the basis of information at time $t-1$, which means they are *predictable* processes. The value of the portfolio at time $t-1$ when it is established is $a_t S_{t-1} + b_t \beta_{t-1}$. This value remains until just before the time t the price S_t is announced. If V_t denotes the *value* of the portfolio (a_t, b_t) just after the time t price is observed, then $V_t = a_t S_t + b_t \beta_t$. The initial value of the portfolio is $V_0 = a_1 S_0 + b_1 \beta_0$. The self-financing portfolio condition is given by $V_t = V_0 + \sum_{i=1}^t (a_i \Delta S_i + b_i \Delta \beta_i)$, $t = 1, 2, \ldots, T$. An alternative condition is obtained by equating the value of the portfolio before rebalancing and immediately after rebalancing, Problem 11.6.

The martingale property of bounded self-financing portfolios is a direct consequence of the self-financing condition, being a stochastic integral with respect to a martingale, see Problem 11.8. In the case of a finite market,

V_t takes finitely many values and the assumption of a bounded portfolio holds.

Options pricing via replication is based on finding a self-financing portfolio replicating claim X, i.e. $X = V_T$. In finite market models, this equation results in a number of linear equations for a_t, b_t, $t = 1, 2 \ldots, T$. Options pricing via an EMM is also based on a self-financing replicating portfolio, where prices are calculated by taking expectations under the EMM.

An important case of models is the binomial model.

In these models, $S_{t+1} = S_t\xi_{t+1}$, with ξ_t's taking two values d or u with $d < u$. The probabilistic setup is given in Problem 11.9. It turns out that the martingale condition implies a unique EMM Q under which all ξ_t's are independent and identically distributed, Problem 11.9. The construction of replicating self-financing portfolios is given in Problem 11.13.

Black–Scholes Model

The Black–Scholes model (also known as the Black–Scholes–Merton model) is one the most famous mathematical models for pricing options in a continuous time setting.

We assume that stock price S_t follows the geometric Brownian motion given by the following stochastic differential equation $dS_t = \mu S_t dt + \sigma S_t dB_t$, where B is a standard Brownian motion starting at $B_0 = 0$. The constant μ is the drift rate, representing the average growth rate or return of the stock. The constant σ is known as the volatility. The solution to the above SDE is given by

$$S_t = S_0 e^{(\mu - \frac{\sigma^2}{2})t + \sigma B_t}.$$

The savings account $\beta_t = e^{rt}$.

The EMM Q always exists in this model, thanks to the Girsanov theorem. $S_t/\beta_t = S_0 e^{(\mu - \frac{\sigma^2}{2} - r)t + \sigma B_t}$ can be made a martingale by changing the drift of B_t. If $c = \frac{\mu - r}{\sigma}$, then Q obtained by $dQ/dP = e^{-cB_T - c^2 T/2}$ is the EMM, which makes the process $B_t + \frac{\mu - r}{\sigma} t$ into a Brownian motion $\hat{B}$. The expression for S_t under Q is $S_t = S_0 e^{(r - \frac{\sigma^2}{2})t + \sigma \hat{B}_t}$, which solves the SDE

$$dS_t = rS_t dt + \sigma S_t d\hat{B}_t.$$

By changing from the real-world measure P to the EMM Q, we have effectively changed the drift term of dS_t from μ to r, while switching to a different Brownian motion. This also has a nice financial interpretation, since in the risk-neutral world (Q), the average growth rate of the stock price should be equal to the risk-free interest rate.

The Black–Scholes model is complete, Problem 11.19. As a consequence, any claim can be priced by a replicating portfolio, and provided some integrability conditions hold, it can be computed by taking expectation under EMM Q, Problem 11.21. For claims $X = g(S_T)$ for a function g, conditional expectations can be evaluated by using the Feynman–Kac formula, see Problem 11.23. Results on some path-dependent (exotic) options are given in Problems 11.30–11.33.

Similar results are available for the Bachelier model, Problems 11.34 and 11.35, however, finding the EMM requires some advanced results on martingale exponential.

Models of stock based on diffusion processes can be treated by and large as the Black–Scholes model, however, closed-form solutions for options prices are rarely available.

Some results on incomplete models are treated in Problems 11.36–11.38.

A change of numeraire is useful technique, when the wealth is measured by the price of stock rather than a savings account. In this case, the EMM Q_1 is such that β_t/S_t is a Q_1 martingale. If S_t/β_t is a martingale EMM Q, and $dQ_1/dQ = \Lambda = \frac{S_T/S_0}{\beta_T/\beta_0}$, then β_t/S_t is a martingale under Q_1. This often yields simplifications in the formulae, especially in the area of bonds.

Bonds, Rates, and Options

A bond issued by Treasury or some other party is an obligation to pay its face value to its holder at time T. A bond holder usually receives a stream of payments, called coupons. However, from a mathematical point of view little is lost when we assume that the face value is \$1 and there are no coupons, and consider the so-called zero-coupon bond. The price of such bond at time $t < T$ is denoted by $P(t,T)$. A model for the bond market consists of a family of bonds maturing at different times T, $T \leq T^*$, as well as a savings accounts β_t. We assume a probability model with a filtration $\mathbb{F} = \{\mathcal{F}_t\}$, $0 \leq t \leq T^*$, and adapted processes $P(t,T)$, $t \leq T \leq T^*$, and β_t. The model does not admit arbitrage between bonds with different maturities and savings accounts if there exists an EMM such that simultaneously for all T the following are martingales $P(t,T)/\beta_t$ under Q. The martingale property yields, by using that $P(T,T) = 1$, the price at time t of T-Bond is given by $P(t,T) = \mathrm{E}_Q(\frac{\beta_t}{\beta_T}|\mathcal{F}_t)$.

Next, we discuss relations to the rates. The spot (short) rate comes into the definition of the savings account $d\beta_t = r_t\beta_t dt$, or $\beta_t = e^{\int_0^t r_s ds}$. Hence, the bond is priced by

$$P(t,T) = \mathrm{E}_Q(e^{-\int_t^T r_s ds}|\mathcal{F}_t).$$

This shows that the bond itself is a derivative of the rate. The rates themselves are not tradable assets; therefore, options on rates are constructed via bonds. In the case of diffusion models for the spot rate, the conditional expectation above can be evaluated by using the Feynman–Kac formula. There are many models based on the spot rate, such as the Merton model, $dr(t) = \mu dt + \sigma dW(t)$, and the Vasicek model, $dr(t) = b(a - r(t))dt + \sigma dW(t)$.

One of the issues with models based on the spot rate is that it is not clear how to find the EMM Q. In fact, the discounted bond process $P(t,T)/\beta_t = \mathrm{E}_Q(\frac{1}{\beta_T}|\mathcal{F}_t)$ is a martingale (of Lévy type) under any measure. To resolve this issue, it is either assumed that the model is already specified under the EMM or the market price of risk quantity is introduced and then used to change the measure. However, this is bypassed by using models based on forward rates, with the rates implied by the bond,

$$f(t,T) = -\frac{\partial \log P(t,T)}{\partial T}.$$

Heath–Jarrow–Morton (HJM) models specify equations for the forward rates and deduce the no-arbitrage condition, which involves only the volatility parameter of the rates. The SDE for forward rates involves adapted processes $\alpha(t,T)$ and volatility $\sigma(t,T)$,

$$df(t,T) = \alpha(t,T)dt + \sigma(t,T)dW(t),$$

and under EMM Q, the drift is changed,

$$df(t,T) = \sigma(t,T)\tau(t,T)dt + \sigma(t,T)dB(t), \quad \text{where } \tau(t,T) = \int_t^T \sigma(t,u)du.$$

One of such models is the LIBOR market model (LMM), also known as the Brace–Gatarek–Musiela (BGM) model, in which rate volatility $\sigma(t,T)$ is deterministic.

Options on bond, such as call and put, correspond to options on the rate, flooring, or capping the rate at which payments on the loan are made. Here, it is convenient to use forward measures by using a T-bond as a numeraire, i.e. when $\beta_t/P(t,T)$, $0 \leq t \leq T$ is a martingale. When such a measure Q_T is used, the price of an option X becomes

$$C_t = \mathrm{E}_Q\left(\frac{\beta_t}{\beta_T}X|\mathcal{F}_t\right) = P(t,T)\mathrm{E}_{Q_T}(X|\mathcal{F}_t).$$

Problems

Problem 11.1 (Payoff functions and diagrams): Graph the following payoffs:

(a) A *straddle* consists of buying a call and a put with the same exercise price and expiration date.
(b) A *butterfly spread* consists of buying two calls with exercise prices K_1 and K_3 and selling a call with exercise price K_2, where $K_1 < K_2 < K_3$.

Solution. A payoff function $f(x)$ for a contract depending on the final price of stock S_T is obtained when the value S_T is replaced by a variable x.

(a) $$\text{Payoff} = (S_T - K)^+ + (K - S_T)^+ = \begin{cases} K - S_T & \text{if} \quad S_T \le K, \\ S_T - K & \text{if} \quad S_T > K, \end{cases}$$

$$f(x) = (K - x)I(x \le K) + (x - K)I(x > K).$$

(b) $$\text{Payoff} = (S_T - K_1)^+ + (S_T - K_3)^+ - (S_T - K_2)^+$$

$$= \begin{cases} 0 \quad \text{if} \quad S_T \le K_1, \\ S_T - K_1 \quad \text{if} \quad K_1 < S_T \le K_2, \\ (S_T - K_1) - (S_T - K_2) = K_2 - K_1 \quad \text{if} \quad K_2 < S_T \le K_3, \\ (S_T - K_1) - (S_T - K_2) + (S_T - K_3) \\ \quad = S_T + K_2 - K_1 - K_3 \quad \text{if} \quad K_3 < S_T, \end{cases}$$

$$\begin{aligned} f(x) &= (x - K_1)I(K_1 < x \le K_2) \\ &\quad + (K_2 - K_1)I(K_2 < x \le K_3) \\ &\quad + (x + K_2 - K_1 - K_3)I(K_3 < x). \end{aligned}$$

Problem 11.2: Let $X \ge 0$. Show that if Q is equivalent to P, then $\mathrm{E}_P(X) > 0$ if and only if $\mathrm{E}_Q(X) > 0$.

Solution. Let $\mathrm{E}_P(X) > 0$. Then, $\mathrm{P}(X > 0) > 0$. This is because if $\mathrm{P}(X > 0) = 0$, then $P(X = 0) = 1$ and $\mathrm{E}_P X = 0$. Since $\mathrm{Q} \sim \mathrm{P}$, it follows that $\mathrm{Q}(X > 0) > 0$. This implies $\mathrm{E}_Q(X) > 0$. Swapping the roles of P and Q, the converse follows.

Problem 11.3: Suppose we can find a self-financing portfolio $(a_t, b_t), t \geq 0$, with wealth process V that replicates an option that pays X at time T, $V_T = X$. Then, the price of this option C_t at time $t < T$ must be given by the value of this portfolio at time t, $C_t = V_t$.

Solution. The proof of this result relies on the assumption that the market does not allow for any arbitrage opportunities.

If $C_t < V_t$, then let us sell the portfolio and buy the option at time t. Our cashflow at time t is then $V_t - C_t > 0$. At time T, the values of the option and portfolio are equal since the portfolio is chosen to replicate the option payoff. So, the cashflow at time T is $C_T - V_T = 0$. Since the portfolio is self-financing, it costs nothing to maintain the portfolio during the time interval $[t, T]$. Thus, we have created a riskless profit, which is equal to $V_t - C_t$ plus any interest earned during $[t, T]$. This is an arbitrage, contradicting our assumption. So, we must rule out $C_t < V_t$.

If $C_t > V_t$, then we can employ the opposite strategy of selling the option and buying the portfolio. Similar to before, we once again create an arbitrage opportunity, which is not allowed. So, we also cannot have $C_t > V_t$. Therefore, the only possibility left is $C_t = V_t$.

Problem 11.4: Show that the existence of EMM implies there is no arbitrage.

Solution. Suppose there is an EMM Q. If there exists an arbitrage opportunity with wealth process V, then $V_0 = 0$, $\mathrm{P}(V_T \geq 0) = 1$, $\mathrm{P}(V_T > 0) > 0$. Since Q is equivalent to P, this implies $V_0 = 0$, $\mathrm{Q}(V_T \geq 0) = 1$, $\mathrm{Q}(V_T > 0) > 0$. In particular, $\mathrm{E}_Q(V_T) > 0$. But that means $\mathrm{E}_Q(V_T)/\beta_T \neq V_0$, so the discounted wealth process cannot be a Q-martingale. This is a contradiction.

Problem 11.5: Show that the expected return on stock under the martingale probability measure Q is the same as on the riskless asset. This is the reason why the martingale measure Q is also called a "risk-neutral" probability measure.

Solution. In discrete time, the statement easily follows from the martingale property of $S(t)/\beta(t)$,

$$\mathrm{E}_Q\left(\frac{S_{t+1}}{\beta(t+1)}\bigg|\mathcal{F}_t\right) = \frac{S_t}{\beta(t)}.$$

Rearranging, using that S_t is $\mathcal{F}_t$ measurable and $\beta(t)$ is deterministic, we obtain for the return S_{t+1}/S_t,

$$\mathrm{E}_Q\left(\frac{S_{t+1}}{S_t}\bigg|\mathcal{F}_t\right) = \frac{\beta(t+1)}{\beta(t)}, \quad \text{and } \mathrm{E}_Q\left(\frac{S_{t+1}}{S_t}\right) = \frac{\beta(t+1)}{\beta(t)},$$

as required.

In continuous time, we have

$$d\left(\frac{S(t)}{\beta(t)}\right) = \frac{dS(t)}{\beta(t)} + S(t)d\left(\frac{1}{\beta(t)}\right).$$

$$\begin{aligned} dR(t) = \frac{dS(t)}{S(t)} &= \frac{\beta(t)}{S(t)}d\left(\frac{S(t)}{\beta(t)}\right) - \beta(t)d\left(\frac{1}{\beta(t)}\right) \\ &= \frac{\beta(t)}{S(t)}d\left(\frac{S(t)}{\beta(t)}\right) + \frac{d\beta(t)}{\beta(t)}. \end{aligned}$$

$$R(t) = \int_0^t \frac{\beta(u)}{S(u)}d\left(\frac{S(u)}{\beta(u)}\right) + \int_0^t \frac{d\beta(s)}{\beta(s)}.$$

The first integral is a martingale, as a stochastic integral with respect to a Q-martingale.

Thus,

$$\mathrm{E}_Q R(t) = \int_0^t \frac{d\beta(s)}{\beta(s)},$$

which is the risk-free return over the interval $[0,t]$.

Problem 11.6: Show that in the multiperiod models, the two self-financing conditions are the same:

$$V_t = V_0 + \sum_{i=1}^{t}(a_i\Delta S_i + b_i\Delta\beta_i), \quad t = 1,2,\ldots,T,$$

and

$$a_{t-1}S_{t-1} + b_{t-1}\beta_{t-1} = a_t S_{t-1} + b_t\beta_{t-1}, \quad t = 1,2,\ldots,T.$$

Solution. By the definition of the value process, $V_t - V_{t-1} = a_tS_t + b_t\beta_t - a_{t-1}S_{t-1} - b_{t-1}\beta_{t-1}$. From the first definition of self-financing, we have $\Delta V_t = V_t - V_{t-1} = a_t(S_t - S_{t-1}) + b_t(\beta_t - \beta_{t-1})$. Taking the difference, we obtain $a_tS_{t-1} - a_{t-1}S_{t-1} + b_t\beta_{t-1} - b_{t-1}\beta_{t-1} = 0$, which is the second equation. On the other hand, if the second definition of self-financing holds, then $\Delta V_t = V_t - V_{t-1} = a_tS_t + b_t\beta_t - a_{t-1}S_{t-1} - b_{t-1}\beta_{t-1} = a_tS_t + b_t\beta_t - a_tS_{t-1} - b_t\beta_{t-1} = a_t(S_t - S_{t-1}) + b_t(\beta_t - \beta_{t-1}) = a_t\Delta S_t + b_t\Delta\beta_t$, which is the first definition.

Problem 11.7: Show that in the multiperiod models with $\beta_t = 1$ for all t if (a_t, b_t) is a self-financing portfolio, with bounded $|a_t|, |b_t| \leq C$ for some constant C, and the discounted stock S_t a martingale under some measure Q, then the discounted value process V_t is also a martingale under Q.

Solution. Since $\beta_t = 1$, we have $\Delta\beta_t = 0$. Hence, $V_t = V_0 + \sum_{i=1}^t a_i\Delta S_i$. Integrability follows by the boundedness of a_t:

$$\mathrm{E}|V_t| \leq |V_0| + \sum_{i=1}^{t} \mathrm{E}|a_i||\Delta S_i| \leq |V_0| + C\sum_{i=1}^{t} \mathrm{E}|S_i| + \mathrm{E}|S_{i-1}| < \infty.$$

We are given that S_t is a martingale under Q, hence V_t is a stochastic integral of a bounded predictable process with respect to a martingale, which itself is a martingale:

$$\begin{aligned}\mathrm{E}(V_{t+1}|\mathcal{F}_t) &= \mathrm{E}(V_0 + \sum_{i=1}^{t+1} a_i\Delta S_i|\mathcal{F}_t) \\ &= V_t + \mathrm{E}(a_{t+1}\Delta S_{t+1}|\mathcal{F}_t) = V_t + a_{t+1}\mathrm{E}(\Delta S_{t+1}|\mathcal{F}_t) = V_t.\end{aligned}$$

Problem 11.8: Show that in the multiperiod models if (a_t, b_t) is a bounded self-financing portfolio, with $|a_t|, |b_t| \leq C$ for some constant C, and the discounted stock S_t/β_t a martingale under some measure Q, then the discounted value process V_t/β_t is also a martingale under Q.

Solution. Integrability follows by boundedness of a_t and b_t just as in the previous problem. The martingale property is verified as follows:

$$\begin{aligned}\mathrm{E}_Q\left(\frac{V_{t+1}}{\beta_{t+1}}|\mathcal{F}_t\right) &= \mathrm{E}_Q\big(a_{t+1}S_{t+1}/\beta_{t+1} + b_{t+1}|\mathcal{F}_t\big) \\ &= a_{t+1}\mathrm{E}_Q\big(S_{t+1}/\beta_{t+1}|\mathcal{F}_t\big) + b_{t+1},\end{aligned}$$

since a_{t+1} and b_{t+1} are $\mathcal{F}_t$ measurable (they are predictable processes).

$$E_Q\left(\frac{V_{t+1}}{\beta_{t+1}}|\mathcal{F}_t\right) = a_{t+1}S_t/\beta_t + b_{t+1} = (a_{t+1}S_t + b_{t+1}\beta_t)/\beta_t.$$

But since (a_t, b_t) is self-financing,

$$a_{t+1}S_t + b_{t+1}\beta_t = a_tS_t + b_t\beta_t.$$

Finally, we obtain

$$\mathrm{E}_Q\left(\frac{V_{t+1}}{\beta_{t+1}}|\mathcal{F}_t\right) = a_tS_t/\beta_t + b_t = \frac{V_t}{\beta_t}.$$

Problem 11.9 (Binomial model EMM): The returns on stock $\xi_{t+1} = S_{t+1}/S_t \in \{d, u\}$, $S_0 = 1$. Savings account $\beta_t = \beta^t$, with $0 < d < \beta < u$. $t = 0, 1, \ldots, n-1$.

(a) Give the probability space for the binomial model, describing the random variables ξ_t's and the process S_t.
(b) Describe the EMM Q, giving the necessary and sufficient conditions for it to exist.
(c) Show that when the EMM Q exists, then it is unique and makes random variables ξ_t's independent and identically distributed.

Solution.

(a) The probability space is given by $(\Omega, \mathcal{F}, \mathbb{F}, \mathrm{P})$:

$$\Omega = \{\omega = (\omega_1, \omega_2, \ldots, \omega_n),\ \omega_t \in \{u, d\},\ t = 1, \ldots, n\}.$$

σ-fields and filtration are given by

$$\mathcal{F} = \mathcal{F}_n = 2^\Omega.$$

$\mathbb{F} = (\mathcal{F}_t)$, $t = 1, \ldots, n$, where $\mathcal{F}_t$ is the field generated by the sets in which the first t coordinates are given and $n - t$ are arbitrary: $A_1 = \{\omega : \omega = (u, u, \ldots, u, \omega_{t+1}, \ldots, \omega_n)\}$, $A_2 = \{\omega : \omega = (d, u, \ldots, u, \omega_{t+1}, \ldots, \omega_n)\}$, etc., $A_{2^t} = \{\omega : \omega = (d, d, \ldots, d, \omega_{t+1}, \ldots, \omega_n)\}$.

The probability P on $\mathcal{F}_n$ gives the joint distribution of all the ξ_t's by assigning probabilities to all 2^n outcomes:

$$\mathrm{P}(\{\omega\}) = \mathrm{P}((\omega_1, \omega_2, \ldots, \omega_n)) = \mathrm{P}(\xi_1 = \omega_1, \xi_2 = \omega_2, \ldots, \xi_n = \omega_n).$$

Note that under P, the random variables ξ_t's need not be independent nor identically distributed.

Provided that all the above probabilities are positive, this probability does not enter the calculations of option prices or replicating portfolios.

The random variable ξ_t is defined by the t coordinate $\xi_t(\omega) = \omega_t$, where $\omega_t \in \{d, u\}$, $t = 1, 2, \ldots, n$. The stochastic process S_t is defined as

$$S_t(\omega) = \prod_{i=1}^{t} \xi_i(\omega) = \prod_{i=1}^{t} \omega_i.$$

(b) We shall show that it is necessary and sufficient for all probabilities $\mathrm{P}(\{\omega\}) > 0$ to be positive for EMM to exist.

Let $\mathrm{P}(\{\omega\}) > 0$ for all $\omega \in \Omega$. Define the number p by $p = \frac{\beta - d}{u - d}$. By the condition on the parameters, $d < \beta < u$, $0 < p < 1$. Next, define the new measure Q by $\mathrm{Q}(\{\omega\}) = p^k(1-p)^{n-k}$, where $k = k(\omega)$ is the number of times u appears in ω (and $n-k$ is the number of d's in ω).

It is clear that $\mathrm{Q}(\{\omega\}) > 0$ for all $\omega \in \Omega$, and it is an easy exercise involving the $Bin(n,p)$ distribution to see that $\sum_{\omega \in \Omega} Q(\omega) = 1$. Hence, Q is indeed a probability. Since P and Q give positive probabilities to all ω's, they are equivalent. This can also be expressed by changing the measure from P to Q using the derivative $d\mathrm{Q}/d\mathrm{P} = \Lambda(\{\omega\}) = \frac{\mathrm{Q}(\{\omega\})}{\mathrm{P}(\{\omega\})}$. Note that since $0 < \Lambda(\{\omega\}) < \infty$ due to all $\mathrm{P}(\{\omega\}) > 0$, Q is equivalent to P.

Next, we show that Q is a martingale measure, namely that the process S_t/β^t is a martingale under Q. To this end, we establish that under Q, all the variables ξ_t's are independent and identically distributed, with

$$\mathrm{Q}(\xi_t = u) = p, \quad \mathrm{Q}(\xi_t = d) = 1 - p.$$

One can recognise in $(\Omega, \mathcal{F}, \mathrm{Q})$ the binomial probability model set up so that it follows that the coordinates are independent and identically distributed. However, we sketch the proof. Let $B_t = \{\omega : \omega_t = u\}$ be the set of outcomes with the tth coordinate u, and B_t^c be its complementary. Then, it is not hard to see by counting the favourable outcomes that $\mathrm{Q}(B_t) = p$ and $\mathrm{Q}(B_t^c) = 1-p$. This can be seen by the following calculation:

$$\begin{aligned}
Q(\xi_t = u) &= \sum_{\omega:\omega_t=u} p^{k(\omega)}(1-p)^{n-k(\omega)} = \sum_{k=1}^{n}\binom{n-1}{k-1}p^k(1-p)^{n-k} \\
&= p\sum_{k=1}^{n}\binom{n-1}{k-1}p^{k-1}(1-p)^{n-k} \\
&= p\sum_{l=0}^{n-1}\binom{n-1}{l}p^l(1-p)^{n-1-l} = p,
\end{aligned}$$

where we used that the sum of binomial probabilities is one.

Moreover, under Q, all these events are independent for different t's, for any choice of these or their complimentary sets, say $D_t = B_t$ or B_t^c,

$$Q(\{\omega\}) = Q(\cap_{t=1}^n D_t) = \prod_{t=1}^{n} Q(D_t).$$

In other words, the coordinates of ω are independent under Q, or all the variables ξ_t's are independent and identically distributed. Once this property of Q is established, the martingale property easily follows. Note that

$$\mathrm{E}_Q(\xi_t) = up + d(1-p) = \beta.$$

$$\begin{aligned}
\mathrm{E}_Q\left(\frac{S_t}{\beta^t}|\mathcal{F}_{t-1}\right) &= \frac{S_{t-1}}{\beta^t}\mathrm{E}_Q(\xi_t|\mathcal{F}_{t-1}) \\
&= \frac{S_{t-1}}{\beta^t}\mathrm{E}_Q(\xi_t) = \frac{S_{t-1}}{\beta^t}\beta = \frac{S_{t-1}}{\beta^{t-1}}.
\end{aligned}$$

The necessity of the condition $\mathrm{P}(\{\omega\}) > 0$ will follow after we show the uniqueness of EMM Q found above.

(c) Uniqueness of EMM Q when it exists: Suppose that EMM Q exists. Then, from the above martingale condition calculation, it follows that

$$\mathrm{E}_Q(S_t|\mathcal{F}_{t-1}) = \beta S_{t-1},\ \mathrm{E}_Q(S_t|\mathcal{F}_{t-2}) = \beta^2 S_{t-2},$$

which imply for all t,

$$\mathrm{E}_Q(\xi_t) = \beta,\ \ \mathrm{E}_Q(\xi_{t-1}\xi_t) = \beta^2.$$

$E_Q(\xi_t) = \beta$ implies that all ξ_t have the same distribution:

$$Q(\xi_t = u) = (\beta - d)/(u - d) = p, \; Q(\xi_1 = d)$$
$$= (u - \beta)/(u - d) = 1 - p.$$

Further, we obtain $E_Q(\xi_{t-1}\xi_t) = E_Q(\xi_{t-1})E_Q(\xi_t)$, so that ξ_{t-1} and ξ_t are uncorrelated. While in general, this does not imply independence, for two-point distributions it does, see Problem 2.34. Similarly, from the martingale condition for $E_Q(S_t) = \beta^t = (E_Q(\xi_1)^t$ so that, for any number of ξ_i's, we obtain that

$$E_Q \prod_{i=1}^{t} \xi_i = \prod_{i=1}^{t} E_Q(\xi_i).$$

Again, due to being a two-point distribution, this implies that ξ_i's are independent.

Thus, Q is exactly the measure found in (b), which means uniqueness. Moreover, it implies that $Q(\{\omega\}) > 0$ for all ω. Since Q and P are equivalent, $P(\{\omega\}) > 0$ for all ω, and necessity is proved.

Problem 11.10: Give the price of stock S_n under the EMM Q in the binomial model, hence derive the price of the call option.

Solution. Since we established that in the n-step binomial model, under EMM Q the distribution of the number of times the stock goes up, i.e. the number of u's amongst in ω has the $Bin(n, p)$ distribution, the Q-distribution of S_n is given by

$$Q(S_n = S_0 u^i d^{n-i}) = \binom{n}{i} p^i (1-p)^{n-i}, \quad i = 0, 1, \ldots, n.$$

Hence, the arbitrage-free price

$$C_0 = \frac{1}{\beta^n} E_Q(S_n - K)^+ = \frac{1}{\beta^n} \sum_{i=0}^{n} \binom{n}{i} p^i (1-p)^{n-i} (u^i d^{n-i} S_0 - K)^+.$$

Problem 11.11: Consider the one-period binomial model, with $S_1 = S_0\xi$, and the contract with payoff X at time 1.

(a) Find the replicating portfolio for this contract.
(b) Find the time 0 price of this contract.
(c) Verify that the price is given by the discounted expectation under the EMM.

Solution.

(a) The value of the replicating portfolio (a, b) $V_1 = aS_1 + b\beta = X$ yields two equations:

$$aS_0u + b\beta = X(u), \quad aS_0d + b\beta = X(d).$$

This gives the solution

$$a = \frac{X(u) - X(d)}{(u-d)S_0}, \quad b = \frac{uX(d) - dX(u)}{(u-d)\beta}.$$

(b) The price of the contract is given by the time 0 of the portfolio

$$C_0 = V_0 = aS_0 + b = \frac{X(u) - X(d)}{(u-d)} + \frac{uX(d) - dX(u)}{(u-d)\beta}.$$

(c) The EMM Q is given by $\mathrm{Q}(S_1 = uS_0) = p = \frac{\beta - d}{u-d}$, $\mathrm{Q}(S_1 = dS_0) = 1 - p = \frac{u-\beta}{u-d}$. Proceeding from (b),

$$\begin{aligned} C_0 &= \frac{X(u) - X(d)}{(u-d)} + \frac{uX(d) - dX(u)}{(u-d)\beta} \\ &= \frac{(\beta - d)X(u) + (u-\beta)X(d)}{(u-d)\beta} = \frac{1}{\beta}\left(\frac{\beta - d}{u-d}X(u) + \frac{u-\beta}{u-d}X(d)\right) \\ &= \frac{1}{\beta}(pX(u) + (1-p)X(d)) = \frac{1}{\beta}\mathrm{E}_Q(X). \end{aligned}$$

Problem 11.12: Consider the one-period binomial model ($S_0 = 1$) and the contract which pays $C_1 = 1/S_1$ at time 1.

(a) Find the time 0 price of this contract.
(b) Find the replicating portfolio for this contract.

Solution.

(a) The price of the contract is given by

$$\begin{aligned} C_0 = \mathrm{E}_Q(1/(\beta S_1)) &= \frac{p/u + (1-p)/d}{\beta} = \frac{pd + (1-p)u}{\beta ud} \\ &= \frac{(\beta - d)d + (u-\beta)u}{\beta ud(u-d)} = \frac{(u^2 - d^2) - \beta(u-d)}{\beta ud(u-d)} \\ &= \frac{u + d - \beta}{ud\beta}. \end{aligned}$$

(b) The replicating portfolio is a pair (a, b) that satisfies $aS_1 + b\beta = D_1$, which corresponds to the equations

$$au + b\beta = 1/u, \quad ad + b\beta = 1/d.$$

The solutions are given by

$$a = \frac{-1}{ud}, \quad b = \frac{u+d}{\beta ud}.$$

Problem 11.13:

(a) Give the equations for a self-financing portfolio, which replicates claim X in the two-period binomial model.
(b) Represent the self-financing portfolio, which replicates claim X by a tree.
(c) Generalise for the multiperiod binomial model.

Solution.

(a) We must find a self-financing portfolio (a_t, b_t), $t = 1, 2$, with value V_t, such that

$$X = V_2 = a_2 S_2 + b_2 \beta_2$$

and also satisfies the self-financing condition

$$(a_2 - a_1)S_1 = -(b_2 - b_1)\beta_1.$$

The replication equation needs to be solved for all four possible outcomes:

$$\omega \in \{\omega_1 = (d, d), \omega_2 = (d, u), \omega_3 = (u, d), \omega_4 = (u, u)\}.$$

The coefficients a_2 and b_2 are not constants, and they depend on the information observed up to time 1 and are functions of S_1. $a_2 = a_2(S_1)$, $b_2 = b_2(S_1)$. Since we assumed $S_0 = 1$, at time 1, we have $S_1(\omega_1) = S_1(\omega_2) = d$ and $S_1(\omega_3) = S_1(\omega_4) = u$. The above equations become

$$X(\omega_1) = a_2(d)d^2 + b_2(d)\beta^2, \; X(\omega_2) = a_2(d)du + b_2(d)\beta^2,$$
$$X(\omega_3) = a_2(u)ud + b_2(u)\beta^2, \; X(\omega_4) = a_2(u)u^2 + b_2(u)\beta^2.$$
$$(a_2(d) - a_1)d = -(b_2(d) - b_1)\beta, \; (a_2(u) - a_1)u = -(b_2(u) - b_1)\beta.$$

The first four equations are replicating the option, while the last two equations are checking the self-financing condition.

These are six equations in six unknowns, $a_1, b_1, a_2(d), a_2(u), b_2(d), b_2(u)$.

Solving these equations we obtain the replicating self-financing portfolio, which allows us to obtain the price of X at times before time $T = 2$.

For $t = 1$, the price of X is $C_1 = V_1 = a_1 S_1 + b_1 \beta_1$.

For $t = 0$, the price of X is $C_0 = V_0 = a_1 S_0 + b_1 \beta_0$.

(b) Denoting by

$$C_{dd} = V_2(\omega_1),\ C_{du} = V_2(\omega_2),\ C_{ud} = V_2(\omega_3),\ C_{uu} = V_2(\omega_4),$$
$$C_u = V_1(\omega_3) = V_1(\omega_4),\ \ C_d = V_1(\omega_1) = V_1(\omega_2),$$

we obtain the following tree for self-financing portfolios:

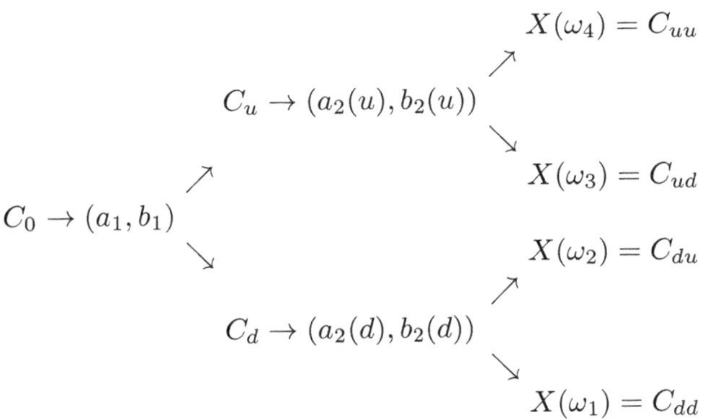

(c) For multiple periods $T = n$, a similar method can be used to find the replicating portfolio. In general, there will be $2^{n+1} - 2$ unknowns and $2^{n+1} - 2$ equations. But to keep things easy to track, it is usually simpler to work backward in time, one branch of the tree at a time. So, start with the payoff $V_n = X$, solve for the portfolio for all branches in the final period (a_n, b_n), then find the value V_{n-1}, then find (a_{n-1}, b_{n-1}) and so on. This effectively breaks one large problem into $2^n - 1$ smaller problems, each one being just like the one-period binomial case.

Problem 11.14: Consider the two-step binomial model, which is specified by the following. The price of one share of stock S_t at time t is given by

$S_0 = 1$, $S_1 = \xi_1 S_0$, $S_2 = \xi_2 S_1$ with random variables ξ_1, ξ_2 taking two values d and u and having the distribution $P(\xi_1 = d, \xi_2 = d) = 1/4$, $P(\xi_1 = d, \xi_2 = u) = 1/8$, $P(\xi_1 = u, \xi_2 = d) = 1/8$, $P(\xi_1 = u, \xi_2 = u) = 1/2$. The savings account is specified by $\beta_t = \beta^t$, $t = 0, 1, 2$. These parameters satisfy $0 < d < \beta < u$.

(a) Give the marginal distributions of ξ_1 and ξ_2 and show that they are not independent under the original (real-world) probability measure P.
(b) Show that the discounted stock price process S_t/β^t, $t = 0, 1, 2$, is not a martingale under P.
(c) Describe the EMM in this model, and give the derivative $\Lambda = dQ/dP$.
(d) Consider the following strategy: start with a portfolio at time $t = 0$, $a_1 = 1$, $b_1 = -1$ (borrow \$1 to buy 1 share), and at time $t = 1$, take $a_2 = 1$, $b_2 = -1$ (do nothing, no rebalancing). Show that this is a self-financing portfolio. Show that the value process is given by $V_0 = 0$, $V_1 = S_1 - \beta$, $V_2 = S_2 - \beta^2$. Which contract X does this portfolio replicate? Is this strategy an arbitrage strategy?
(e) Start with portfolio at time $t = 0$, $a_1 = 1$, $b_1 = -1$ (borrow \$1 to buy 1 share). At time $t = 1$, let $a_2 = a_2(S_1) = 0$ if $S_1 = u$ and $a_2 = a_2(S_1) = 2$ if $S_1 = d$; in other words, $a_2 = I(\xi_1 = u) + 2I(\xi_1 = d)$. Find $b_2 = b_2(S_1)$, so that the self-financing condition holds. Give the value of the portfolio at times $t = 1$ and $t = 2$. Which contract X does this portfolio replicate?
(f) Let $X = S_2/S_1^2$. Give the price of X at times $t = 0, 1$ first by the EMM method of pricing and second by the portfolio method of pricing.
(g) Explain why the portfolio replicating the claim needs rebalancing; in other words, why do we need to change it at time $t = 1$ after it is established at time $t = 0$. What can happen if we don't rebalance it?

Solution.

(a) The marginals are given by $P(\xi_1 = d) = P(\xi_1 = d, \xi_2 = d) + P(\xi_1 = d, \xi_2 = u) = 1/4 + 1/8 = 3/8$, $P(\xi_1 = u) = 5/8$,

$$P(\xi_2 = d) = 1/4, \ P(\xi_2 = u) = 3/4.$$

Since $P(\xi_1 = d, \xi_2 = d) = 1/4 \neq P(\xi_1 = d)P(\xi_2 = d) = 3/8 \times 1/4$ the rvs. ξ_1 and ξ_2 are not independent under probability P.

(b) Writing the martingale condition for times $t = 1$ and $t = 2$, we obtain the following equations: $\mathrm{E}_P(S_1) = \mathrm{E}_P(\xi_1) = (3d+5u)/8 = \beta$,

$$\mathrm{E}_P(S_2|S_1) = \mathrm{E}_P(S_1\xi_2|S_1) = S_1\mathrm{E}_P(\xi_2) = S_1(d+3u)/4 = S_1\beta.$$

It now follows from these equations that $d = u$, which is a contradiction with the assumption $d < u$.

(c) $dQ/dP = \Lambda$ is as follows:
$\Lambda((d,d)) = Q((d,d))/P((d,d)) = 4(1-p)^2$; similarly,

$$\Lambda((d,u)) = 8p(1-p),\ \Lambda((u,d)) = 8p(1-p),\ \Lambda((u,u)) = 2p^2.$$

(d) Since $\Delta a_{t+1} = a_{t-1} - a_t = 0 = \Delta b_{t+1} = b_{t+1} - b_t$, the portfolio is self-financing.

As $V_t = a_t S_t + b_t\beta^t$, with $V_0 = a_1 S_0 + b_0$, we have

$$V_0 = S_0 - 1 = 0. \quad V_1 = a_1 S_1 + b_1\beta = S_1 - \beta.$$
$$V_2 = a_2 S_2 + b_2\beta^2 = S_2 - \beta^2.$$

This portfolio replicates the claim $X = V_2 = S_2 - \beta^2$.

The portfolio does not represent an arbitrage strategy because $Q(V_2 < 0) \geq Q(\xi_1 = d, \xi_2 = d) = (1-p)^2 > 0$.

Hence, it follows from the equivalence of Q and P that $P(V_2 < 0) > 0$.

(e) From the self-financing condition,

$$\begin{aligned}(a_2(S_1) - a_1)S_1 &= -(b_2(S_1) - b_1)\beta,\\ b_2(S_1) &= -(a_2(S_1) - a_1)S_1/\beta - b_1\\ &= -(a_2(S_1) - 1)S_1/\beta + 1 = I(S_1 = u)(u+\beta)/\beta\\ &\quad + I(S_1 = d)(\beta - d)/\beta.\end{aligned}$$

The portfolio replicates $X = V_2 = a_2 S_2 + b_2\beta^2$. $X = b_2\beta^2$ if $S_1 = u$ and $X = 2S_2 + b_2\beta^2$ if $S_1 = d$.

(f) The EMM method:

$$C_1 = \frac{1}{\beta}\mathrm{E}_Q(S_2/S_1^2|S_1) = \frac{1}{\beta}\frac{1}{S_1^2}E_Q(S_2|S_1) = \frac{1}{\beta}\frac{1}{S_1^2}(S_1\beta) = \frac{1}{S_1},$$

$$C_0 = \frac{1}{\beta^2}E_Q(X) = \frac{1}{\beta^2}\mathrm{E}_Q(S_2/S_1^2) = \frac{1}{\beta^2}\mathrm{E}_Q\mathrm{E}_Q(S_2/S_1^2|S_1)$$
$$= \frac{1}{\beta^2}\mathrm{E}_Q\left(\frac{1}{S_1^2}E_Q(S_2|S_1)\right) = \frac{1}{\beta^2}\mathrm{E}_Q\left(\frac{1}{S_1^2}S_1\beta\right)$$
$$= \frac{1}{\beta}\mathrm{E}_Q\left(\frac{1}{S_1}\right) = \frac{1}{\beta}\left(\frac{\beta-d}{u(u-d)} + \frac{u-\beta}{d(u-d)}\right) = \frac{u+d-\beta}{ud\beta}.$$

Replicating the portfolio method,

$$V_2 = a_2(S_1)S_2 + b_2(S_1)\beta^2 = X(S_1, S_2).$$

Since $S_2 = S_1\xi_2$, we have

$$a_2(S_1)S_1\xi_2 + b_2(S_1)\beta^2 = X(S_1, S_1\xi_2) = S_2/S_1^2 = \xi_2/S_1.$$

Letting $\xi_2 = u$ and d, we have two linear equations with solutions

$$a_2(S_1) = \frac{(u/S_1) - (d/S_1)}{(u-d)S_1} = \frac{1}{S_1^2},$$
$$b_2(S_1) = \frac{u(d/S_1) - d(u/S_1)}{(u-d)\beta^2} = 0.$$

Clearly, $V_2 = a_2S_2 + b_2\beta^2 = a_2S_2 = S_2/S_1^2$. Next, the rebalancing condition gives

$$V_1 = a_1S_1 + b_1\beta = a_2(S_1)S_1 + b_2\beta = a_2(S_1)S_1 = \frac{1}{S_1}.$$

This gives $C_1 = V_1$, as above.

Letting $S_1 = u$ and d, we have two linear equations with solutions

$$a_1 = \frac{(1/u) - (1/d)}{u-d} = -\frac{1}{ud}, \quad b_1 = \frac{(u/d) - (d/u)}{(u-d)\beta} = \frac{u+d}{ud\beta},$$
$$V_0 = a_1S_0 + b_1 = -\frac{1}{ud} + \frac{u+d}{ud\beta} = \frac{u+d-\beta}{ud\beta}.$$

This gives $C_0 = V_0$, as above.

(g) If the replicating portfolio is not rebalanced, then there is a positive probability that it won't match the claim at the end, resulting in a loss.

Problem 11.15: Consider the two-period binomial model. Show that if $\xi_2 = \xi_1$, then there is no EMM. Hence, deduce that there are arbitrage strategies in the model and find an arbitrage strategy.

Solution. If $\xi_2 = \xi_1$, then $S_2 = \xi_2\xi_1 = \xi_1^2$. There is no measure Q that makes $1, \xi_1/\beta, \xi_1^2/\beta^2$ into a martingale because for any measure, $\mathrm{E}(\xi_1^2/\beta^2|S_1) = \xi_1^2/\beta^2 \neq \xi_1/\beta$. It is not a martingale for any choice of distribution of ξ_1, equivalent to the original distribution on the two points d and u. In this model, there is zero probability for outcomes (d,u) and (u,d), cf. Problem 11.9. An arbitrage strategy is given by the following. After step 1, we observe ξ_1. Since ξ_2 is exactly the same, we sell the stock if $\xi_1 = u$ and invest in the savings account. If $\xi_1 = d$, then we borrow d and buy the stock. This results in arbitrage profit.

Problem 11.16: Consider the one-period model $S_0 = 1$, $S_1 = \xi$, where $\xi \in \{d, 1, u\}$, and savings account $\beta_t = \beta^t$, $d < 1 < \beta < u$.

(a) Describe the space of EMMs.
(b) Let X be a claim that can be replicated by a portfolio. Show that $\mathrm{E}_Q(X)$ is the same for any EMM Q, and the time 0 price of X is given by $C_0 = \mathrm{E}_Q(X)/\beta$.
(c) Describe the set of claims X which cannot be replicated by a portfolio. Show that for such claims, $\mathrm{E}_Q(X)$ depends on Q.

Solution.

(a) The EMM is given by the vector of probabilities (p_d, p_n, p_u), such that $dp_d + p_n + up_u = \beta$. There are infinitely many EMMs.
(b) Introduce the vectors of payoffs of stock $w = (d, 1, u)$ and savings account $v = (\beta, \beta, \beta)$. Then, a claim X can be replicated if there are constants a and b such that $X = aS_1 + b\beta$. Writing this in vector form yields that the payoff $(X(d), X(1), X(u))$ is a linear combination of vectors w and v. In other words, it lies in the span of w, v. So, $Span(w, v)$ is the set of all attainable claims.

   Since the vectors w and v are not collinear, the replicating portfolio (a, b) is unique. The time 0 price of X is the value of the portfolio $V_0 = aS_0 + b$.

   Let Q be an EMM. Then,

$$\mathrm{E}_Q(X) = a\mathrm{E}_Q(S_1) + b = aS_0\beta + b\beta,$$

   which is the same for any Q. Moreover, $C_0 = aS_0 + b = \mathrm{E}_Q(X)/\beta$.
(c) Clearly, any claim with a payoff vector not in $Span(w, v)$ cannot be replicated. This space has dimension two. Since the dimension of the space of all claims $\mathbb{R}^3$ is three, there are many claims that cannot be replicated. An example is the claim that pays \$1 when

the stock goes up and nothing in any other case with payoff vector $(0, 0, 1)$. One way to know that this vector is not in the span of the other two is to see that the following matrix has a non-zero determinant: $\begin{pmatrix} d & 1 & 0 \\ 1 & 1 & 0 \\ u & 1 & 1 \end{pmatrix}$. $\mathrm{E}_Q(X) = p_u$ clearly depends on the choice of p_u, i.e. the martingale measure Q. More generally, any claim with a payoff which can be written as a linear combination of the orthogonal component of $Span(w, v)$ and possibly w and v cannot be replicated.

Problem 11.17: Consider a general one-period model with n states and a finite number of assets. Let the $n \times m$ matrix M have its columns represent the growth rate of m assets (stocks and savings account) for the n possible states of the economy at time 1. Show that the rank of M is the dimension of the space of attainable claims (payoffs that can be replicated), while the nullity of M is the dimension of the space of EMMs. Deduce that a complete market is equivalent to the rank being n, while a unique EMM is equivalent to the nullity being 0.

Solution. Let the vector $a = (a_1, a_2, \ldots, a_m)$ denote a portfolio of assets $S_1, S_2, \ldots, S_m$. Let vector $X = (X(e_1), X(e_2), \ldots, X(e_n))$ be the payoff of claim X, where $X(e_j)$ is the payoff when the economy is in the state j, and similarly define vectors $S_1, S_2, \ldots, S_m$, $S_i(e_j)$ as the payoff of asset i when the economy is in the state e_j, $i = 1, 2, \ldots, m$, $j = 1, 2 \ldots, n$. (This is a slight abuse of notation whereby the payoff vectors have the same names as the assets.) The mth asset is the savings account. The vectors $S_1, S_2, \ldots, S_m$ are the columns of the matrix M. Then, the claim is attainable if there is an a such that $Ma = X$, in other words, when X belongs to $Span(S_1, S_2, \ldots, S_m)$. Since the rank of M is the dimension of this space, it is also the dimension of the space of attainable claims.

Let now the vector $p = (p_1, p_2, \ldots, p_n)$ denote a probability vector, which represents an EMM. Then, the equations for p can be written using the martingale condition as the inner product of vectors S_i and p:

$$\mathrm{E}_Q(S_i) = \sum_{j=1}^{n} S_i(e_j)p_j = (S_i p) = \beta.$$

This can be written as $M^T p = (\beta, \beta, \ldots, \beta)$. This equation has a unique solution if M is invertible or of the full rank. The null space of M is the space of vectors x orthogonal to the column space, $M^T x = 0$. Nullity is the dimension of this space. The nullity is 0 if and only if M is invertible. Clearly, the difference of two EMM vectors $p^1 - p^2$ belongs to the null space. This is how nullity enters the description of the space of EMMs. The rank–nullity theorem states that $Rank(M) + Nullity(M) = dim(Domain(M))$.

Problem 11.18: Let (B_t) denote a Brownian motion under the real-world measure with $B_0 = 0$. Does each of the following market models have an EMM? If not, find an arbitrage strategy:

(a) $\beta_t = 1, S_t = 2$.
(b) $\beta_t = 1, S_t = 2\sin t$.
(c) $\beta_t = 2 + \sin t, S_t = 4 + 2\sin t$.
(d) $\beta_t = 1, S_t = t + B_t$.
(e) $\beta_t = 1, S_t = e^{B_t}$.
(f) $\beta_t = 1, S_t = 1 + e^{B_t}$.
(g) $\beta_t = e^t, S_t = 1$.
(h) $\beta_t = e^t, S_t = e^{2t}$.
(i) $\beta_t = e^t, S_t = e^{2t+B_t}$.
(j) $\beta_t = e^{B_t}, S_t = e^{2B_t}$.

Solution.

(a) The discounted stock price $S_t/\beta_t = 2$, which is a constant. So, there is an EMM.
(b) The discounted stock price $S_t/\beta_t = 2\sin t$, which is deterministic but not constant. So, there is no EMM. An arbitrage would be buying the stock at time 0 and selling it at time $\pi/2$.
(c) The discounted stock price $S_t/\beta_t = 2$, which is a constant. So, there is an EMM.
(d) The discounted stock price $S_t/\beta_t = t + B_t$. By the Girsanov theorem, we can find a new measure so that $S_t/\beta_t = \hat{B}_t$, which is a martingale. So, there is an EMM.
(e) This is the Black–Scholes model, so there is an EMM.
(f) The discounted stock price $S_t/\beta_t = 1 + e^{B_t}$. By the Girsanov theorem, we can find a new measure so that $S_t/\beta_t = 1 + e^{-t/2+\hat{B}_t}$, which is a martingale. So, there is an EMM.
(g) The discounted stock price $S_t/\beta_t = e^{-t}$, which is deterministic but not constant. So, there is no EMM. An arbitrage would involve selling the stock at time 0 and buying it back at any later time.
(h) The discounted stock price $S_t/\beta_t = e^t$, which is deterministic but not constant. So, there is no EMM. An arbitrage would involve buying the stock at time 0 and selling it at any later time.

(i) This is the Black–Scholes model, so there is an EMM.

(j) The discounted stock price $S_t/\beta_t = e^{B_t}$. By the Girsanov theorem, we can find a new measure so that $S_t/\beta_t = e^{-t/2+\hat{B}_t}$, which is a martingale. So, there is an EMM.

Problem 11.19: Show that EMM Q in the Black–Scholes model is unique.

Solution. Let B_t, $0 \le t \le T$, be a Brownian motion under the Wiener measure P, and Q be equivalent to P. Then, there is a predictable process H_t, so that $dQ/dP = e^{\int_0^T H_t dB_t - \frac{1}{2}\int_0^T H_t^2 dt}$; moreover $B_t = \hat{B}_t + \int_0^t H_s ds$, for a Brownian motion $\hat{B}_t$ under Q (Corollary 10.21 in Klebaner's book).

Since $S_t e^{-rt} = S_0 e^{(\mu-r)t - \frac{\sigma^2}{2}t + \sigma B_t}$, and any continuous change of measure $Q \sim P$ results in adding a drift to B_t, we have

$$S_t e^{-rt} = S_0 e^{(\mu-r)t - \frac{\sigma^2}{2}t + \sigma\hat{B}_t + \sigma\int_0^t H_s ds}.$$

If it is a martingale under the new measure, then

$$\frac{(\mu-r)}{\sigma}t + \int_0^t H_s ds = 0.$$

This implies that a.e. $H_t = -\frac{(\mu-r)}{\sigma} = -c$, and $dQ/dP = e^{-cB_T - c^2T/2}$, proving the uniqueness of Q.

Problem 11.20: Let (a_t, b_t) be a portfolio in the Black–Scholes model with value $V_t = a_t S_t + b_t\beta_t$. Let $Z_t = S_t/\beta_t$.

(a) Show that the portfolio is self-financing if and only if $d(V_t/\beta_t) = a_t dZ_t$.

(b) Suppose that V_t is a self-financing portfolio, and a_t is known. Find b_t.

Solution.

(a) If the portfolio is self-financing, we have $dV_t = a_t dS_t + b_t d\beta_t$. Hence,

$$\begin{aligned} d(V_t/\beta_t) &= \beta_t^{-1}dV_t - \beta_t^{-2}V_t d\beta_t \\ &= a_t\beta_t^{-1}dS_t + b_t\beta_t^{-1}d\beta_t - \beta_t^{-2}(a_tS_t + b_t\beta_t)d\beta_t \\ &= a_t\beta_t^{-1}dS_t - \beta_t^{-2}a_tS_t d\beta_t = a_t dZ_t, \end{aligned}$$

since $dZ_t = \beta_t^{-1}dS_t - \beta_t^{-2}S_t d\beta_t$.

Conversely, suppose that $d(V_t/\beta_t) = a_t dZ_t$. Then,

$$\begin{aligned} dV_t &= \beta_t d(V_t/\beta_t) + \beta_t^{-1} V_t d\beta_t \\ &= \beta_t a_t dZ_t + \beta_t^{-1}(a_t S_t + b_t\beta_t) d\beta_t \\ &= a_t dS_t - \beta_t^{-1} a_t S_t d\beta_t + \beta_t^{-1} a_t S_t d\beta_t + b_t d\beta_t = a_t dS_t + b_t d\beta_t. \end{aligned}$$

(b) From the definition of the portfolio value, solving for b_t yields $b_t = (V_t - a_t S_t)/\beta_t = V_t/\beta_t - a_t Z_t$.

Problem 11.21: Let X be a payoff of contract at time T in the Black–Scholes model, such that $\mathrm{E}X^2 < \infty$.

(a) Find the replicating self-financing portfolio for X by using the predictable representation property of Brownian motion.
(b) Give C_t, the price of X at time $t < T$, as the value of the replicating portfolio. Show that it is discounted the expected payoff under the EMM.

Solution.

(a) Since under Q the martingale $Z_t = S_t e^{-rt} = S_0 e^{\sigma B_t - \frac{\sigma^2}{2}t}$ for a Q Brownian motion B_t (we dropped hat), it is the stochastic exponential of σB_t, and $Z_t = S_0 + \int_0^t \sigma Z_u dB_u$. Since X^2 is integrable under P and

$$\mathrm{E}_Q|X| = \mathrm{E}_P(|X|e^{-cB_T - c^2T/2}) \leq \left(\mathrm{E}_P X^2 \mathrm{E}_P \Lambda^2\right)^{1/2} < \infty,$$

X is integrable under Q.

Note that all the following expectations are under Q.

Thus, the random variable Xe^{-rT} can be written as an integral of a predictable process H_t by the predictable representation property of Q Brownian motion B_t,

$$Xe^{-rT} = e^{-rT}\mathrm{E}X + \int_0^T H_t dB_t.$$

Hence we can write, using $dB_t = dZ_t/(\sigma Z_t)$,

$$Xe^{-rT} = e^{-rT}\mathrm{E}X + \int_0^T (H_t/(\sigma Z_t)) dZ_t.$$

Let now

$$M_t = \mathrm{E}(Xe^{-rT}|\mathcal{F}_t).$$

Then, M_t is a martingale and from the above representation for Xe^{-rT}, it has the form

$$M_t = e^{-rT}\mathrm{E}X + \int_0^t (H_u/(\sigma Z_u))dZ_u.$$

(This martingale is in fact the discounted portfolio value process.) The replicating self-financing portfolio is given by (read a_t from above and b_t from Problem 11.20)

$$a_t = H_t/(\sigma Z_t), \;\; b_t = e^{-rT}\mathrm{E}X + \int_0^t a_u dZ_u - H_t/\sigma.$$

Indeed, we can see that it is a replicating portfolio:

$$\begin{aligned} V_T = a_T S_T + b_T e^{rT} &= H_T e^{rT}/\sigma \\ &\quad + \mathrm{E}X + e^{rT}\int_0^T a_u dZ_u - e^{rT}H_T/\sigma = X. \end{aligned}$$

Check the self-financing property for $V_t = a_t S_t + b_t e^{rt}$, use Problem 11.20:

$$\begin{aligned} d(V_t/\beta_t) = d(a_t Z_t + b_t) &= d(H_t/\sigma + \int_0^t a_u dZ_u - H_t/\sigma) \\ &= e^{-rT}a_t dZ_t. \end{aligned}$$

(b) $C_t = V_t = a_t S_t + b_t\beta_t = e^{-r(T-t)}\mathrm{E}X + e^{rt}\int_0^t a_u dZ_u$. Hence,

$$C_t = e^{rt}M_t = e^{-r(T-t)}\mathrm{E}(X|\mathcal{F}_t).$$

Problem 11.22: Let the price process S_t be a martingale in the Black–Scholes model, and savings account $\beta_t = 1$. Let the claim have payoff $X = S_1 - \int_0^1 S_u du$ at time $T = 1$.

(a) Find the price of this claim at time $t < 1$.

(b) Give the replicating portfolio for this claim, and show that it is self-financing.

Solution.

(a) First, check the conditions of Problem 11.21. Note that by integration by parts $\int_0^1 S_u du = S_1 - \int_0^1 u dS_u$. Hence, $X = \int_0^1 u dS_u$, and

$$\mathrm{E}X^2 = \mathrm{E}\int_0^1 u^2 d[S,S]_u.$$

Using $dS_t = \sigma S_t dB_t$ and that $S_t = S_0 e^{\sigma B_t - \frac{\sigma^2}{2}t}$,

$$\mathrm{E}X^2 = \mathrm{E}\int_0^1 u^2\sigma^2 S_u^2 du = \sigma^2 \int_0^1 u^2 \mathrm{E}S_u^2 du$$
$$= \sigma^2 S_0^2 \int_0^1 u^2 e^{\sigma^2 u} du < \infty.$$

We can apply Problem 11.21, with the discounted stock process $Z_t = S_t$, and $\mathrm{Q} = \mathrm{P}$:

$$C_t = \mathrm{E}(X|\mathcal{F}_t) = \mathrm{E}\left(\int_0^1 u dS_u|\mathcal{F}_t\right) = \int_0^t u dS_u.$$

(b) From above, we have immediately $a_t = t$, $b_t = -\int_0^t S_u du$. Clearly, as $\beta_t = 1$, and using integration by parts, $V_t = tS_t - \int_0^t S_u du = \int_0^t u dS_u = C_t$. In particular, $V_1 = \int_0^1 u dS_u = X$. It is also self-financing,

$$dV_t = tdS_t + S_t dt - S_t dt = tdS_t = a_t dS_t + b_t d\beta_t.$$

Problem 11.23: Consider pricing an option on stock in the Black–Scholes model, with the payoff $X = g(S_T)$ at time T, where g is a given function, and $\mathrm{E}X^2 < \infty$. Let C_t be the price of this option at time $t \leq T$.

(a) Show that $C_t = C(S_t, t)$, for a function $C(x, t)$.
(b) Give the partial differential equation for $C(x, t)$ and the boundary conditions.
(c) Find the replicating self-financing portfolio for X.

Solution.

(a) Since the conditions of Problem 11.21 hold,

$$C_t = e^{-r(T-t)}\mathrm{E}_Q(g(S_T)|\mathcal{F}_t) = e^{-r(T-t)}\mathrm{E}_Q(g(S_T)|S_t)$$

because S_t is a Markov process. But the latter equals $C(S_t, t)$ with the function

$$C(x, t) = e^{-r(T-t)}\mathrm{E}_Q(g(S_T)|S_t = x).$$

(b) S_t is a diffusion process under Q with the generator $L = \frac{1}{2}\sigma^2 x^2 \frac{\partial^2}{\partial x^2} + rx\frac{\partial}{\partial x}$, and by the Feynman–Kac formula, $C(x, t)$ solves the PDE (the Black–Scholes PDE) $LC + \frac{\partial C}{\partial t} = rC$, or explicitly,

$$\frac{1}{2}\sigma^2 x^2 \frac{\partial^2 C}{\partial x^2} + rx\frac{\partial C}{\partial x} + \frac{\partial C}{\partial t} - rC = 0, \;\; C(x, T) = g(x).$$

(c) To find the replicating portfolio, we use Problem 11.20, by representing V_te^{-rt} as an integral of a_t with respect to $Z_t = S_te^{-rt}$. The discounted value of the replicating portfolio by (a) equals to $V_te^{-rt} = e^{-rt}C(S_t, t)$. Hence, by using Itô's formula,

$$d(e^{-rt}C(S_t,t)) = e^{-rt}\left(\frac{\partial C}{\partial x}dS_t\right)$$
$$+ e^{-rt}\left(\frac{\partial C}{\partial t} + \frac{1}{2}\sigma^2 S_t^2 \frac{\partial^2 C}{\partial x^2} - rC\right)dt,$$

where all the derivatives are evaluated at $(x, t) = (S_t, t)$. However, by the PDE in (b), the expression in the second bracket equals to $-rS_t\frac{\partial C}{\partial x}$. Therefore,

$$d(e^{-rt}C(S_t,t)) = \frac{\partial C}{\partial x}\left(e^{-rt}dS_t - re^{-rt}S_tdt\right) = \frac{\partial C}{\partial x}dZ_t$$

because $dZ_t = d(e^{r-t}S_t) = e^{-rt}dS_t - re^{-rt}S_tdt$. Thus, by Problem 11.20,

$$a_t = \frac{\partial C}{\partial x}(S_t,t), \quad b_t = e^{-rt}(C(S_t,t) - a_tS_t)$$
$$= e^{-rt}r^{-1}\left(\frac{\partial C}{\partial t} + \frac{1}{2}\frac{\partial^2 C}{\partial x^2}\sigma^2 S_t^2\right),$$

where again the PDE in (b) is used for b_t.

Problem 11.24: Consider pricing a call option and a put option on stock in the Black–Scholes model. The call option has payoff $X = (S_T - K)^+$ at time T, and the put option has payoff $X = (K - S_T)^+$ at time T. Obtain explicit results (price and replicating portfolios) for these options by using Problem 11.23.

(a) Give the solution to the Black–Scholes PDE for $C(x, t)$.
(b) Find the replicating self-financing portfolio for X.
(c) Give the Black–Scholes formula for the put option.

Solution.

(a) $C_t = C(S_t, t)$, where $C(x, t)$ solves the Black–Scholes PDE with the boundary condition $C(x, T) = (x - K)^+$. The Black–Scholes PDE is usually solved by transforming it into a standard heat equation which has a known solution. We, however, solve it by using the probability method, by computing expectations, Problem 11.26. The Black–Scholes option pricing formula for the time t

price of a call option with strike K and maturity T (so the payoff is $C_T = (S_T - K)^+$) is given by $C(S_t, t)$, where

$$C(x,t) = x\Phi(h_t) - Ke^{-r(T-t)}\Phi(h_t - \sigma\sqrt{T-t}),$$

where

$$h_t = \frac{\ln(x/K) + (r + \frac{1}{2}\sigma^2)(T-t)}{\sigma\sqrt{T-t}}$$

and Φ is the cumulative distribution function of a standard normal distribution.

(b) The replicating self-financing portfolio for a call option in the Black–Scholes model is given by

$$\begin{aligned} a_t &= \frac{\partial C}{\partial x} = \Phi(h_t), \quad b_t = \beta_t^{-1}(C_t - a_t x) \\ &= -Ke^{-rT}\Phi(h_t - \sigma\sqrt{T-t}). \end{aligned}$$

The term $a_t = \frac{\partial C}{\partial x}$ is often called the *delta* of the option. It represents both the sensitivity of the option price with respect to changes in the stock price and the amount of stocks to buy when hedging the option.

(c) The Black–Scholes option pricing formula for a put option with strike K and maturity T is

$$C_t = e^{-r(T-t)}K\Phi(\sigma\sqrt{T-t} - h_t) - S_t\Phi(-h_t),$$

where h_t is the same as in the call option.

Problem 11.25: Let Z be a standard normal distribution. Show that

$$E(e^{u+vZ} - K)^+ = e^{u+v^2/2}\Phi\left(\frac{u+v^2-\log K}{v}\right) - K\Phi\left(\frac{u-\log K}{v}\right),$$

where Φ is the cumulative distribution of the standard normal distribution.

Solution. Let us rewrite the expectation using indicator functions:

$$\begin{aligned} E(e^{u+vZ} - K)^+ &= E\big((e^{u+vZ} - K)I(e^{u+vZ} - K > 0)\big) \\ &= E\big(e^{u+vZ}I(e^{u+vZ} - K > 0)\big) \\ &\quad - KE\big(I(e^{u+vZ} - K > 0)\big) \\ &= E\big(e^{u+vZ}I(e^{u+vZ} - K > 0)\big) - KP(e^{u+vZ} - K > 0). \end{aligned}$$

The second term is easy to compute since

$$P(e^{u+vZ} - K > 0) = P\Big(Z > \frac{\log K - u}{v}\Big) = \Phi\Big(\frac{u - \log K}{v}\Big).$$

Note that we have used the fact that $P(Z > c) = 1 - \Phi(c) = \Phi(-c)$.

The first term requires some algebraic manipulations. In particular, we have to complete the squares in the exponent inside the integral:

$$\begin{aligned} E\big(e^{u+vZ} I(e^{u+vZ} - K > 0)\big) &= \int_{\log K}^{\infty} e^{z} \frac{1}{\sqrt{2\pi}v} e^{-\frac{(z-u)^2}{2v^2}} dz \\ &= e^{u+v^2/2} \int_{\log K}^{\infty} \frac{1}{\sqrt{2\pi}v} e^{-\frac{(z-(u+v^2))^2}{2v^2}} dz. \end{aligned}$$

Since the integrand on the right-hand side is actually the density function of the normal distribution $N(u + v^2, v^2)$, which has the same distribution as $u + v^2 + vZ$, we can simplify further:

$$\begin{aligned} E\big(e^{u+vZ} I(e^{u+vZ} - K > 0)\big) &= e^{u+v^2/2} P(u + v^2 + vZ > logK) \\ &= e^{u+v^2/2} P\Big(Z > \frac{\log K - (u + v^2)}{v}\Big) \\ &= e^{u+v^2/2} \Phi\Big(\frac{u + v^2 - \log K}{v}\Big). \end{aligned}$$

Problem 11.26: Consider pricing a call option in the Black–Scholes model. Use Problem 11.25 to find C_t, the price at time $t < T$. Hence, solve the Black–Scholes PDE in Problem 11.24.

Solution. Since $C_t = e^{-r(T-t)} \mathrm{E}_Q((S_T - K)^+ | \mathcal{F}_t) = e^{-r(T-t)} \mathrm{E}_Q((S_T - K)^+ | S_t)$, we need to know the conditional distribution of S_T given S_t under Q. It is obtained from the solution of the SDE for S_t under Q with a Q-Brownian motion $\hat{B}_t$, solved on the interval $[t, T]$,

$$S_T = S_t e^{(r - \frac{\sigma^2}{2})(T-t) + \sigma(\hat{B}_T - \hat{B}_t)}.$$

So, the conditional distribution of S_T given S_t under Q is normal with parameters $u = \log S_t + (r - \frac{\sigma^2}{2})(T - t)$ and $v^2 = \sigma^2(T - t)$.

Applying Problem 11.25 above, we arrive at the Black–Scholes option pricing formula:

$$C_t = S_t\Phi(h_t) - Ke^{-r(T-t)}\Phi(h_t - \sigma\sqrt{T-t}),$$

$$h_t = \frac{\log(S_t/K) + (r + \frac{\sigma^2}{2})(T-t)}{\sigma\sqrt{T-t}}.$$

This gives the solution of the Black–Scholes PDE for the call option at the point $(x, t) = (S_t, t)$, i.e. the function $C(x, t)$ after replacing S_t by x.

Problem 11.27: Consider the Black–Scholes model for the stock price, $dS_t = -2S_t\,dt + 2S_t\,dB_t$. $S_0 = 1$, and the savings account is given by $\beta_t = e^{2t}$.

(a) Find C_t, the price at time $t \leq 1$ of the call option on this stock with exercise price $K = 1$ and expiration date $T = 1$.
(b) Consider the portfolio given by $a_t = -1$ (stock) and $b_t = S_te^{-2t}$ (savings account). Show that this portfolio has value 0 for all t but it is not self-financing.

Solution.

(a) Substituting $T = 1, t = \frac{1}{2}, K = 1, r = 2$ and $\sigma = 2$ into the Black–Scholes formula, we have

$$C_{1/2} = S_{1/2}\Phi\left(\frac{\ln(S_{1/2}) + 2}{\sqrt{2}}\right) - e^{-1}\Phi\left(\frac{\ln(S_{1/2})}{\sqrt{2}}\right).$$

(b) $V_t = a_tS_t + b_te^{2t} = -S_t + S_t = 0$. The portfolio is clearly not self-financing because a_t is constant, but $b_t = S_te^{-2t}$ is continuously changing, thus requiring either withdrawal (if the price of the stock goes down) or injection (if the price of the stock goes up) of funds into the savings account. Formally, $dV_t = d(0) = 0$, but

$$\begin{aligned} a_tdS_t + b_td\beta_t &= (-1) \times (-2S_tdt + 2S_tdB_t) + S_te^{-2t}(2e^{2t}dt) \\ &= 2S_tdt - 2S_tdB_t + 2S_tdt = 2S_t(2dt - dB_t) \neq 0. \end{aligned}$$

Problem 11.28: Assume the Black–Scholes model on time interval $0 \leq t \leq T$. Suppose that the number of shares processed in a self-financing portfolio a_t is of finite variation.

(a) Show then that the self-financing condition is given by $S_t da_t = -\beta_t db_t$, $0 \leq t \leq T$.
(b) Find the other process b_t by using the self-financing condition in (a).
(c) Consider the portfolio V_t for $t \leq 1$, such that $a_0 = 1$, $b_0 = 0$, and at time t, $a_t = 1 - t$. Find the amount to be held in the savings account b_t so that this portfolio is self-financing. Give the claim X this portfolio replicates at time $T = 1$ and the price of this claim at time $t = 0$.

Solution.

(a) Since a portfolio satisfies $V_t = a_t S_t + b_t \beta_t$,

$$dV_t = d(a_t S_t + b_t \beta_t) = a_t dS_t + S_t da_t + b_t d\beta_t + \beta_t db_t.$$

Here, we used that due to a_t having finite variation, $da_t dS_t = 0$, and since β_t has finite variation, $db_t d\beta_t = 0$. The portfolio is self-financing if $dV_t = a_t dS_t + b_t d\beta_t$. Taking the difference, we can write the self-financing condition as $S_t da_t = -\beta_t db_t$.

(b) From here, we can obtain b_t if we know a_t:

$$b_t = b_0 - \int_0^t (S_u/\beta_u) da_u.$$

(c) Using $da_t = d(1 - t) = -dt$, we have $b_t = \int_0^t \frac{S_u}{\beta_u} du$. Therefore, the self-financing portfolio is

$$a_t = 1 - t, \;\; b_t = \int_0^t \frac{S_u}{\beta_u} du$$

with value $V_t = (1-t)S_t + \beta_t \int_0^t \frac{S_u}{\beta_u} du$. Hence, $V_1 = \beta_1 \int_0^1 \frac{S_u}{\beta_u} du$. This portfolio replicates the claim $X = V_1 = \beta_1 \int_0^1 \frac{S_u}{\beta_u} du$, the stock price average. The time zero price of this claim is $C_0 = V_0 = a_0 S_0 + b_0 = S_0$. Alternatively, we obtain the same by using the pricing by EMM formula:

$$C_0 = (1/\beta_1)\mathrm{E}_Q(X) = \mathrm{E}_Q\left(\int_0^1 \frac{S_u}{\beta_u} du\right) = \int_0^1 \mathrm{E}_Q\left(\frac{S_u}{\beta_u}\right) du = S_0,$$

where we interchanged integrals by the Fubini theorem and used that $\frac{S_u}{\beta_u}$ is a Q-martingale.

Problem 11.29: Let $U(t)$ be the value of a foreign asset in foreign currency. Assume that it satisfies the Black–Scholes SDE. Let $X(t)$ be the exchange rate. A call option on an asset in the foreign market pays at time T, $X(T)(U(T)-K)^+$ in the domestic currency, and its time t price is

$$C(t) = e^{-r_d(T-t)}\mathrm{E}_{Q_d}(X(T)(U(T)-K)^+|\mathcal{F}_t).$$

Taking the numeraire based on the Q_d-martingale $X(t)e^{-(r_d-r_f)t}$, obtain the formula for $C(t)$.

Solution. Take $\Lambda(T) = \frac{d\mathrm{Q}_f}{d\mathrm{Q}_d} = e^{-T(r_d-r_f)}\frac{X(T)}{X(0)}$. Then,

$$\begin{aligned} C(t) &= e^{-r_d(T-t)}\mathrm{E}_{Q_d}(X(T)(U(T)-K)^+|\mathcal{F}_t) \\ &= e^{-r_d(T-t)}\frac{\mathrm{E}_{Q_f}(\frac{1}{\Lambda(T)}X(T)(U(T)-K)^+|\mathcal{F}_t)}{\mathrm{E}_{Q_f}(\frac{1}{\Lambda(T)}|\mathcal{F}_t)} \\ &= e^{-r_f(T-t)}X(t)\mathrm{E}_{Q_f}((U(T)-K)^+|\mathcal{F}_t). \end{aligned}$$

Under Q_f, $U(t)$ satisfies the Black–Scholes SDE

$$dU(t) = r_f U(t)dt + \sigma_U U(t)dB_f(t).$$

Thus, $e^{-r_f(T-t)}\mathrm{E}_{Q_f}((U(T)-K)^+|\mathcal{F}_t)$ can be obtained from the Black–Scholes formula, and we have

$$C(t) = X(t)U(t)\Phi(h(t)) - X(t)Ke^{-r_f(T-t)}\Phi(h(t)-\sigma_U\sqrt{T-t}),$$

$$\text{where} \quad h(t) = \frac{\ln\frac{U(t)}{K} + (r_f + \frac{1}{2}\sigma_U^2)(T-t)}{\sigma_U\sqrt{T-t}}.$$

Problem 11.30: A lookback call pays off $S(T) - S_*$ at maturity T, where $S_* = \inf_{0\le t\le T} S(t)$. Derive the price of a lookback call.

Solution. The price of a lookback call is given by

$$\begin{aligned} C &= e^{-rT}\mathrm{E}_Q(S(T)-S_*) = e^{-rT}\mathrm{E}_Q(S(T)) - e^{-rT}\mathrm{E}_Q(S_*) \\ &= S(0) - e^{-rT}\mathrm{E}_Q(S_*), \end{aligned}$$

since $e^{-rT}\mathrm{E}_Q(S(T)) = S(0)$ due to $S_t e^{-rt}$ being a Q-martingale. Under this EMM,

$$S(t) = S(0)e^{\sigma B(t)+(r-\sigma^2/2)t} = S(0)e^{\sigma\left(B(t)+\frac{r-\sigma^2/2}{\sigma}t\right)} = S(0)e^{\sigma W(t)},$$

where $W(t) = B(t) + vt$ with $v = \frac{r-\sigma^2/2}{\sigma}$. Since $\mathrm{E}_Q(S_*) = S(0)\mathrm{E}_Q(e^{\sigma W_*})$, it reduces to computing $\mathrm{E}_Q(e^{\sigma W_*})$. We do it by another change of measure to remove the drift in W_t.

Define a measure Q_1 by $dQ_1/dQ = e^{-vB(T)-\frac{1}{2}v^2T}$. Then, $W(t)$ is a Q_1-Brownian motion.

Let q_{W,W_*} denote the joint density of $(W(T), W_*)$ under Q and q^1_{W,W_*} under Q_1.

Then, by the reflection principle for Brownian motion and Theorem 3.21,

$$q^1_{W,W_*}(x,y) = q^1_{W,W^*}(-x,-y) = \sqrt{\frac{2}{\pi}}\frac{(x-2y)}{T^{3/2}}e^{-\frac{(x-2y)^2}{2T}}, \ x \ge y, \ y \le 0.$$

Since $dQ/dQ_1 = e^{vW(T)-\frac{1}{2}v^2T}$, the joint density under Q is obtained as (Problem 10.10)

$$q_{W,W_*}(x,y) = e^{vx-\frac{1}{2}v^2T}q^1_{W,W_*}(x,y) = \sqrt{\frac{2}{\pi}}\frac{(x-2y)}{T^{3/2}}e^{vx-\frac{1}{2}v^2T-\frac{(x-2y)^2}{2T}}.$$

Hence, the density of W_* under Q is

$$f^Q_{W_*}(y) = \int q_{W,W_*}(x,y)dx = \int_y^\infty \sqrt{\frac{2}{\pi}}\frac{(x-2y)}{T^{3/2}}e^{vx-\frac{1}{2}v^2T-\frac{(x-2y)^2}{2T}}dx.$$

Therefore,

$$\begin{aligned} E_Q(e^{\sigma W_*}) &= \int_{-\infty}^0 e^{\sigma y} f^Q_{W_*}(y)dy \\ &= \left(1-\frac{\sigma^2}{2r}\right)\Phi\left(\frac{\sqrt{T}(r-\sigma^2/2)}{\sigma}\right) \\ &\quad + e^{rT}\left(1+\frac{\sigma^2}{2r}\right)\left(1-\Phi\left(\frac{\sqrt{T}(r+\sigma^2/2)}{\sigma}\right)\right). \end{aligned}$$

Hence, we obtain

$$\begin{aligned} C &= S(0)(1-e^{-rT}E_Q(e^{\sigma W_*})) \\ &= S(0)\left(-\frac{\sigma^2}{2r} + \left(1+\frac{\sigma^2}{2r}\right)\Phi\left(\frac{\sqrt{T}(r+\sigma^2/2)}{\sigma}\right)\right. \\ &\quad \left. -e^{-rT}\left(1-\frac{\sigma^2}{2r}\right)\Phi\left(\frac{\sqrt{T}(r-\sigma^2/2)}{\sigma}\right)\right). \end{aligned}$$

This formula should replace the formula in (11.65), which has an error.

Problem 11.31: Show that the price of a down-and-in call is given by

$$e^{-rT}\left(\frac{H}{S(0)}\right)^{2r/\sigma^2-1}\left(F\Phi\left(\frac{\ln(F/K)+\frac{\sigma^2 T}{2}}{\sigma\sqrt{T}}\right)-K\Phi\left(\frac{\ln(F/K)-\frac{\sigma^2 T}{2}}{\sigma\sqrt{T}}\right)\right),$$

where $F = e^{rT}H^2/S(0)$, if the barrier $H < K$.

Solution. The payoff of a down-and-in call is $(S_T - K)^+ I(S_* \leq H)$. Thus, its price is

$$\begin{aligned} C &= e^{-rT}\mathrm{E}_Q\left((S(T)-K)^+ I(S_* \leq H)\right) \\ &= e^{-rT}\mathrm{E}_Q\left((S(T)-K) I(S(T) > K) I(S_* \leq H)\right). \end{aligned}$$

Under the EMM Q,

$$S(t) = S(0)e^{\sigma B(t)+(r-\sigma^2/2)t} = S(0)e^{\sigma W(t)},$$

where $W(t) = B(t) + vt$ with $v = \frac{r-\sigma^2/2}{\sigma}$. Then, we have

$$C = e^{-rT}\mathrm{E}_Q\big(\big(S(0)e^{\sigma W(T)} - K\big) I\big(S(0)e^{\sigma W(T)} > K\big) I\big(S(0)e^{\sigma W_*} \leq H\big)\big).$$

Using change of measure (see Exercise 11.15), we can show that the joint density of $(W(T), W_*)$ is given by

$$q_{W,W_*}(x,y) = \sqrt{\frac{2}{\pi}}\frac{(x-2y)}{T^{3/2}}e^{vx-\frac{1}{2}v^2T-\frac{(x-2y)^2}{2T}}, \quad x \geq y, y \leq 0.$$

Therefore, the expectation above is

$$\begin{aligned} &\mathrm{E}_Q\left(\left(S(0)e^{\sigma W(T)} - K\right) I\left(W(T) > \frac{1}{\sigma}\ln\frac{K}{S(0)}\right) I\right. \\ &\quad \times \left.\left(W_* \leq \frac{1}{\sigma}\ln\frac{H}{S(0)}\right)\right) = \int_{\frac{1}{\sigma}\ln\frac{K}{S(0)}}^{\infty}\int_{-\infty}^{\frac{1}{\sigma}\ln\frac{H}{S(0)}} \\ &\quad \times \left(S(0)e^{\sigma x} - K\right)\sqrt{\frac{2}{\pi}}\frac{(x-2y)}{T^{3/2}}e^{vx-\frac{1}{2}v^2T-\frac{(x-2y)^2}{2T}}\,dydx \\ &= \left(\frac{H}{S(0)}\right)^{\frac{2r}{\sigma^2}-1}\left(F\Phi\left(\frac{\ln(F/K)+\frac{\sigma^2 T}{2}}{\sigma\sqrt{T}}\right)\right. \\ &\quad \left.-K\Phi\left(\frac{\ln(F/K)-\frac{\sigma^2 T}{2}}{\sigma\sqrt{T}}\right)\right), \end{aligned}$$

where $F = e^{rT}H^2/S(0)$. Multiplying by e^{-rT}, we obtain the answer.

Problem 11.32: Derive the formula for the price of a down-and-in call option when the barrier $H > K$.

Solution. Proceeding with the calculations as in Exercise 11.16, we obtain that

$$\begin{aligned} C &= e^{-rT}\mathrm{E}_Q\left((S(T)-K)I(S(T)>K)I(S_* \le H)\right) \\ &= e^{-rT}\mathrm{E}_Q\Big(\big(S(0)e^{\sigma W(T)}-K\big)I\big(S(0)e^{\sigma W(T)}>K\big) \\ &\quad \times I\big(S(0)e^{\sigma W_*}\le H\big)\Big). \end{aligned}$$

Using the joint density of $W(T)$ and W_*,

$$q_{W,W_*}(x,y) = \sqrt{\frac{2}{\pi}}\frac{(x-2y)}{T^{3/2}}e^{vx-\frac{1}{2}v^2T-\frac{(x-2y)^2}{2T}}, \quad x \ge y, y \le 0,$$

where $v = \frac{1}{\sigma}(r-\frac{1}{2}\sigma^2)$, we obtain for $H > K$, the expectation is

$$\int_{\frac{1}{\sigma}\ln\frac{H}{S(0)}}^{\infty}\int_{-\infty}^{\frac{1}{\sigma}\ln\frac{H}{S(0)}}\left(S(0)e^{\sigma x}-K\right)q_{W,W_*}(x,y)dxdy$$
$$+\int_{\frac{1}{\sigma}\ln\frac{K}{S(0)}}^{\frac{1}{\sigma}\ln\frac{H}{S(0)}}\int_{-\infty}^{x}\left(S(0)e^{\sigma x}-K\right)q_{W,W_*}(x,y)dxdy,$$

which gives the option price

$$\begin{aligned} C &= e^{-rT}K\left(\frac{H}{S(0)}\right)^{1-\frac{2r}{\sigma^2}}\Phi\left(\frac{\ln(H/S(0))+rT-\frac{\sigma^2T}{2}}{\sigma\sqrt{T}}\right) \\ &\quad + S(0)\left(\frac{H}{S(0)}\right)^{1+\frac{2r}{\sigma^2}}\Phi\left(\frac{\ln(H/S(0))+rT+\frac{\sigma^2T}{2}}{\sigma\sqrt{T}}\right) \\ &\quad + e^{-rT}K\left(\Phi\left(\frac{\ln(K/S(0))-rT+\frac{\sigma^2T}{2}}{\sigma\sqrt{T}}\right)\right. \\ &\quad \left. -\Phi\left(\frac{\ln(H/S(0))-rT+\frac{\sigma^2T}{2}}{\sigma\sqrt{T}}\right)\right) \\ &\quad + S(0)\left(\Phi\left(\frac{\ln(H/S(0))-rT-\frac{\sigma^2T}{2}}{\sigma\sqrt{T}}\right)\right. \\ &\quad \left. + -\Phi\left(\frac{\ln(K/S(0))-rT-\frac{\sigma^2T}{2}}{\sigma\sqrt{T}}\right)\right). \end{aligned}$$

Problem 11.33: Derive the formula for the price of an up-and-in call option when the barrier $H > K$.

Solution. This can be an option using a similar calculation to Exercise 11.16. The price of an up-and-in call option is

$$C = e^{-rT}\mathrm{E}_Q\left((S(T)-K)I(S(T)>K)I(S^* \le H)\right),$$

where $S^* = \sup_{0\le t\le T} S(t)$. From Theorem 3.21, the joint density function of a Brownian motion at time T and its supremum over $[0,T]$ is

$$p_{W,W^*}(x,y) = \sqrt{\frac{2}{\pi}}\frac{2y-x}{T^{3/2}}e^{-\frac{(2y-x)^2}{2T}}, \quad x \le y,\ y \ge 0.$$

With a change of measure, the expectation term above becomes

$$\int_{\frac{1}{\sigma}\ln\frac{K}{S(0)}}^{\frac{1}{\sigma}\ln\frac{H}{S(0)}}\int_{\frac{1}{\sigma}\ln\frac{H}{S(0)}}^{\infty}(S(0)e^{\sigma x}-K)\sqrt{\frac{2}{\pi}}\frac{2y-x}{T^{3/2}}e^{vx-\frac{1}{2}v^2T-\frac{(2y-x)^2}{2T}}\,dxdy$$

$$+\int_{\frac{1}{\sigma}\ln\frac{H}{S(0)}}^{\infty}\int_{x}^{\infty}(S(0)e^{\sigma x}-K)\sqrt{\frac{2}{\pi}}\frac{2y-x}{T^{3/2}}e^{vx-\frac{1}{2}v^2T-\frac{(2y-x)^2}{2T}}\,dxdy,$$

which gives the option price

$$C = e^{-rT}K\left(\frac{H}{S(0)}\right)^{\frac{2r}{\sigma^2}-1}\left(\Phi\left(\frac{\ln\frac{KS(0)}{H^2}-rT-\frac{1}{2}\sigma^2T}{\sigma\sqrt{T}}\right)\right.$$

$$\left.-\Phi\left(\frac{\ln\frac{S(0)}{H}-rT+\frac{1}{2}\sigma^2T}{\sigma\sqrt{T}}\right)\right)$$

$$+\ S(0)\left(\frac{H}{S(0)}\right)^{\frac{2r}{\sigma^2}+1}\left(\Phi\left(\frac{\ln\frac{H^2}{KS(0)}+rT+\frac{1}{2}\sigma^2T}{\sigma\sqrt{T}}\right)\right.$$

$$\left.-\ \Phi\left(\frac{\ln\frac{H}{S(0)}+rT+\frac{1}{2}\sigma^2T}{\sigma\sqrt{T}}\right)\right)$$

$$-\ e^{-rT}K\Phi\left(\frac{\ln\frac{S(0)}{H}+rT-\frac{1}{2}\sigma^2T}{\sigma\sqrt{T}}\right)$$

$$+\ S(0)\Phi\left(\frac{\ln\frac{S(0)}{H}+rT+\frac{1}{2}\sigma^2T}{\sigma\sqrt{T}}\right).$$

Problem 11.34 (Bachelier model): Let $S_t = S_0 + \mu t + \sigma B_t$, and $\beta_t = 1$ $(r = 0)$. Let X be a payoff of contract at time T, such that $\mathrm{E}X^2 < \infty$.

(a) Find the EMM in this model and show that it is unique.
(b) Find the replicating self-financing portfolio for X by using the predictable representation property of Brownian motion.
(c) Give C_t, the price of X at time $t < T$ as the value of the replicating portfolio. Show that it is discounted the expected payoff under the EMM.
(d) Let $X = g(S_T)$ for some function g. Show that $C_t = C(x,t) = \mathrm{E}(g(S_T)|S_t = x)$. Give the PDE for $C(x,t)$. Give the replicating portfolio for such an option.
(e) Solve the PDE in (d) above, and find the price C_t.
(f) Specify the solution for $g(x) = (x - K)^+$.
(g) Give the Bachelier formula for the price of a call option and the self-financing replicating portfolio.

Solution.

(a) By the Girsanov theorem, $\mu t + \sigma B_t = \sigma(B_t + (\mu/\sigma)t) = \sigma \hat{B}_t$ for a Q-Brownian motion $\hat{B}_t$, with $d\mathrm{Q}/d\mathrm{P} = \Lambda = e^{-cB_T - c^2T/2}$ for $c = \mu/\sigma$. Uniqueness follows from the result of the absolute continuous change of the Wiener measure (Corollary 10.21), in the same way as in Problem 11.19.

(b) Since X^2 is integrable under P and

$$\mathrm{E}_Q|X| = \mathrm{E}_P(|X|e^{-cB_T - c^2T/2}) \leq \left(\mathrm{E}_P X^2 \mathrm{E}_P \Lambda^2\right)^{1/2} < \infty,$$

X is integrable under Q. Note that all the expectations in the following are under Q.

The random variable X can be written as an integral of a predictable process H_t by the predictable representation property of Q-Brownian motion B_t (we dropped the hat),

$$X = \mathrm{E}X + \int_0^T H_t dB_t.$$

Since under Q the martingale $Z_t = S_t = S_0 + \sigma B_t$, we can write using $dB_t = dZ_t/\sigma$,

$$X = \mathrm{E}X + \int_0^T (H_t/\sigma)dZ_t.$$

Let now $M_t = \mathrm{E}(X|\mathcal{F}_t)$. Then, M_t is a martingale and from the above representation for X, it has the form

$$M_t = \mathrm{E}X + \int_0^t (H_u/\sigma)dZ_u.$$

(This martingale is in fact the discounted portfolio value process.) The replicating self-financing portfolio is given by

$$a_t = H_t/\sigma, \quad b_t = \mathrm{E}X + \int_0^t a_u dZ_u - H_t S_t/\sigma.$$

Indeed, we can see that it is a replicating portfolio:

$$V_T = a_T S_T + b_T = H_T S_T/\sigma + \mathrm{E}X + \int_0^T a_u dZ_u - H_T S_T/\sigma = X.$$

Check the self-financing property for $V_t = a_t S_t + b_t$,

$$\begin{aligned} d(V_t) &= d(a_t S_t + b_t) \\ &= d(H_t S_t/\sigma + \mathrm{E}X + \int_0^t a_u dZ_u - H_t S_t/\sigma) = a_t dS_t. \end{aligned}$$

(c) $C_t = V_t = a_t S_t + b_t = \mathrm{E}X + \int_0^t a_u dZ_u$. Hence, $C_t = M_t = \mathrm{E}(X|\mathcal{F}_t)$.

(d) It follows from the above, by the Markov property of Brownian motion, that

$$C_t = \mathrm{E}(X|\mathcal{F}_t) = \mathrm{E}(g(S_T)|\mathcal{F}_t) = \mathrm{E}(g(S_T)|S_t).$$

The function $C(x,t) = \mathrm{E}(g(S_T)|S_t = x)$ solves the backward PDE $LC + \frac{\partial}{\partial t}C = 0$ with the boundary condition $C(x,T) = g(x)$, where L is the diffusion operator of S_t under Q, $L = \frac{1}{2}\sigma^2 \frac{\partial^2}{\partial x^2}$:

$$\frac{1}{2}\sigma^2 \frac{\partial^2}{\partial x^2} C(x,t) + \frac{\partial}{\partial t} C(x,t) = 0, \quad C(x,T) = g(x).$$

Since on the one hand, by replicating the self-financing portfolio, $dC_t = a_t dS_t$, and on the other hand by using Itô's formula with the above PDE,

$$dC_t = dC(S_t,t) = \frac{\partial C}{\partial x} dS_t + \left(\frac{1}{2}\sigma^2 \frac{\partial^2}{\partial x^2} C(x,t) + \frac{\partial}{\partial t} C(x,t)\right) dt = \frac{\partial C}{\partial x} dS_t,$$

$$a_t = \frac{\partial C}{\partial x}(S_t,t). \; b_t = C_t - a_t S_t.$$

(e) The above PDE is the heat equation with the normal probability density as the fundamental solution. First, find $P(S_T \le y | S_t = x)$, where calculations are done under Q. Since $S_T = S_t + \sigma(B_T - B_t)$, the conditional distribution of S_T given $S_t = x$ is normal with mean x and variance $\sigma^2(T-t)$. It has the density $p_t(x,y) = \frac{1}{\sigma\sqrt{2\pi(T-t)}} e^{-\frac{(y-x)^2}{2\sigma^2(T-t)}}$. Thus,

$$C(x,t) = \mathrm{E}(g(S_T)|S_t = x) = \int_{-\infty}^{\infty} g(y)p_t(x,y)dy$$
$$= \mathrm{E}g(N(x, \sigma^2(T-t)),$$
$$C_t = C(S_t, t).$$

(f) If $g(x) = (x-K)^+$, then by the above result, $C(x,t) = \mathrm{E}(N(x,v^2) - K)^+$ with $v^2 = \sigma^2(T-t)$. Let $Z \sim N(0,1)$, then

$$C(x,t) = \mathrm{E}(vZ + x - K)^+ = \mathrm{E}(vZ + x - K)I(vZ + x - K > 0)$$
$$= \mathrm{E}(vZ + x - K)I(vZ + x - K > 0)$$
$$= \mathrm{E}(vZ)I(vZ + x - K > 0) + (x-K)P(vZ + x - K > 0)$$
$$= (x-K)\Phi\left(\frac{x-K}{\sigma\sqrt{T-t}}\right) + \sigma\sqrt{T-t}\phi\left(\frac{x-K}{\sigma\sqrt{T-t}}\right).$$

We have used that $\mathrm{E}Z1_{Z>a} = \phi(a)$, by direct integration, since $\phi'(x) = -x\phi(x)$

(g) The price of a call option, see above,

$$C_t = (S_t - K)\Phi\left(\frac{S_t - K}{\sigma\sqrt{T-t}}\right) + \sigma\sqrt{T-t}\phi\left(\frac{S_t-K}{\sigma\sqrt{T-t}}\right).$$

The replicating portfolio $a_t = \frac{\partial C}{\partial x}(S_t, t) = \Phi\left(\frac{S_t-K}{\sigma\sqrt{T-t}}\right)$, by direct differentiation, using that the derivative of the normal density $\phi(x)' = -x\phi(x)$. $b_t = C_t - a_t S_t = \sigma\sqrt{T-t}\phi\left(\frac{S_t-K}{\sigma\sqrt{T-t}}\right) - K\Phi\left(\frac{S_t-K}{\sigma\sqrt{T-t}}\right)$.

Problem 11.35 (Bachelier model with interest rate): Let $S_t = S_0 + \mu t + \sigma B_t$, and $\beta_t = e^{rt}$.

(a) Assuming there is an EMM Q, find the SDE for S_t under the EMM Q and solve it on the interval $[t, T]$.

(b) Give the conditional distribution of S_T given S_t for $t < T$ under the EMM.
(c) Give the price of the option with payoff $X = g(S_T)$, and specify for the call option.
(d) Using Problem 10.27 justify the EMM in (a).

Solution.

(a) Using that $dS_t = \mu dt + \sigma dB_t$, we have

$$\begin{aligned} d(S_t e^{-rt}) &= e^{-rt} dS_t + S_t d(e^{-rt}) \\ &= e^{-rt}(\mu dt + \sigma dB_t) - re^{-rt} S_t dt \\ &= \sigma e^{-rt} \left(\frac{\mu - rS_t}{\sigma} dt + dB_t \right). \end{aligned}$$

Assuming that the EMM Q exists, it follows that there is a Brownian motion $\hat{B}_t$ under Q, so that

$$\frac{\mu - rS_t}{\sigma} dt + dB_t = d\hat{B}.$$

Therefore, by writing dB_t in terms of $d\hat{B}$, we obtain the SDE for S_t under Q:

$$dS_t = rS_t dt + \sigma d\hat{B}_t.$$

Thus, under Q, the drift term μ is replaced by rS_t, yielding the Ornstein–Uhlenbeck SDE. Its solution is obtained by looking at $d(S_t e^{-rt})$ and is given by

$$S_T e^{-rT} = S_t e^{-rt} + \sigma \int_t^T e^{-ru} d\hat{B}_u,$$

$$S_T = S_t e^{r(T-t)} + \sigma e^{rT} \int_t^T e^{-ru} d\hat{B}_u.$$

(b) Since the Itô integral $\int_t^T e^{-ru} d\hat{B}_u$ is independent of S_t (and of $\mathcal{F}_t$) and has a normal distribution with mean 0 and variance $\int_t^T e^{-2ru} du = \frac{1}{2r}(e^{-2rt} - e^{-2rT})$, the conditional distribution of S_T given S_t is normal with mean $u = S_t e^{r(T-t)}$ and variance $v^2 = \frac{\sigma^2}{2r}(e^{2r(T-t)} - 1)$.

(c) The price of the option is then

$$C_t = e^{-r(T-t)} \mathrm{E}_Q(g(S_T)|S_t) = e^{-r(T-t)} \mathrm{E} g(N(u, v^2)).$$

Using the previous Problem 11.34, we can easily obtain the price of the option with $g(x) = (x - K)^+$, just by replacing the appropriate variables with new values of the mean and the variance of the normal distribution. After some simplification using $K' = Ke^{-r(T-t)}$ and $v' = (\sigma/2r)\sqrt{1 - e^{-2r(T-t)}}$,

$$\begin{aligned} C_t &= e^{-r(T-t)}\mathrm{E}(N(u, v^2) - K)^+ = e^{-r(T-t)}\mathrm{E}(N(u - K, v^2))^+ \\ &= (S_t - K')\Phi\left(\frac{S_t - K'}{v'}\right) + v'\phi\left(\frac{S_t - K'}{v'}\right). \end{aligned}$$

(d) Let $J_t = \frac{\mu - rS_t}{\sigma}$ and consider $\Lambda_t = \mathcal{E}(\int J_s dB_s)_t$. To show that the appropriate change of measure exists comes to the question of verifying that $\mathrm{E}\Lambda_T = 1$. Then, $\Lambda_T = d\mathrm{Q}/d\mathrm{P}$ gives the desired change of measure in (a). To check this condition is not so easy because the Kazamaki condition (and therefore the Novikov condition) fails. However, it can be done along the lines of Problem 10.27.

Problem 11.36 (Pricing in incomplete markets, when EMM Q is not unique, i.e. $M_t = S_t/\beta_t$ is a martingale under different Q's):

(a) Show that if M_t, $0 \le t \le T$, is a martingale under two different equivalent probability measures Q and P, then for $s < t$, $\mathrm{E}_Q(M_t|\mathcal{F}_s) = \mathrm{E}_P(M_t|\mathcal{F}_s)$ almost surely. Moreover, if M_0 is non-random, then $\mathrm{E}_P M_t = \mathrm{E}_Q M_t$.

(b) Let $\Lambda = d\mathrm{Q}/d\mathrm{P}$ and $\Lambda_t = E_P(\Lambda|\mathcal{F}_t)$. Show that both processes M_t and $\Lambda_t M_t$ are P-martingales.

(c) Show that the price C_t of an attainable claim X, with $\mathrm{E}(X^2) < \infty$, at time t is the same for all EMM.

Solution.

(a) As M is a martingale under Q and P,

$$\begin{aligned} \mathrm{E}_Q(M_t|\mathcal{F}_s) &= M_s, \text{ Q-a.s,} \\ \mathrm{E}_P(M_t|\mathcal{F}_s) &= M_s, \text{ P-a.s.} \end{aligned}$$

Since $\mathrm{Q} \sim \mathrm{P}$, $\mathrm{E}_Q(M_t|\mathcal{F}_s) = \mathrm{E}_P(M_t|\mathcal{F}_s)$ a.s. P and Q. Take $s = 0$, then $\mathrm{E}_Q(M_t) = M_0 = \mathrm{E}_P(M_t)$.

(b) By the generalised Bayes, formula with $\Lambda = d\mathrm{Q}/d\mathrm{P}$,

$$\mathrm{E}_Q(M_t|\mathcal{F}_s) = \frac{\mathrm{E}_P(\Lambda M_t|\mathcal{F}_s)}{\mathrm{E}_P(\Lambda|\mathcal{F}_s)}.$$

But $\mathrm{E}_Q(M_t|\mathcal{F}_s) = \mathrm{E}_P(M_t|\mathcal{F}_s)$, and $\mathrm{E}_P(\Lambda|\mathcal{F}_s) = \Lambda_s$, and by conditioning on $\mathcal{F}_t$, $\mathrm{E}_P(\Lambda M_t|\mathcal{F}_s) = \mathrm{E}_P(\Lambda_t M_t|\mathcal{F}_s)$, so that

$$\mathrm{E}_P(M_t|\mathcal{F}_s)\mathrm{E}_P(\Lambda|\mathcal{F}_s) = \mathrm{E}_P(\Lambda M_t|\mathcal{F}_s), \text{ and } \Lambda_s M_s = \mathrm{E}_P(\Lambda_t M_t|\mathcal{F}_s).$$

Hence, we have that M_t is a Q-martingale if and only if $\Lambda_t M_t$ is a P-martingale. Thus, both processes are martingales under P, M, and ΛM.

(c) As X is attainable, it can be replicated by a self-financing replicating portfolio with $V_T = X$. The price of the claim at time t is $C_t = V_t$. Since $V_t = a_t S_t + b_t \beta_t$ does not depend on probability measure, so is C_t. But for an EMM Q by the martingale property of V_t/β_t $C_t = \beta_t \mathrm{E}_Q(X/\beta_T|\mathcal{F}_t)$, implying it is the same for all Q.

Problem 11.37 (Non-uniqueness of EMM in mixed models): In this problem, the price of an asset is modelled as a sum of a diffusion and a jump process.

(a) Let $X_t = B_t + N_t$, with Brownian motion B and Poisson process N. Give at least two equivalent probability measures Q_1 and Q_2, such that X_t is a Q_i-martingale, $i = 1, 2$ (see Problem 10.22).
(b) Consider the market model of Bachelier type $S_t = 1 + B_t + N_t$ and $\beta_t = 1$. Show that there are infinitely many EMMs.

Solution. For any $a > 0$, define Q_a by

$$\frac{d\mathrm{Q}_a}{d\mathrm{P}} = e^{N(T)\ln a - aB(T) + (1-a-\frac{1}{2}a^2)T}.$$

Then, X_t is a Q_a-martingale. See Problem 10.22. It follows that for the model in (b) $S_t/\beta_t = S_t$ is a martingale under any Q_a.

Problem 11.38 (An incomplete market model with a non-traded asset): Consider a continuous time model on $0 \le t \le 1$ in which the savings account is given by $\beta_t = e^{rt}$ and the stock price is given by $S_t = S_0 e^{rt}$. The model also includes another random variable X that takes values from $\{0, 1\}$, indicating whether an earthquake has occurred at time 1. It is known at time 0 that the probability of the earthquake occurring at time 1 is strictly between 0 and 1. Only the stock and the savings account are traded.

(a) Does this model have arbitrage opportunities?
(b) Find two different EMMs for this model. Using the fundamental theorems of asset pricing, what can you conclude about the model?

(c) Show that any self-financing portfolio is deterministic, and find it.
(d) Find a contract that cannot be replicated in the market.

Solution.

(a) The discounted stock price is given by $S_t/\beta_t = S_0$, which is a constant. Since constants are always martingales, there exists at least one EMM (e.g. the real-world measure). By the first fundamental theorem of asset pricing, the existence of an EMM implies there is no arbitrage in this model.

(b) Since S_t and β_t are deterministic, we only have to assign probabilities to the random variable X. For example, we can choose $Q(X = 1) = p$ and $Q(X = 0) = (1 - p)$ for any real number p with $0 < p < 1$. Any two choices of such p would suffice. By the second fundamental theorem of asset pricing, since the EMM is not unique, the market model is incomplete.

(c) Since X cannot be traded, any self-financing portfolio (a_t, b_t) must satisfy

$$\begin{aligned} dV_t &= a_t dS_t + b_t d\beta_t = a_t r S_0 e^{rt} dt + b_t r e^{rt} dt \\ &= r(a_t S_t + b_t \beta_t) dt = r V_t dt. \end{aligned}$$

So, V_t is also deterministic with $V_t = V_0 e^{rt}$.

(d) Consider any contract whose payoff is a non-constant function of X. For example, consider a contract that simply pays X at time 1. It is impossible to replicate the payoff X with V_1 because X is random but V_1 is deterministic.

Problem 11.39 (Change of numeraire): Assume $S(t)$ evolves according to the Black–Scholes model. Show that under the EMM Q_1, when $S(t)$ is a numeraire, $d(e^{rt}/S(t)) = \sigma(e^{rt}/S(t))dW_t$, where $W(t) = B(t) - \sigma t$ is a Q_1-Brownian motion. Give the likelihood dQ_1/dQ. Give the SDE for $S(t)$ under Q_1.

Solution. Recall that $dS(t) = rS(t)dt + \sigma S(t)dB(t)$ for a Q-Brownian motion. Under Q_1 when $S(t)$ is numeraire, $e^{rt}/S(t)$ is martingale:

$$\begin{aligned} d(e^{rt}/S(t)) &= e^{rt} dS(t) + re^{rt} dt/S(t) \\ &= e^{rt}\Big(-\frac{1}{S^2(t)} dS(t) + \frac{1}{S^3(t)} dS(t)^2 \Big) + re^{rt} \frac{1}{S(t)} dt \\ &= e^{rt}\Big(-\frac{\sigma}{S(t)} dB(t) + \frac{\sigma^2}{S(t)} dt \Big) = -\sigma e^{rt} \frac{1}{S(t)} dW(t) \end{aligned}$$

for a Q_1-Brownian motion $W(t) = B(t) - \sigma t$. Since $-W(t)$ is also a Q_1-Brownian motion, we can remove the minus sign in the SDE and write

$$d(e^{rt}/S(t)) = \sigma e^{rt}\frac{1}{S(t)}dW(t).$$

The likelihood $dQ_1/dQ = e^{\sigma B_T - \sigma^2 T/2}$.

Under Q_1, $dS(t) = (r + \sigma^2)S(t)dt + \sigma S(t)dW(t)$.

Problem 11.40: Derive the Black–Scholes formula by using the stock price as the numeraire.

Solution. Let $\Lambda(T) = \frac{dQ_1}{dQ} = \frac{S(T)/S(0)}{\beta(T)/\beta(0)}$, where $\beta(t) = e^{rt}$. According to Theorem 11.17, with $S(t)$ being numeraire (under Q_1),

$$\begin{aligned}C(t) &= \mathrm{E}_{Q_1}\left(\frac{S(t)}{S(T)}(S(T) - K)^+|\mathcal{F}_t\right)\\ &= S(t)\mathrm{E}_{Q_1}\left(\left(1 - \frac{K}{S(T)}\right)I(S(T) > K)|\mathcal{F}_t\right)\\ &= S(t)Q_1(S(T) > K|\mathcal{F}_t) - KS(t)\mathrm{E}_{Q_1}\left(\frac{1}{S(T)}I(S(T) > K)|\mathcal{F}_t\right).\end{aligned}$$

The two expressions above can be evaluated using the distribution of $S(T)$ given $S(t)$ under Q_1; however, the second term is easier to calculate using the change of measure:

$$\begin{aligned}C(t) &= S(t)Q_1(S(T) > K|\mathcal{F}_t)\\ &\quad - KS(t)\mathrm{E}_Q\left(\frac{\Lambda(T)}{\Lambda(t)}\frac{1}{S(T)}I(S(T) > K)|\mathcal{F}_t\right)\\ &= S(t)Q_1(S(T) > K|\mathcal{F}_t) - K\frac{\beta(t)}{\beta(T)}Q(S(T) > K|\mathcal{F}_t).\end{aligned}$$

Under Q, $S(T) = S(t)e^{\sigma(B_T - B_t) + (r - \sigma^2/2)(T-t)}$; whereas under Q_1, $dS(t) = (r + \sigma^2)S(t)dt + \sigma S(t)dW(t)$ gives $S(T) = S(t)$ $e^{\sigma(W_T - W_t) + (r + \sigma^2/2)(T-t)}$. Evaluating the probabilities, we obtain the

Black–Scholes formula:

$$
\begin{aligned}
C(t) &= S(t)\mathrm{Q}_1\big(\ln S(T) > \ln K|\mathcal{F}_t\big) - Ke^{-r(T-t)}\mathrm{Q}\big(\ln S(T) > \ln K|\mathcal{F}_t\big) \\
&= S(t)\mathrm{Q}_1\left(Z > \frac{\ln K - (r+\sigma^2/2)(T-t) - \ln S(t)}{\sigma\sqrt{T-t}}\right) \\
&\quad - Ke^{-r(T-t)}\mathrm{Q}\Big(Z > \frac{\ln K - (r-\sigma^2/2)(T-t) - \ln S(t)}{\sigma\sqrt{T-t}}\Big) \\
&= S(t)\Phi(h_t) - Ke^{-r(T-t)}\Phi(h_t - \sigma\sqrt{T-t}),
\end{aligned}
$$

where $h_t = \frac{\ln(S(t)/K)+(r+\sigma^2/2)(T-t)}{\sigma\sqrt{T-t}}$.

Problem 11.41: One way to derive a PDE for the function $C(x,t)$ when the option price is of the form $C(S_t,t)$ is based on the fact that $C(S_t,t)e^{-rt} = V(t)e^{-rt}$ is a Q-local martingale, which implies that the coefficient of dt in $d(C(S(t),t)e^{-rt})$ equals to zero. Derive the PDE for the price of the option in Heston's model, which is under Q (from (11.48*)):

$$
\begin{aligned}
dS(t) &= rS(t)dt + \sqrt{v(t)}S(t)dB(t) \\
dv(t) &= \alpha(\mu - v(t))dt + \delta\sqrt{v(t)}dW(t),
\end{aligned}
$$

where B and W are correlated. Let ρ be the correlation of B and W.

Solution.

$$
\begin{aligned}
d(C(S,v,t)e^{-rt}) &= e^{-rt}d(C(S,v,t)) - re^{-rt}C(S,v,t)dt \\
&= e^{-rt}\bigg(\frac{\partial C}{\partial S}dS(t) + \frac{\partial C}{\partial v}dv(t) + \frac{\partial C}{\partial t}dt + \frac{1}{2}\frac{\partial^2 C}{\partial S^2}dS(t)^2 \\
&\quad + \frac{1}{2}\frac{\partial^2 C}{\partial v^2}dv(t)^2 + \frac{\partial^2 C}{\partial S\partial v}dS(t)dv(t) - rCdt\bigg) \\
&= e^{-rt}\bigg(\frac{\partial C}{\partial S}(rS(t)dt + \sqrt{v(t)}S(t)dB(t)) \\
&\quad + \frac{\partial C}{\partial v}(\alpha(\mu - v(t))dt + \delta\sqrt{v(t)}dW(t)) \\
&\quad + \frac{\partial C}{\partial t}dt + \frac{1}{2}\frac{\partial^2 C}{\partial S^2}v(t)S^2(t)dt \\
&\quad + \frac{1}{2}\frac{\partial^2 C}{\partial v^2}\delta^2 v(t)dt + \frac{\partial^2 C}{\partial S\partial v}\delta v(t)S(t)\rho dt - rCdt\bigg).
\end{aligned}
$$

As $C(S(t),v(t),t)e^{-rt}$ is a Q-local martingale, the dt term above is zero, and we have

$$rS(t)\frac{\partial C}{\partial S}+\alpha(\mu-v(t))\frac{\partial C}{\partial v}+\frac{\partial C}{\partial t}+\frac{1}{2}v(t)S^2(t)\frac{\partial^2 C}{\partial S^2}$$
$$+\frac{1}{2}\delta^2 v(t)\frac{\partial^2 C}{\partial v^2}+\delta\rho v(t)S(t)\frac{\partial^2 C}{\partial S\partial v}-rC=0.$$

Problem 11.42: Suppose that S_T has a continuous distribution. Show that prices of call options for all strikes $K\in\mathbb{R}$, namely the knowledge of the function $C(T,K)$ as a function of K, determine the marginal distribution of the stock S_T under the EMM Q.

Solution. Let $h(x,K)=(x-K)^+$. Then, the price of the call with strike K is given by $C(T,K)=e^{-rT}\mathrm{E}_Q h(S_T,K)$. Now, $\frac{\partial h}{\partial K}(x,K)=-I(x>K)$ at any point $x\neq K$. Hence, $\frac{\partial h}{\partial K}(S_T,K)=-I(S_T>K)$, provided $S_T\neq K$. But this event has zero probability, $\mathrm{Q}(S_T=K)=0$ since S_T has a continuous distribution. Next, by changing the expectation and derivative (using dominated convergence), we obtain that

$$\begin{aligned}\frac{\partial C}{\partial K}(T,K)&=e^{-rT}\frac{\partial}{\partial K}\mathrm{E}_Q h(S_T,K)\\&=e^{-rT}\mathrm{E}_Q\frac{\partial}{\partial K}h(S_T,K)\\&=-e^{-rT}\mathrm{E}_Q I(S_T>K)=-e^{-rT}\mathrm{Q}(S_T>K).\end{aligned}$$

Problem 11.43 (Option as a function of the initial stock price $S_0=S$ and strike K): Assume that in the model for stock price $S(T)/S$ does not depend on S, where $S(T)$ is the price of stock at time T. Let T be the exercise time and K the exercise price of the call option. Show that the price of this option $C=C(S,K)$ satisfies the following PDE:

$$C=S\frac{\partial C}{\partial S}+K\frac{\partial C}{\partial K}.$$

You may assume all the necessary differentiability. Hence, show that the delta of the option $\frac{\partial C}{\partial S}$ in the Black–Scholes model is given by $\Phi(h(t))$ with $h(t)$ given by formula (11.36).

Solution. The price of a call option is

$$C=e^{-rT}\mathrm{E}_Q(S(T)-K)^+=e^{-rT}S\mathrm{E}_Q\left(\frac{S(T)}{S}-\frac{K}{S}\right)^+.$$

Let

$$F(y) = e^{-rT} \mathrm{E}_Q \left(\frac{S(T)}{S} - y \right)^+.$$

Then, we have $C = SF(K/S)$. This gives

$$\frac{\partial C}{\partial S} = F(K/S) + SF'(K/S)(-K/S^2) = F(K/S) - (K/S)F'(K/S)$$
$$\frac{\partial C}{\partial K} = SF'(K/S)(1/S) = F'(K/S).$$

It follows that

$$C = SF(K/S) = S\frac{\partial C}{\partial S} + K\frac{\partial C}{\partial K}.$$

For the Black–Scholes model, we have

$$C = S\Phi(h) - Ke^{-rT}\Phi(h - \sigma\sqrt{T}),$$

where $h = \frac{\ln(S/K)+(r+\sigma^2/2)T}{\sigma\sqrt{T}}$. Comparing this to the PDE for C as a function of S and K, we deduce that

$$\frac{\partial C}{\partial S} = \Phi(h), \quad \text{and} \quad \frac{\partial C}{\partial K} = -e^{-rT}\Phi(h - \sigma\sqrt{T}).$$

Note that while h also depends on S and K, differentiation of $C(S, K)$ directly (albeit tediously) confirms this.

The same argument applied to the price of the option at time t (since t is fixed) gives the Black–Scholes formula for C_t.

Problem 11.44: The next three problems concern the applications of Kalman–Bucy filter (observation of a constant.) Let the signal be a constant $X(t) = c$, for all t, and the observation process satisfies $dY(t) = X(t)dt + dW(t)$. Give the Kalman–Bucy filter and find $\widehat{X}(t)$.

Solution. Note that $dX(t) = 0$. Applying Theorem 14.6 with $a(t) = b(t) = 0$ and $A(t) = B(t) = 1$, we have $d\widehat{X}(t) = -v(t)\widehat{X}(t)dt + v(t)dY(t)$, where $v(t)$ satisfies the Riccati equation $dv(t) = -v^2(t)dt$. Solving the differential equation, we obtain $v(t) = 1/t$. Hence, $d\widehat{X}(t) = -(1/t)\widehat{X}(t)dt + (1/t)dY(t)$, which gives $\widehat{X}(t) = Y(t)/t$.

Problem 11.45 (Observation of Brownian motion): Let the signal be a Brownian motion $X(t) = W_1(t)$, and the observation process satisfies $dY(t) = X(t)dt + dW_2(t)$. Give the Kalman–Bucy filter and find $\widehat{X}(t)$.

Solution. Here, $dX(t) = dW_1(t)$. Applying Theorem 14.6 with $a(t) = 0$, $b(t) = 1$, and $A(t) = B(t) = 1$, we have $d\widehat{X}(t) = -v(t)\widehat{X}(t)dt + v(t)dY(t)$ and $v(t)$ satisfies the Riccati equation $\frac{dv(t)}{dt} = 1 - v^2(t)$, which gives $v(t) = \frac{e^{2t}-1}{e^{2t}+1}$ from Equation (14.21*).

Problem 11.46 (Filtering of indirectly observed stock prices): Let the signal follow the Black–Scholes model $X(t) = X(0)\exp(\sigma W_1(t) + \mu t)$, and the observation process satisfies $dY(t) = X(t)dt + dW_2(t)$. Give the Kalman–Bucy filter and find $\widehat{X}(t)$.

Solution. Here, $dX(t) = (\mu + \frac{1}{2}\sigma^2)X(t)dt + \sigma X(t)dB(t)$. Applying Theorem 14.6 with $a(t) = \mu + \sigma^2/2$, $b(t) = \sigma$ and $A(t) = B(t) = 1$, we have

$$d\widehat{X}(t) = (\mu + \frac{1}{2}\sigma^2 - v(t))\widehat{X}(t)dt + v(t)dY(t)$$

and

$$\frac{dv(t)}{dt} = (2\mu + \sigma^2)v(t) + \sigma^2 - v^2(t).$$

Problem 11.47: Assume that the spot rate r_t, specified under the EMM, follows the Merton model and is given by $r_t = 0.1 + 0.1t + 0.1B_t$.

(a) Give the conditional distribution of r_s given r_t for $s > t$.
(b) Find the price of the bond with maturity at T at time t, $P(t,T)$.
(c) Find the forward rates implied by this model. Deduce that for two terms $T_1 < T_2$, $f(t,T_1)$ and $f(t,T_2)$ have correlation one.

Solution.

(a) $\{r_t\}$ is a Gaussian process as a linear function of Brownian motion B_t:

$$r_s = r_t + 0.1(s-t) + 0.1(B_s - B_t).$$

Thus, the conditional distribution of r_s given r_t is normal with mean $r_t + 0.1(s-t)$ and variance $0.01(s-t)$.

(b) The price of the bond at time t is given by

$$\begin{aligned} P(t,T) &= \mathrm{E}(e^{-\int_t^T r_s ds}|\mathcal{F}_t) = \mathrm{E}(e^{-\int_t^T (r_t + 0.1(s-t) + 0.1(B_s - B_t))ds}|\mathcal{F}_t) \\ &= \mathrm{E}(e^{-r_t(T-t) - 0.1[\frac{s^2}{2} - st]_t^T - 0.1\int_t^T (B_s - B_t))ds}|\mathcal{F}_t) \\ &= e^{-r_t(T-t) - 0.1(\frac{1}{2})(T-t)^2}\mathrm{E}(e^{-0.1\int_t^T (B_s - B_t))ds}). \end{aligned}$$

The last expectation is evaluated by using integration by parts and the fact that $\int_0^a h(t)dB_t$ with non-random h has normal distribution with mean 0 and variance $\int_0^a h^2(t)dt$, we obtain $\int_t^T (B_s - B_t)ds$ is normally distributed with mean 0 and variance $\frac{1}{3}(T-t)^3$.

Now, it follows that $\mathrm{E}(e^{-0.1\int_t^T (B_s - B_t))ds}) = e^{\frac{1}{2}0.1^2\frac{(T-t)^3}{3}}$ and hence

$$P(t,T) = e^{-r_t(T-t)-0.1(\frac{1}{2})(T-t)^2+\frac{0.1^2(T-t)^3}{6}}.$$

(c) The forward rates are given by $f(t,T) = -\frac{\partial \log P(t,T)}{\partial T} = r_t + 0.01(T-t) - 0.005(T-t)^2$.

Therefore, $f(t,T_2) - f(t,T_1) = D(T_1,T_2,t)$, since r_t cancels out, leaving a deterministic expression. Since random variables that differ by a constant have correlation one, $f(t,T_1)$ and $f(t,T_2)$ have correlation one.

Problem 11.48: $P(t,T)$ stands for the price at time t of \$1 at time T. Assume that under the EMM, the spot rate follows SDE $dr_t = \mu(r_t)dt + \sigma(r_t)dW_t$, and savings account is given by $\beta_t = e^{\int_0^t r_u du}$. Derive the PDE for the $P(t,T)$.

Solution. Since $P(t,T) = \mathrm{E}(e^{-\int_t^T r_s ds}|\mathcal{F}_t)$ where the expectation is taken under the EMM, and the process r_t is a diffusion and therefore Markov, $P(t,T) = \mathrm{E}(e^{-\int_t^T r_s ds}|r_t)$. Hence, $P(t,T)$ is a function of r_t (as well as of t and T). Letting $C(x,t)$ denote this function (and dropping T from notations), we have

$$C(x,t) = \mathrm{E}(e^{-\int_t^T r_s ds}|r_t = x).$$

Now, by the Feynman–Kac formula, this function solves the PDE

$$LC(x,t) + \frac{\partial}{\partial t}C(x,t) - xC(x,t) = 0, \quad C(x,T) = 1.$$

The term xC is because in the discount integral the function of the underlying process r_t is the identity x, and where L is the diffusion operator of r_t. Therefore, the PDE for the price of bond is given by

$$\frac{1}{2}\sigma^2(x)\frac{\partial^2}{\partial x^2}C(x,t) + \mu(x)\frac{\partial}{\partial x}C(x,t) - xC(x,t) = 0, \quad C(x,T) = 1,$$

where the boundary condition is due to $P(T,T) = 1$.

Of course, the same PDE is obtained if one writes $d(P(t,T)/\beta_t)$ and equates the coefficient of dt to zero.

Problem 11.49: In the Vasicek model, the spot rate satisfies the stochastic differential equation

$$dr_t = b(a - r_t)dt + \sigma dB_t, \ \ 0 \le t \le T, \ \ r_0 > 0,$$

where B_t is a P-Brownian motion. Let $\Lambda = e^{\lambda B_T - \lambda^2 T/2}$, for some $\lambda > 0$, and the measure Q defined by $dQ/dP = \Lambda$. Show that the equation for r_t under Q results in the change of a,

$$dr_t = b(\tilde{a} - r_t)dt + \sigma dW_t, \ \ 0 \le t \le T, \ \ r_0 > 0,$$

where $\tilde{a} = a + \sigma\lambda/b$, and W_t is a Q-Brownian motion.

Solution. Under Q, $B_t - \lambda t = W_t$ is a BM. So, $dB_t = \lambda dt + dW_t$. Therefore, under Q,

$$dr_t = b(a + \sigma\lambda/b - r_t)dt + \sigma dW_t, \ \ 0 \le t \le T, \ \ r_0 > 0.$$

Problem 11.50: Assume the Vasicek model under the EMM. Take all the parameters equal to 1, $a = b = \sigma = 1$. Show that the price of the bond is

$$P(t,T) = \exp\left((1 - e^{-(T-t)}) \left(\frac{1}{2} - r_t \right) - \frac{1}{2}(T-t) - \frac{1}{4}(1 - e^{-(T-t)})^2 \right).$$

Give the partial differential equation including the boundary condition for the bond price, and then give its solution.

Solution. The price of bond is given by

$$P(t,T) = E\left[e^{-\int_t^T r_s ds} | \mathcal{F}_t \right].$$

Taking $a = b = \sigma = 1$, we have

$$r_s = 1 - e^{-(s-t)}(1 - r_t) + e^{-s} \int_t^s e^u dB_u.$$

Hence,

$$\int_t^T r_s ds = \int_t^T [1 - e^{-(s-t)}(1 - r_t)]ds + \int_t^T e^{-s} \int_t^s e^u dB_u ds.$$

Change of variables gives

$$= (T-t) + (1-r_t)(e^{-(T-t)} - 1) + \int_0^{T-t} e^{-(v+t)} \int_0^v e^{w+t} d\tilde{B}_w dv.$$

Proceeding, by using the stochastic Fubini theorem, swap integrals to have the Itô integral:

$$\begin{aligned}\int_t^T r_s ds &= (T-t) + (1-r_t)(e^{-(T-t)} - 1)\\ &\quad + \int_0^{T-t}\int_0^v e^{-v} e^w d\tilde{B}_w dv\\ &= (T-t) + (1-r_t)(e^{-(T-t)} - 1)\\ &\quad + \int_0^{T-t}\int_w^{T-t} e^{-v} e^w dv d\tilde{B}_w\\ &= (T-t) + (1-r_t)(e^{-(T-t)} - 1)\\ &\quad + \int_0^{T-t} (e^{-w} - e^{-(T-t)}) e^w d\tilde{B}_w\\ &= (T-t) + (1-r_t)(e^{-(T-t)} - 1)\\ &\quad + \int_0^{T-t} 1 - e^{-(T-t)+w} d\tilde{B}_w,\end{aligned}$$

where we used the change of variables $s - t = v$ and $u - t = w$ and $\tilde{B}_w = B_{t+w}$, a Brownian motion. The Ito integral is a Gaussian process by Theorem 4.11 in the text, and so, conditioned on $\mathcal{F}_t$, $-\int_t^T r_s ds$ is normally distributed with mean,

$$\mu_1 = -(T-t) - (1-r_t)(e^{-(T-t)} - 1)$$

and variance

$$\begin{aligned}\sigma_1^2 &= Var\left(\int_0^{T-t} 1 - e^{-(T-t)+w} d\tilde{B}_w\right) = \int_0^{T-t} (1 - e^{-(T-t)+w})^2 dw\\ &= (T-t) - 2(1 - e^{-(T-t)}) + \frac{1}{2}(1 - e^{-2(T-t)}).\end{aligned}$$

Alternatively, σ_1^2 can be found using

$$\sigma_1^2 = \mathrm{Cov}\left(\int_t^T r_s ds, \int_t^T r_u du\right) = \int_t^T \int_t^T \mathrm{Cov}(r_s, r_u) ds du.$$

Finally, using the moment-generating function of normal distribution,

$$P(t,T) = e^{\mu_1 + \frac{1}{2}\sigma_1^2}$$

gives the answer.

To give the PDE, note that the bond price $P(t,T)$ is a function of the rate r_t,

$$P(t,T) = C(r_t, t).$$

Using the Feynmann–Kac formula, we have for the function $C(x,t)$,

$$\partial_t C + (1-x)\partial_x C + \frac{1}{2}\partial_x^2 C = xC.$$

The boundary condition $C(x,T) = 1$. The solution to this PDE is seen from the above price of bond $P(t,T)$:

$$C(x,t) = \exp\left((1 - e^{-(T-t)})\left(\frac{1}{2} - x\right) - \frac{1}{2}(T-t) - \frac{1}{4}(1 - e^{-(T-t)})^2\right).$$

Problem 11.51: Show that a European call option on the T-bond is given by

$$C(t) = P(t,T)\mathrm{Q}_T(P(S,T) > K \mid \mathcal{F}_t) - KP(t,S)\mathrm{Q}_s(P(S,T) > K \mid \mathcal{F}_t),$$

where S is the exercise time of the option and Q_S, Q_T are S and T-forward measures.

Solution. For $t \le S \le T$, the price of the option is given by

$$\begin{aligned} C(t) &= \mathrm{E}_Q\left(\frac{\beta(t)}{\beta(S)} X \middle| \mathcal{F}_t\right) = \mathrm{E}_Q\left(\frac{\beta(t)}{\beta(S)}(P(S,T) - K)^+ \middle| \mathcal{F}_t\right) \\ &= \mathrm{E}_Q\left(\frac{\beta(t)}{\beta(S)} P(S,T) I(P(S,T) > K) \middle| \mathcal{F}_t\right) \\ &\quad - K\mathrm{E}_Q\left(\frac{\beta(t)}{\beta(S)} I(P(S,T) > K) \middle| \mathcal{F}_t\right). \end{aligned}$$

Define the T-forward measure Q_T by $\frac{d\mathrm{Q}_T}{d\mathrm{Q}} = \frac{P(S,T)/P(0,T)}{\beta(S)/\beta(0)} = \frac{P(S,T)}{\beta(S)P(0,T)}$. Then (cf. Theorem 10.10),

$$\begin{aligned}
&\mathrm{E}_Q\left(\frac{\beta(t)}{\beta(S)}P(S,T)I(P(S,T)>K)|\mathcal{F}_t\right)\\
&\quad= \mathrm{E}_{Q_T}\left(\frac{d\mathrm{Q}}{d\mathrm{Q}_T}\frac{\beta(t)}{\beta(S)}P(S,T)I(P(S,T)>K)|\mathcal{F}_t\right)\Big/\mathrm{E}_{Q_T}\left(\frac{d\mathrm{Q}}{d\mathrm{Q}_T}|\mathcal{F}_t\right)\\
&\quad= P(t,T)\mathrm{E}_{Q_T}\left(I(P(S,T)>K)|\mathcal{F}_t\right)\\
&\quad= P(t,T)\mathrm{Q}_T\left(P(S,T)>K|\mathcal{F}_t\right).
\end{aligned}$$

Also, define the S-forward measure Q_S by $\frac{d\mathrm{Q}_S}{d\mathrm{Q}} = \frac{P(S,S)/P(0,S)}{\beta(S)/\beta(0)} = \frac{1}{\beta(S)P(0,S)}$. Then,

$$\begin{aligned}
&\mathrm{E}_Q\left(\frac{\beta(t)}{\beta(S)}I(P(S,T)>K)|\mathcal{F}_t\right)\\
&\quad= \mathrm{E}_{Q_S}\left(\frac{d\mathrm{Q}}{d\mathrm{Q}_S}\frac{\beta(t)}{\beta(S)}I(P(S,T)>K)|\mathcal{F}_t\right)\Big/\mathrm{E}_{Q_S}\left(\frac{d\mathrm{Q}}{d\mathrm{Q}_S}|\mathcal{F}_t\right)\\
&\quad= P(t,S)\mathrm{E}_{Q_S}\left(I(P(S,T)>K)|\mathcal{F}_t\right)\\
&\quad= P(t,T)\mathrm{Q}_S\left(P(S,T)>K|\mathcal{F}_t\right).
\end{aligned}$$

Thus, putting these together, we obtain

$$C(t) = P(t,T)\mathrm{Q}_T\left(P(S,T)>K|\mathcal{F}_t\right) - KP(t,T)\mathrm{Q}_S\left(P(S,T)>K|\mathcal{F}_t\right).$$

Problem 11.52: Show that a European call option on the bond in the Merton model is given by

$$\begin{aligned}
&P(t,T)\Phi\left(\frac{\ln\frac{P(t,T)}{KP(t,S)} + \frac{\sigma^2(T-S)^2(S-t)}{2}}{\sigma(T-S)\sqrt{S-t}}\right)\\
&\qquad - KP(t,S)\Phi\left(\frac{\ln\frac{P(t,T)}{KP(t,S)} - \frac{\sigma^2(T-S)^2(S-t)}{2}}{\sigma(T-S)\sqrt{S-t}}\right).
\end{aligned}$$

Solution. The spot rate in the Merton model satisfies the SDE $dr(t) = (\mu+\sigma q)dt + \sigma dB(t)$ under Q, and (cf. Equation (12.59*))

$$P(S,T) = \frac{P(t,T)}{P(t,S)}e^{-\int_t^S(\tau(u,T)-\tau(u,S))dB(u) - \frac{1}{2}\int_t^S(\tau^2(u,T)-\tau^2(u,S))du},$$

where $\tau(t,T) = \sigma(T-t)$. Since $dB^T(t) = dB(t) + \tau(t,T)dt$ for a Q_T-Brownian motion, after some algebra we obtain

$$P(S,T) = \frac{P(t,T)}{P(t,S)} e^{-\sigma(T-S)(B^T(S)-B^T(t))+\frac{1}{2}\sigma^2(T-S)^2(S-t)}.$$

Therefore, the distribution of $P(S,T)$ conditional on $\mathcal{F}_t$ under Q_T is

$$LN\left(\ln\frac{P(t,T)}{P(t,S)} + \frac{1}{2}\sigma^2(T-S)^2(S-t), \sigma^2(T-S)^2(S-t)\right).$$

Similarly, $dB^S(t) = dB(t) + \tau(t,S)dt$ for a Q_S-Brownian motion and we have

$$P(S,T) = \frac{P(t,T)}{P(t,S)} e^{-\sigma(T-S)(B^S(S)-B^S(t))-\frac{1}{2}\sigma^2(T-S)^2(S-t)}.$$

Therefore, the distribution of $P(S,T)$ conditional on $\mathcal{F}_t$ under Q_S is

$$LN\left(\ln\frac{P(t,T)}{P(t,S)} - \frac{1}{2}\sigma^2(T-S)^2(S-t), \sigma^2(T-S)^2(S-t)\right).$$

Hence, the price of the option is

$$C(t) = P(t,T)Q_T\left(P(S,T) > K|\mathcal{F}_t\right) - KP(t,T)Q_S\left(P(S,T) > K|\mathcal{F}_t\right),$$

where

$$\begin{aligned} &Q_T\left(P(S,T) > K|\mathcal{F}_t\right) \\ &= 1 - \Phi\left(\frac{\ln K - \ln\frac{P(t,T)}{P(t,S)} - \frac{1}{2}\sigma^2(T-S)^2(S-t)}{\sigma(T-S)\sqrt{S-t}}\right) \\ &= \Phi\left(\frac{\ln\frac{P(t,T)}{KP(t,S)} + \frac{1}{2}\sigma^2(T-S)^2(S-t)}{\sigma(T-S)\sqrt{S-t}}\right), \end{aligned}$$

$$Q_S\left(P(S,T) > K|\mathcal{F}_t\right) = \Phi\left(\frac{\ln\frac{P(t,T)}{KP(t,S)} - \frac{1}{2}\sigma^2(T-S)^2(S-t)}{\sigma(T-S)\sqrt{S-t}}\right).$$

Problem 11.53: (Stochastic Fubini theorem.) Let $H(t,s)$, $0 \le t, s \le T$, be continuous, and for any fixed s, $H(t,s)$ as a process in t, $0 \le t \le T$ is adapted to the Brownian motion filtration $\mathcal{F}_t$. Assume $\int_0^T H^2(t,s)dt < \infty$ so that for each s, the Itô integral $X(s) = \int_0^T H(t,s)dW(t)$ is defined. Since $H(t,s)$ is continuous, $Y(t) = \int_0^T H(t,s)ds$ is defined and it is continuous and adapted. Assume

$$\int_0^T \mathrm{E}\left(\int_0^T H^2(t,s)dt\right) ds < \infty.$$

(a) Show that $\int_0^T \mathrm{E}|X(s)|ds \le \int_0^T \left(\mathrm{E}\int_0^T H^2(t,s)dt\right)^{1/2} ds < \infty$, consequently $\int_0^T X(s)ds$ exists.

(b) If $0 = t_0 < t_1 < \cdots < t_n = T$ is a partition of $[0,T]$, show that

$$\int_0^T \left(\sum_{i=0}^{n-1} H(t_i,s)\big(W(t_{i+1}) - W(t_i)\big)\right) ds$$
$$= \sum_{i=0}^{n-1}\left(\int_0^T H(t_i,s)ds\right)\big(W(t_{i+1}) - W(t_i)\big).$$

(c) By taking the limits as the partition shrinks, show that

$$\int_0^T X(s)ds = \int_0^T Y(t)dW(t),$$

in other words, the order of integration can be interchanged:

$$\int_0^T \left(\int_0^T H(t,s)dW(t)\right) ds = \int_0^T \left(\int_0^T H(t,s)ds\right) dW(t).$$

Solution.

(a) From Jensen's inequality $\mathrm{E}|X| \le \sqrt{\mathrm{E}(X^2)}$, we have

$$\int_0^T \mathrm{E}|X(s)|ds \le \int_0^T \left(\mathrm{E}(X^2(s))\right)^{1/2} ds$$
$$= \int_0^T \left(\mathrm{E}\int_0^T H^2(t,s)dt\right)^{1/2} ds < \infty.$$

(b) By additivity of the integral,

$$\int_0^T \left(\sum_{i=0}^{n-1} H(t_i,s)\big(W(t_{i+1}) - W(t_i)\big)\right) ds$$
$$= \sum_{i=0}^{n-1}\left(\int_0^T H(t_i,s)ds\right)\big(W(t_{i+1}) - W(t_i)\big).$$

(c) Let $X_n(s) = \sum_{i=0}^{n-1} H(t_i,s)\big(W(t_{i+1}) - W(t_i)\big)$. Then, $X_n(s)$ is an approximation to the Itô integral $X(s) = \int_0^T H(t,s)dW(t)$. $\mathrm{E}X^2(s) = \int_0^T \mathrm{E}H^2(t,s)dt < \infty$, and under the stated conditions, $\int_0^T \mathrm{E}(X_n(s) - X(s))^2 ds$ converges to zero. This implies converges in L^2 and in the probability of $\int_0^T X_n(s)ds$ to $\int_0^T X(s)ds$. Also, the right-hand side $\sum_{i=0}^{n-1} \left(\int_0^T H(t_i,s)ds\right)\left(W(t_{i+1}) - W(t_i)\right)$ is an approximation of the Itô integral $\int_0^T Y(t)dW(t)$, and converges to it in probability.

Problem 11.54 (One-factor HJM): Show that the correlation between the forward rates $f(t,T_1)$ and $f(t,T_2)$ in the HJM model with deterministic volatilities $\sigma(t,T)$ is given by

$$\rho(T_1,T_2) = Corr(f(t,T_1), f(t,T_2)) = \frac{\int_0^t \sigma(s,T_1)\sigma(s,T_2)ds}{\sqrt{\int_0^t \sigma^2(s,T_1)ds \int_0^t \sigma^2(s,T_2)ds}}.$$

Give the class of volatilities for which the correlation is one.

Solution. For the one-factor HJM model (see (12.35*)),

$$df(t,T) = \alpha(t,T)dt + \sigma(t,T)dW(t),$$

and

$$f(t,T) = f(0,T) + \int_0^t \alpha(s,T)ds + \int_0^t \sigma(s,T)dW(s).$$

Thus, $Var(f(t,T)) = \int_0^t \sigma^2(s,T)ds$ and

$$\mathrm{Cov}(f(t,T_1), f(t,T_2)) = \int_0^t \sigma(t,T_1)\sigma(t,T_2)ds.$$

Therefore, we obtain

$$\begin{aligned}\rho(T_1,T_2) &= \frac{\mathrm{Cov}(f(t,T_1), f(t,T_2))}{\sqrt{Var(f(t,T_1))Var(f(t,T_2))}} \\ &= \frac{\int_0^t \sigma(s,T_1)\sigma(s,T_2)ds}{\sqrt{\int_0^t \sigma^2(s,T_1)ds \int_0^t \sigma^2(s,T_2)ds}}.\end{aligned}$$

The correlation is one when $\sigma(t,T) = \sigma(t)$ does not depend on T. More generally, we can use the Cauchy–Schwarz inequality to find the general case:

$$\left(\int_0^t h(s)g(s)ds\right)^2 \le \int_0^t h^2(s)ds \int_0^t g^2(s)ds,$$

with equality if and only if h and g are linearly related, $g(s) = ah(s)+b$.

Problem 11.55: Find the forward rates $f(t,T)$ in Vasicek's model. Give the price of a cap.

Solution. Vasicek's model: $dr(t) = b(a - r(t))dt + \sigma dW(t)$. Assume that the market price of risk is constant q and let $R(\infty) = a + \frac{\sigma q}{b} - \frac{\sigma^2}{2b^2}$. Then, the bond price for this model is given in Equation (12.32*) as

$$P(t,T) = e^{\frac{1}{b}(1-e^{-b(T-t)})(R(\infty)-r(t))-(T-t)R(\infty)-\frac{\sigma^2}{4b^3}(1-e^{-b(T-t)})^2}.$$

Thus, the forward rate is

$$\begin{aligned} f(t,T) &= -\frac{\partial \ln P(t,T)}{\partial T} \\ &= -e^{-b(T-t)}(R(\infty) - r(t)) + R(\infty) \\ &\quad + \frac{\sigma^2}{2b^2}e^{-b(T-t)}(1 - e^{-b(T-t)}) \\ &= r(t)e^{-b(T-t)} + R(\infty)(1 - e^{-b(T-t)}) \\ &\quad + \frac{\sigma^2}{2b^2}e^{-b(T-t)}(1 - e^{-b(T-t)}). \end{aligned}$$

A cap is a sum of caplets and its price is $\text{Cap}(t) = \sum_{i=1}^n \text{Caplet}_i(t)$, where

$$\begin{aligned} \text{Caplet}_i(t) &= P(t,T_i)\text{E}_{Q_{T_i}}((f_{i-1} - k)^+\delta|\mathcal{F}_t) \\ &= P(t,T_i)\text{E}_{Q_{T_i}}\left(\left(\frac{1}{P(T_{i-1},T_i)} - 1 - k\delta\right)^+ |\mathcal{F}_t\right). \end{aligned}$$

For the evaluation of this price, see Problem 11.56, where the pricing is done in a more general context.

Problem 11.56 (Caps pricing market formula): Show that in the HJM model with a deterministic $\sigma(t,T)$, the price of a cap with trading dates $T_i = T + i\delta$, $i = 1,\ldots,n$, and strike rate k is given by

$$\sum_{i=1}^n (P(t,T_{i-1})\Phi(-h_{i-1}(t)) - (1+k\delta)P(t,T_i)\Phi(-h_{i-1}(t) - \gamma_{i-1}(t))),$$

where $\gamma_{i-1}^2(t) = Var(\ln P(T_{i-1}, T_i)) = \int_t^{T_{i-1}} |\tau(s, T_i) - \tau(s, T_{i-1})|^2 ds$, with $\tau(t, T) = \int_t^T \sigma(t, s) ds$ and $h_{i-1}(t) = \frac{1}{\gamma_{i-1}(t)} \left(\ln \frac{(1+k\delta)P(t,T_i)}{P(t,T_{i-1})} - \frac{1}{2}\gamma_{i-1}^2(t) \right)$.

Solution. $\text{Cap}(t) = \sum_{i=1}^n \text{Caplet}_i(t)$ and

$$\begin{aligned}
\text{Caplet}_i(t) &= P(t, T_i) \mathrm{E}_{Q_{T_i}} \left((f_{i-1} - k)^+ \delta | \mathcal{F}_t \right) \\
&= P(t, T_i) \mathrm{E}_{Q_{T_i}} \left(\left(\frac{1}{P(T_{i-1}, T_i)} - (1 + k\delta) \right)^+ | \mathcal{F}_t \right) \\
&= P(t, T_i) \mathrm{E}_{Q_{T_i}} \left(\frac{1}{P(T_{i-1}, T_i)} I \left(\frac{1}{P(T_{i-1}, T_i)} > 1 + k\delta \right) | \mathcal{F}_t \right) \\
&\quad - P(t, T_i)(1 + k\delta) \mathrm{E}_{Q_{T_i}} \left(I \left(\frac{1}{P(T_{i-1}, T_i)} > 1 + k\delta \right) | \mathcal{F}_t \right).
\end{aligned}$$

The conditional distribution under Q_{T_i} of $P(T_{i-1}, T_i)$ given $\mathcal{F}_t$ is

$$LN \left(\ln \frac{P(t, T_i)}{P(t, T_{i-1})} + \frac{1}{2} \gamma_{i-1}^2(t), \gamma_{i-1}^2(t) \right),$$

as given in Corollary 12.8. Note that conditioned on $\mathcal{F}_t$, the distribution of $\frac{1}{P(T_{i-1}, T_i)}$ is $LN \left(\ln \frac{P(t, T_{i-1})}{P(t, T_i)} - \frac{1}{2}\gamma_{i-1}^2(t), \gamma_{i-1}^2(t) \right)$.

The second expectation in the equation of $\text{Caplet}_i(t)$ above is

$$\begin{aligned}
&Q_{T_i} \left(\frac{1}{P(T_{i-1}, T_i)} > 1 + k\delta | \mathcal{F}_t \right) \\
&= Q_{T_i} \left(\ln P(T_{i-1}, T_i) < - \ln(1 + k\delta) | \mathcal{F}_t \right) \\
&= Q_{T_i} \left(N(0, 1) < \frac{1}{\gamma_{i-1}(t)} \left(- \ln(1 + k\delta) - \ln \frac{P(t, T_i)}{P(t, T_{i-1})} - \frac{1}{2} \gamma_{i-1}^2(t) \right) \right) \\
&= \Phi \left(\frac{1}{\gamma_{i-1}(t)} \left(- \ln \frac{(1 + k\delta) P(t, T_i)}{P(t, T_{i-1})} - \frac{1}{2} \gamma_{i-1}^2(t) \right) \right) \\
&= \Phi(-h_{i-1}(t) - \gamma_{i-1}(t)).
\end{aligned}$$

The first expectation in the $\text{Caplet}_i(t)$ equation can be computed by changing the measure.

Suppose that $X \sim LN(\mu, \gamma^2)$ under measure Q and introduce an equivalent measure $\widetilde{Q}$ by $\frac{d\widetilde{Q}}{dQ} = X/E_Q 10(X) = X e^{-\mu - \frac{1}{2}\gamma^2}$. Then, the distribution of X under $\widetilde{Q}$ is $LN(\mu+\gamma^2,\gamma^2)$. Indeed,

$$\mathrm{E}_{\widetilde{Q}}\left(e^{u\ln X}\right)=\mathrm{E}_Q\left(Xe^{-\mu-\frac{1}{2}\gamma^2}e^{u\ln X}\right)=\mathrm{E}_Q\left(e^{(u+1)\ln X-\mu-\frac{1}{2}\gamma^2}\right)$$
$$=e^{(u+1)\mu+\frac{1}{2}(u+1)^2\gamma^2-\mu-\frac{1}{2}\gamma^2}=e^{u(\mu+\gamma^2)+\frac{1}{2}u^2\gamma^2}.$$

We also have

$$\mathrm{E}_Q\big(XI(X>c)\big)=\mathrm{E}_{\widetilde{Q}}\big((1/X)e^{\mu+\frac{1}{2}\gamma^2}XI(X>c)\big)$$
$$=e^{\mu+\frac{1}{2}\gamma^2}\widetilde{Q}(X>c)=e^{\mu+\frac{1}{2}\gamma^2}\left(1-\Phi\left(\frac{\ln c-\mu-\gamma^2}{\gamma}\right)\right)$$
$$=e^{\mu+\frac{1}{2}\gamma^2}\Phi\left(\frac{\mu+\gamma^2-\ln c}{\gamma}\right).$$

Therefore, with X being a random variable having the distribution of $\frac{1}{P(T_{i-1},T_i)}$ conditioned on $\mathcal{F}_t$ with $\mu=\ln\frac{P(t,T_{i-1})}{P(t,T_i)}-\frac{1}{2}\gamma_{i-1}^2(t)$ and $\gamma^2=\gamma_{i-1}^2(t)$, we obtain that

$$\mathrm{E}_{Q_{T_i}}\left(\frac{1}{P(T_{i-1},T_i)}I\Big(\frac{1}{P(T_{i-1},T_i)}>1+k\delta\Big)\big|\mathcal{F}_t\right)$$
$$=e^{\ln\frac{P(t,T_{i-1})}{P(t,T_i)}-\frac{1}{2}\gamma_{i-1}^2(t)+\frac{1}{2}\gamma_{i-1}^2(t)}\Phi$$
$$\times\left(\frac{\ln\frac{P(t,T_{i-1})}{P(t,T_i)}-\frac{1}{2}\gamma_{i-1}^2(t)+\gamma_{i-1}^2(t)-\ln(1+k\delta)}{\gamma_{i-1}(t)}\right)$$
$$=\frac{P(t,T_{i-1})}{P(t,T_i)}\Phi\left(\frac{\ln\frac{P(t,T_{i-1})}{P(t,T_i)}+\frac{1}{2}\gamma_{i-1}^2(t)-\ln(1+k\delta)}{\gamma_{i-1}(t)}\right)$$
$$=\frac{P(t,T_{i-1})}{P(t,T_i)}\Phi\left(\frac{1}{\gamma_{i-1}(t)}\left(\ln\frac{P(t,T_{i-1})}{(1+k\delta)P(t,T_i)}+\frac{1}{2}\gamma_{i-1}^2(t)\right)\right)$$
$$=\frac{P(t,T_{i-1})}{P(t,T_i)}\Phi(-h_{i-1}(t)).$$

Finally, putting all the terms together, we obtain the price of a caplet,

$$\mathrm{Caplet}_i(t)=P(t,T_{i-1})\Phi(-h_{i-1}(t))$$
$$-P(t,T_i)(1+k\delta)\Phi(-h_{i-1}(t)-\gamma_{i-1}(t)),$$

and the price of a cap as required.

Problem 11.57 (Two-factor and higher HJM models): Two-factor HJM is given by the SDE

$$df(t,T) = \alpha(t,T)dt + \sigma_1(t,T)dW_1(t) + \sigma_2(t,T)dW_2(t),$$

where W_1 and W_2 are independent Brownian motions.

(a) Give the stochastic differential equation for the log of the bond prices, and show that

$$\frac{P(t,T)}{\beta(t)} = P(0,T)e^{-\int_0^t A_1(u,T)du - \int_0^t \tau_1(u,T)dW_1(u) - \int_0^t \tau_2(u,T)dW_2(u)},$$

with $\tau_i(t,T) = \int_t^T \sigma_i(t,s)ds$, $i = 1, 2$, and $A_1(t,T) = \int_t^T \alpha(t,s)ds$.

(b) Using the same proof as in the one-factor model, show that the no-arbitrage condition is given by

$$\alpha(t,T) = \sigma_1(t,T)\int_t^T \sigma_1(t,s)ds + \sigma_2(t,T)\int_t^T \sigma_2(t,s)ds.$$

Solution.

(a) The bond price is $P(t,T) = e^{-\int_t^T f(t,s)ds}$. It follows from Example 12.1 that

$$\begin{aligned} d(\ln P(t,T)) &= -d\left(\int_t^T f(t,s)ds\right) \\ &= r(t)dt - A_1(t,T)dt - \tau_1(t,T)dW_1(t) \\ &\quad - \tau_2(t,T)dW_2(t). \end{aligned}$$

Thus,

$$\begin{aligned} P(t,T) &= P(0,T) \\ &\quad \times e^{\int_0^t r(u)du - \int_0^t A_1(u,T)du - \int_0^t \tau_1(u,T)dW_1(u) - \int_0^t \tau_2(u,T)dW_2(u)}. \end{aligned}$$

With $\beta(t) = e^{\int_0^t r(u)du}$, we have

$$\frac{P(t,T)}{\beta(t)} = P(0,T)e^{-\int_0^t A_1(u,T)du - \int_0^t \tau_1(u,T)dW_1(u) - \int_0^t \tau_2(u,T)dW_2(u)}.$$

(b) Writing $X(t) = \ln P(t,T)$, from Equation (12.40*),

$$d\left(\frac{P(t,T)}{\beta(t)}\right) = \frac{P(t,T)}{\beta(t)}\left(X(t) + \frac{1}{2}d[X,X](t) - r(t)dt\right)$$
$$= \frac{P(t,T)}{\beta(t)}\Big(- A_1(t,T)dt - \tau_1(t,T)dW_1(t)$$
$$-\tau_2(t,T)dW_2(t) + \frac{1}{2}\tau_1^2(t,T)dt + \frac{1}{2}\tau_2^2(t,T)dt\Big).$$

For the no-arbitrage condition, the dt term must be zero, i.e.

$$\int_t^T \alpha(t,s)ds = \frac{1}{2}\left(\int_t^T \sigma_1(t,s)ds\right)^2 + \frac{1}{2}\left(\int_t^T \sigma_2(t,s)ds\right)^2.$$

Differentiating in T gives

$$\alpha(t,T) = \sigma_1(t,T)\int_t^T \sigma_1(t,s)ds + \sigma_2(t,T)\int_t^T \sigma_2(t,s)ds.$$

Problem 11.58: Show that a swap can be written as

$$\text{Swap}(t,T_0,k) = P(t,T_0) - P(t,T_n) - k\delta\sum_{i=1}^n P(t,T_i).$$

Solution. At T_i the following exchange is made: the amount received is $f_{i-1}(T_i - T_{i-1})$ and paid out is $k(T_i - T_{i-1})$. Using $1/P(T_{i-1},T_i) = 1 + f_{i-1}(T_i - T_{i-1})$, the resulting amount at time T_i is $1/P(T_{i-1},T_i) - 1 - k\delta$, where $\delta = T_i - T_{i-1}$. Thus, the value at time t of the swap is

$$\text{Swap}(t,T_0,k) = \sum_{i=1}^n \text{E}_Q\left(\frac{\beta(t)}{\beta(T_i)}\left(\frac{1}{P(T_{i-1},T_i)} - 1 - k\delta\right) \mid \mathcal{F}_t\right).$$

Using the martingale property of the discounted bonds $\frac{P(t,T)}{\beta(t)}$ under Q,

$$\text{Swap}(t,T_0,k)$$
$$= \sum_{i=1}^n \text{E}_Q\left(\text{E}_Q\left(\frac{\beta(t)}{\beta(T_i)}\left(\frac{1}{P(T_{i-1},T_i)} - 1 - k\delta\right)|\mathcal{F}_{T_{i-1}}\right)|\mathcal{F}_t\right)$$
$$= \sum_{i=1}^n \text{E}_Q\left(\beta(t)\frac{P(T_{i-1},T_i)}{\beta(T_{i-1})}\left(\frac{1}{P(T_{i-1},T_i)} - 1 - k\delta\right)|\mathcal{F}_t\right)$$

$$= \sum_{i=1}^{n} \mathrm{E}_Q \left(\frac{\beta(t)}{\beta(T_{i-1})} - \frac{P(T_{i-1},T_i)}{\beta(T_{i-1})} \beta(t)(1+k\delta) \big| \mathcal{F}_t \right)$$

$$= \sum_{i=1}^{n} \left(P(t,T_{i-1}) - P(t,T_i)(1+k\delta) \right)$$

$$= P(t,T_0) - P(t,T_n) - k\delta \sum_{i=1}^{n} P(t,T_{i-1}).$$

Problem 11.59: Denote by $b(t) = \delta \sum_{i=1}^{n} P(t,T_i)$. Show that for $0 < t \le T_0$, $\mathrm{Swap}(t,T_0,k) = P(t,T_0) - P(t,T_n) - kb(t)$ and that the swap rate

$$k(t) = \frac{P(t,T_0) - P(t,T_n)}{b(t)}.$$

Solution. This follows immediately from the previous exercise and that the swap rate $k(t)$ solves $\mathrm{Swap}(t,T_0,k) = 0$.

Problem 11.60 (Jamshidian (1996) swaptions pricing formula): Assume that the swap rate $k(t) > 0$ and that $v^2(t) = \int_t^T \frac{1}{k^2(s)} d[k,k](s)$ is deterministic. Show that

$$\mathrm{Swaption}(t) = b(t) \left(\alpha_+(t)k(t) - k\alpha_-(t) \right),$$

where

$$\alpha_{\pm}(t) = \Phi \left(\frac{\ln k(t)/k}{v(t)} \pm \frac{v(t)}{2} \right).$$

Solution. Consider the portfolio of bonds that at any time $t < T$ is long $\alpha_+(t)$ of the T_0-bond, short $\alpha_+(t)$ of the T_n-bond, and for each i, $1 \le i \le n$, short $\alpha_-(t)\delta k$ of the T_i-bond. The value of this portfolio at time $t < T$ is $C(t) = \alpha_+(t)(P(t,T_0) - P(t,T_n)) - k\alpha_-(t)b(t)$, with $b(t) = \delta \sum_{i=1}^{n} P(t,T_i)$. By using the expression for the swap rate (see Exercise 12.10),

$$C(t) = b(t)(\alpha_+(t)k(t) - k\alpha_-(t)).$$

It can be seen that this portfolio has the correct final value and is self-financing, i.e. $dC(t) = \alpha_+(t)d\left(P(t,T_0) - P(t,T_n)\right) - k\alpha_-(t)db(t)$.

Chapter 12

Applications in Biology

Populations developing in time often have uncertain growth factors, be it through random variation in reproduction or environmental factors. Models for developing populations are given by diffusion processes, such as Feller branching diffusion, Wright–Fisher model, birth–death processes and branching processes. Such models generalise classical population models described by differential equations. Often, the classical models are recovered from stochastic ones in the limit, as for example, is the stochastic Lotka–Volterra model for competition of species.

Feller branching diffusion is a continuous time and continuous state space analogue of a branching process. It follows the diffusion equation

$$dX_t = \mu X_t dt + \sigma\sqrt{X_t}dB_t, \quad X_0 > 0.$$

Using this equation, an extinction probability and the growth rate on the complimentary set can be determined.

The Wright–Fisher and Moran models describe the development of an allele frequency in a population and are important in population genetics. These models are used to calculate and predict, among other questions of interest, fixation probabilities of alleles. The Wright–Fisher diffusion is given by the SDE

$$dX_t = g(X_t)dt + \sqrt{X_t(1-X_t)}dB_t, \quad 0 < X_0 < 1.$$

The function $g(x)$ is linear or quadratic, depending on the biological context, such as the presence of mutation ($g(x) = ax + b$)) and selection ($g(x) = sx(1-x)$).

In a birth–death process, when population size is x, each particle lives for an exponential length of time with parameter $a(x)$ and at the end of life splits into two particles with probability $p(x)$ or leaves no offspring with probability $1 - p(x)$. The change in the population occurs only at the death of a particle, and from state x the process jumps to $x + 1$ if a split occurs or to $x - 1$ if a particle dies without splitting. Thus, the jump variables $\xi(x)$ take values 1 and -1 with probabilities $p(x)$ and $1 - p(x)$, respectively. Using the fact that the minimum of exponentially distributed random variables is exponentially distributed, we can see that the process stays at x for an exponentially distributed length of time with parameter $\lambda(x) = xa(x)$. Instead of specifying the probability of split $p(x)$ or the lifespan of a particle $a(x)$, a birth–death process is described in terms of population birth and death rates, i.e. in a population of size x, a particle is born at rate $b(x)$ and dies at the rate $d(x)$. Then, the process stays at x for an exponential time with parameter $\lambda(x) = b(x) + d(x)$ and has a jump from x according to

$$\xi(x) = \begin{cases} +1 & \text{with probability } \dfrac{b(x)}{b(x)+d(x)}, \\ -1 & \text{with probability } \dfrac{d(x)}{b(x)+d(x)}. \end{cases}$$

If $X(t)$ denotes the population size at time t, then it can be represented as

$$X(t) = X(0) + \int_0^t N_1(b(X(s))ds - \int_0^t N_2(d(X(s))ds,$$

where N_1 and N_2 are independent Poisson processes with rate one. Compensating the Poisson processes above, we obtain a decomposition into a drift term and a martingale term.

This process has the following representation akin to a stochastic differential equation:

$$X_t = X_0 + \int_0^t (b(X_s) - d(X_s))ds + M_t,$$

where M_t is a martingale with predictable quadratic variation:

$$\langle M, M \rangle_t = \int_0^t (b(X_s) + d(X_s))ds.$$

In many stochastic models, such as positively recurrent models, a stationary distribution exists. For diffusions, it is given in Chapter 6. For birth–death processes with birth and death population rates $b(x)$ and $d(x)$, respectively, $x = 0, 1, 2 \ldots$, it is given by ($b(-1) = 0$, $\prod_{x=0}^{-1} = 1$):

$$\pi(j) = \pi(0) \prod_{x=0}^{j-1} \frac{b(x)}{d(x+1)}, \quad \pi(0) = \left(\sum_{i=0}^{\infty} \prod_{x=0}^{i-1} \frac{b(x)}{d(x+1)} \right)^{-1}.$$

There is a large part of mathematical biology that uses models of differential equations to study biological phenomena such as populations of genes, bacteria, and animals, populations competing for resources, and predator–prey populations. While studying such phenomena on a microscopic level, a question arises as to how to introduce noise in such models. There are two ways to do it. The first way is to perturb the coefficients in the models by (a white) noise. This approach leads to diffusion models. There is a question when introducing noise into equations as to which stochastic integrals should be used, Itô or Stratanovich. It is known that if we introduce noise such that its correlation function tends to Dirac's δ-function (the correlation function of white noise), then stochastic differential equations in the Stratonovich form result.

The second way, more appropriate for modelling with discrete state spaces, is that of birth–death processes. By converting the given dynamics to rates, we obtain stochastic models which in the limit (as noise goes to zero) recover the deterministic dynamical models. Both such approaches are illustrated in the Lotka–Volterra model for competition between prey and predator. Another example is that of a model for ant foraging.

Deterministic differential equations models are used, for example, to find fixed points of the system, which allow for conclusions about biological populations. In a stochastic setting, these points are replaced by stationary distributions. The mode of the stationary distribution indicates the most probable state of the system.

Problems

Problem 12.1: Let $dX(t) = -X(t)dt + X(t)^r dB(t)$, where $r < \frac{1}{2}$. Find the expected time for $X(t)$ to hit zero.

Solution. We first consider the expected exit time of $X(t)$ from an interval $(0, b)$ and then take the limit $b \to \infty$. From Theorem 6.16

and Exercise 6.8, with $\mu(x) = -x$ and $\sigma^2(x) = x^{2r}$, we have $G(x) = \exp(\int_0^x \frac{2s}{s^{2r}} ds) = \exp(\frac{x^{2-2r}}{1-r})$ and

$$\begin{aligned} \mathrm{E}_x(T_0 \wedge T_b) = & -2\int_0^x \exp\left(\frac{y^{2-2r}}{1-r}\right)\int_0^y s^{-2r}\exp\left(-\frac{s^{2-2r}}{1-r}\right) dsdy \\ & +2\int_0^b \exp\left(\frac{y^{2-2r}}{1-r}\right)\int_0^y s^{-2r}\exp\left(-\frac{s^{2-2r}}{1-r}\right) dsdy \\ & \times \frac{\int_0^x \exp\left(\frac{s^{2-2r}}{1-r}\right) ds}{\int_0^b \exp\left(\frac{s^{2-2r}}{1-r}\right) ds}. \end{aligned}$$

Note that

$$\begin{aligned} & \lim_{b\to\infty} \frac{\int_0^b \exp\left(\frac{y^{2-2r}}{1-r}\right)\int_0^y s^{-2r}\exp\left(-\frac{s^{2-2r}}{1-r}\right) dsdy}{\int_0^b \exp\left(\frac{s^{2-2r}}{1-r}\right) ds} \\ & = \lim_{b\to\infty} \frac{\exp\left(\frac{b^{2-2r}}{1-r}\right)\int_0^b s^{-2r}\exp\left(-\frac{s^{2-2r}}{1-r}\right) ds}{\exp\left(\frac{b^{2-2r}}{1-r}\right)} \\ & = \int_0^\infty s^{-2r}\exp\left(-\frac{s^{2-2r}}{1-r}\right) ds. \end{aligned}$$

Thus,

$$\begin{aligned} \mathrm{E}_x(T_0) = & -2\int_0^x \exp\left(\frac{y^{2-2r}}{1-r}\right)\int_0^y s^{-2r}\exp\left(-\frac{s^{2-2r}}{1-r}\right) dsdy \\ & +2\int_0^\infty s^{-2r}\exp\left(-\frac{s^{2-2r}}{1-r}\right) ds\int_0^x \exp\left(\frac{y^{2-2r}}{1-r}\right) dy \\ = & \, 2\int_0^x \exp\left(\frac{y^{2-2r}}{1-r}\right)\int_y^\infty s^{-2r}\exp\left(-\frac{s^{2-2r}}{1-r}\right) dsdy \\ = & \, (1-r)^{-\frac{1}{2-2r}}\int_0^x \exp\left(\frac{y^{2-2r}}{1-r}\right)\int_{\frac{y^{2-2r}}{1-r}}^\infty u^{-\frac{1}{2-2r}}\exp(-u) dudy \\ = & \, (1-r)^{-\frac{1}{2-2r}}\int_0^x \exp\left(\frac{y^{2-2r}}{1-r}\right)\Gamma\left(\frac{1-2r}{2-2r}, \frac{y^{2-2r}}{1-r}\right) dy, \end{aligned}$$

where $\Gamma(z,v) = \int_v^\infty u^{z-1}e^{-u} du$ is the upper incomplete Gamma function.

Problem 12.2: Let $X(t)$ be a Feller branching diffusion satisfying SDE $dX_t = \mu X_t dt + \sigma\sqrt{X_t}dB_t$, $X_0 > 0$, and $\mu > 0$.

(a) Let $c = 2\mu/\sigma^2$. Show that e^{-cX_t} is a martingale.

(b) Let T be the time to extinction, $T = \inf\{t : X_t = 0\}$, where $T = \infty$ if $X_t > 0$ for all $t \geq 0$, and let $q_t(x) = \mathrm{P}_x(T \leq t)$ be the probability that extinction occurs by time t when the initial population size is x. Prove that $q_t(x) \leq e^{-cx}$.

Solution.

(a) Using Itô's formula,

$$\begin{aligned} d(e^{-cX(t)}) &= -ce^{-cX(t)}dX(t) + \frac{1}{2}c^2e^{-cX(t)}d[X,X](t) \\ &= -ce^{-cX(t)}\sigma\sqrt{X}(t)dB(t). \end{aligned}$$

By Theorem 13.1, $X(t) \geq 0$ for all t. Thus, the function $e^{-cx}\sqrt{x}$ is bounded for $x \geq 0$. Therefore, $e^{-cX(t)}$ is a martingale as an Itô integral of a bounded process.

(b) Let $\tau = T \wedge t$, then τ is a bounded stopping time. Applying optional stopping theorem to martingale $e^{-cX(t)}$, we have

$$\begin{aligned} e^{-cx} &= \mathrm{E}e^{-cX(0)} = \mathrm{E}e^{-cX(\tau)} \\ &= \mathrm{E}(e^{-cX(T)}I(T \leq t)) + \mathrm{E}(e^{-cX(t)}I(T > t)) \\ &\geq \mathrm{E}(e^{-cX(T)}I(T \leq t)) = \mathrm{P}(T \leq t). \end{aligned}$$

Problem 12.3: A model for population growth is given by the SDE $dX(t) = 2X(t)dt + \sqrt{X}(t)dB(t)$, and $X(0) = x > 0$. Find the probability that the population doubles its initial size x before it becomes extinct. Show that when the initial population size x is small then this probability is approximately $1/2$, but when x is large this probability is nearly one.

Solution. Let $T_a = \inf\{t : X_t = a\}$ and $\tau = T_0 \wedge T_{2x}$. To find $p = \mathrm{P}_x(T_{2x} < T_0)$, we apply the optional stopping theorem to martingale $M_t = e^{-4X_t}$ with the bounded stopping time τ,

$$e^{-4x} = \mathrm{E}(e^{-4X_\tau}) = e^{-4(0)}(1-p) + e^{-4(2x)}(p) = 1 - p + pe^{-8x},$$

which gives $\mathrm{P}_x(T_{2x} < T_0) = \frac{1-e^{-4x}}{1-e^{-8x}}$. Alternatively, apply Theorem 6.17. When x is small, $e^x \approx 1 + x$, thus $\mathrm{P}_x(T_{2x} < T_0) = \frac{1-e^{-4x}}{1-e^{-8x}} \approx \frac{1}{2}$, but when x is large, $\mathrm{P}_x(T_{2x} < T_0) = \frac{1-e^{-4x}}{1-e^{-8x}} \approx 1$.

Problem 12.4: Derive the stationary distribution for the Wright–Fisher diffusion.

Solution. Wright–Fisher diffusion:

$$dX_t = (-\gamma_1 X_t + \gamma_2(1 - X_t))dt + \sqrt{X_t(1 - X_t)}dB_t, \quad 0 < X_0 < 1.$$

The stationary distribution is given by formula (6.69*), that is,

$$\begin{aligned}
\pi(x) &= \frac{C}{\sigma^2(x)} \exp\left(\int_{x_0}^{x} \frac{2\mu(y)}{\sigma^2(y)} dy\right) \\
&= \frac{C}{x(1-x)} \exp\left(\int_{x_0}^{x} \frac{-2\gamma_1 y + 2\gamma_2(1-y)}{y(1-y)} dy\right) \\
&= \frac{C}{x(1-x)} \exp\left([2\gamma_1 \ln(1-y) + 2\gamma_2 \ln(y)]_{x_0}^{x}\right) \\
&= C(1-x)^{2\gamma_1 - 1} x^{2\gamma_2 - 1} (1 - x_0)^{-2\gamma_1} x_0^{-2\gamma_2} \\
&= C'(1-x)^{2\gamma_1 - 1} x^{2\gamma_2 - 1},
\end{aligned}$$

where C' is such that $\int_0^1 \pi(x)dx = 1$. Since $1/C' = \int_0^1 (1 - x)^{2\gamma_1 - 1} x^{2\gamma_2 - 1} dx = \frac{\Gamma(2\gamma_1)\Gamma(2\gamma_2)}{\Gamma(2\gamma_1 + 2\gamma_2)}$, we obtain that

$$\pi(x) = \frac{\Gamma(2\gamma_1 + 2\gamma_2)}{\Gamma(2\gamma_1)\Gamma(2\gamma_2)} (1-x)^{2\gamma_1 - 1} x^{2\gamma_2 - 1}.$$

Indeed, one can verify that $L^*\pi(x) = \frac{1}{2}\frac{\partial^2}{\partial x^2}\{\sigma^2(x)\pi(x)\} - \frac{\partial}{\partial x}\{\mu(x)\pi(x)\} = 0$.

Problem 12.5: Let a deterministic growth model be given by the differential equation

$$dx(t) = g(x(t))dt, \ \ x_0 > 0, \ \ t \geq 0,$$

for a positive function $g(x)$ and consider its stochastic analogue

$$dX(t) = g(X(t))dt + \sigma(X(t))dB(t), \ \ X(0) > 0.$$

One way to analyze the stochastic equation is by comparison with the deterministic solution.

(a) Find $G(x)$ such that $G(x(t)) = G(x(0)) + t$ and consider $Y(t) = G(X(t))$. Find $dY(t)$.

(b) Let $g(x) = x^r$, $0 \le r < 1$, and $\sigma(x)/x^r \to 0$ as $x \to \infty$. Give conditions on $g(x)$ and $\sigma^2(x)$ so that the low of large numbers (LLN) holds for $Y(t)$, that is, $Y(t)/t \to 1$, as $t \to \infty$ on the set $\{Y(t) \to \infty\}$.

Solution.

(a) $G(x(t)) - G(x(0)) = t$ is equivalent to $dG(x(t)) = dt$. But $dG(x(t)) = G'(x(t))dx(t) = G'(x(t))g(x(t))dt$. Thus, we must have $G'(x) = \frac{1}{g(x)}$ or $G(x) = \int_a^x \frac{1}{g(u)}du$.

$Y(t) = G(X(t))$. Using Itô's formula, we obtain

$$\begin{aligned} dY(t) &= G'(X(t))dX(t) + \frac{1}{2}G''(X(t))d[X,X]_t \\ &= \frac{1}{g(X(t))}(g(X(t))dt + \sigma(X(t))dB(t)) \\ &\quad - \frac{1}{2}\frac{g'(X(t))}{g^2(X(t))}\sigma^2(X(t))dt \\ &= dt - \frac{g'(X(t))\sigma^2(X(t))}{2g^2(X(t))}dt + \frac{\sigma(X(t))}{g(X(t))}dB(t). \end{aligned}$$

(*Note*: "$\frac{1}{2}$" is missing in the solution given in the book.)

(b) For $g(x) = x^r$, $dY(t) = dt - \frac{rX^{r-1}(t)\sigma^2(X(t))}{2X^{2r}(t)}dt + \frac{\sigma(X(t))}{X^r(t)}dB(t)$ and $Y(t) = Y(0) + t - \int_0^t \frac{r\sigma^2(X(s))}{2X^{r+1}(s)}ds + \int_0^t \frac{\sigma(X(s))}{X^r(s)}dB(s)$.

For $Y(t)/t \to 1$, we need to have $\frac{1}{t}\int_0^t \frac{\sigma(X(s))}{X^r(s)}dB(s) - \frac{1}{t}\int_0^t \frac{r\sigma^2(X(s))}{2X^{r+1}(s)}ds \to 0$. If $\frac{\sigma^2(x)}{g^2(x)} = \frac{\sigma^2(x)}{x^{2r}} < C$ for some constant C, then $\mathrm{E}(\int_0^t \frac{\sigma(X(s))}{2X^r(s)}dB(s))^2 = \int_0^t \mathrm{E}(\frac{\sigma^2(X(s))}{4X^{2r}(s)})ds \le Ct$ and $\mathrm{E}((\frac{1}{t}\int_0^t \frac{\sigma(X(s))}{2X^r(s)}dB(s))^2) \to 0$, which implies $\frac{1}{t}\int_0^t \frac{\sigma(X(s))}{2X^r(s)}dB(s) \to 0$ in probability. Also, using L'Hopital's rule, $\frac{1}{t}\int_0^t \frac{\sigma^2(X(s))r}{2X^{r+1}(s)}ds \to \lim_{t\to\infty} \frac{\sigma^2(X(t))r}{2X^{r+1}(t)} = 0$ on the set $\{X(t) \to \infty\}$. The LLN for $Y(t)$ now follows.

Problem 12.6 (Individual and population rates): Let $L_1, L_2, \ldots, L_x$ be independent exponentially distributed random variables with parameter $a(x)$, (they represent the lifespans of particles in the population of size x). Show that $\min(L_1, L_2, \ldots, L_x)$ has an exponential distribution with parameter $xa(x)$. Application of this is in birth–death (and other Markov population) processes, where the change in the population size occurs when a particle dies. So, if each particle lives for exponential time with parameter

$a(x)$ when there are x particles, then the time between the change in the population is $\min(L_1, L_2, \ldots, L_x)$.

Solution.

$$\begin{aligned}\mathrm{P}(\min(L_1, L_2, \ldots, L_x) > t) &= \mathrm{P}(L_1 > t, L_2 > t, \ldots, L_x > t)\\ &= (\mathrm{P}(L_1 > t))^x = (e^{-a(x)t})^x = e^{-xa(x)t}.\end{aligned}$$

Problem 12.7 (Birth–death processes mean exit time): Let X_t be a birth–death process with birth and death rates $b(i)$ and $d(i)$, respectively, $i = 1, 2, \ldots$.

(a) Find the expected time to exit from (a, b) by using Theorem 13.8.
(b) Find the expected time to extinction by taking $a = 0$ and $b \to \infty$ in the result above.

Solution. Let $\tau = \inf\{t \geq 0 : X(t) \notin (a, b)\}$ denote the exit time from (a, b) and $v(x) = \mathrm{E}_x\tau$ the expected time to exit from (a, b) with $X(0) = x$. Then, v satisfies $Lv = -1$ with boundary conditions $v(a) = v(b) = 0$. By Theorem 13.8,

$$v(x) = \sum_{l=m}^{x-1}\left[w\prod_{i=m}^{l}\frac{d(i)}{b(i)} - \sum_{j=m}^{l}\frac{1}{b(j)}\prod_{i=j+1}^{l}\frac{d(i)}{b(i)}\right],$$

where

$$w = \sum_{l=m}^{n-1}\sum_{j=m}^{l}\frac{1}{b(j)}\prod_{i=j+1}^{l}\frac{d(i)}{b(i)}\Big/\sum_{l=m}^{n-1}\prod_{i=m}^{l}\frac{d(i)}{b(i)}.$$

Problem 12.8: Find the variance of the proportion of type 1 particles $X(t)/N$ in the Moran model. The Moran model in discrete time with neutral drift is given by

$$\mathrm{P}\big(X(t+1) = X(t) + 1|X(t)\big) = \Big(1 - \frac{X(t)}{N}\Big)\frac{X(t)}{N}$$

$$\mathrm{P}\big(X(t+1) = X(t) - 1|X(t)\big) = \frac{X(t)}{N}\Big(1 - \frac{X(t)}{N}\Big)$$

$$\mathrm{P}\big(X(t+1) = X(t)|X(t)\big) = 1 - 2\frac{X(t)}{N}\Big(1 - \frac{X(t)}{N}\Big).$$

Solution. Note that $X(t)$ is a martingale and $\mathrm{E}(X(t)) = X(0)$. Computing the second moment, we have

$$\begin{aligned}\mathrm{E}(X(t+1)^2) &= \mathrm{E}(X(t+1)^2|X(t)) \\ &= 2\mathrm{E}\left(\frac{X(t)}{N}\left(1-\frac{X(t)}{N}\right)\right) + \mathrm{E}(X(t)^2).\end{aligned}$$

This gives us a recursive relation:

$$\mathrm{E}(X(t+1)^2) = 2\frac{X(0)}{N} + \mathrm{E}(X(t)^2)\Big(1-\frac{1}{N^2}\Big).$$

By evaluating the first few terms, we see that

$$\begin{aligned}\mathrm{E}(X(t)^2) &= X(0)^2\Big(1-\frac{1}{N^2}\Big)^t + 2\frac{X(0)}{N}\sum_{k=0}^{t-1}\Big(1-\frac{1}{N^2}\Big)^k \\ &= X(0)^2\Big(1-\frac{1}{N^2}\Big)^t + 2X(0)N\left(1-\Big(1-\frac{1}{N^2}\Big)^t\right).\end{aligned}$$

Therefore,

$$\begin{aligned}Var(X(t)) = \mathrm{E}(X(t)^2) - (\mathrm{E}(X(t)))^2 &= \big(2X(0)N - X(0)^2\big) \\ &\times \left(1-\Big(1-\frac{1}{N^2}\Big)^t\right)\end{aligned}$$

and

$$Var\Big(\frac{X(t)}{N}\Big) = \frac{1}{N^2}Var(X(t)) = \Big(2\frac{X(0)}{N} - \frac{X(0)^2}{N^2}\Big)\Big(1-\Big(1-\frac{1}{N^2}\Big)^t\Big).$$

Problem 12.9: Find the mean and variance of the proportion of type 1 particles $X(t)/N$ in the Moran model with mutation. The Moran model in discrete time with mutation is given by

$$\begin{aligned}&\mathrm{P}\big(X(t+1) = X(t)+1|X(t)\big) \\ &\qquad = \left(1-\frac{X(t)}{N}\right)\frac{X(t)}{N}(1-\gamma_1) + \left(1-\frac{X(t)}{N}\right)^2\gamma_2\end{aligned}$$

$$\mathrm{P}\big(X(t+1) = X(t)-1|X(t)\big) = \frac{X(t)}{N}\left(1-\frac{X(t)}{N}\right)(1-\gamma_2) + \left(\frac{X(t)}{N}\right)^2\gamma_1$$

$$\mathrm{P}\big(X(t+1)=X(t)|X(t)\big)=1-2\frac{X(t)}{N}\left(1-\frac{X(t)}{N}\right)(2-\gamma_1-\gamma_2)$$
$$-\left(\frac{X(t)}{N}\right)^2\gamma_1-\left(1-\frac{X(t)}{N}\right)^2\gamma_2,$$

where γ_i, $i=1,2$, are the probabilities of mutation of type i, $i=1,2$.

Solution. Computing the conditional mean and simplifying, we have

$$\mathrm{E}\big(X(t+1)|X(t)\big)=X(t)+\frac{X(t)}{N}(\gamma_1+\gamma_2)+\gamma_2.$$

Thus, $\mathrm{E}(X(t)))$ satisfies the recursion relation

$$\mathrm{E}(X(t+1))=\gamma_2+\mathrm{E}(X(t))\Big(1-\frac{\gamma_1+\gamma_2}{N}\Big).$$

Solving this gives

$$\mathrm{E}(X(t))=X(0)\left(1-\frac{\gamma_1+\gamma_2}{N}\right)^t+\sum_{k=0}^{t-1}\gamma_2\left(1-\frac{\gamma_1+\gamma_2}{N}\right)^k$$
$$=X(0)\left(1-\frac{\gamma_1+\gamma_2}{N}\right)^t+\frac{N}{\gamma_1}\left(1-\left(1-\frac{\gamma_1+\gamma_2}{N}\right)^t\right).$$

Again, with direct calculation and simplification, the conditional second moment is

$$\mathrm{E}\big(X(t+1)^2|X(t)\big)=\gamma_2+2\gamma_2X(t)+\frac{X(t)}{N}(2-\gamma_1-3\gamma_2)$$
$$+\ X(t)^2-2X(t)\frac{X(t)}{N}(\gamma_1+\gamma_2)-2\left(\frac{X(t)}{N}\right)^2$$
$$\times\ (1-\gamma_1-\gamma_2).$$

With $\mathrm{E}(X(t))$, this gives

$$\mathrm{E}\big(X(t+1)^2\big)=\gamma_2+\mathrm{E}(X(t))\left(2\gamma_2+\frac{1}{N}(2-\gamma_1-3\gamma_2)\right)$$
$$+\,\mathrm{E}\big(X(t)^2\big)\left(1-\frac{2}{N}(\gamma_1+\gamma_2)-\frac{2}{N^2}(1-\gamma_1-\gamma_2)\right)$$

$$= \gamma_2 + \left(\frac{N}{\gamma_1} + \left(X(0) - \frac{N}{\gamma_1}\right)\left(1 - \frac{\gamma_1+\gamma_2}{N}\right)^t\right)$$
$$\times \left(2\gamma_2 + \frac{1}{N}(2 - \gamma_1 - 3\gamma_2)\right)$$
$$+ \mathrm{E}(X(t)^2)\left(1 - \frac{2}{N}(\gamma_1+\gamma_2) - \frac{2}{N^2}(1 - \gamma_1 - \gamma_2)\right).$$

Solving this, we have

$$\mathrm{E}(X(t)^2) = A\sum_{k=0}^{t-1} C^k + B\sum_{k=0}^{t-1} C^k\left(-\frac{\gamma_1+\gamma_2}{N}\right)^{t-k-1} + CX(0)^2,$$

where

$$A = \gamma_2 + 2N\frac{\gamma_2}{\gamma_1} + \frac{1}{\gamma_1}(2 - \gamma_1 - 3\gamma_2),$$
$$B = \left(X(0) - \frac{N}{\gamma_1}\right)\left(2\gamma_2 + \frac{1}{N}(2 - \gamma_1 - 3\gamma_2)\right),$$
$$C = 1 - \frac{2}{N}(\gamma_1+\gamma_2) - \frac{2}{N^2}(1 - \gamma_1 - \gamma_2).$$

Then,

$$Var\left(\frac{X(t)}{N}\right) = \frac{1}{N^2}Var(X(t)) = \frac{1}{N^2}(\mathrm{E}(X(t)^2) - (\mathrm{E}(X(t)))^2).$$

Problem 12.10: Find the mean and variance of the proportion of type 1 particles $X(t)/N$ in the Moran model with mutation in continuous time. The Moran model in continuous time is given by a birth–death process with population rates

$$b(x) = N\left[\left(1 - \frac{x}{N}\right)\frac{x}{N}(1 - \gamma_1) + \left(1 - \frac{x}{N}\right)^2\gamma_2\right],$$
$$d(x) = N\left[\frac{x}{N}\left(1 - \frac{x}{N}\right)(1 - \gamma_2) + \left(\frac{x}{N}\right)^2\gamma_1\right].$$

Note that all states $0 < x < N$ have positive rates (unless $\gamma_1 = 1 - \gamma_2 = 0$ or $\gamma_2 = 1 - \gamma_1 = 0$), and if $\gamma_1, \gamma_2 > 0$, the states 0 and N are reflecting, $b(0) > 0$, $d(0) = 0$, $b(N) = 0$, $d(N) > 0$.

Solution. Using the stochastic representation of birth–death processes,

$$X(t) = X(0) + N\int_0^t \left(\gamma_2 - (\gamma_1+\gamma_2)\frac{X(s)}{N}\right) ds + M(t),$$

where $M(t)$ is a martingale with

$$\begin{aligned}\langle M, M\rangle(t) = N\int_0^t \Bigg[&(2-\gamma_1-\gamma_2)\frac{X(s)}{N}\left(1-\frac{X(s)}{N}\right)\\ &+\gamma_2\left(1-\frac{X(s)}{N}\right)^2 + \gamma_1\left(\frac{X(s)}{N}\right)^2\Bigg] ds.\end{aligned}$$

First, let's compute $\mathrm{E}(X(t))$. Taking expectation in the equation for $X(t)$ gives

$$\mathrm{E}(X(t)) = X(0) + N\int_0^t \left(\gamma_2 - (\gamma_1+\gamma_2)\frac{\mathrm{E}(X(s))}{N}\right) ds.$$

Let $y_t = \mathrm{E}(X(t))$, we then have $y_t = X_0 + N\gamma_2 t - (\gamma_1+\gamma_2)\int_0^t y_s ds$, or

$$y_t' + (\gamma_1+\gamma_2)y_t = N\gamma_2.$$

Solving this differential equation $y_0 = X(0)$ gives

$$\mathrm{E}(X(t)) = y_t = X(0)e^{-(\gamma_1+\gamma_2)t} + N\frac{\gamma_2}{\gamma_1+\gamma_2}(1-e^{-(\gamma_1+\gamma_2)t}).$$

Now, we compute $E(X(t)^2)$. Note that, by Itô's formula,

$$\begin{aligned}d(X(t)^2) &= 2X(t)N\left(\gamma_2 - (\gamma_1+\gamma_2)\frac{X(t)}{N}\right)dt\\ &\quad + 2X(t)dM(t) + d\langle M\rangle(t)\\ &= \Bigg(2N\gamma_2 X(t) - 2(\gamma_1+\gamma_2)X(t)^2 + (2-\gamma_1-\gamma_2)X(t)\\ &\quad\times\left(1-\frac{X(t)}{N}\right) + N\gamma_2\left(1-\frac{X(t)}{N}\right)^2\\ &\quad + N\gamma_1\left(\frac{X(t)}{N}\right)^2\Bigg) dt + 2X(t)dM(t).\end{aligned}$$

Taking expectation and rearranging, we have

$$\begin{aligned}
&\mathrm{E}(X(t)^2)\\
&= X(0)^2 + \int_0^t \left(N\gamma_2 + \left(2\frac{\gamma_1+\gamma_2-1}{N} - 2(\gamma_1+\gamma_2) \right) \mathrm{E}(X(s)^2) \right.\\
&\quad \left. + \big(2\gamma_2(N-1) + (2-\gamma_1-\gamma_2)\big)\mathrm{E}(X(s)) \right) ds\\
&= X(0)^2 + \int_0^t \left(N\gamma_2 + \left(2\frac{\gamma_1+\gamma_2-1}{N} - 2(\gamma_1+\gamma_2) \right) \mathrm{E}(X(s)^2) \right.\\
&\quad + \left(2\gamma_2(N-1) + (2-\gamma_1-\gamma_2)\right)\\
&\quad \left. \times \left(X(0)e^{-(\gamma_1+\gamma_2)s} + \frac{N\gamma_2}{\gamma_1+\gamma_2}(1-e^{-(\gamma_1+\gamma_2)s}) \right) \right) ds.
\end{aligned}$$

With $y_t = \mathrm{E}(X(t)^2)$, we obtain an ordinary differential equation for y_t, solving which gives

$$\begin{aligned}
&e^{(2(\gamma_1+\gamma_2)-\frac{2(\gamma_1+\gamma_2-1)}{N})t} y_t\\
&\quad = \int_0^t \left(N\gamma_2 + (2N\gamma_2 + 2 - \gamma_1 - 3\gamma_2) \right.\\
&\quad\quad \times \left(\frac{N\gamma_2}{\gamma_1+\gamma_2} + \left(X(0) - \frac{N\gamma_2}{\gamma_1+\gamma_2} \right) e^{-(\gamma_1+\gamma_2)s} \right)\\
&\quad\quad \left. \times\, e^{(2(\gamma_1+\gamma_2)-\frac{2(\gamma_1+\gamma_2-1)}{N})s} \right) ds + X(0)^2
\end{aligned}$$

or

$$\begin{aligned}
\mathrm{E}(X(t)^2) = y_t &= X(0)^2 e^{(-2(\gamma_1+\gamma_2)+\frac{2(\gamma_1+\gamma_2-1)}{N})t}\\
&\quad + \frac{N^2\gamma_2 + \frac{N^2\gamma_2}{\gamma_1+\gamma_2}(2N\gamma_2 + 2 - \gamma_1 - 3\gamma_2)}{2N(\gamma_1+\gamma_2) - 2(\gamma_1+\gamma_2-1)}\\
&\quad \times \left(1 - e^{(-2(\gamma_1+\gamma_2)+\frac{2(\gamma_1+\gamma_2-1)}{N})t} \right)\\
&\quad + \frac{(X(0) - \frac{N\gamma_2}{\gamma_1+\gamma_2})N(2N\gamma_2 + 2 - \gamma_1 - 3\gamma_2)}{N(\gamma_1+\gamma_2) - 2(\gamma_1+\gamma_2-1)}\\
&\quad \times \left(e^{-(\gamma_1+\gamma_2)t} - e^{(-2(\gamma_1+\gamma_2)+\frac{2(\gamma_1+\gamma_2-1)}{N})t} \right).
\end{aligned}$$

This together with $\mathrm{E}(X(t))$ gives $Var(X(t))$. Finally, we get

$$\begin{aligned}
Var\left(\frac{X(t)}{N}\right) &= \frac{1}{N^2}Var(X(t)) = \left(\frac{X(0)}{N}\right)^2 e^{\frac{2(\gamma_1+\gamma_2-1)}{N}t} \\
&+ \frac{\gamma_2\left(1+\frac{2N\gamma_2+2-\gamma_1-3\gamma_2}{\gamma_1+\gamma_2}\right)}{2N(\gamma_1+\gamma_2)-2(\gamma_1+\gamma_2-1)} \\
&\times \left(1-e^{\left(-2(\gamma_1+\gamma_2)+\frac{2(\gamma_1+\gamma_2-1)}{N}\right)t}\right) \\
&+ \frac{\left(\frac{X(0)}{N}-\frac{\gamma_2}{\gamma_1+\gamma_2}\right)(2N\gamma_2+2-\gamma_1-3\gamma_2)}{N(\gamma_1+\gamma_2)-2(\gamma_1+\gamma_2-1)} \\
&\times \left(e^{-(\gamma_1+\gamma_2)t}-e^{\left(-2(\gamma_1+\gamma_2)+\frac{2(\gamma_1+\gamma_2-1)}{N}\right)t}\right) \\
&- 2\frac{X(0)}{N}\frac{\gamma_2}{\gamma_1+\gamma_2}\left(e^{-(\gamma_1+\gamma_2)t}-e^{-2(\gamma_1+\gamma_2)t}\right) \\
&- \left(\frac{\gamma_2}{\gamma_1+\gamma_2}\right)^2\left(1-e^{-(\gamma_1+\gamma_2)t}\right)^2.
\end{aligned}$$

Problem 12.11 (Birth–death processes: Stochastic representation): Let $N_k^\lambda(t)$ and $N_k^\mu(t)$, $k \geq 1$, be two independent sequences of independent Poisson processes with rates λ and μ. Let $X(0) > 0$ and for $t > 0$, $X(t)$ satisfies

$$X(t) = X(0) + \int_0^t \sum_{k\geq 1} I(X(s-) \geq k)dN_k^\lambda(s) - \int_0^t \sum_{k\geq 1} I(X(s-) \geq k)dN_k^\mu(s).$$

(a) Show that $X(t)$ is a birth–death process and identify the rates.
(b) Give the semimartingale decomposition for $X(t)$.

Solution.

(a) A birth–death process stays at x for an exponential time and then increases or decreases by jumps of size one. Here, the jumps of $X(t)$ are governed by

$$\pi_1(t) = \int_0^t \sum_{k\geq 1} I(X(s-) \geq k)dN_k^\lambda(s),$$

$$\pi_2(t) = \int_0^t \sum_{k\geq 1} I(X(s-) \geq k)dN_k^\mu(s).$$

Their jumps are of size one. The jumps are also disjoint since $N_k^\lambda(t)$ and $N_k^\mu(t)$ are independent Poisson processes.

Next, we identify the compensators of $\pi_1(t)$ and $\pi_2(t)$. The process

$$A_1(t) = \int_0^t \sum_{k\geq 1} I(X(s-) \geq k)\lambda ds$$

is adapted and continuous. Hence, it is predictable. Observe that

$$\begin{aligned}
&\pi_1(t) - A_1(t)\\
&\quad = \int_0^t \sum_{k\geq 1} I(X(s-) \geq k) dN_k^\lambda(s) - \int_0^t \sum_{k\geq 1} I(X(s-) \geq k)\lambda ds\\
&\quad = \int_0^t \sum_{k\geq 1} I(X(s-) \geq k) d(N_k^\lambda(s) - \lambda s)
\end{aligned}$$

is an integral with respect to a martingale. Hence, it is a local martingale. Thus, the compensator of $\pi_1(t)$ is

$$\begin{aligned}
A_1(t) &= \int_0^t \sum_{k\geq 1} I(X(s-) \geq k)\lambda ds\\
&= \int_0^t \sum_{k\geq 1} I(X(s) \geq k)\lambda ds = \int_0^t \lambda X(s) ds,
\end{aligned}$$

since $\sum_{k\geq 1} I(X(s) \geq k) = X(s)$. Similarly, $\pi_2(t)$ has compensator $A_2(t) = \int_0^t \mu X(s)ds$. It follows that $X(t)$ is a linear birth–death process with birth rate λx and death rate μx.

(b) Since $\pi_1(t) - A_1(t)$ and $\pi_2(t) - A_2(t)$ are (local) martingales and $A_1(t)$ and $A_2(t)$ are processes of finite variation, the semimartingale decomposition for $X(t)$ is given by

$$X(t) = X(0) + A_1(t) - A_2(t) + \pi_1(t) - A_1(t) - \pi_2(t) + A_2(t).$$

Problem 12.12: Nonlinear birth–death processes: stochastic representation as time change in a Poisson process. This is a most useful representation for modelling in applications, as well as for theoretical analysis.

Let X_t be a birth–death process with population rates $\lambda(x)$ and $\mu(x)$. Let $N_1(t)$ and $N_2(t)$ be independent Poisson processes with rate 1. Show that

$$X_t = X_0 + N_1\left(\int_0^t \lambda(X_s)ds\right) - N_2\left(\int_0^t \mu(X_s)ds\right).$$

Solution. The jumps of N_1 and N_2 are of size one. The jumps are also disjoint since N_1 and N_2 are independent Poisson processes. Thus, X_t has jumps of size 1 and -1. The compensator of X_t is

$$A_t = \int_0^t (\lambda(X_s) - \mu(X_s))ds.$$

The sharp bracket process is given by

$$\langle M, M\rangle_t = \int_0^t (\lambda(X_s) + \mu(X_s))ds.$$

One can also verify that X_t is a Markov process with the same generator as for the birth–death process.

Problem 12.13 (A stochastic model for ants foraging): A deterministic model for ant foraging is given by

$$\frac{dx_t}{dt} = (\alpha + \beta x_t)(n - x_t) - \frac{sx_t}{K + x_t},$$

where x_t is the number of ants on the trail to the food source, n is the total number of ants available to forage, α is the rate at which ants randomly find the feeders, β is the strength of recruitment, and $sx/(K + x)$ is a saturating function which determines the rate at which individual ants lose the pheromone trail. s is the maximum rate at which ants lose the trail when the trail is saturated with ants, and K is the number of ants on a trail that gives a rate of loss of $s/2$. Formulate a stochastic model for ant foraging and give its stationary distribution.

Solution. The number of ants on the trail X_t is modelled by a birth–death process with state space $\{0, 1, \ldots, n\}$, with birth and death rates

$$b(x) = (\alpha + \beta x)(n - x), \quad d(x) = \frac{sx}{K + x}.$$

Using $b(x) - d(x) = (\alpha + \beta x)(n - x) - x/(1 + x)$, we obtain a representation that captures stochastic dynamics:

$$X_t = X_0 + \int_0^t \left((\alpha + \beta X_u)(n - X_u) - \frac{sX_u}{K + X_s}\right) ds + M_t,$$

where M_t is a martingale with predictable quadratic variation,

$$\langle M, M\rangle_t = \int_0^t (b(X_u) + d(X_u))du.$$

This process has a stationary distribution, indicating the likelihood of having j ants $(0 \le j \le n)$ on the trail. It is given by the formula in the introduction to the chapter. It can be studied numerically for given values of parameters.

Problem 12.14 (A stochastic Lotka–Volterra model): The classical Lotka–Volterra predator–prey model is given by a system of two nonlinear differential equations for prey x_t and predator y_t densities (with constants $a, b, c, e > 0$):

$$\frac{dx_t}{dt} = ax_t - bx_t y_t,$$
$$\frac{dy_t}{dt} = -cy_t + ex_t y_t.$$

Taking this model as a guide, we specify the rates in a nonlinear birth–death process for a stochastic Lotka–Volterra model. Each population changes by adding 1 (birth) or subtracting 1 (death). Each of these occurs according to independent Poisson processes with random, population-dependent rates, described as follows. Each population is non-negative, with state zero being absorbing. The prey population has a survival rate per individual a (birth rate minus death rate) but, in addition, dies proportional to the number of predators. The predator population has a negative survival rate c and a birth rate proportional to the number of prey. Using the construction in the previous question, we arrive at the following equations for the numbers of the prey population X_t and predator population Y_t:

$$X_t = X_0 + N_1\left(\int_0^t aX_s ds\right) - N_2\left(\int_0^t \frac{b}{K}X_s Y_s ds\right),$$
$$Y_t = Y_0 - N_3\left(\int_0^t cY_s ds\right) + N_4\left(\int_0^t \frac{e}{K}X_s Y_s ds\right).$$

Parameter K plays a role of carrying capacity. Assuming $X_0 = x_0 K$, $Y_0 = y_0 K$, obtain equations for the process of densities, $x_t^K = X_t/K$ and $y_t^K = Y_t/K$. Identify the drift terms and the martingales. Show that the martingales converge to zero as $K \to \infty$. (It can be shown that the limit as $K \to \infty$ of the density process is the classical system of Lotka–Volterra equations.)

Solution. Dividing the defining equations by K, we obtain that the densities x_t, y_t satisfy the following stochastic equations:

$$x_t^K = x_0^K + \frac{1}{K} N_1 \left(\int_0^t K a x_s^K ds \right) - \frac{1}{K} N_2 \left(\int_0^t K b x_s^K y_s^K ds \right),$$

$$y_t^K = y_0^K - \frac{1}{K} N_3 \left(\int_0^t K c y_s^K ds \right) + \frac{1}{K} N_4 \left(\int_0^t K e x_s^K y_s^K ds \right).$$

The compensator of $N_1\left(\int_0^t aKx_s^K ds\right)$ is $\int_0^t aKx_s^K ds$, i.e.

$$M_t^1 = N_1 \left(\int_0^t aKx_s^K ds \right) - \int_0^t aKx_s^K ds$$

is a martingale. It has predictable quadratic variation $\langle M^1 \rangle_t = \int_0^t aKx_s^K ds$. After dividing the equations by K, K disappears from the drift term but remains in the martingale term because the predictable quadratic variation

$$\left\langle \frac{1}{K} M^1 \right\rangle_t = \frac{1}{K^2} \langle M^1 \rangle_t = \frac{1}{K} \int_0^t a x_s^K ds.$$

Compensating all other Poisson processes, in a similar way we obtain the following equations:

$$x_t^K = x_0^K + \int_0^t a x_s^K ds - \int_0^t b x_s^K y_s^K ds + m_t^1 - m_t^2,$$

$$y_t^K = y_0^K - \int_0^t c y_s^K ds + \int_0^t e x_s^K y_s^K ds - m_t^3 + m_t^4,$$

where the martingales

$$m_t^1 = \frac{1}{K} M_t^1, \; m_t^2 = \frac{1}{K} M_t^2, \; m_t^3 = \frac{1}{K} M_t^3, \; m_t^4 = \frac{1}{K} M_t^4$$

have predictable quadratic variation:

$$\langle m^1 \rangle_t = \left\langle \frac{1}{K} M^1 \right\rangle_t = \frac{1}{K} \int_0^t a x_s^K ds,$$

$$\langle m^2 \rangle_t = \left\langle \frac{1}{K} M^2 \right\rangle_t = \frac{1}{K} \int_0^t b x_s^K y_s^K ds,$$

$$\langle m^3 \rangle_t = \left\langle \frac{1}{K} M^3 \right\rangle_t = \frac{1}{K} \int_0^t c y_s^K ds,$$

$$\langle m^4 \rangle_t = \left\langle \frac{1}{K} M^4 \right\rangle_t = \frac{1}{K} \int_0^t e x_s^K y_s^K ds.$$

Note that the quadratic variation of martingales converges to zero as $K \to \infty$. This yields by Doob's inequality (Theorem 7.32) with any T:

$$\mathrm{P}\left(\sup_{t\le T}|m_t^1| > \varepsilon\right) \le \frac{1}{\varepsilon^2}E(m_T^1)^2 = \frac{1}{\varepsilon^2}E\left\langle m^1\right\rangle_T = \frac{1}{\varepsilon^2 K}E\int_0^T ax_s^K\,ds.$$

Using the first equation and neglecting the negative term, after taking expectation, we obtain

$$Ex_t^K \le x_0^K \int_0^t aEx_s^K\,ds.$$

Now, by Gronwall's inequality for $t \le T$,

$$Ex_t^K \le x_0^K e^{aT}.$$

This shows that

$$\mathrm{P}\left(\sup_{t\le T}|m_t^1| > \varepsilon\right) \le \frac{1}{\varepsilon^2 K}x_0^K aTe^{aT}.$$

Thus, $\sup_{t\le T}|m_t^1|$ converges to zero in probability as $K \to \infty$ on any finite interval $[0, T]$.

Next, from the first equation,

$$\int_0^t bEx_s^K y_s^K\,ds = x_0^K + \int_0^t aEx_s^K\,ds - Ex_t^K \le x_0^K + \int_0^t aEx_s^K\,ds \le x_0^K + aTe^{aT}.$$

This implies that the quadratic variation of m_t^2 converges to zero, and we have in the same way as above that $\sup_{t\le T}|m_t^2|$ and $\sup_{t\le T}|m_t^4|$ converges to zero in probability as $K \to \infty$. Neglecting the negative term in the second equation, we obtain that $Ey_t^K \le y_0^K + x_0^K + aTe^{aT}$. This in turn implies that $\sup_{t\le T}|m_t^3|$ converges to zero in probability as $K \to \infty$. Thus, all the martingale vanish in the limit. Taking (at this stage) the formal limit as $K \to \infty$ and interchanging it with the integrals, we obtain the classical Lotka–Volterra system of equations.

Problem 12.15 (A stochastic Lotka–Volterra model): The classical Lotka–Volterra predator–prey model is given by a system of two nonlinear differential equations for prey x_t and predator y_t densities, given in the previous problem. Derive a stochastic model by perturbing the birth of prey and the death of predator coefficients, a and c, by independent white noises with intensities $\varepsilon\sigma_1^2$, and $\varepsilon\sigma_2^2$, where $\varepsilon > 0$ is a small parameter. Consider perturbations in the sense of Itô and Stratanovich. Using the first integral of the deterministic system $r(x_t, y_t) = \text{const.}$, find an SDE for the process $R_t = r(X_t, Y_t)$.

Solution. Writing perturbations formally by using independent white noises $\dot{B}_1$ and $\dot{B}_2$, we obtain

$$\frac{dX_t}{dt} = (a + \sqrt{\varepsilon}\sigma_1 \dot{B}_1)X_t - bX_tY_t,$$
$$\frac{dY_t}{dt} = (-c + \sqrt{\varepsilon}\sigma_2 \dot{B}_2)Y_t + eX_tY_t.$$

Converting these to a well-defined system of two stochastic equations, we have

$$dX_t = X_t(a - bY_t)dt + \sqrt{\varepsilon}\sigma_1 X_t dB_1(t),$$
$$dY_t = Y_t(-c + eX_t) + \sqrt{\varepsilon}\sigma_2 Y_t dB_2(t),$$

where stochastic integrals are treated in the sense of Itô. When we treat them in the sense of Stratanovich, we have the system

$$dX_t = X_t(a - bY_t)dt + \sqrt{\varepsilon}\sigma_1 X_t \circ dB_1(t),$$
$$dY_t = Y_t(-c + eX_t) + \sqrt{\varepsilon}\sigma_2 Y_t \circ dB_2(t).$$

It follows from the properties of Stratanovich integral that this system is equivalent to the following Itô equations:

$$dX_t = X_t(a + \varepsilon\sigma_1^2/2 - bY_t)dt + \sqrt{\varepsilon}\sigma_1 X_t dB_1(t),$$
$$dY_t = Y_t(-c + \varepsilon\sigma_2^2/2 + eX_t) + \sqrt{\varepsilon}\sigma_2 Y_t dB_2(t).$$

It is intuitively clear and can be shown rigorously, similar to the previous question by showing that martingales converge to zero, that as the noise intensity ε converges to zero, we obtain in the limit the classical Lotka–Volterra system of equations on any interval $[0, T]$.

It is easy to check that the trajectories of the deterministic Lotka–Volterra system in the phase space x, y are closed curves described by the first integral (a function $r(x, y)$, such that for any t, $r(x_t, y_t) = r(x_0, y_0) = \text{const.}$),

$$r(x, y) = ex - c + c\ln\left(1 + \frac{ex - c}{c}\right) + by - a - a\ln\left(1 + \frac{by - a}{a}\right).$$

We choose to write the first integral so that $r(x, y) \geq 0$ and $r(x, y) = 0$ if and only if $x = c/e$ and $y = a/b$, the only fixed point of the deterministic system in a positive quadrant.

By Itô formula, $R_t = r(X_t, Y_t)$ solves

$$dR_t = \frac{\varepsilon}{2}(e\sigma_1^2 X_t + b\sigma_2^2 Y_t)dt + \sqrt{\varepsilon}(\sigma_1(eX_t - c)dB_1(t) + \sigma_2(bY_t - a)dB_2(t)).$$

Whereas $dr(x_t, y_t) = 0$ on the solutions of the deterministic system, on the solutions of the stochastic system, $dr(X_t, Y_t)$ is not zero but small, since its drift and diffusion coefficients are of order ε. This is called a slowly changing component.

The above representation is used to establish the convergence of $R(t/\varepsilon)$ to a diffusion process as $\varepsilon \to 0$.

Index

$\{A_n\ ev.\}$, 17
$\{A_n\ i.o.\}$, 17
σ-field, 15
σ-field generated by $\mathcal{G}$, 16
σ-field generated by $\{\mathcal{F}_n\}$, 16

A

absorbing, 284
adapted, 24
adaptedness, 104
adjoint differential operator, 150
allele frequency, 445
almost surely, 18
ants foraging, 460
arbitrage opportunities, 382
arbitrage pricing theory, 381
average rate options, 381

B

Bachelier model, 385
Bachelier model with interest rate, 420
backward and forward equations, 150
backward equation, 176
Bayes' theorem, 350
bilinear form, 3
binomial model, 384
birth–death process, 285
birth–death processes:
  stochastic representation, 458
Black–Scholes model, 384
Black–Scholes PDE, 407
bonds, 385
Borel σ-field, 16
Borel–Cantelli lemma, 17
Brownian motion (BM), 61, 199
Burkholder–Davis–Gundy inequality, 230

C

càdlàg, 1
calculation of expectations, 176
call options, 381
canonical decomposition, 261
capping, 386
caps pricing market formula, 438
Cauchy criteria, 19
Cauchy sequence in L^2, 102
Cauchy–Schwarz inequality, 3
Cauchy–Schwarz inequality for integrals, 31
CDF, 17
Chan–Koralyi–Longstaff–Sanders (CKLS) SDE, 216
change of measure for point processes, 352
change of measure for processes, 350
change of measure for random variables, 349
change of numeraire, 385
change of probability measure, 349

change of time, 231, 243
characteristic functions, 18
class E_T, 101
class P_T, 101
class of simple processes S_T, 101
compensator, 236, 259, 283
complete, 383
conditional distribution, 78–79
conditional expectation, 22
continuous martingale component, 262
continuous probability model, 15
convergence, 18
convergence in L^2, 102
convergence sets of a submartingale, 286
covariance, 117
covariance function, 62
covariation, 104

D

DDS, 243
derivative contracts, 381
diffusion on an interval, 183
diffusion process, 144
dirichlet class (D), 228
discrete multi-period models, 383
distribution, 18
dominated convergence, 21
Doob's inequality, 229
Doob–Lévy martingales, 24
Dynkin's formula, 175

E

$\mathrm{E}(X, |\mathcal{G})$, 22
EMM, 382
Engelbert and Schmidt, 179
European call option on the T-bond, 433
existence and uniqueness of solutions, 147
exit times, 180
expectation, 19
explosion, 204, 303

F

$f(B(t))$ is not a semimartingale, 264
fake BM, 92
Fatou's lemma, 21
Feller diffusion, 247
Feller process, 303
Feynman–Kac formula, 178
filtering, 429
filtration, 24
finite variation (FV), 1
first fundamental theorem, 382
flooring, 386
forward measures, 386
forward rates, 429
fractional Brownian motion, 111

G

gambler's ruin, 37
Gaussian diffusions, 152
Gaussian martingale, 118
Gaussian process, 62, 152
Gaussian vector, 45
general Bayes' formula, 350
generator, 149
Girsanov's theorem, 351
gradient, 122
Gronwall inequality, 13

H

Hölder's inequality, 230
hedge, 382
hedging, 262
higher dimensions, 109
hitting time, 64, 79, 284

I

independent, 17
independent increment, 24
inner product, 3
integrable, 22
integration by parts formula, 4
interchanging expectation and derivative, 41
isometry, 104

Itô integral, 100
Itô processes, 106
Itô's formula for $f(B_1(t), \ldots, B_n(t))$, 122
Itô's formula for BM, 105
Itô's formula for Itô processes, 108

J

Jensen's inequality, 118
joint distributions, 64
jumps, 6

K

Kalman–Bucy filter, 428
Kazamaki condition, 260
Kolmogorov probability model, 15

L

$\liminf A_n$, 17
$\limsup A_n$, 17
Langevin equation, 146
Lebesgue decomposition, 349
Lebesgue integral, 19
likelihood functions, 353
linearity, 3
Loève criterion, 19
local martingales, 227
local time, 259
logarithm, 145

M

Markov jump processes, 284
Markov process, 24
martingale, 24
martingale convergence, 224
martingale differences, 253
martingale inequalities, 229
maximum, 64
mean exit time, 452
measurable and adapted processes, 101
measurable sets, 15
measurable space, 15
measure of jumps, 262
Merton model, 429
Meyer–Tanaka formula, 259
minimum, 64
models of stock based on diffusion processes, 385
moment-generating function, 39, 43
moment-generating function of the multivariate normal distribution, 70
monotone convergence, 21
Moran model, 445
multidimensional SDE, 186
mutation, 453
mutually exclusive, 25

N

n-th moment, 40
natural filtration, 287
nonlinear birth–death processes, 459
normal number, 27
Novikov's condition, 260

O

observation of Brownian motion, 428
occupation time formula, 260
one-factor HJM, 437
optional stopping, 225
Ornstein–Uhlenbeck process, 168

P

payoff, 381
Poisson process, 283
polarisation, 3
positive local martingale, 228
positivity, 3
predictable representation, 383
predictable representation property, 262
probability, 18
probability measure, 16
probability of ruin, 84
pure jump point process, 283
put options, 381

Q

quadratic covariation, 2
quadratic variation, 2, 104
quadratic variation of martingales, 228

R

$\mathbb{R}^n$-valued Itô integral, 110
r'th mean, 18
Radon–Nikodym theorem, 22
random integral, 77
random measure of jumps, 261
random variable, 17, 20
random walk hits b before it hits a, 35
rates, 385
real-world probability measure, 382
recurrence and transience, 182
remaining life, 300
renewal process, 283
representation of solutions, 200
Riemann integral, 3, 20
Riemann–Stieltjes integral, 3
risk management, 381
ruin probability in insurance, 241

S

$sign(B_t)$, 112
sample path continuity, 104
scale function, 199
second fundamental theorem, 383
self-financing portfolio, 381
semigroup of operators, 189
semimartingale, 259
simple, 20
size of increments of BM, 66
square-root process, 246
standard normal distribution, 31
stationary distribution, 184, 450
stationary increments, 61
stochastic calculus for poisson processes, 287
stochastic differential equations (SDEs), 99, 143
stochastic differentials, 106
stochastic exponential, 145
stochastic exponential of BM, 225
stochastic Lotka–Volterra model, 445
stochastic processes, 23
stopping time, 25
Stratonovich integral, 131
strong solution, 144
strong uniqueness, 144
submartingale, 24
sufficient conditions for explosion, 285
symmetry, 3
symmetric difference, 56

T

T-bond, 386
telegraphic signal, 293
terminal value problem, 177
three martingales of BM, 63
time-changed Brownian motion $B(f(t))$, 246
time-homogeneous, 24
towering/smoothing, 23
transition probabilities, 62
triplet of predictable characteristics, 262
two-factor and higher HJM models, 441

U

uniformly integrable, 224
uniqueness of solutions under a weaker assumption, 171

V

Vallée–Poussin theorem, 224
variation of a function, 1
Vasicek model, 431

W

weak convergence, 18
weak solution, 148
Wiener, Norbert, 125
Wright–Fisher model, 445

X

$X^+ = \max(X, 0)$, 20
$X^- = \max(-X, 0)$, 20

Y

Yamada–Watanabe uniqueness, 172

Z

zero-mean, 104
zeros of BM, 64

www.ingramcontent.com/pod-product-compliance
Lightning Source LLC
Chambersburg PA
CBHW070744220326
41598CB00026B/3732